T0199362

Optical Angular Momentum

L Allen
Universities of Glasgow, Strathclyde and Sussex

Stephen M Barnett
University of Strathclyde

Miles J Padgett
University of Glasgow

CRC Press
Taylor & Francis Group
Boca Raton London New York

CRC Press is an imprint of the
Taylor & Francis Group, an **informa** business

First published 2003 by IOP Publishing Ltd

Published 2020 by CRC Press
Taylor & Francis Group
6000 Broken Sound Parkway NW, Suite 300
Boca Raton, FL 33487-2742

First issued in paperback 2020

© 2003 by Taylor & Francis Group, LLC
CRC Press is an imprint of Taylor & Francis Group, an Informa business

No claim to original U.S. Government works

ISBN 13: 978-0-367-57853-4 (pbk)
ISBN 13: 978-0-7503-0901-1 (hbk)

This book contains information obtained from authentic and highly regarded sources. Reasonable efforts have been made to publish reliable data and information, but the author and publisher cannot assume responsibility for the validity of all materials or the consequences of their use. The authors and publishers have attempted to trace the copyright holders of all material reproduced in this publication and apologize to copyright holders if permission to publish in this form has not been obtained. If any copyright material has not been acknowledged please write and let us know so we may rectify in any future reprint.

Except as permitted under U.S. Copyright Law, no part of this book may be reprinted, reproduced, transmitted, or utilized in any form by any electronic, mechanical, or other means, now known or hereafter invented, including photocopying, microfilming, and recording, or in any information storage or retrieval system, without written permission from the publishers.

For permission to photocopy or use material electronically from this work, please access www. copyright.com (http://www.copyright.com/) or contact the Copyright Clearance Center, Inc. (CCC), 222 Rosewood Drive, Danvers, MA 01923, 978-750-8400. CCC is a not-for-profit organization that provides licenses and registration for a variety of users. For organizations that have been granted a photocopy license by the CCC, a separate system of payment has been arranged.

Trademark Notice: Product or corporate names may be trademarks or registered trademarks, and are used only for identification and explanation without intent to infringe.

Visit the Taylor & Francis Web site at
http://www.taylorandfrancis.com

and the CRC Press Web site at
http://www.crcpress.com

British Library Cataloguing-in-Publication Data

A catalogue record for this book is available from the British Library.

Library of Congress Cataloging-in-Publication Data are available

Cover Design: Frédérique Swist

Typeset by Academic + Technical, Bristol

Contents

Preface

It has been recognised for a long time that a photon has spin angular momentum, observable macroscopically in a light beam as polarisation. It is less well known that a beam may also carry orbital angular momentum linked to its phase structure. Although both forms of angular momentum have been identified in electromagnetic theory for very many years, it is only over the past decade that orbital angular momentum has been the subject of intense theoretical and experimental study. The concepts combine neatly into optical angular momentum.

This book is designed to be an accessible survey of the current status of optical angular momentum. It reproduces 44 original papers arranged in eight sections. Each section has a brief introduction to set the reproduced papers in the context of a wider range of related work.

It is a pleasure to thank, in the same order as their papers appear here, G Nienhuis, MV Berry, JM Vaughan, C Tamm, JP Woerdman, L Torner, MS Soskin, NR Heckenberg, A Ashkin, H Rubinsztein-Dunlop, SJ van Enk, DV Petrov, BA Garetz, I Bialynicki-Birula, WJ Firth, M Segev and A Zeilinger for being willing to have their work and that of their co-authors reproduced.

We are pleased, too, to acknowledge the publishers who allowed us to reproduce papers originally published in their journals: the *Royal Society* for paper **1.1**; the *American Physical Society* for papers **1.2, 2.1, 2.3, 2.4, 2.7, 3.6, 3.10, 4.2, 4.3, 5.1, 5.3, 6.4, 6.6, 7.1, 7.2, 7.3, 8.2** and **8.3**; the *Optical Society of America* for papers **3.2, 3.3, 4.1, 4.4, 6.1** and **6.3**; *SPIE* – the *International Society for Optical Engineering* for **2.5**; *Elsevier Science B.V* for papers **2.2, 2.6, 2.9, 3.4, 3.5, 3.11, 3.12, 6.2** and **6.5**; *Kluwer Academic Publishers* for **3.9**; *Nature* for paper **8.1**; *EDP Sciences* for **2.8**; the *American Association of Physics Teachers* and the *American Institute of Physics* for **3.1** and **3.8**; while four papers were published by the *Institute of Physics Publishing,* namely **1.3, 2.10, 3.7** and **5.2**.

We wish to express, too, our gratitude to the many friends, too numerous to list here, with whom we have enjoyed exploring the fascinating topic of optical angular momentum.

Optical angular momentum is a new area of physics, but one for which the foundations have been firmly established. It will be very interesting to see how it develops over the next few years.

L Allen
Stephen M Barnett
Miles J Padgett

August 2002

Acknowledgments and Copyright Information

We are grateful to the copyright holders listed below for granting permission to reprint materials that are the core of this book. Section numbers refer to the contents list.

1.1 "The wave motion of a revolving shaft, and a suggestion as to the angular momentum in a beam of circularly polarised light" by JH Poynting. Reprinted by permission from *Proc. Roy. Soc. London* Ser. A **82** 560–567 (1909)

1.2 "Mechanical detection and measurement of the angular momentum of light" by RA Beth. Reprinted by permission from *Phys. Rev.* **50** 115–125 (1936)

1.3 "Introduction to the atoms and angular momentum of light special issue" by L Allen. Reprinted by permission from *J. Opt. B: Quantum Semiclass. Opt.* **4** S1–6 (2002)

2.1 "Orbital angular momentum of light and the transformation of Laguerre–Gaussian laser modes" by L Allen, MW Beijersbergen, RJC Spreeuw and JP Woerdman. Reprinted by permission from *Phys. Rev. A* **45** 8185–8189 (1992)

2.2 "Eigenfunction description of laser beams and orbital angular momentum of light" by SJ van Enk and G Nienhuis. Reprinted by permission from *Opt. Commun.* **94** 147–158 (1992)

2.3 "Paraxial wave optics and harmonic oscillators" by G Nienhuis and L Allen. Reprinted by permission from *Phys. Rev. A* **48** 656–665 (1993)

2.4 "Matrix formulation for the propagation of light beams with orbital and spin angular momenta" by L Allen, J Courtial and MJ Padgett. Reprinted by permission from *Phys. Rev. E* **60** 7497–7503 (1999)

2.5 "Paraxial beams of spinning light" by MV Berry. Reprinted by permission from *Singular Optics* (Eds. MS Soskin and MV Vasnetsov) SPIE 3487, 6–11 (1998)

2.6 "The Poynting vector in Laguerre–Gaussian beams and the interpretation of their angular momentum density" by L Allen and MJ Padgett. Reprinted by permission from *Opt. Commun.* **184** 67–71 (2000)

2.7 "Intrinsic and extrinsic nature of the orbital angular momentum of a light beam" by AT O'Neil, I MacVicar, L Allen and MJ Padgett. Reprinted by permission from *Phys. Rev. Lett.* **88** 053601-1-4 (2002)

2.8 "Spin and orbital angular momentum of photons" by SJ van Enk and G Nienhuis. Reprinted by permission from *Europhys. Lett.* **25** 497–501 (1994)

2.9 "Orbital angular momentum and nonparaxial light beams" by SM Barnett and L Allen. Reprinted by permission from *Opt. Commun.* **110** 670–678 (1994)

2.10 "Optical angular-momentum flux" by SM Barnett. Reproduced by permission from *J. Opt. B: Quantum Semiclass. Opt.* **4** S7–16 (2002)

3.1 "An experiment to observe the intensity and phase structure of Laguerre–Gaussian laser modes" by M Padgett, J Arlt, N Simpson and L Allen. Reprinted by permission from *Am. J. Phys.* **64** 77–82 (1996)
3.2 "Temporal and interference fringe analysis of $TEM_{01}{}^{*}$ laser modes" by JM Vaughan and DV Willets. Reprinted by permission from *J. Opt. Soc. A* **73** 1018–1021 (1983)
3.3 "Bistability and optical switching of spatial patterns in a laser" by C Tamm and CO Weiss. Reprinted by permission from *J. Opt. Soc. Am. B* **7** 1034–1038 (1990)
3.4 "Optical helices and spiral interference fringes" by M Harris, CA Hill and JM Vaughan. Reprinted by permission from *Opt. Commun.* **106** 161–166 (1994)
3.5 "Astigmatic laser mode converters and transfer of orbital angular momentum" by MW Beijersbergen, L Allen, HELO van der Veen and JP Woerdman. Reprinted by permission from *Opt. Commun.* **96** 123–132 (1993)
3.6 "Observation of the dynamical inversion of the topological charge of an optical vortex" by G Molina-Terriza, J Recolons, JP Torres, L Torner and EM Wright. Reprinted by permission from *Phys. Rev. Lett.* **87** 023902-1-4 (2001)
3.7 "Orbital angular momentum exchange in cylindrical-lens mode converters" by MJ Padgett and L Allen. Reprinted by permission from *J. Opt B. Quantum Semiclass. Opt.* **4** S17–19 (2002)
3.8 "Laser beams with screw dislocations in their wavefronts" by V Yu Bazhenov, MV Vasnetsov and MS Soskin. Reprinted by permission from *JETP Lett.* **52** 429–431. (1990)
3.9 "Laser beams with phase singularities" by NR Heckenberg, R McDuff, CP Smith, H Rubinsztein-Dunlop and MJ Wegener. Reprinted by permission from *Opt. Quantum Electron.* **24** S951–962 (1992)
3.10 "Topological charge and angular momentum of light beams carrying optical vortices" by MS Soskin, VN Gorshkov, MV Vasnetsov, JT Malos and NR Heckenberg. Reprinted by permission from *Phys Rev A* **56** 4064–4075 (1997)
3.11 "Helical-wavefront laser beams produced with a spiral phaseplate" by MW Beijersbergen, RPC Coerwinkel, M Kristensen and JP Woerdman. Reprinted by permission from *Opt. Commun.* **112** 321–327 (1994)
3.12 "The generation of free-space Laguerre–Gaussian modes at millimetre-wave frequencies by use of a spiral phaseplate" by GA Turnbull, DA Robertson, GM Smith, L Allen and MJ Padgett, *Opt. Commun.* **127** 183–188 (1996)
4.1 "Observation of a single-beam gradient force optical trap for dielectric particles" by A Ashkin, JM Dziedzic, JE Bjorkholm and S Chu. Reprinted by permission from *Opt. Lett.* **11** 288–290 (1986)
4.2 "Direct observation of transfer of angular momentum to absorptive particles from a laser beam with a phase singularity" by H He, MEJ Friese, NR Heckenberg and H Rubinsztein-Dunlop. Reprinted by permission from *Phys. Rev. Lett.* **75** 826–829 (1995)
4.3 "Optical angular-momentum transfer to trapped absorbing particles" by MEJ Friese, J Enger, H Rubinsztein-Dunlop and NR Heckenberg. Reprinted by permission from *Phys. Rev. A* **54** 1593–1596 (1996)
4.4 "Mechanical equivalence of spin and orbital angular momentum of light: an optical spanner" by NB Simpson, K Dholakia, L Allen and MJ Padgett. Reprinted by permission from *Opt. Lett.* **22** 52–54 (1997)
5.1 "Atom dynamics in multiple Laguerre–Gaussian beams" by L Allen, M Babiker, WK Lai and VE Lembessis. Reprinted by permission from *Phys. Rev. A* **54** 4259–4270 (1996)

5.2 "Selection rules and centre-of-mass motion of ultracold atoms" by SJ van Enk. Reprinted by permission from *Quantum Opt.* **6** 445–457 (1994)
5.3 "Optical pumping of orbital angular momentum of light in cold cesium atoms" by JWR Tabosa and DV Petrov. Reprinted by permission from *Phys. Rev. Lett.* **83** 4967–4970 (1999)
6.1 "Angular Doppler effect" by BA Garetz. Reprinted by permission from *J. Opt. Soc. Am.* **71** 609–611 (1981)
6.2 "Doppler effect induced by rotating lenses" by G Nienhuis. Reprinted by permission from *Opt. Commun.* **132** 8–14 (1996)
6.3 "Poincaré-sphere equivalent for light beams containing orbital angular momentum" by MJ Padgett and J Courtial. Reprinted by permission from *Opt. Lett.* **24** 430–432 (1999)
6.4 "Rotational frequency shift" by I Bialynicki-Birula and Z Bialynicka-Birula. Reprinted by permission from *Phys. Rev. Lett.* **78** 2539–2542 (1997)
6.5 "Azimuthal Doppler shift in light beams with orbital angular momentum" by L Allen, M Babiker and WL Power. Reprinted by permission from *Opt. Commun.* **112** 141–144 (1994)
6.6 "Rotational frequency shift of a light beam" by J Courtial, DA Robertson, K Dholakia, L Allen and MJ Padgett. Reprinted by permission from *Phys. Rev. Lett.* **81** 4828–4830 (1998)
7.1 "Second-harmonic generation and the conservation of orbital angular momentum with high-order Laguerre–Gaussian modes" by J Courtial, K Dholakia, L Allen and MJ Padgett. Reprinted by permission from *Phys. Rev. A* **56** 4193–4196 (1997)
7.2 "Optical solitons carrying orbital angular momentum" by WJ Firth and DV Skryabin. Reprinted by permission from *Phys. Rev. Lett.* **79** 2450–2453 (1997)
7.3 "Integer and fractional angular momentum borne on self-trapped necklace-ring beams" by M Soljačić and M Segev. Reprinted by permission from *Phys. Rev. Lett.* **86** 420–423 (2001)
8.1 "Entanglement of the orbital angular momentum states of photons" by A Mair, A Vaziri, G Weihs and A Zeilinger. Reprinted by permission from *Nature* **412** 313–316 (2001)
8.2 "Two-photon entanglement of orbital angular momentum states" by S Franke-Arnold, SM Barnett, MJ Padgett and L Allen. Reprinted by permission from *Phys. Rev. A* **65** 033823-1-6 (2002)
8.3 "Measuring the orbital angular momentum of a single photon" by J Leach, MJ Padgett, SM Barnett, S Franke-Arnold and J Courtial. Reprinted by permission from *Phys. Rev. Lett.* **88** 257901-1-4 (2002)

Section 1

INTRODUCTION

That light should have mechanical properties has been known, or at least suspected, since Kepler proposed that the tails of comets were due to radiation pressure associated with light from the sun. A quantitative theory of such effects became possible only after the development of Maxwell's unified theory of electricity, magnetism and optics. However, although his treatise on electromagnetism (1.1) contains a calculation of the radiation pressure at the earth's surface, there is little more on the mechanical effects of light. It was Poynting who quantified the momentum and energy flux associated with an electromagnetic field (1.2). In modern terms, the momentum per unit volume associated with an electromagnetic wave is given by $\varepsilon_0 \mathbf{E} \times \mathbf{B}$. The angular momentum density is, naturally enough, the cross product of this with position, that is $\mathbf{r} \times \varepsilon_0(\mathbf{E} \times \mathbf{B})$ (1.3).

Poynting reasoned that circularly polarised light must carry angular momentum (1.4, **Paper 1.1**). His argument proceeded by analogy with the wave motion associated with a line of dots marked on a rotating cylindrical shaft. His calculation showed that $E\lambda/2\pi$ is the angular momentum transmitted through a plane in unit time, per unit area, where E is Poynting's notation for the energy per unit volume and λ is the wavelength. When the energy of each photon crossing the surface is associated with $\hbar\omega$, we obtain the result that circularly polarised photons each carry $\hbar$ units of angular momentum.

Poynting's paper concludes with a proposal for measuring the angular momentum associated with circularly polarised light. His idea was that circularly polarised light passing through a large number of suspended quarter-wave plates, and so becoming linearly polarised, should give up all its angular momentum and so induce a rotation in the suspension. He concludes, however, that "my present experience of light-forces does not give me much hope that the effect could be detected". The effect was detected, however, about twenty years after Poynting's death by Beth (1.5, **Paper 1.2**) who used a single quarter wave plate, together with a mirror which sent the light back through the plate enhancing the torque on the suspension. He showed that the same quantitative result is obtained for the classical torque as for that which arises from the assumption that each photon carried an angular momentum $\hbar$.

Careful examination of $\varepsilon_0 \mathbf{r} \times (\mathbf{E} \times \mathbf{B})$ shows that polarisation does not account for all of the angular momentum that can be carried by the electromagnetic field. The part associated with polarisation is known as spin, but in addition there is also an orbital contribution. Until recently, however, the discussion of spin and orbital angular momenta for light was largely restricted to textbooks and related to non-specific forms for the electric field. Current research activity in this area originated with the realisation that physically realisable light beams, familiar from the paraxial optics of laser theory, can carry a well-defined quantity of orbital angular momentum for each photon (1.6, **Paper 2.1**). An extensive review of orbital angular momentum was published in 1999 (1.7). This was followed, in 2002, by the publication of a special issue of *Journal of Optics B* devoted to atoms and angular momentum of light. The introduction to that special issue provides a brief overview of the current status of the field and is reprinted here as the natural introduction to this book (1.8, **Paper 1.3**).

REFERENCES

1.1 JC Maxwell, 1891, *A Treatise on Electricity and Magnetism* (Oxford: Clarendon Press) Art. 793. These volumes are available in the series *Oxford Classic Texts in the Physical Sciences* (Oxford: Oxford University Press).

1.2 JH Poynting, 1884, *Phil. Trans.* **174**, 343. This paper is also available in 1920, *Collected Scientific Papers by John Henry Poynting* (Cambridge: Cambridge University Press).

1.3 JD Jackson, 1999, *Classical Electrodynamics* 3rd edn (New York: Wiley) p. 608.

1.4 JH Poynting, 1909, *Proc. Roy. Soc. A* **82** 560.

1.5 RA Beth, 1936, *Phys. Rev.* **50** 115.

1.6 L Allen, MW Beijersbergen, RJC Spreeuw and JP Woerdman, 1992, *Phys. Rev. A* **45** 8185.

1.7 L Allen, MJ Padgett and M Babiker, 1999, *Prog. Opt.* **39** 291.

1.8 L Allen, 2002, *J. Opt. B: Quantum Semiclass. Opt.* **4** S1.

The Wave Motion of a Revolving Shaft, and a Suggestion as to the Angular Momentum in a Beam of Circularly Polarised Light

By JH Poynting, ScD, FRS
(Received June 2,—Read June 24, 1909)

When a shaft of circular section is revolving uniformly, and is transmitting power uniformly, a row of particles originally in a line parallel to the axis will lie in a spiral of constant pitch, and the position of the shaft at any instant may be described by the position of this spiral.

Let us suppose that the power is transmitted from left to right, and that as viewed from the left the revolution is clockwise. Then the spiral is a left-handed screw. Let it be on the surface, and there make an angle ε with the axis. Let the radius of the shaft be a, and let one turn of the spiral have length λ along the axis. We may term λ the wave-length of the spiral. We have $\tan\varepsilon = 2\pi a/\lambda$. If the orientation of the section at the origin at time t is given by $\theta = 2\pi \mathrm{N}t$, where N is the number of revolutions per second, the orientation of the section at x is given by

$$\theta = 2\pi \mathrm{N}t - \frac{x}{a}\tan\varepsilon = \frac{2\pi}{\lambda}(\mathrm{N}\lambda t - x), \tag{1}$$

which means movement of orientation from left to right with velocity $\mathrm{N}\lambda$.

The equation of motion for twist waves on a shaft of circular section is

$$\frac{d^2\theta}{dt^2} = \mathrm{U}_n{}^2\frac{d^2\theta}{dx^2}, \tag{2}$$

where $\mathrm{U}_n{}^2$ = modulus of rigidity/density = n/ρ.

Though (1) satisfies (2), it can hardly be termed a solution for $d^2\theta/dt^2$, and $d^2\theta/dx^2$ in (1) are both zero. But we may adapt a solution of (2) to fit (1) if we assume certain conditions in (1).

The periodic value

$$\theta = \Theta \sin\frac{2\pi}{l}(\mathrm{U}_n t - x)$$

satisfies (2), and is a wave motion with velocity U_n and wave-length l. Make l so great that for any time or for any distance under observation $U_n t/l$ and x/l are so small that the angle may be put for the sine. Then

$$\theta = \Theta \frac{2\pi}{l}(U_n t - x). \tag{3}$$

This is uniform rotation. It means that we only deal with the part of the wave near a node, and that we make the wave-length l so great that for a long distance the "displacement curve" obtained by plotting θ against t coincides with the tangent at the node. We must distinguish, of course, between the wave-length l of the periodic motion and the wave-length λ of the spiral.

We can only make (1) coincide with (3) by putting

$$\Theta/l = 1/\lambda \quad \text{and} \quad N\lambda = U_n.$$

Then it follows that for a given value of N, the impressed speed of uniform rotation, there is only one value of λ or one value of ε for which the motion may be regarded as part of a natural wave system, transmitted by the elastic forces of the material with velocity $= \sqrt{(n/\rho)}$. There is therefore only one "natural" rate of transmission of energy.

The value of ε is given by

$$\tan\varepsilon = 2\pi a/\lambda = 2\pi a N/N\lambda = 2\pi a N/U_n = 2\pi a N\sqrt{(\rho/n)}.$$

Suppose, for instance, that a steel shaft with radius $a = 2\,\text{cm}$, density $\rho = 7.8$, and rigidity $n = 10^{12}$ is making $N = 10$ revs. per sec. We may put $\tan\varepsilon = \varepsilon$, since it is very small. The shaft is twisted through 2π in length λ or through $2\pi/\lambda$ per centimetre, and the torque across a section is

$$G = \tfrac{1}{2}n\pi a^4 2\pi/\lambda = n\pi^2 a^4 N\sqrt{(\rho/n)},$$

since

$$\lambda = \frac{U_n}{N} = \frac{1}{N}\sqrt{\frac{n}{\rho}}.$$

The energy transmitted per second is

$$2\pi NG = 2\pi^3 a^4 N^2 \sqrt{(n\rho)}.$$

Putting 1 H.P. $= 746 \times 10^7$ ergs per second, this gives about 38 H.P.

But a shaft revolving with given speed N can transmit any power, subject to the limitation that the strain is not too great for the material. When the power is not that "naturally" transmitted, we must regard the waves as "forced." The velocity of transmission is no longer U_n, and forces will have to be applied from outside in addition to the internal elastic forces to give the new velocity.

Let H be the couple applied per unit length from outside. Then the equation of motion becomes

$$\frac{d^2\theta}{dt^2} = U_n{}^2 \frac{d^2\theta}{dx^2} + \frac{2H}{\pi a^4},$$

where $\frac{1}{2}\pi a^4$ is the moment of inertia of the cross section. Assuming that the condition travels on with velocity U unchanged in form,

$$\frac{d\theta}{dt} = -\mathrm{U}\frac{d\theta}{dx} \quad \text{and} \quad \mathrm{H} = \tfrac{1}{2}\pi a^4(\mathrm{U}_n{}^2)\frac{d^2\theta}{dx^2},$$

or H has only to be applied where $d^2\theta/dx^2$ has value, that is where the twist is changing.

The following adaptation of Rankine's tube method of obtaining wave velocities* gives these results in a more direct manner. Suppose that the shaft is indefinitely extended both ways. Any twist disturbance may be propagated unchanged in form with any velocity we choose to assign, if we apply from outside the distribution of torque which, added to the torque due to strain, will make the change in twist required by the given wave motion travelling at the assigned speed.

Let the velocity of propagation be U from left to right, and let the displacement at any section be θ, positive if clockwise when seen from the left. The twist per unit length is

$$\frac{d\theta}{dx} = -\frac{1}{\mathrm{U}}\frac{d\theta}{dt} = -\frac{\dot{\theta}}{\mathrm{U}}.$$

The torque across a section from left to right in clockwise direction is

$$-\tfrac{1}{2}n\pi a^4\frac{d\theta}{dx} = \frac{n\pi a^4}{2\mathrm{U}}\,.\,\dot{\theta}.$$

Let the shaft be moved from right to left with velocity U; then the disturbance is fixed in space, and if we imagine two fixed planes drawn perpendicular to the axis, one, A, at a point where the disturbance is θ and the other, B, outside the wave system, where there is no disturbance, the condition between A and B remains constant, except that the matter undergoing that condition is changing. Hence the total angular momentum between A and B is constant. But no angular momentum enters at B, since the shaft is there untwisted and has merely linear motion. At A, then, there must be on the whole no transfer of angular momentum from right to left. Now, angular momentum is transferred in three ways:—

1. By the carriage by rotating matter. The angular momentum per unit length is $\frac{1}{2}\rho\pi a^4\dot{\theta}$, and since length U per second passes out at A, it carries out $\frac{1}{2}\rho\pi a^4\dot{\theta}\mathrm{U}$.

2. By the torque exerted by matter on the right of A on matter on the left of A. This takes out $-n\pi a^4\dot{\theta}/2\mathrm{U}$.

3. By the stream of angular momentum by which we may represent the forces applied from outside to make the velocity U instead of U_n.

If H is the couple applied per unit length, we may regard it as due to the flow of angular momentum L along the shaft from left to right, such that $\mathrm{H} = -d\mathrm{L}/dx$. There is then angular momentum L flowing out per second from right to left. Since the total flow due to (1), (2), and (3) is zero,

$$\tfrac{1}{2}\rho\pi a^4\dot{\theta}\mathrm{U} - n\pi a^4\dot{\theta}/2\mathrm{U} - \mathrm{L} = 0,$$

and

$$\mathrm{L} = \frac{\pi a^4\dot{\theta}}{2}\left(\rho\mathrm{U} - \frac{n}{\mathrm{U}}\right) = \frac{\rho\pi a^4\dot{\theta}}{2\mathrm{U}}(\mathrm{U}_n{}^2) = -\frac{\rho\pi a^4}{2}\frac{d\theta}{dx}(\mathrm{U}^2 - \mathrm{U}_n{}^2),$$

* 'Phil. Trans.,' 1870, p. 277.

and

$$H = -\frac{dL}{dx} = \frac{\rho\pi a^4}{2}\frac{d^2\theta}{dx^2}(U_n^{\ 2}).$$

If $H = 0$, either $U^2 = U_n^{\ 2}$ when the velocity has its "natural value," or $d^2\theta/dx^2 = 0$, and the shaft is revolving with uniform twist in the part considered.

Now put on to the system a velocity U from left to right. The motion of the shaft parallel to its axis is reduced to zero, and the disturbance and the system H will travel on from left to right with velocity U. A "forced" velocity does not imply *transfer* of physical conditions by the material with that velocity. We can only regard the conditions as reproduced at successive points by the aid of external forces. We may illustrate this point by considering the incidence of a wave against a surface. If the angle of incidence is i and the velocity of the wave is V, the line of contact moves over the surface with velocity $v = V/\sin i$, which may have any value from V to infinity. The velocity v is not that of transmission by the material of the surface, but merely the velocity of a condition impressed on the surface from outside.

Probably in all cases of transmission with forced velocity, and certainly in the case here considered, the velocity depends upon the wave-length, and there is dispersion.

With a shaft revolving N times per second $U = N\lambda$, and it is interesting to note that the group velocity $U - \lambda dU/d\lambda$ is zero. It is not at once evident what the group velocity signifies in the case of uniform rotation. In ordinary cases it is the velocity of travel of the "beat" pattern, formed by two trains of slightly different frequencies. The complete "beat" pattern is contained between two successive points of agreement of phase of the two trains. In our case of superposition of two strain spirals with constant speed of rotation, points of agreement of phase are points of intersection of the two spirals. At such points the phases are the same, or one has gained on the other by 2π. Evidently as the shaft revolves these points remain in the same cross-section, and the group velocity is zero.

With deep water waves the group velocity is half the wave velocity, and the energy flow is half that required for the onward march of the waves.* The energy flow thus suffices for the onward march of the group, and the case suggests a simple relation between energy flow and group velocity.

But the simplicity is special to unforced trains of waves. Obviously, it does not hold when there are auxiliary working forces adding or subtracting energy along the waves. For the revolving shaft the simple relation would give us no energy flow, whereas the strain existing in the shaft implies transmission of energy at a rate given as follows.

The twist per unit length is $d\theta/dx$, and therefore the torque across a section is $-\frac{1}{2}n\pi a^4\, d\theta/dx$, or $\frac{1}{2}n\pi a^4\dot{\theta}/U$, since $d\theta/dx = -\dot{\theta}/U$. The rate of working or of energy flow across the section is $\frac{1}{2}n\pi a^4\dot{\theta}^2/U$.

The relation of this to the strain and kinetic energy in the shaft is easily found. The strain energy per unit length being $\frac{1}{2}$ (couple $\times$ twist per unit length) is $\frac{1}{4}n\pi a^4(d\theta/dx)^2$, which is $\frac{1}{4}n\pi a^4\dot{\theta}^2/U^2$. The kinetic energy per unit length is $\frac{1}{2}\rho\pi a^4\dot{\theta}^2$, or, putting $\rho = n/U_n^{\ 2}$, is $\frac{1}{4}n\pi a^4\dot{\theta}^2/U_n^{\ 2}$.

In the case of natural velocity, for which no working forces along the shaft are needed, when $U = U_n = \sqrt{(n/\rho)}$, the kinetic energy is equal to the strain energy at every point and the energy transmitted across a section per second is that contained in length U_n.

*O. Reynolds, 'Nature,' August 23, 1877; Lord Rayleigh, 'Theory of Sound,' vol. 1, p. 477.

But if the velocity is forced this is no longer true,* and it is easily shown that the energy transferred is that in length

$$\frac{2U}{1+U^2/U_n{}^2},$$

which is less than U if $U > U_n$, and is greater than U if $U < U_n$.

It appears possible that always the energy is transmitted along the shaft at the speed U_n. If the forced velocity $U > U_n$, we may, perhaps, regard the system in a special sense as a natural system with a uniform rotation superposed on it.

Let us suppose that the whole of the strain energy in length U_n is transferred per second while only the fraction μ of the kinetic energy is transferred, the fraction $1-\mu$ being stationary.

The energy transferred : strain energy in U_n : kinetic energy in $U_n = 1/U : U_n/2U^2 : U_n/2U_n{}^2$.

Put $U = pU_n$, and our supposition gives

$$\frac{1}{pU_n} = \frac{1}{2p^2U_n} + \frac{\mu}{2U_n} \quad \text{or} \quad \mu = \frac{2}{p} - \frac{1}{p^2} = 1 - \left(1 - \frac{1}{p}\right)^2.$$

If the forced velocity $U < U_n$, we may regard the system as a natural one, with a uniform stationary strain superposed on it.

We now suppose that the whole of the kinetic energy is transferred, but only a fraction ν of the strain energy, and we obtain

$$\frac{1}{pU_n} = \frac{\nu}{2p^2U_n} + \frac{1}{2U_n} \quad \text{or} \quad \nu = 2p - p^2 = 1 - (1-p)^2.$$

It is perhaps worthy of note that a uniform longitudinal flow of fluid may be conceived as a case of wave motion in a manner similar to that of the uniform rotation of a shaft.

A SUGGESTION AS TO THE ANGULAR MOMENTUM IN A BEAM OF CIRCULARLY POLARISED LIGHT

A uniformly revolving shaft serves as a mechanical model of a beam of circularly polarised light. The expression for the orientation θ of any section of the shaft distant x from the origin, $\theta = 2\pi\lambda^{-1}(Ut - x)$, serves also as an expression for the orientation of the disturbance, whatever its nature, constituting circularly polarised light.

For simplicity, take a shaft consisting of a thin cylindrical tube. Let the radius be a, the cross-section of the material s, the rigidity n, and the density ρ. Let the tube make N revolutions per second, and let it have such twist on it that the velocity of transmission of the spiral indicating the twist is the natural velocity $U_n = \sqrt{(n/\rho)}$.

Repeating for this special case what we have found above, the strain energy per unit length is $\frac{1}{2}n\varepsilon^2 s$, or, since $\varepsilon = a\,d\theta/dx = -a\dot{\theta}/U_n$, the strain energy is $\frac{1}{2}na^2 s\dot{\theta}^2/U_n{}^2 = \frac{1}{2}\rho a^2 s\dot{\theta}^2$.

*In the Sellmeier model illustrating the dispersion of light, the particles may be regarded as outside the material transmitting the waves and as applying forces to the material which make the velocity forced. The simple relation between energy flow and group velocity probably does not hold for this model.

But the kinetic energy per unit length is also $\frac{1}{2}\rho a^2 s\dot{\theta}^2$, so that the total energy in length U_n is $\rho a^2 s\dot{\theta}^2 U_n$. The rate of working across a section is

$$n\varepsilon sa\dot{\theta} = na^2 s\dot{\theta}^2/U_n = \rho a^2 s\dot{\theta}^2 U_n,$$

or the energy transferred across a section is the energy contained in length U_n.

If we put E for the energy in unit volume and G for the torque per unit area, we have $Gs\dot{\theta} = EsU_n$, whence

$$G = EU_n/\dot{\theta} = EN\lambda/2\pi N = E\lambda/2\pi.$$

The analogy between circularly polarised light and the mechanical model suggests that a similar relation between torque and energy may hold in a beam of such light incident normally on an absorbing surface. If so, a beam of wave-length λ containing energy E per unit volume will give up angular momentum $E\lambda/2\pi$ per second per unit area. But in the case of light waves $E = P$, where P is the pressure exerted. We may therefore put the angular momentum delivered to unit area per second as

$$P\lambda/2\pi.$$

In the 'Philosophical Magazine,' 1905, vol. 9, p. 397, I attempted to show that the analogy between distortional waves and light waves is still closer, in that distortional waves also exert a pressure equal to the energy per unit volume. But as I have shown in a paper on "Pressure Perpendicular to the Shear Planes in Finite Pure Shears, etc." (*ante*, p. 546), the attempt was faulty, and a more correct treatment of the subject only shows that there is probably a pressure. We cannot say more as to its magnitude than that if it exists it is of the order of the energy per unit volume.

When a beam is travelling through a material medium we may, perhaps, account for the angular momentum in it by the following considerations. On the electromagnetic theory the disturbance at any given point in a circularly polarised beam is a constant electric strain or displacement f uniformly revolving with angular velocity $\dot{\theta}$. In time dt it changes its direction by $d\theta$.

This may be effected by the addition of a tangential strain $fd\theta$; or the rotation is produced by the addition of tangential strain $f\dot{\theta}$ per second, or by a current $f\dot{\theta}$ along the circle described by the end of f. We may imagine that this is due to electrons drawn out from their position of equilibrium so as to give f, and then whirled round in a circle so as to give a circular convection current $f\dot{\theta}$. Such a circular current of electrons should possess angular momentum.

Let us digress for a moment to consider an ordinary conduction circuit as illustrating the possession of angular momentum on this theory. Let the circuit have radius a and cross-section s, and let there be N negative electrons per unit volume, each with charge e and mass m, and let these be moving round the circuit with velocity v. If i is the total current, $i = Nsve$. The angular momentum will be

$$Ns2\pi a\,.\,mva = 2\pi a^2 im/e = 2Aim/e = 2Mm/e,$$

where A is the area of the circuit and M is the magnetic moment. This is of the order of $2M/10^7$.

It is easily seen that this result will hold for any circuit, whatever its form if A is the projection of the circuit on a plane perpendicular to the axis round which the moment is taken and if $M = Ai$. If we suppose that a current of negative electrons flows round the circuit in this way and that the reaction while their momentum is being established is on the material of the conductor, then at make of current there should be an impulse on the conductor of moment $2M/10^7$. If the circuit could be suspended so that it lay in a horizontal plane and was able to turn about a vertical axis in a space free from any magnetic field, we might be able to detect such impulse if it exists. But it is practically impossible to get a space free from magnetic intensity. If the field is H, the couple in the circuit due to it is proportional to HM. It would require exceedingly careful construction and adjustment of the circuit to ensure that the component of the couple due to the field about the vertical axis was so small that its effect should not mask the effect of the impulsive couple. The electrostatic forces, too, might have to be considered as serious disturbers.

Returning to a beam of circularly polarised light, supposed to contain electrons revolving in circular orbits in fixed periodic times, the relations between energy and angular momentum are exactly the same as those in a revolving shaft or tube, and the angular momentum transmitted per second per square centimetre is $E\lambda/2\pi = P\lambda/2\pi$, where P is the pressure of the light per square centimetre on an absorbing surface.

The value of this in any practical case is very small. In light pressure experiments, P is detected by the couple on a small disc, of area A say, at an arm b and suspended by a fibre. What we observe is the moment APb. If the same disc is suspended by a vertical fibre attached at its centre and the same beam circularly polarised in both cases is incident normally upon it, according to the value suggested the torque is $AP\lambda/2\pi$.

The ratio of the two is $\lambda/2\pi b$. Now b is usually of the order of 1 cm. Put $\lambda = 6 \times 10^{-5}$, or, say, $2\pi/10^{-5}$, and the ratio becomes 10^{-5}.

It is by no means easy to measure the torque APb accurately, and it appears almost hopeless to detect one of a hundred-thousandth of the amount. The effect of the smaller torque might be multiplied to some extent, as shown in accompanying diagram.

Let a series of quarter wave plates, $p_1, p_2, p_3, \ldots$, be suspended by a fibre above a Nicol prism N, through which a beam of light is transmitted upwards, and intermediate between these let a series of quarter wave plates, $q_1, q_2, q_3, \ldots$, be fixed, each with a central hole for the free passage of the fibre. The beam emerges from N plane polarised. If N is placed so that the beam after passing through p_1 is circularly polarised, it has gained angular momentum, and therefore tends to twist p_1 round. The next plate q_1 is to be arranged so that the beam emerges from it plane polarised and in the original plane. It then passes through p_2, which is similar to p_1, and again it is circularly polarised and so exercises another torque. The process is repeated with q_2 and p_3, and so on till the beam is exhausted. By revolving N through a right angle round the beam, the effect is reversed. But, even with such multiplications, my present experience of light forces does not give me much hope that the effect could be detected, if it has the value suggested by the mechanical model.

PAPER 1.2

Mechanical Detection and Measurement of the Angular Momentum of Light

RICHARD A. BETH,* *Worcester Polytechnic Institute, Worcester, Mass. and Palmer Physical Laboratory, Princeton University*

(Received May 8, 1936)

The electromagnetic theory of the torque exerted by a beam of polarized light on a doubly refracting plate which alters its state of polarization is summarized. The same quantitative result is obtained by assigning an angular momentum of $\hbar$ $(-\hbar)$ to each quantum of left (right) circularly polarized light in a vacuum, and assuming the conservation of angular momentum holds at the face of the plate. The apparatus used to detect and measure this effect was designed to enhance the moment of force to be measured by an appropriate arrangement of quartz wave plates, and to reduce interferences. The results of about 120 determinations by two observers working independently show the magnitude and sign of the effect to be correct, and show that it varies as predicted by the theory with each of three experimental variables which could be independently adjusted.

ELECTROMAGNETIC FIELD THEORY

THE moment of force or torque exerted on a doubly refracting medium by a light wave passing through it arises from the fact that the dielectric constant **K** is a tensor. Consequently the electric intensity **E** is, in general, not parallel to the electric polarization **P** or to the electric displacement

$$\mathbf{D}=\mathbf{KE}=\mathbf{E}+4\pi\mathbf{P}$$

in the medium. The torque per unit volume produced by the action of the electric field on the polarization of the medium is

$$\mathbf{l}=\mathbf{P}\times\mathbf{E}=(\mathbf{D}\times\mathbf{E})/4\pi. \tag{1}$$

Following Sadowsky and Epstein[1] we may calculate this torque for a simple case as follows: Assume a doubly refracting medium of permeability unity and take as x, y, and z axes the principal axes of the tensor **K** for the light frequency in question. Denote the principal values of **K** by n_x^2, n_y^2 and n_z^2 so that n_x, n_y and n_z will be the principal indices of refraction. Let the electric components of a plane light wave propagated in the $+z$ direction be given by

$$\mathbf{E}=A\cos\theta\cos(Z_1+\Delta),\quad A\sin\theta\cos(Z_1-\Delta),\quad 0,$$

$$\mathbf{D}=\mathbf{KE}=n_x^2A\cos\theta\cos(Z_1+\Delta),\quad n_y^2A\sin\theta\cos(Z_1-\Delta),\quad 0, \tag{2}$$

where $Z_1=\omega(t-nz/c)$; $n=(n_y+n_x)/2$,

$$\Delta=\pi z(n_y-n_x)/\lambda;\quad \omega=2\pi c/\lambda.$$

For any value of z (in particular $z=0$) for which the wave plate thickness Δ/π is a whole number the light is plane polarized and E makes an angle $+\theta$ with the x axis.

The torque per unit volume at any point in the crystal is calculated from (2) according to (1) $\mathbf{l}=0,\ 0,$

$$(A^2/8\pi)(n_x^2-n_y^2)\sin 2\theta\cos(Z_1+\Delta)\cos(Z_1-\Delta)$$

and the time average value of the z component is

$$(A/4\pi)^2(n_x^2-n_y^2)\sin 2\theta\cos 2\Delta.$$

Integrating from z_1 to z_2 we get the torque per unit area on the crystal between these values:

$$L=-(A/4\pi)^2 n\lambda\sin 2\theta(\sin 2\Delta_2-\sin 2\Delta_1). \tag{3}$$[2]

This torque, which has hitherto been generally considered too small for experimental detection, was detected and measured in the present experiment.[3]

* Member of the Physics Department, Worcester Polytechnic Institute. Preliminary work and final calculations were carried on at Worcester. Experimental work was done on leave-of-absence from Worcester as Research Associate at Palmer Physical Laboratory, Princeton University. Additional measurements were made by Mr. W. Harris at Princeton, as will be explained later.

[1] A. Sadowsky, Acta et Commentationes Imp. Universitatis Jurievensis **7**, No. 1–3 (1899); **8**, No. 1–2 (1900). P. S. Epstein, Ann. d. Physik **44**, 593 (1914). The derivation is repeated here because the first articles are in Russian and relatively inaccessible, while in the last there seems to be a misprint in the result. I wish to thank Dr. Boris Podolsky for translating parts of Sadowsky's articles for me from the Russian. I am also deeply indebted to Professor A. Einstein not only for his advice in checking Professor Epstein's calculation but also for several interesting discussions about the experimental part of the work here reported.

[2] J. H. Poynting, Proc. Roy. Soc. **A82**, 560 (1909), infers from an ingenious mechanical analogy that circularly polarized light should exert a torque equal to the light energy per unit volume times $\lambda/2\pi$ on unit area of a quarter-wave plate which makes the light plane polarized. This result, which neglects surface reflections, is contained in (3).

[3] R. A. Beth, Phys. Rev. **48**, 471 (1935); also Annual Meeting of the Am. Phys. Soc., St. Louis, Missouri, January 1, 1936. A. H. S. Holbourn, Nature **137**, 31 (1936), also reports successful measurement of the effect.

Quantum Theory

It is worth observing that (3) may be derived from the quantum theory by assigning an angular momentum or spin of $\hbar(-\hbar)$[4] to each photon of left (right) circularly polarized light, the spin axis being in the direction of propagation of the light.[5]

Consider an elliptically polarized light wave, propagated in the $+z$ direction in a vacuum, whose components in the directions of the principal axes of the ellipse (phase difference $=\pi/2$) have amplitudes X_0 and Y_0. Let the components of the same wave, when resolved along an arbitrary pair of perpendicular x and y axes in the plane of the ellipse, be given by

$$\mathbf{E}=X\cos(Z+\Delta),\quad Y\cos(Z-\Delta),\quad 0, \tag{4}$$

where $Z=\omega(t-z/c)$ and 2Δ is the phase angle by which the y component lags behind the x component of the wave. It may be shown that[6]

$$XY\sin 2\Delta=X_0Y_0,\quad X^2+Y^2=X_0^2+Y_0^2. \tag{5}$$

If we now regard the wave (4) as the superposition of left and right circularly polarized components of amplitudes L and R, respectively, calculation shows that

$$L^2-R^2=XY\sin 2\Delta,\quad L^2+R^2=(X^2+Y^2)/2.$$

The number of left circularly polarized photons transmitted per unit area per second is the Poynting energy flow $cL^2/4\pi$ divided by the energy per photon $2\pi\hbar\nu=2\pi\hbar c/\lambda$, or $\lambda L^2/8\pi^2\hbar$. Multiplying by $\hbar$ gives $\lambda L^2/8\pi^2$, the angular momentum transmitted per unit area per second by the left circularly polarized component. Taking account of the similar expression for the right circularly polarized component, the angular momentum transmitted per unit area per second by the wave (4) is

$$M=\lambda(L^2-R^2)/8\pi^2=\lambda XY\sin 2\Delta/8\pi^2$$
$$=\lambda X_0Y_0/8\pi^2. \tag{6}$$

It is well known that the linear momentum transmitted per unit area per second is the energy flow divided by c, in this case:

$$(L^2+R^2)/4\pi=(X^2+Y^2)/8\pi=(X_0^2+Y_0^2)/8\pi.$$

Thus the angular and linear momenta are proportional to the natural invariants (5) of the wave (4). For $X_0=Y_0$ their ratio is $\lambdabar=\lambda/2\pi$, which is another form of Poynting's result.[2]

The value (6) of the angular momentum, which is independent of the quantum constant $\hbar$, can be derived from the linear momentum (Poynting vector divided by c) in a finite beam, but the elementary method given is more descriptive in bringing out the fact that the quantum theory leads to the same value (3) as the wave theory for the torque on a section of a crystal.

To show this equivalence we have now to obtain the angular momentum for a wave (2) *in the crystal* from the expression (6) for a wave *in a vacuum*, by placing the face of the crystal at $z=z_1$ (vacuum for $z<z_1$) and using the conservation of angular momentum and Fresnel's expressions for the amplitudes of the reflected and transmitted waves. In the reflected wave the y component of amplitude $Y(n_y-1)/(n_y+1)$ lags behind the x component of amplitude $X(n_x-1)/(n_x+1)$ by the phase angle 2Δ. We get the angular momentum transmitted through the face $z=z_1$ by calculating (6) for the direct and reflected waves in the vacuum (taking account of direction of transmission of each) and assuming the conservation of angular momentum at the face

$$M_1=XY(n_x+n_y)\lambda\sin 2\Delta/4\pi^2(n_x+1)(n_y+1).$$

Using Fresnel's expressions for the amplitudes, we identify the transmitted wave *in the crystal* with (2) at $z=z_1$ by choosing

$$A\cos\theta=2X/(n_x+1);\quad A\sin\theta=2Y/(n_y+1);$$
$$\Delta=\Delta_1=\pi z_1(n_y-n_x)/\lambda.$$

Hence the angular momentum transmitted by (2) at $z=z_1$ is

$$M_1=(A/4\pi)^2 n\lambda\sin 2\theta\sin 2\Delta_1. \tag{7}$$

It can now be seen directly that the torque (3) is equal to the excess of the angular momentum (7) per unit area per second flowing into a section of the crystal at $z=z_1$ over that flowing out at $z=z_2$. Hence both the field and the

---

[4] $\hbar=h/2\pi$.

[5] A. E. Ruark and H. C. Urey, Proc. Nat. Acad. Sci. **13**, 763 (1927); Harnwell and Livingood, *Experimental Atomic Physics* (1933), p. 81; Brillouin, *Les Statistiques Quantiques*, Chap. III (1930).

[6] See, e.g., Schuster and Nicholson, *Theory of Optics*, third edition, pp. 14, 15 (1924), especially Eqs. (15) and (17).

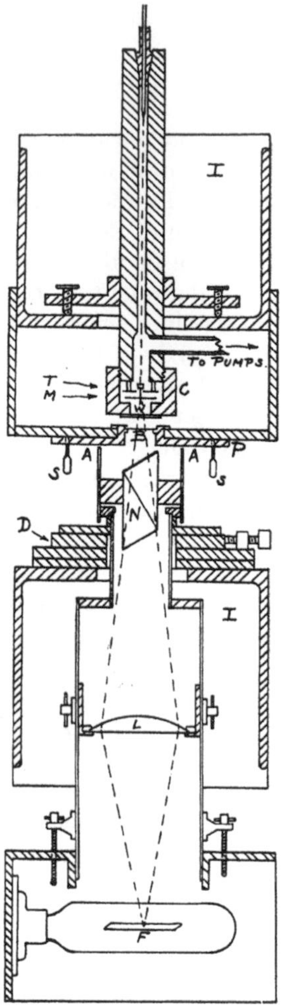

FIG. 1. Diagram of apparatus.

FIG. 2. Photograph of apparatus.

quantum theories predict the same value for the effect which has been measured in the present experiment.

APPARATUS

The basic idea used in detecting and measuring the effect[7] is to observe the deflection of a quartz wave plate hung from a fine quartz fiber when suitably polarized light is sent through the plate. In order to increase the effect to be measured and to avoid as far as possible interfering effects due to mechanical and electrostatic disturbances, stray heating effects, radiometer and gas effects, and variations in light pressure, the apparatus was constructed as shown in Figs. 1, 2, 3, and 4. Some of the precautions taken to avoid these interferences may have been unnecessary, but they were sufficient.

The whole apparatus is supported on the heavy cast iron brackets I in Fig. 1, each 12″×12″, which are bolted to a brick pier to avoid mechanical disturbances, especially torsional vibrations of a period comparable to that

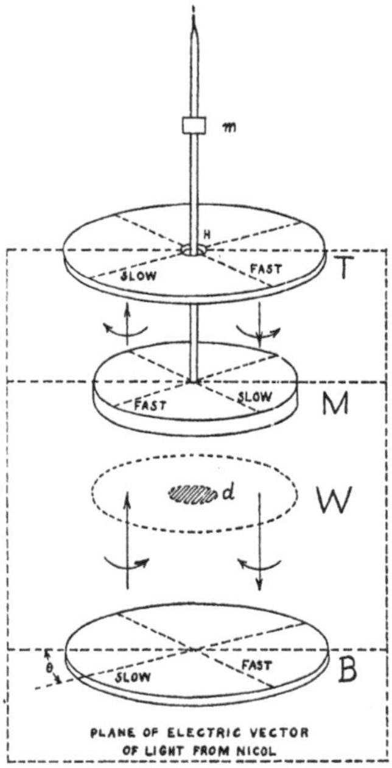

FIG. 3. Wave plate arrangement.

[7] See Poynting, reference 2, A. Kastler, Société des Sciences physiques et naturelles de Bordeaux, Jan. 28, 1932. R. A. Beth, abstract, Boston Meeting, American Physical Society, Phys. Rev. **45**, 296 (1934).

FIG. 4. Detail showing method by which light beam could be given lateral adjustment.

of the torsional pendulum formed by the one-inch circular wave plate M hanging at the bottom of the quartz fiber (about ten minutes). The fiber is about 25 centimeters long and hangs in a quarter inch hole bored lengthwise through a copper cylinder two inches in diameter. The upper end of the fiber is attached with a bit of flake shellac to a conical copper plug which was carefully ground in place and sealed with Apiezon grease. Thus the zero position as well as the amplitude of the torsional pendulum can be changed while the chamber is evacuated.

The hanging plate and the fixed plate above it, Fig. 3, are located in the cylindrical vacuum chamber C which is of copper, three inches in diameter with walls one-half inch thick. The heavy copper chamber is intended to eliminate unequal heating effects from the room as well as to shield off stray light and electrostatic disturbances.[8]

The chamber was evacuated through wide tubing by a high speed diffusion pump and three liquid-air traps. A two-stage diffusion pump and forepump provided the forevacuum for the high speed pump. According to the rated air calibration of the Western Electric ionization gauge, pressures below 10^{-6} mm of mercury existed in the chamber while the effect was being measured; no noticeable fluctuations appeared in the actual measurements which might be ascribed to radiometer or gas effects.

Light from a three-millimeter tungsten ribbon filament F is focused by the fused quartz lens L, 10.7 centimeters in diameter, through a large Nicol prism N, the lower plate B (Fig. 3), the fused quartz window W, and the hanging plate M, on the reflecting layer of aluminum on the top of the upper plate T in the chamber. The maximum deviation of the light from the vertical was about 10°. W has a circular aperture one inch in diameter. A round one-quarter inch copper disk d fastened in the middle of W prevents light from passing through the hole H in the top plate T. With this arrangement no light energy reaches the fiber and most of the energy is reflected out of the vacuum chamber altogether, thus minimizing fluctuations in the position of the pendulum due to unequal heating of the fiber and light pressure on the small mirror m. These undesirable effects are found to be present if the shield d is removed.

The lower plate B is placed outside the vacuum chamber in a brass ring bushing held so it is free to turn in a brass frame attached to the upper cast iron bracket. The position of the lower wave plate may be read on a circular scale by means of a pointer attached to the bushing. A circular brass plate P attached to the under side of the bushing carries adjustable stops S, and the plate may be rotated from one stop to the other while observing the swing of the pendulum (with a telescope placed in front of the apparatus) by means of cords running in grooves on the edge of the plate P. Frame, plate P, cords, stops, pointer and circular scale are visible at the top of Fig. 4.

The ideas underlying the wave plate arrangement, Fig. 3, may be qualitatively understood as follows. The light coming upward through the plates is reflected by the aluminum player on the top side of T, passing downward again through the plates. For a certain wave-length λ_0, B and T are quarter wave plates with their axes as shown at 90° to those of the hanging half-wave plate M. If light of wave-length λ_0 from the Nicol, plane polarized at $\theta=45°$, enters the bottom plate B, both direct and reflected beams will be circularly polarized in each space between the plates in the directions indicated by the

[8] I wish to thank Professor G. P. Harnwell for this and many other helpful suggestions concerning the mechanical and optical design of the apparatus.

curved arrows. Thus the angular momentum delivered per second to the half-wave plate M is almost four times what could be obtained by simply absorbing the same amount of circularly polarized light.

For wave-lengths λ different from λ_0 it will be seen that, because of the alternate advancing and retarding of each component with respect to the other, light in both direct and reflected beams is plane polarized at the initial angle θ at the bottom of B, the middle of M, and the top of T. If the plates vary somewhat from the exact specifications these levels of plane polarization will be correspondingly shifted. In general the light will be elliptically polarized in each space between the plates. Careful consideration shows that the angular momentum delivered to M is in the same sense as that for λ_0 for all wave-lengths λ for which

$$(n_y - n_x)_\lambda / \lambda < 2(n_y - n_x)_{\lambda_0} / \lambda_0$$
$$\text{or, approximately, } \lambda > \lambda_0/2.$$

If the lower plate B is rotated 90° from the position shown, the torque due to light of wave-length λ_0 will be just reversed. For this position the contributions to the torque of other wave-lengths is not easily considered qualitatively, but detailed calculation shows, as might be expected, that the integrated torque for all values of λ used is smaller in magnitude and opposite in direction compared to the torque similarly calculated for the position shown in Fig. 3. The torque is a continuous function of the position angle of the bottom plate, and its value can, with some labor, be calculated for each position.

The design wave-length λ_0 is determined on the basis of (3) by noting that A^2 is practically proportional to the energy in the light at each wave-length, and that λ enters as a factor. The energy distribution has a rather sharp peak in the region just above 1.0 μ for the tungsten filament temperatures to be considered. By a series of trial calculations $\lambda_0 = 1.2\ \mu$ was chosen to make the light torque a maximum.

The actual plates used, of course, vary somewhat from the ideal specifications. The most accurate plates obtained were made to within a few percent of the specified thickness by the Bryden Company of Waltham, Massachusetts, whom I wish to thank for this excellent cooperation. For the purposes of the exact calculations the plates were measured by means of a Babinet type compensator with sodium light. From the known values of the refractive indices of quartz[9] the retardation was then calculated for each plate for all wave-lengths contributing to the effect.

Besides almost quadrupling the torque compared to what would be obtained if the light were simply absorbed, and furthermore reducing heating, radiometer, and gas effects in the vacuum chamber, this plate arrangement greatly reduced difficulties due to radiation pressure. The disk M cannot be mounted exactly at right angles to the fiber axis, nor can the light be projected exactly in a vertical direction. Thus the resultant light pressure integrated over the disk will not be quite vertical and will produce a "light pressure torque" unless its line of action lies in a vertical plane containing the fiber axis. Without special precautions this torque might very easily mask the effect to be detected and measured.[10]

A tungsten filament light source was chosen, at a sacrifice of light intensity compared to an arc, in order to keep the resultant light pressure on the disk M as steady as possible, both in position and magnitude, and to give a steady and reproducible spectral energy distribution for purposes of calculation. The undesirable resultant light pressure on the disk M is decreased in the plate arrangement described, first, because most of the energy is allowed to pass through the disk and, secondly, because the pressure due to surface reflections and absorption for the upward beam is largely cancelled by the similar pressure for the beam reflected downward.

Finally, the moment arm for the residual light pressure can be varied by shifting the entire light beam parallel to itself. This is done by mounting the entire optical system up to $A-A$ in Fig. 1 on the double slide D, which in turn is supported in the lower of the two cast iron brackets I. The detail view, Fig. 4, shows the two micrometer screws by which the lateral

[9] *International Critical Tables*, VI, 341–342.

[10] Besides the smallness of the effect itself, the light pressure torque is perhaps the greatest difficulty in this experiment. I wish to thank Professor A. Kastler of the University of Bordeaux for a personal communication emphasizing this fact.

motion in two coordinates is produced. These are calibrated in thousandths of an inch so that any particular setting is easily repeated at a later time. By using this arrangement and leaving the lower wave plate B in a fixed position, the change in the sum of light torque, and light pressure, radiometer, and gas effect torque produced by a ten-percent change in filament current was measured for various settings of one of the double slide screws, the other being kept fixed. The results when plotted give a smooth curve crossing the zero torque line when the filament image lies across the middle of the window W as nearly as the eye can judge. This indicates that the spurious torques are probably of the nature anticipated and, in any event, can be held steady and made very small by appropriate setting of the double slide. In making the measurements of the light torque effect itself, the filament current was held constant to better than one percent, and the filament image was placed across the middle of the window. Care was also taken to use the same lengthwise portion of the filament in all cases.

The entire optical system can be rotated in its support on the double slide. Changes in the angle θ so produced are read on the circular scale visible in Fig. 4 on top of the double slide and below the Nicol housing N.

As described below, changes in the light torque effect corresponding to definite rotations of the bottom plate were measured and found to be reproducible for the same settings to within a few percent. When the bottom plate was removed, or when it was replaced by a fused quartz plate, rotation in the same manner produced no measurable effect. Hence it may be assumed that rotating the lower plate leaves the light energy, light pressure, heating, radiometer and gas effects sensibly unchanged. In other words, the apparatus used successfully eliminates the interfering effects, and enhances the light torque effect to be measured.

Measurements

The apparatus constructed according to the plan described showed the effect in the predicted direction and of about the predicted magnitude at the first trial on July 10, 1935. With this, the first aim, that of *detecting* the effect, was accomplished. The second aim, that of *measuring* the effect, was carried out by a resonance method (see reference 7). One of the first series of measurements taken is shown in Fig. 5, which illustrates both the definiteness of the effect observed and the kind of minor fluctuations encountered.

At the top of Fig. 5 are given the readings $a_\nu(\nu=0, 1, 2, \cdots)$ in centimeters on the scale (338 centimeters from the fiber axis) corresponding to the end points of the swing of the torsional pendulum. Tenths of millimeters on the scale could be estimated by means of the cross hairs in the telescope. The bottom plate B was in the position shown in Fig. 3 ($\theta=45°$, right stop R) while the readings a_0 to a_4 were being taken, but was turned 90° to the left (to $\theta=-45°$, to left stop L) at the instant at which the end point a_5 was read. The change in light torque produced by this rotation of B was such as to *oppose* the swing of the pendulum from a_5 to a_6. At a_6 the plate B was turned back to R, the torque change again opposing the ensuing swing from a_6 to a_7. The plate was turned similarly in "antiresonance" at a_7 and a_8 as indicated, but was then left in the position R when a_9, a_{10} and a_{11} were

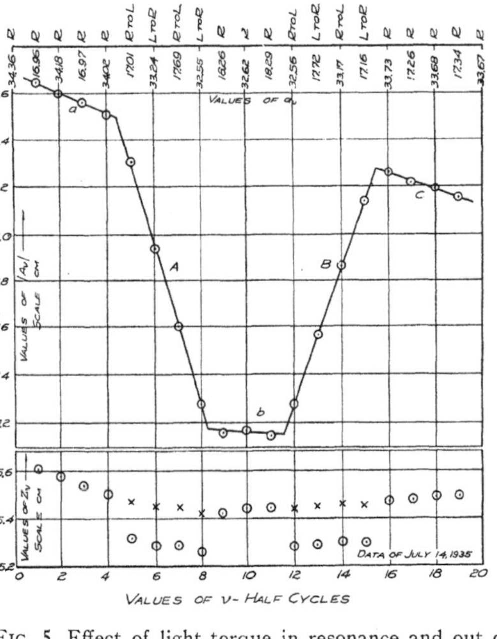

Fig. 5. Effect of light torque in resonance and out of resonance on the vibration amplitude of the torsion pendulum.

read. At a_{12} the plate was turned R to L, the change in torque now *aiding* the swing of the pendulum between a_{12} and a_{13}. At a_{13} the plate was turned L to R. This "resonance" turning was continued at a_{14} and a_{15}, and then the plate was again left in the position R for the drift readings a_{16} to a_{20}. For most of the readings taken another "antiresonance" series was added to the schedule described, and, for the readings taken by Mr. Harris, the "drift" periods were shortened while the "resonance" series in the middle of the schedule was lengthened to six half-cycles.

From the end-point readings taken according to such a schedule $4A_\nu = -a_{\nu-1} + 2a_\nu - a_{\nu+1}$ and $4Z_\nu = a_{\nu-1} + 2a_\nu + a_{\nu+1}$ were calculated. The absolute values of the "amplitudes" A_ν are plotted in the upper part of Fig. 5 and the "zeros" Z_ν (circles) in the lower part. For small values of damping (see "drift" slopes) the arithmetic means used differ by much less than the experimental error from the ideally correct geometric means for the quantities calculated.

From each such set of readings we wish to determine the value of the change in light torque on the hanging plate M when the bottom plate B is turned R to L or L to R. Since the torsional amplitude of the pendulum was at most a few degrees, we may assume that the light torque on M was a constant for a given position of the bottom plate B. We may consider a change in the position of the bottom plate to be equivalent to a change in the equilibrium position about which the torsional pendulum executes its oscillations. Let x be the corresponding change in scale reading. Careful consideration shows that, leaving damping out of account, the change in A_ν per *cycle* must then be $2x$ during "resonance" or "antiresonance" sequences. With a small amount of damping, x is therefore very nearly equal to the average of the rates of change of A_ν per *half-cycle* during "resonance" and "antiresonance." For example, in Fig. 5, the magnitudes of the slopes A and B, by least squares, are 0.341 and 0.289 centimeter per half-cycle, respectively. Hence $x = 0.315$ centimeter.

This method of taking readings and calculating affords several checks on the self-consistency of the sequence of readings a_ν used to determine one value of x. For example, the damping slopes a, b, and c should be approximately equal to $(A-B)/2$ in magnitude. Furthermore $Z_\nu + x/2$ during the "resonance" and "antiresonance" intervals (plotted as crosses) should form a smooth sequence with the values of Z_ν during the drift intervals. Slight disturbances, such as between a_9 and a_{10}, may be taken into account in evaluating these diagrams.

The quartz fibers were made in a gas-oxygen blast flame. Of those which seemed straight and uniform over a length of a foot or so, the *finest* (*small* K)[11] was taken. K for the first pendulum, with which the measurements of Fig. 5 were taken, was 8.52×10^{-6} dyne-cm/radian, or 1.26×10^{-8} dyne-cm/scale centimeter. Hence the change in light torque in Fig. 5 was 3.97×10^{-9} dyne-cm. This first fiber was accidentally broken and most of the measurements were taken with the second fiber for which K was 10.1×10^{-6} dyne-cm/radian or 1.50×10^{-8} dyne-cm/scale centimeter. The period in the latter case was about nine minutes, so that most of the determinations of x required consecutive observation of end-points for practically two hours. I made about 80 determinations during July, August and September, 1935, and Mr. Wilbur Harris made about 40 determinations during October, November and December, 1935, after I had left Princeton. Each of us worked in the absence of the other, and each made determinations of the three types to be described. Mr. Harris used a different bottom plate from the one

[11] With regard to the choice of the torsion constant K (in dyne-cemtimeters per radian) for the fiber, the following is of interest. During a "resonance" interval the amplitude (neglecting damping) increases by an amount $2Nx$ in N cycles. For a given torque change, x is inversely proportional to K. The number of cycles in a given time is directly proportional to $\sqrt{K}$, assuming the moment of inertia of the pendulum is a constant. Hence the change in amplitude $2Nx$ in *a given time* is inversely proportional to $\sqrt{K}$. In other words, the time required to achieve a given increase in amplitude by the resonance method varies directly as $\sqrt{K}$, other factors being constant. *Increasing* K in this experiment would therefore involve an increase in the time required for each measurement, a disadvantage which was not present in the corresponding case of the Einstein-de Haas experiment, in which the *impulse* per half cycle was independent of the length of the cycle. Nevertheless, Professor Einstein has pointed out that it would also be worth while in this experiment to *increase* K considerably, in spite of the increase in time required to achieve a given change in amplitude, because then the ratio of the energy in the effect desired to the energy of random disturbances would become much more favorable. Furthermore, it is an advantage to have the end-point readings a_ν come at shorter intervals insofar as the possible segregation of the larger random disturbances is concerned.

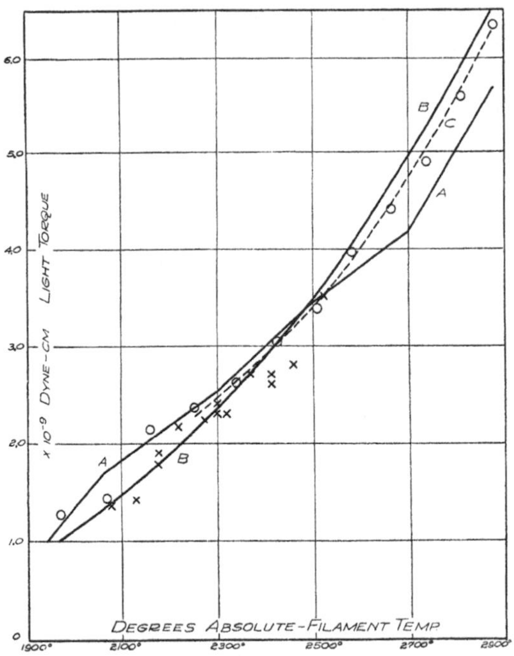

FIG. 6. Comparison of theory and experiment. Type I measurements: light torque as a function of filament temperature.

which I used, which seems to eliminate the possibility that the results, which were throughout consistent, could have been influenced by irregular patches of "infrared dirt" on the clear crystal quartz plates. Mr. Harris also investigated very thoroughly whether any effects were obtainable in the absence of the bottom plate, and found absolutely nothing. These checks were invaluable in substantiating the results of the present experiment, and Mr. Harris deserves much credit for the thoroughness and enthusiasm with which he carried out the tedious observations.

Three types of measurements, presented in Figs. 6, 8 and 9, were taken to show that the effect varies as required by the theory. The crosses represent my determinations and the circles those of Mr. Harris. Each point is derived from observations over a period of one and one-half to two hours as described in connection with Fig. 5. The smooth curves give the corresponding theoretical variation in each case.

For type I (Fig. 6) the torque change was measured for various filament temperatures as determined by the current through the lamp. The bottom plate was turned between stops at $\theta=45°$ and $\theta=-45°$, and the plane of polarization of the Nicol was at 45° to the axes of the plates as in Fig. 3. Mr. Harris' values (circles), taken consecutively, without shifting the position of the lamp, form the smoothest sequence obtained. The scattering of my values (crosses) taken with a different bottom plate is partly explained by the fact that I did not make these determinations consecutively, but shifted the lamp position a number of times for other readings. The total amount of light entering the plate system changes rather sharply with the position of the lamp in its socket and the setting of the double slide because of the presence of the small light shield d in the middle of the window W. The measurements previously reported[12] of this type, were made with a different lamp and fiber, and were made with a somewhat larger light shield on the window W. The values were therefore somewhat smaller than those shown in Fig. 6, but the magnitudes increased in a similar way with temperature.

The torque to be expected at various filament temperatures was calculated from (3) and the spectral energy distributions as follows. From the Poynting vector, the energy flow in the plate M is $A^2(n_x \cos^2\theta + n_y \sin^2\theta)c/8\pi$ or, practically, $A^2nc/8\pi = 2\pi c(A/4\pi)^2 n$ ergs/cm²-sec. Hence, if $J_\lambda' d\lambda$ is the total energy flow (integrated over the area of the plate) in M in the wave-length range λ to $\lambda+d\lambda$, the net torque on the plate from the energy flow in this wave-length range will be $L_\lambda' d\lambda$ dyne-centimeters, where

$$L_\lambda' = \sin 2\theta_\lambda'(\sin 2\Delta_1 - \sin 2\Delta_2)\lambda J_\lambda'/2\pi c.$$

Here $\tan\theta_\lambda'$ is, in general, the ratio of the amplitudes of the components of the light in the directions of the axes of the plates T and M. $\theta_\lambda' = \theta$ only when the position of the bottom plate B is such that the directions of its axes coincide with those for M and T. If the total energy flow entering the plate system in this wave-length range is $J_\lambda d\lambda$ ergs/sec., we set $J_\lambda' = J_\lambda F_{\lambda\eta}$ for each of the light beams contributing to the effect (enumerated by the index η), and compute $F_{\lambda\eta}$ using Edison Pettit's values for the reflectivity of aluminum (top surface of T)[13] and the re-

[12] R. A. Beth, Phys. Rev. **48**, 471 (1935).
[13] Edison Pettit, Pub. Astronom. Soc. Pacific **46**, 27 (1934).

fractive indices of quartz, which enter into Fresnel's expressions for the transmitted and reflected intensities at the plate surfaces. The absorption in the quartz plates may be neglected because the light has been filtered through the fused quartz lens L and the calcite and Canada balsam of the Nicol prism. From the theory for elliptically polarized light, the angles θ_λ' and the phases 2Δ may be calculated for each beam and each wave-length and for any position of the bottom plate, because the incoming light from the Nicol is plane-polarized, and the plate thicknesses, as measured by the compensator in wave plates for sodium light, were: $T=0.649$, $M=1.051$, $B_1=0.535$ (my torque measurements), $B_2=0.454$ (Harris' torque measurements). The whole torque on M is then

$$L=\int L_\lambda J_\lambda d\lambda \text{ dyne-centimeters,} \qquad (8)$$

where the torque factors L_λ are given by

$$L_\lambda=\sum_\eta \lambda F_{\lambda\eta}\{\sin 2\theta_\lambda'(\sin 2\Delta_1-\sin 2\Delta_2)\}_{\lambda\eta}/2\pi c \text{ seconds.} \qquad (9)$$

The torque factors L_λ were calculated for the sum η of the effects of the main beam going upward through M, the main beam reflected downward from the aluminized top surface of T, and for the reflections of these beams at the surfaces of B, W, M, and T, repeated surface reflections being neglected. These values of (9) are shown in Fig. 7, the angles denoting the azimuthal position of the bottom plate; at 45° the direction of the "fast" axis of B coincides with the direction of the "slow" axis of M and the "fast" axis of T (see Fig. 3). The full curve refers to the calculations for B_2, the dotted curve to those for B_1.

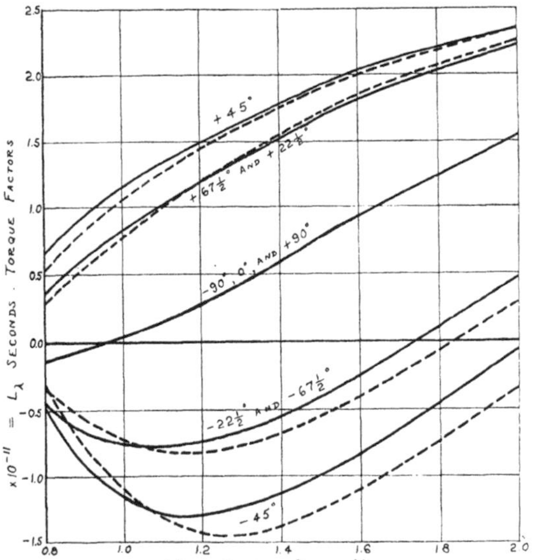

FIG. 7. Torque factor as a function of wave-length.

Using a rocksalt spectrometer and thermocouple arrangement Mr. Harris compared the amount of radiation from a black body at a known temperature (radiation from the interior of a furnace made out of an alundum cylinder $2\frac{1}{2}$ inches in diameter and a foot long at 766°C and 784°C was used) with the amount of radiation coming through the Nicol in the same solid angle at intervals of 0.1 μ for each of six different lamp currents. The energy factors J_λ were computed in absolute terms from these measurements using calculated values of black body radiation, Lambert's law, and constants of the optical system to determine the solid angle of the cone of light entering the plate system from the Nicol.

Replacing the integral in (8) by the corresponding sum for 0.1 μ intervals from 1.0 μ to 2.0 μ inclusive, the change in total torque from $\theta=45°$ to $\theta=-45°$ for each of the six filament currents and for the bottom plate B_2 was obtained. These theoretical torque values are plotted against the temperatures, 1948°, 2069°, 2297°, 2506°, 2701° and 2879° absolute, corresponding to the filament currents used, in Fig. 6 and connected by straight lines A. The irregularity of this broken line A indicates the difficulties encountered in making the energy determinations. The principal sources of error are: The spectrometer was not set up especially for this purpose, but was originally designed for somewhat longer wave-lengths; the solid angle of the light cone is difficult to determine with accuracy; the neglected effect below 1.0 μ is probably not exactly compensated by assuming that the Nicol actually polarizes all the light out to 2.0 μ. There is reason to believe, from the energy measurements themselves, that the energies computed for the upper two temperatures are somewhat too low. The root-mean-square deviation of the measured torques (circles) from the computed values A is 0.29×10^{-9} dyne-centimeter.

A second attempt was made to get the values of the energy factors J_λ in absolute terms by extrapolating the tungsten emissivities given by Forsythe and Worthing[14] (assuming the emissivity to be a linear function of temperature at each wave-length), multiplying by the black body radiation at the temperature of the filament, taking account of the solid angle of radiation collected from each point on the filament by the lens L, and making an allowance for the reflections from the surfaces of the bulb of the lamp, the lens, and the Nicol prism. The change in torque at each temperature was calculated from (8) as before, the integral in this case being replaced by the corresponding sum for 0.2 μ intervals from 0.8 μ to 2.0 μ inclusive. The results must be reduced by about 15 percent to give the curve B in Fig. 6. This 15 percent may be caused by the absorption of the Canada balsam in the Nicol, by nonpolarization of the light near 2.0 μ, and by inaccuracies in the emissivities used. The root-mean-square deviation of the circles from the curve B *as drawn* is. 0.24×10^{-9} dyne-centimeter. It should be emphasized that the uncertainties in B leave the *absolute value* in considerably greater doubt than in the case of A.

An important conclusion from Fig. 6 is, however, that the *slope* and *shape* of the theoretically determined curves A and B agrees within the limits of error with the trend of the measured torque values. The trend of B relative to the torque values (measured for certain lamp *currents*) depends on the current-temperature calibration of the lamp which was extrapolated from one value at 2941° absolute by comparison with the calibration for a similar lamp, while the trend of A relative to the torque values measured does not depend on this calibration, the energy distributions having been measured for given lamp *currents* also. The dotted smooth curve C, empirically fitted to the upper nine of Harris' torque measurements, appears as a natural compromise between the trends of A and B. Half of the circles lie within one percent of this curve. Their root-mean-square deviation from the curve is 0.054×10^{-9} dyne-centimeter.

[14] Forsythe and Worthing, Astrophys. J. **61**, 151 (1925).

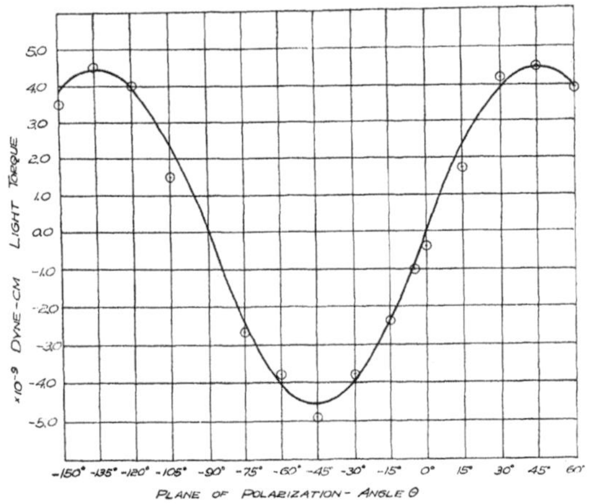

FIG. 8. Type II measurements: variation of light torque with angle between plane of polarization of light and the axes of the plates.

If we choose to think of C as an empirical calibration curve, this indicates that we may determine the temperature of the lamp in this range with a probable error of about 7° by measuring the torque with the accuracy attained by Mr. Harris, the lamp position being left unchanged.

The type I measurements may be regarded as a test of the factors $\lambda J_\lambda/2\pi c$ in the calculation of the torque from (8) and (9).

For the type II measurements (Fig. 8) the lamp current, and therefore the light intensity, was maintained constant and the plate B_2 was rotated between the same stops as in type I, but the plane of polarization of the light was rotated to various angles θ with the axes of the plates. These measurements are intended to test the factor $\sin 2\theta$ in the formula for the torque. The root-mean-square deviation of the measured values from the computed sine curve shown is 0.34×10^{-9} dyne-centimeter. A small systematic deviation from the sine curve may be due to the difficulty encountered in lining up the plate axes accurately in setting up the apparatus. In setting the angle θ the whole lamp housing and optical system, including the Nicol, was rotated and the double slide reset as described. Because of the presence of the light shield d this causes a greater scattering of the measurements than would be obtained if the lamp could be left fixed in position.

The measurements shown were made by Mr. Harris. In my measurements of this type only a

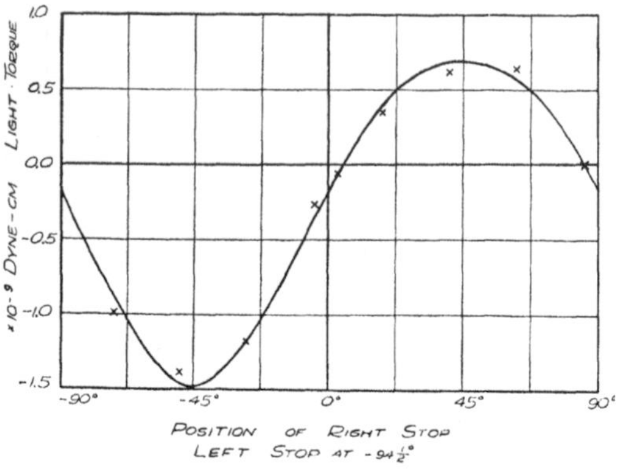

FIG. 9. Type III measurements: light torque as a function of position of the right stop.

quarter cycle of a sine curve was covered, but at 15° intervals. These points also form a smooth segment of a sine curve; though the root-mean-square deviation is considerably less, the test is not as critical and the result not as interesting as the full cycle curve shown.

For the type III measurements (Fig. 9) the light intensity was again maintained constant and the plane of polarization of the optical system was again put at 45° to the directions of the axes of plates M and T as in type I. The left stop was fixed at $-94\frac{1}{2}°$ and measurements made for different positions of the right stop. Since the torque for B_1 at the left stop was constant, the curve shows how the light torque in this plate system varies with the position of the bottom plate. The elliptical polarization produced by the bottom plate depends on its position, and these measurements are, therefore, in the main a test of the phase factors $(\sin 2\Delta_1 - \sin 2\Delta_2)$ in the formula. From the torque factors of Fig. 7 the smooth curve shown in Fig. 9 was calculated from the energy distribution. The root-mean-square deviation of the measured values of the torque from this curve is 0.073×10^{-9} dyne-centimeter. This very good agreement is again to be expected because the lamp position was left fixed. Mr. Harris also made measurements of this type, but did not cover a whole cycle. They are in similar agreement with the predicted values.

CONCLUSION

The change in the angular momentum of polarized light in passing through a crystal plate has been detected and measured with the apparatus described. This conclusion rests on the fact that the measured effect agrees in sign and, within the limits of error, in magnitude with the effect as predicted both by the wave and by the quantum theories of light. In particular it has been shown that the effect varies as required by the theory with each of three experimental variables studied: filament temperature (light intensity), azimuth of the plane of polarization, and position of the bottom plate.

I am deeply indebted to Professor A. W. Duff of the Worcester Polytechnic Institute who first suggested this problem to me and who personally helped me with the purchase of apparatus in the preliminary stages of the investigation. I wish to thank the National Research Council for a Grant-in-Aid for the purchase of apparatus and the Palmer Physical Laboratory of Princeton for providing all the specially made apparatus and technical assistance as well as for the hospitality of the Laboratory. In particular I wish also to thank Professors Harnwell, Einstein, Condon, Robertson, Shenstone, Bleakney, and Dr. Barnes of Princeton for their willing helpfulness in discussing the work with me on numerous occasions. Dr. S. N. Van Voorhis and Messrs. C. W. Curtis and C. W. Lampson of Princeton helped me both with suggestions about experimental details and in actually taking readings. Mr. Curtis shared the work of taking readings with me for many weeks. The very valuable part of the work done by Mr. W. Harris has been noted in detail above. Finally, I want to thank Mr. A. H. S. Holbourn of the Clarendon Laboratory, Oxford, for his interesting correspondence with me about the problem.

PAPER 1.3

Introduction to the atoms and angular momentum of light special issue

L Allen

Optics and Applications Group, Department of Physics and Astronomy, Kelvin Building, University of Glasgow, Glasgow G12 8QQ, UK and
University of Sussex, Falmer, Brighton BN1 9RH, UK

Received 8 January 2002
Published 28 March 2002
Online at stacks.iop.org/JOptB/4/S1

Abstract
A very brief comment concerning the literature relating to the angular momentum of light is followed by a précis of a review article on the orbital angular momentum of light published in 1999. An outline is then given of the key developments since 1999 in the study of the angular momentum of light and of the related topic of optical vortices.

Keywords: Orbital angular momentum of light, optical vortices, atoms

1. Introduction

There is a vast physics literature related to the angular momentum of light. The great majority of it concerns spin angular momentum, discussed in the context of polarization, virtually all of it developed for idealized polarized plane waves. Sometimes a formal theoretical examination of the interaction of light and matter, particularly with a single atom, is written over a sum of plane waves. But, very often, the summation is something of an illusion and the final results are for a single plane wave of well defined wavevector $\boldsymbol{k}$. A great proportion of the published work relates to the polarization selection rules of atomic transitions and, with the advent of laser trapping and cooling, to their exploitation for specific ends.

This does not mean that physicists believed that plane waves really existed. The approximation of a plane wave was used because it worked very well for most purposes. It has been known for a very long time, too, that a calculation of the angular momentum of a free electromagnetic field produced contributions which did not depend on photon spin. Attributing the term *orbital angular momentum* to those parts of the angular momentum of a light beam which did not arise from spin rarely proved necessary. Occasionally the insistence of angular momentum conservation, particularly in nuclear physics, dictated that a photon almost notionally carried off a certain number of units of $\hbar$, but in general the concept has not been much used.

A discussion of the behaviour of waves which were patently not plane, other than in the context of diffraction theory, did not really begin until the 1970s and early 1980s with the work of Nye and Berry [1, 2]. Their work on singularities and dislocations in wavefronts has led to a very considerable literature. For screw dislocations, or phase singularities, this leads to an examination of the behaviour of *optical vortices*. Light beams possessing such a property can, of course, be generated in the laboratory using laser beams and they were first generated by Vaughan and Willetts in 1983 [3]. Even more recently, 1992, came the realization [4] that laboratory-generated laser beams could exist with a unique integer amount of orbital angular momentum, $\ell\hbar$ per photon. Only since then has serious thought been given in optical physics to the possibility of investigating such a property.

The literatures of optical vortices and of orbital angular momentum have stayed resolutely separate. At the lowest level this was because of the difference in the chosen language of the two areas of optical physics. It was also because the concept of angular momentum *per se* was not for some time strongly associated with optical vortices and, indeed, not all dislocations have an angular momentum: although angular momentum is conserved in a noninteracting light beam, it is not necessarily conserved in its vortex. Nevertheless, the discussion of stable beams consisting of a circularly symmetric field such as higher-order Laguerre–Gaussian modes lends itself rather readily to the use of orbital angular momentum, even though the beam does contain an optical vortex. When such a beam is polarized it lends itself even more so to the concept of *total angular momentum*. Generalizations about the separation of these two areas of interest are likely to be too severe, because the same effects are, as ever in physics, describable in more than one way. For example, in many cases the main property of a Laguerre–Gaussian beam exploited in an atom guiding or trapping experiment may well not be its degree of orbital angular momentum.

It was a particular pleasure for me, having been strongly identified with the work on orbital angular momentum, to learn that an Institute of Physics journal was to have a special issue on the subject of atoms and the angular momentum of light together with the role of vortices. A wide range of interests has been welcomed in the hope that the issue might play the role, in some senses at least, of an update of more formal reviews.

The editor of *Journal of Optics B: Quantum and Semiclassical Optics* suggested that I might care to write this introduction and that a possible form it might take would be to bring the *Progress in Optics* [5] review on orbital angular momentum, published in 1999 and co-authored by myself, Padgett and Babiker, up to date. However, I feel that the detailed underpinning required for a review article is inappropriate here. In consequence, what follows is an outline of published work since 1999 rather than a coherent exposition of it. It includes, where appropriate, papers on vortices, although my emphasis will inevitably be on angular momentum.

At the time of the *Progress in Optics* review, much had already been done. The essential paraxial approximation theory [4] of orbital angular momentum had been devised, as had a quantum mechanical equivalent [6]. The concept had been shown not to be an artefact of the paraxial approximation in a rigorous examination of an unapproximated solution to an arbitrary free field [7], although certain important problems still remained. A general way of producing Laguerre–Gaussian beams from a laser operating in high-order Gaussian modes by means of a cylindrical lens mode-convertor [8] had already been devised, although this approach is often supplanted by their production with a hologram [9, 10].

Experimental evidence was produced of the existence of orbital angular momentum [11] and for the case of a Laguerre–Gaussian beam of $\ell = 1$ shown to be mechanically equivalent [12] to a circularly polarized beam where the spin angular momentum is by necessity unity, $\sigma = 1$. This experiment, of course, was the equivalent of Beth's famous determination of $\hbar$ from the spin angular momentum of a circularly polarized beam. This work extended that of Ashkin *et al* [13] on optical tweezers to what may be called *optical spanners*.

The frequency shift known variously as the angular Doppler shift [14], the rotational frequency shift [15] or the shift arising from a rotating mode-converter [16] has been experimentally observed both for orbital and for spin and orbital angular momentum combined [17, 18]. The same shift was predicted as an azimuthal Doppler shift in the interaction of Laguerre–Gaussian laser light with atoms [19], in a body of theoretical work concerned with atoms interacting with such light beams [20–23].

Other work on atoms with such modes had also begun with the possibility of dark field excitation previously achieved by blocking the light [24], and the construction of novel optical traps [25]. Second-harmonic generation had been observed [26] with the frequency-doubled beam shown to possess twice the angular momentum of the pump beam, $2\ell\hbar$. The first theoretical description of the conservation of orbital angular momentum in soliton motion in nonlinear media had also been completed [27] essentially using the language of vortices, following the earlier work of Swartzlander and Law [28] on soliton propagation.

All of these aspects of the properties and behaviour of orbital angular momentum have been revisited and extended, in the time between the 1999 review and now. The topic, which had a slow and halting start, has undergone a rapid expansion. The areas of interest can be categorized as general theory, mode converters, optical tweezers, atoms, Bose–Einstein condensates and entanglement.

2. General theory

A series of analogues now exists which shows how similar orbital angular momentum is to spin angular momentum. Allen *et al* [29] devised the equivalent of Jones matrices, which describe the orbital angular momentum as it passes through an system of mode-order preserving optical elements in terms of the constituent Hermite–Gaussian modes and developed the matrices to account for light that has both spin and orbital angular momentum. Immediately [30], a similar matrix was described in terms of Laguerre–Gaussian modes and the approach was used to construct a variable-phase mode converter. This strong analogy between spin and orbital angular momentum enabled Padgett and Courtial [31] to propose a sphere for light beams in orbital angular momentum states which is precisely equivalent to the Poincaré sphere, positions on which represent the polarization state of the light. This equivalent sphere was used to interpret the rotational frequency shift of light beams with orbital angular momentum.

An $SU(2)$ basis for the proposed analogue of the Poincaré sphere was presented by Agarwal [32], who generalized the work to include quantized as well as partially coherent beams with orbital angular momentum. He also demonstrated that the $SU[\mathrm{N}]$ version of the theory could be used for higher-order Laguerre–Gaussian beams. Ponomarenko [33] introduced a new class of partially coherent beams which carry optical vortices. He showed that any such beam can be represented as an incoherent superposition of fully coherent Laguerre–Gaussian modes of arbitrary order with the same azimuthal index, the same family examined by Agarwal. He investigated their free-space propagation properties. Serna and Movilla [34] extended the consideration of coherent beams with orbital angular momentum to partially coherent beams in terms of two elements of the beam matrix. They considered a family of more general fields and the relationship between the twist of the beam and the orbital angular momentum. Simon and Agarwal [35] present a phase-space description to derive a closed form expression for the Wigner function for Laguerre–Gaussian modes and to show that it is as compact as the one for Hermite–Gaussian modes.

The principle of operation of a new class of optical devices operating in quadratic nonlinear media, that mix wavefront dislocations in focused beams and produce certain patterns of bright spatial solitons, was described by Torner *et al* [36]. They argued that the orbital angular momentum of the light beam is central to this behaviour. Self-trapped necklace-ring beams that carry and conserve angular momentum were demonstrated by Soljacic and Segev [37], who showed that such beams can have a fractional ratio of angular momentum to energy.

The local linear momentum density in Laguerre–Gaussian beams has been used [38] to investigate classically the trajectory of the Poynting vector for light of any polarization, while some difficulties of interpretation of the local angular momentum density are discussed. The local angular momentum density plays a vital role in the determination of what value of angular momentum determines the nature of an interaction with matter. Spin angular momentum is always intrinsic and no matter how, or where, an interaction takes place the angular momentum involved is $\hbar$ per photon. Berry [39] showed that, contrary to popular belief, orbital angular momentum is independent of the beam axis provided it is reasonable and possible to stipulate a direction for which the transverse momentum current is exactly zero. It is tempting, therefore, to say that orbital angular momentum is also always intrinsic, but O'Neill *et al* [40] have shown experimentally that it may be both extrinsic and intrinsic.

The well known spin-dependent linear and angular transverse shifts of the centre of gravity of partially and refracted light beams have been shown theoretically to have a non-spin-dependent analogue by Fedoseyev [41]. For a Laguerre–Gaussian beam the linear transverse shift is interpreted as a transformation of the orbital angular momentum at reflection and refraction: a parallel is drawn between this shift and the Goos–Hanchen shift.

Under the broad heading of general theory we may also include several other papers which appear in this issue. Desyatnikov and Kivshar (pages S58–S65) describe ring-profile optical solitons in nonlinear media whose propagation behaviour depends upon their orbital angular momentum. Piccarillo *et al* (pages S20–S24) investigate the competition between spin and orbital angular momentum in liquid crystals both theoretically and experimentally. What I believe to be the first totally QED approach to this topic has been applied by Dávila Romero *et al* (pages S66–S72) to represent the photonics of beams with orbital angular momentum interacting with a nonlinear medium. They apply the theory to second-harmonic generation as an illustration. Tiwari (pages S39–S46) maps out a philosophical and rather speculative view of relativity, entanglement and the reality of photons and proposes a laboratory experiment to test the consequences which would arise from a vortex structure for the photon.

Finally a very significant contribution to the general theory of this area is to be found in this issue. Barnett (pages S7–S16) introduces the use of angular-momentum flux as a natural description of angular momentum carried by light and solves a number of outstanding problems. Most important is the fact that the angular-momentum flux of a light beam about its direction of propagation can be separated into spin and orbital parts and that this separation is not only gauge invariant, but does not rely on the paraxial approximation. Remarkably, this gauge invariant separation and the fluxes of both spin and orbital angular momentum reduce to the known paraxial forms in the appropriate limit. The title of his paper suggests the most functional title in the future for this area of physics, namely *optical angular momentum*.

3. Mode converters

Perhaps not surprisingly, given the effectiveness of cylindrical lens and hologram mode converters, this area of the subject has not seen a very strong degree of activity. However, Clifford *et al* [42] generated high-order Laguerre–Gaussian modes from a simple external-cavity diode laser and the detailed performance of a cylindrical lens mode converter has been analysed by examining the output beam for imperfections in the implementation of the converter [43]. Nevertheless this issue of *J. Opt. B: Quantum Semiclass. Opt.* has a paper by Chávez-Cerda *et al* (pages S52–S57), who use a holographic technique to produce high-order Mattieu beams, the orbital angular momentum content of which is discussed in terms of Bessel beams.

4. Optical tweezers

The possibility of revolving a small particle by means of a light beam and of developing light-driven micromachines is creating a distinct literature of its own. Friese *et al* [44] pointed out that the difficulty of earlier measurements with orbital angular momentum was that it was difficult to develop high power because of overheating and unwanted axial forces, but that this could be overcome by the use of transparent particles. They showed that an optical torque can be induced on microscopic birefringent particles of calcite held by optical tweezers. The particles either became aligned with the polarization, or spun with constant rotation frequency. Because the particles were transparent they could be held in three-dimensional optical traps at very high power at rotation rates of more than 350 rps. Higurashi *et al* [45, 46] also achieved optically induced mechanical alignment of trapped birefringent micro-objects by use of linearly polarized light. Trapping and rotation were achieved by the same Gaussian beam. The micro-objects were aligned about the laser beam axis and their angular position could be smoothly controlled by rotating the plane of the electric field of the light.

Luo *et al* [47] describe an optical micromotor where the spinning particle consists of a two-bead linkage. The large bead is trapped in optical tweezers and rotates about the laser beam axis. The small bead is coated with gold or palladium and generates a rotational torque through the change of momentum from the gradient radiation pressure of the same laser. A laser power of 29 mW generated a rotation of $\sim$160 rpm. They suggest that, apart from developing optically operated mechanical devices, such a system could be used to twist macromolecules or to generate shear forces in a medium at the nano-level.

O'Neil and Padgett [48] reported the trapping of metallic particles in inverted optical tweezers. The particles were loosely confined in three dimensions to an annular region just below the beam waist where gravity is counterbalanced by the scattering forces. Trapping efficiency was improved by the use of a Laguerre–Gaussian beam and off-axis particle rotation was observed, which is induced by the orbital angular momentum of the beam. The polarization of the light did not influence the motion of the particles. The same workers also show [49] that axial trapping is improved in optical tweezers when a Laguerre–Gaussian beam is used rather than a Gaussian. However, contrary to earlier suggestions, the lateral efficiency is unchanged. Friese *et al* [50] also visited this area in an approach which invoked the force constants of a trap modelled as a harmonic oscillator.

Galajda and Ormos [51] show how light-induced polymerization of light curing resins may be applied to generate rotating particles several microns in size. They produce mechanical devices, consisting of multiple moving parts, driven by these rotors. Friese *et al* [52] also show how the torque they previously described may be harnessed to drive an optically trapped microfabricated structure.

Finally, Nieminen *et al* [53] argue that systematic application of electromagnetic scattering theory can provide a general theory of laser trapping and yield results missing from existing theory. This is particularly important for wavelength scale particles. They present calculations of forces and torques on trapped particles obtained from this approach.

In this issue there is evidence of further activity in papers by Volke-Sepulveda *et al* (pages S82–S89) on the mechanical transfer of orbital angular momentum from a high-order Bessel beam to trapped particles, and on the rotation of microscopic propellers in laser tweezers by Galajda and Ormos (pages S78–S81).

5. Atoms

The intrinsic shape of the intensity distribution of Laguerre–Gaussian modes, as well as their orbital angular momentum, make them ideal for interactions with atoms. It is not surprising therefore that there is seemingly a fast-growing literature associated with them. Morsch and Meacher [54] proposed an all-optical method for channelling and cooling a cloud of pre-cooled atoms making use of the light shifts induced by the laser mode tuned close to the atomic resonance. They showed that optical pumping and inelastic collisions of the atoms with the optical walls of the trap gave rise to cooling. Because this cooling offsets the heating due to random scattering of photons, the trap did not suffer any lifetime limitation. They presented the results of a semi-classical Monte Carlo simulation for a three-dimensional model. Simultaneously, Kuppens *et al* [55] presented experimental and theoretical studies of polarization-gradient cooling of neon atoms at the dark centre of a TEM01* mode which, of course, has no angular momentum. The transverse motion within the annular mode is cooled by two-dimensional optical molasses. The results were found to be in qualitative agreement with a one-dimensional quantum Monte Carlo simulation of cooling in the presence of an external light-shift potential. They found that the focusing properties of Laguerre–Gaussian modes were not altered in the presence of the cooling beams. The same group reported [56] on the application of the mode as a strong dipole potential to guide a slow-atom beam. Here polarization gradient cooling generates a bimodal momentum distribution.

Lembessis [57] considered an atom moving in a linearly polarized mode and included the Roentgen term in the electric dipole interaction. He displayed terms which involve the coupling of the photon angular momentum with the atomic particle angular momentum. An investigation was carried out by Liu and Milburn [58] into the classical two-dimensional nonlinear dynamics of cold atoms in far-off-resonant Laguerre–Gaussian beams. They showed that chaotic developments exist provided $\ell > 1$, when the beam is periodically modulated. The atoms are predicted to accumulate on several ring regions when the system enters a regime of global chaos.

A fascinating experiment on optical pumping in cold caesium atoms [59] presented experimental results on the transfer of orbital angular momentum from one transition to another at a quite different frequency. This transfer was observed using a nondegenerate four-wave mixing process as an indirect tool. The fact that the orbital angular momentum was not only preserved but conserved, suggests that radiation processes of higher order than dipole should demonstrate mode conversion as both orbital and spin selection rules become invoked.

Truscott *et al* [60] produced an optically written waveguide in an atomic vapour using Laguerre–Gaussian modes and these experimental results were subsequently explained by use of a full density matrix approach by Kapoor and Agarwal [61]. Anderson *et al* [62] returned to the problem more recently and considered a five-level system in detail. A high degree of theoretical sophistication has been applied [63] to study the effect of phase fluctuations of a driving laser field on the dissipative and dipolar forces of a two-level atom by means of the phase diffusion model. The effect of the phase fluctuations on orbital angular momentum transfer to atoms was investigated for both Lagurerre-Gaussian and Bessel beams.

6. Bose–Einstein condensates

The degree of activity involving atoms makes it no surprise that Bose–Einstein condensates (BECs) should also be addressed. Bolda and Walls [64] proposed a method of simultaneously taking a ground state to a different atomic hyperfine state and to a vortex trap state. This would be accomplished by use of one laser having a Laguerre–Gaussian mode profile. It is the vortex of the mode rather than the orbital angular momentum which would play the key role here just as in the work of Wright *et al* [65], who numerically investigated red-detuned Laguerre–Gaussian beams of various mode-index to produce suitable optical dipole toroidal traps for two-dimensional atomic BECs. The efficient optical generation of vortices in trapped Bose–Einstein condensates [66] was demonstrated numerically, while the creation of Bose–Einstein condensates was demonstrated in Rb in a specially designed hybrid, optical dipole and magnetic trap [67]. This trap allows the coherent transfer of matter waves into a pure optical potential waveguide, based again on the TEM 01* beam. It is also the spatial distribution of the mode which is dominant in the work of Tsurumi and Wadati [68]. They investigated ground state properties of toroidal condensates through the variational method. It was found that the condensate is attracted towards the axis of symmetry of the torus.

In this issue Alexander *et al* (pages S33–S38) analyse the structure and stability of vortices in hybrid atomic–molecular BECs. They predict new types of topological vortex state and show that their dynamics in the presence of losses are nontrivial. Also here is the work of Boussiakou *et al* (pages S25–S32), who develop a theory to determine the electric and magnetic properties of vortex states in BECs. The Lagrangian is symmetric in the electric and magnetic aspects of the problem and includes both the Aharanov–Casher and Roentgen interaction terms. Specific field distributions are evaluated for electric and magnetic dipole active BECs in a vortex state.

7. Entanglement

Perhaps the most exciting new aspect of orbital angular momentum and the most promising for fundamental work in the future is that of entanglement. The state of one particle in a two-particle entangled state defines the state of the second particle, while neither particle has a well defined state before measurement. An enormous literature exists for the entanglement of two-state quantum systems, in particular those involving the two orthogonal polarization states of photons. Mair *et al* [69] have demonstrated, by means of a parametric down-conversion experiment, that modes carrying orbital angular momentum can also show entanglement. As Laguerre–Gaussian modes can be used to define an infinite-dimensional discrete Hilbert space, this approach offers the possibility of entanglement involving many orthogonal quantum states. Consequently, it may be possible to resolve some aspects of long-standing arguments and discussion concerning the results on polarization states. Franke-Arnold *et al* [70] have investigated theoretically the orbital angular momentum correlation of a photon pair created in a spontaneous parametric down-conversion process. They show how the conservation of orbital angular momentum results from phase-matching in the nonlinear crystal. The same workers [71] calculate the anticipated correlation between measurements of the orbital angular momentum of the signal and idler beams of parametric down-conversion. The orbital angular momentum is measured by use of a hologram and displacement of the hologram with respect to the beam axis allows the measurement of superpositions of Laguerre–Gaussian modes. The correlations between such superposition modes show entanglement and could be used for Bell-type tests of nonlocality. In this issue Vaziri *et al* (pages S47–S51) examine superpositions of the orbital angular momentum for application in entanglement experiments in a not dissimilar way.

In an interesting paper by Arnaut and Barbosa [72] the orbital and spin angular momentum of single photons and entangled pairs of photons generated by parametric down-conversion was investigated. However, Eliel *et al* [73] show that the crucial role played by crystal symmetry in their work cannot be validated.

An even newer aspect of entanglement is given in this issue in the paper by Muthukrishnan and Stroud (pages S73–S77), who consider the exchange of spin and orbital angular momenta between a circularly polarized Laguerre–Gaussian beam and a single atom trapped in a two-dimensional harmonic potential. Spin and orbital angular momentum are individually conserved and result in the entanglement of the internal and external degrees of freedom of the atom.

8. Conclusions

The concepts of orbital angular momentum and of optical vortices have extended over a large number of aspects of optical physics. To what extent this can be expected to continue, and to what extent it will ultimately prove to be important, remains to be seen. Over a decade, the foundations have been put into place. My guess is that the subject will continue to grow, although it may become increasingly technical, and that we are still a long way from appreciating its full impact. I should like to think that this special issue would make a positive contribution to that process. There would seem to be no reason why not. It contains a mixed bag of papers, some inevitably purely transitory, but nevertheless including several which I suspect will pass the test of time. If so, that is probably more than anyone could have reasonably hoped.

Amusingly enough for me I find, almost by accident, that following 'the observation of the dynamical inversion of the topical charge of an optical vortex' by Molina-Terriza *et al* [74], Padgett and I have cited their reference and the word 'vortex' in our short paper published here. If such inveterate users of the formalism of orbital angular momentum can begin to combine the separate literatures, however frail the link, perhaps there is some possibility that the separate literatures will begin to coalesce.

References

[1] Nye J F and Berry M V 1974 *Proc. R. Soc.* London A **336** 165–90
[2] Nye J F 1981 *Proc. R. Soc.* London A **378** 219–39
[3] Vaughan J M and Willets D V 1983 *J. Opt. Soc. Am.* **73** 1018–21
[4] Allen L, Beijersbergen M W, Spreeuw R J C and Woerdman J P 1992 *Phys. Rev.* A **45** 8185–9
[5] Allen L, Padgett M J and Babiker M 1999 *Prog. Opt.* **39** 291–372
[6] van Enk S J and Nienhuis G 1992 *J. Opt. Commun.* **94** 147–58
[7] Barnett S M and Allen L 1994 *J. Opt. Commun.* **110** 670–8
[8] Beijersbergen M W, Allen L, van der Veen H E L O and Woerdman J P 1993 *J. Opt. Commun.* **96** 123–32
[9] Bazhenov V Yu, Soskin M S and Vasnetsov M V 1990 *JETP Lett.* **52** 429–30
[10] Heckenberg N R, McDuff R, Smith C P, Rubinsztein-Dunlop H and Wegener M J 1992 *J. Opt. Commun.* **24** S951–62
[11] He H, Friese M E J, Heckenberg N R and Rubinsztein-Dunlop H 1995 *Phys. Rev. Lett.* **75** 826–9
[12] Simpson N B, Dholakia K, Allen L and Padgett M J 1997 *Opt. Lett.* **22** 52–4
[13] Ashkin A, Dziedzic J M, Bjorkholm J E and Chu S 1986 *Opt. Lett.* **11** 288–90
[14] Garetz B A and Arnold S 1979 *J. Opt. Commun.* **31** 1–3
[15] Bialynicki-Birula I and Bialynicka-Birula Z 1997 *Phys. Rev. Lett.* **78** 2539–42
[16] Nienhuis G 1996 *J. Opt. Commun.* **132** 8–14
[17] Courtial J, Dholakia K, Robertson D A, Allen L and Padgett M J 1998 *Phys. Rev. Lett.* **80** 3217–19
[18] Courtial J, Robertson D A, Dholakia K, Allen L and Padgett M J 1998 *Phys. Rev. Lett.* **81** 4828–30
[19] Allen L, Babiker M and Power W L 1994 *J. Opt. Commun.* **42** 141–4
[20] Babiker M, Power W L and Allen L 1994 *Phys. Rev. Lett.* **73** 1239–42
[21] Power W L, Allen L, Babiker M and Lembessis V E 1995 *Phys. Rev.* A **52** 479–88
[22] Allen L, Babiker M, Lai W K and Lembessis V E 1996 *Phys. Rev.* A **54** 4259–70
[23] Lai W K, Babiker M and Allen L 1997 *J. Opt. Commun.* **133** 487–94
[24] Andersen M H, Petrich W, Ensher J R and Cornell E A 1994 *Phys. Rev.* A **50** R3597–600
[25] Kuga T, Torii Y, Shiokawa N and Hirano T 1997 *Phys. Rev. Lett.* **78** 4713–16
[26] Dholakia K, Simpson N B, Padgett M J and Allen L 1996 *Phys. Rev.* A **54** R3742–5
[27] Firth W J and Skryabin D V 1997 *Phys. Rev. Lett.* **79** 2450–3
[28] Swartzlander G A Jr and Law C T 1992 *Phys. Rev. Lett.* **69** 2503–6

[29] Allen L, Courtial J and Padgett M J 1999 *Phys. Rev.* E **60** 7497–503
[30] O'Neil A T and Courtial J 2000 *J. Opt. Commun.* **181** 35–45
[31] Padgett M J and Courtial J 1999 *J. Opt. Lett.* **24** 430–2
[32] Agarwal G S 1999 *J. Opt. Soc. Am.* A **16** 2914–16
[33] Ponomarenko S A 2001 *J. Opt. Soc. Am.* A **18** 150–6
[34] Serna J and Movilla J M 2001 *J. Opt. Lett.* **26** 405–7
[35] Simon R and Agarwal G S 2000 *J. Opt. Lett.* **25** 1313–15
[36] Torner L, Torres J P, Petrov D V and Soto-Crespo J M 1998 *J. Opt. Quantum Electron.* **30** 809–27
[37] Soljacic M and Segev M 2001 *Phys. Rev. Lett.* **86** 420–3
[38] Allen L and Padgett M J 2000 *J. Opt. Commun.* **184** 67–71
[39] Berry M V 1998 *Proc. SPIE* **3487** 6–11
[40] O'Neil A T, MacVicar I, Allen L and Padgett M J 2002 *Phys. Rev. Lett.* at press
[41] Fedoseyev V G 2001 *J. Opt. Commun.* **193** 9–18
[42] Clifford M A, Arlt J, Courtial J and Dholakia K 1998 *J. Opt. Commun.* **156** 300–6
[43] Courtial J and Padgett M J 1999 *J. Opt. Commun.* **159** 13–18
[44] Friese M E J, Nieminen T A, Heckenberg N R and Rubinsztein-Dunlop H 1998 *Nature* **394** 348–50
[45] Higurashi E, Sawarda R and Ito T 1999 *Appl. Phys. Lett.* **72** 2951–3
[46] Higurashi E, Sawarda R and Ito T 1999 *Phys. Rev.* E **59** 3676–81
[47] Luo Z P, Sun Y L and An K N 2000 *Appl. Phys. Lett.* **76** 1779–81
[48] O'Neil A T and Padgett M J 2000 *J. Opt. Commun.* **185** 139–43
[49] O'Neil A T and Padgett M J 2001 *J. Opt. Commun.* **193** 45–50
[50] Friese M E, Rubinsztein-Dunlop H, Heckenberg N R and Dearden E W 1996 *J. Appl. Opt.* **35** 7112–6
[51] Galajda P and Ormos P 2001 *Appl. Phys. Lett.* **78** 249–51
[52] Friese M E J, Rubinsztein-Dunlop H, Gold J, Hagberg P and Hanstrop D 2001 *Appl. Phys. Lett.* **78** 547–9
[53] Nieminen T A, Rubinsztein-Dunlop H and Heckenberg N R 2001 *J. Quant. Spectrosc. Radiat. Transfer* **70** 627–37
[54] Morsch O and Meacher D R 1998 *J. Opt. Commun.* **148** 49–53
[55] Kuppens S, Rauner M, Schiffer M, Sengstock K, Ertmer W, van Dorsselaer F E and Nienhuis G 1998 *Phys. Rev.* A **58** 3068–79
[56] Schiffer M, Rauner M, Kuppens S, Zinner M, Sengstock K and Ertmer W 1998 *J. Appl. Phys.* B **67** 705–8
[57] Lembessis V E 1999 *J. Opt. Commun.* **159** 243–7
[58] Liu X M and Milburn G 1999 *Phys. Rev.* E **59** 2842–5
[59] Tabosa J W R and Petrov D V 1999 *Phys. Rev. Lett.* **83** 4967–70
[60] Truscott A G, Friese M E J, Heckenberg N R and Rubinsztein-Dunlop H 1999 *Phys. Rev. Lett.* **82** 1438–41
[61] Kapoor R and Agarwal G S 2000 *Phys. Rev.* A **61** 053818–24
[62] Anderson J A, Friese M E J, Truscott A G, Ficek Z, Drummond P D, Heckenberg N R and Rubinsztein-Dunlop H 2001 *Phys. Rev.* A **63** 023820–9
[63] Lawande S V and Panat P V 2000 *Mod. Phys. Lett.* B **14** 631–7
[64] Bolda E L and Walls D F 1998 *Phys. Lett.* A **246** 32–6
[65] Wright E M, Arlt J and Dholakia K 2000 *Phys. Rev.* **63** 013608–17
[66] Dobrek L, Gajda M, Lewenstein M, Sengstock K, Birkl G and Ertmer W 1999 *Phys. Rev.* A **60** R3381–4
[67] Bongs K, Burger S, Dettmer S, Hellweg D, Arlt J, Ertmer W and Sengstock K 2001 *Phys. Rev.* A **63** 311602–4
[68] Tsurumi T and Wadati M 2001 *J. Phys. Soc. Japan* **70** 1512–18
[69] Mair A, Vaziri A, Weihs G and Zeilinger A 2001 *Nature* **412** 313–16
[70] Franke-Arnold S, Barnett S M, Padgett M J and Allen L 2002 *Phys. Rev.* A at press
[71] Padgett M J, Courtial J, Allen L, Franke-Arnold S and Barnett S M 2002 *J. Mod. Opt.* at press
[72] Arnaut H H and Barbosa G A 2000 *Phys. Rev. Lett.* **85** 286–9
[73] Eliel E R, Dutra S M, Nienhuis G and Woerdman J P 2001 *Phys. Rev. Lett.* **86** 5208–9
[74] Molina-Terriza G, Recolons J, Torres J P and Torner L 2001 *Phys. Rev. Lett.* **87** 023902/1–4

Section 2

SPIN AND ORBITAL ANGULAR MOMENTUM

The modern study of optical angular momentum can be said to have started with the paper of Allen *et al.* (2.1, **Paper 2.1**). This work showed that any beam with the amplitude distribution $u(r, \phi, z) = u_0(r, z) \exp il\phi$, carried angular momentum about the beam axis. Moreover, this angular momentum could be separated into orbital and spin components. The orbital contribution is determined solely by the azimuthal phase dependence and is equivalent to $l\hbar$ per photon. The spin angular momentum is determined by the polarisation and has the value $\sigma_z\hbar$ per photon, where $-1 \leq \sigma_z \leq 1$, with the extremal values corresponding to pure circular polarisation. A physically realisable example of light with this phase distribution is a Laguerre–Gaussian beam, familiar from paraxial optics (2.2).

The identification of an orbital angular momentum is strongly suggested by a powerful analogy between paraxial optics and quantum mechanics (2.3). Here the Schrödinger wave equation is identical to the paraxial form of the wave equation with t replaced by z, while the operator corresponding to the z-component of orbital angular momentum can be represented in the form $L_z = -i\hbar\partial/\partial\phi$. The analogy allows much of paraxial optics, including orbital angular momentum, to be studied using the formalism of quantum mechanics. This is has been amply demonstrated by van Enk and Nienhuis (2.4, **Paper 2.2**) who introduce an eigenfunction description of laser beams and by Nienhuis and Allen (2.5, **Paper 2.3**) who compare paraxial modes and their properties to those of the quantum harmonic oscillator. The propagation of spin and orbital angular momenta through optical elements can usefully be described without reference to the spatial form of the light beam, by a generalisation of the Jones matrix description of optical polarisation (2.6, **Paper 2.4**). This abstraction is analogous to separating the angular and radial wavefunctions in quantum mechanics.

The separation of optical angular momentum into spin and orbital parts is not, however, as straightforward as the analogy with quantum mechanics might suggest. There are two principal difficulties. The first arises from problems in distinguishing between the concepts of spin and orbital angular momentum and of the idea of intrinsic and extrinsic angular momentum. The second concerns the difficulty in identifying meaningful spin and orbital angular momenta within full Maxwellian electromagnetism. A number of publications have dealt with each of these problems.

It is helpful, in approaching the meaning and interpretation of optical orbital and spin angular momenta, to keep in mind the corresponding problem in the mechanics of a set of material bodies. These might be a set of electrons and nuclei, a gas of molecules or even a system of planets and a star. Each body may have its own spin and orbital angular momentum, the latter arising as a consequence of its linear momentum. For such a system of bodies, labelled by the subscript α, we may write the total angular momentum as

$$\mathbf{J} = \sum_{\alpha} [(\mathbf{r}_{\alpha} \times \mathbf{p}_{\alpha}) + \mathbf{s}_{\alpha}].$$

Here, $\mathbf{r}_{\alpha}$, $\mathbf{p}_{\alpha}$ and $\mathbf{s}_{\alpha}$ denote the position, linear momentum and spin of body α. We can identify the total spin and orbital angular momenta of the system as the sum of the individual quantities for each of the particles. Hence our total spin $\mathbf{S}$ and orbital angular momenta $\mathbf{L}$ are

$$\mathbf{S} = \sum_{\alpha} \mathbf{s}_{\alpha},$$

$$\mathbf{L} = \sum_{\alpha} \mathbf{r}_{\alpha} \times \mathbf{p}_{\alpha}.$$

Intrinsic and extrinsic angular momenta are often identified with the spin and orbital angular momenta. It is more meaningful, however, to define these terms by the manner in which they change if we displace the origin of the axes of rotation. Hence, the external angular momentum $\mathbf{J}_{\text{ext}}$ will be the cross product of the position of the centre of mass with the total linear momentum, $\mathbf{P}$

$$\mathbf{J}_{\text{ext}} = \mathbf{R} \times \mathbf{P}$$

where

$$\mathbf{R} = \sum_{\alpha} \frac{m_{\alpha}}{M} \mathbf{r}_{\alpha}; \quad M = \sum_{\alpha} m_{\alpha}; \quad \mathbf{P} = \sum_{\alpha} \mathbf{p}_{\alpha}.$$

The intrinsic angular momentum comprises the spin angular momentum together with the orbital angular momentum *relative to* the centre of mass. If we make simple use of the fact that momentum depends on velocity, it follows that

$$\mathbf{J}_{\text{int}} = \sum_{\alpha} m_{\alpha} (\mathbf{r}_{\alpha} - \mathbf{R}) \times (\dot{\mathbf{r}}_{\alpha} - \dot{\mathbf{R}}) + \mathbf{S}.$$

Clearly, displacement of the axes of rotation will change the extrinsic angular momentum but leave the intrinsic angular momentum unchanged. If there is only a single body then the orbital and spin angular momenta are the same as the extrinsic and intrinsic angular momenta respectively. In general, however, they are different quantities.

Many of the properties of the mechanical system can be usefully applied to optical angular momentum. It is clear from the definition of the external angular momentum that its component parallel to the total momentum must be zero. This means that the angular momentum carried by a Laguerre–Gaussian beam about its axis, or parallel to it, is entirely intrinsic (2.7, **Paper 2.5**) as the total linear momentum points along the beam axis. It is interesting to ask how the orbital and spin angular momenta are distributed within the beam (2.8, **Paper 2.6**). If only part of the beam is considered then the balance between the orbital contributions to the intrinsic and extrinsic angular momenta will change. In the mechanical system this arises because, if only a subset of the bodies is considered, the centre of mass will in general be different. If we consider only a part of a Laguerre–Gaussian beam then an extrinsic angular momentum (2.9, **Paper 2.7**) is found. The spin angular momentum, as it is wholly intrinsic, remains unchanged. The spin of a macroscopic body can, of course, be re-interpreted in terms of the orbital angular momentum of its constituent particles. It is interesting to note that a similar procedure is possible, even for a single electron, if its spin is associated with an orbital angular momentum for its wave field (2.10). However, for light fields, the association of optical spin angular momentum with polarisation is the most useful approach.

Our previous discussion has focused on optical angular momentum within the paraxial approximation and by analogy with quantum theory. It is highly desirable, however, to be able to ground such a fundamental concept within electromagnetic theory. We can write the total angular momentum very simply in terms of the momentum density ($\varepsilon_0 \mathbf{E} \times \mathbf{B}$) as:

$$\mathbf{J} = \int \varepsilon_0 \mathbf{r} \times (\mathbf{E} \times \mathbf{B})\, d^3x.$$

Much has been written about suitable spin and orbital parts of this angular momentum. Yilmaz identifies a spin or intrinsic angular momentum, but notes that it is not in general gauge invariant (2.11). Gauge invariant quantities can be derived in terms of the gauge invariant transverse part of the vector potential $\mathbf{A}^{\perp}$ (2.12, 2.13). No distinction is made between spin and orbital, or extrinsic and intrinsic, angular momenta but the total angular momenta is written as a the sum of "spin" and "orbital" parts:

$$\mathbf{S} = \int \varepsilon_0 \mathbf{E}^{\perp} \times \mathbf{A}^{\perp}\, d^3x,$$

$$\mathbf{L} = \sum_l \int \varepsilon_0 E_l^{\perp} (\mathbf{r} \times \nabla) A_l^{\perp}\, d^3x.$$

The spin part gives the difference between the numbers of right and left circularly polarised photons (2.13), but it has been suggested that neither quantity is itself physically observable (2.12). A more dramatic problem arises from the fact that neither of these quantities is, by itself, an angular momentum (2.14, 2.15, **Paper 2.8**). These difficulties question the validity of the simple picture obtained within the paraxial approximation. A study of exact, or non-paraxial, beams with $\exp(-il\phi)$ dependence revealed that the ratio of the angular momentum per unit length to the energy per unit length is only approximately $(l + \sigma_z)/\omega$ (2.16, **Paper 2.9**). It is more correct, however, to analyse the properties of optical beams in terms of fluxes. The recent introduction of the optical angular momentum flux has resolved some of the outstanding problems with the electromagnetic description of optical spin and orbital angular momentum (2.17, **Paper 2.10**). In particular, the angular momentum flux for a beam about its axis can be separated into well-behaved spin and orbital components and the ratio of these to the energy flux are precisely σ_z/ω and l/ω respectively.

REFERENCES

2.1 L Allen, MW Beijersbergen, RJC Spreeuw and JP Woerdman, 1992, *Phys. Rev. A* **45** 8185.

2.2 AE Siegman, 1986, *Lasers* (Mill Valley, CA: University Science).

2.3 D Marcuse, 1972, *Light Transmission Optics* (New York: Van Nostrand).

2.4 SJ van Enk and G Nienhuis, 1992, *Opt. Commun.* **94** 147.

2.5 G Nienhuis and L Allen, 1993, *Phys. Rev. A* **48** 656.
2.6 L Allen, J Courtial and MJ Padgett, 1999, *Phys. Rev. E* **60** 7497.
2.7 MV Berry, in *Singular optics* (ed MS Soskin) Frunzenskoe, Crimea 1998, *SPIE* **3487** 1.
2.8 L Allen and MJ Padgett, 2000, *Opt. Commun.* **184** 67.
2.9 AT O'Neil, A MacVicar, L Allen and MJ Padgett, 2002, *Phys. Rev. Lett.* **88** 053601.
2.10 HC Ohanian, 1986, *Am. J. Phys.* **54** 500.
2.11 H Yilmaz, 1965, *Introduction to the theory of relativity and the principles of modern physics* (New York: Blaisdell).
2.12 C Cohen-Tannoudji, J Dupont-Roc and G Grynberg, 1989, *Photons and atoms* (New York: Wiley).
2.13 L Mandel and E Wolf, 1995, *Optical coherence and quantum optics* (Cambridge: Cambridge University Press).
2.14 SJ van Enk and G Nienhuis, 1994, *J. Mod. Opt.* **41** 963.
2.15 SJ van Enk and G Nienhuis, 1994, *Europhys. Letts.* **25** 497.
2.16 SM Barnett and L Allen, 1994, *Opt. Commun.* **110** 670.
2.17 SM Barnett, 2002, *J. Opt. B: Quantum Semiclass. Opt.* **4** S1.

PAPER 2.1

Orbital angular momentum of light and the transformation of Laguerre-Gaussian laser modes

L. Allen, M. W. Beijersbergen, R. J. C. Spreeuw, and J. P. Woerdman
Huygens Laboratory, Leiden University, P.O. Box 9504, 2300 RA Leiden, The Netherlands
(Received 6 January 1992)

Laser light with a Laguerre-Gaussian amplitude distribution is found to have a well-defined orbital angular momentum. An astigmatic optical system may be used to transform a high-order Laguerre-Gaussian mode into a high-order Hermite-Gaussian mode reversibly. An experiment is proposed to measure the mechanical torque induced by the transfer of orbital angular momentum associated with such a transformation.

PACS number(s): 42.50.Vk

I. INTRODUCTION

It is well known from Maxwell's theory that electromagnetic radiation carries both energy and momentum. The momentum may have both linear and angular contributions; angular momentum has a spin part associated with polarization [1] and an orbital part associated with spatial distribution [2]. Any interaction between radiation and matter is inevitably accompanied by an exchange of momentum. This often has mechanical consequences some of which are related to radiation pressure. Although an experimental demonstration of the mechanical torque created by the transfer of angular momentum of a circularly polarized light beam was performed over 50 years ago [1], the work associated with the mechanical influence of light beams on atoms and matter has been almost exclusively concerned with linear momentum [3–5].

Beth [1] made the first observation of the angular momentum of light following Poynting [6], who inferred from a mechanical analogy that circularly polarized light should exert a torque on a birefringent plate and that the ratio of angular to linear momentum is equal to $\lambda/2\pi$. In his experiment a half-wave plate was suspended by a fine quartz fiber [see Fig. 1(a)]. A beam of light, circularly polarized by a fixed quarter-wave plate, passed through the plate which transformed right-handed circularly polarized light into left-handed circularly polarized light and transferred $2\hbar$ of spin angular momentum for each photon to the birefringent plate. It was found that the measured torque agreed in sign and magnitude with that predicted by both wave and quantum theories of light. The ratio of the angular momentum of N photons in the beam, $J=\pm N\hbar$, to their energy, $W=N\hbar\omega$, is $\pm 1/\omega$ and Beth's measurement is sometimes referred to as the measurement of the spin angular momentum of the photon.

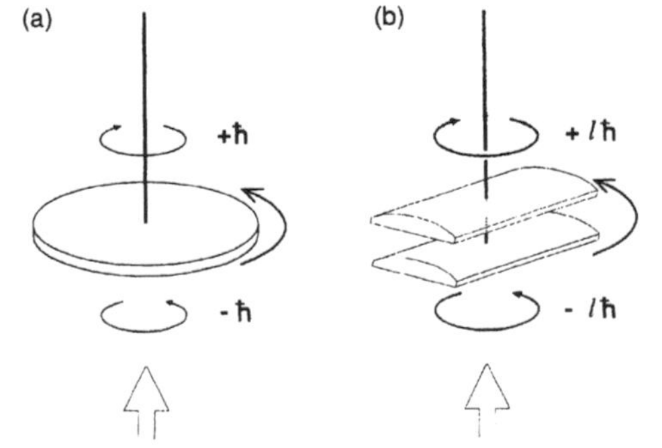

FIG. 1. (a) A suspended $\lambda/2$ birefringent plate undergoes torque in transforming right-handed into left-handed circularly polarized light. (b) Suspended cylindrical lenses undergo torque in transforming a Laguerre-Gaussian mode of orbital angular momentum $-l\hbar$ per photon, into one with $+l\hbar$ per photon.

The purpose of this paper is to investigate whether a Gaussian mode may be said to possess orbital angular momentum and to propose a study of the mechanical consequences which might then arise. The amplitude of a Laguerre-Gaussian mode has an azimuthal angular dependence of $\exp(-il\phi)$, where l is the azimuthal mode index. Analogy between quantum mechanics and paraxial optics [7] suggests that such modes are the eigenmodes of the angular momentum operator L_z and carry an orbital angular momentum of $l\hbar$ per photon. The transverse amplitude distribution of laser light is usually described in terms of a product of Hermite polynomials $H_n(x)H_m(y)$ and associated with TEM_{nmq} modes. Laguerre polynomial distributions of amplitude, TEM_{plq} modes, are also possible but occur less often in actual lasers. This allows the analogy with quantum mechanics to be stressed even more strongly. It is well known [8] that the one-dimensional quantum-mechanical harmonic oscillator has a solution in the form of a Hermite polynomial and that in two dimensions the solutions may be written as Laguerre polynomials with energy $(n+m+1)\hbar\omega$, while the eigenvalue of the two-dimensional angular momentum operator is $(n-m)\hbar$. It seems likely, therefore, that TEM_{plq} modes possess well-defined orbital angular momenta.

II. ORBITAL ANGULAR MOMENTUM OF A LAGUERRE-GAUSSIAN MODE

The angular momentum density associated with the transverse electromagnetic field may be shown [2] to be

$$\mathbf{M}=\epsilon_0\mathbf{r}\times(\mathbf{E}\times\mathbf{B}) \tag{1}$$

while the total angular momentum of the field is

$$\mathbf{J}=\epsilon_0\int \mathbf{r}\times(\mathbf{E}\times\mathbf{B})d\mathbf{r} \; . \tag{2}$$

In atomic physics it is normal to expect that

$$\mathbf{J}=\mathbf{L}+\mathbf{S} \; , \tag{3}$$

where the first term is identified with the orbital angular momentum **L**, and the second with the spin **S**. But there is doubt as to whether **L** and **S** are, in general, separately physically observable [9] for vector fields.

Clearly the linear momentum of a transverse plane wave, $\mathbf{E}\times\mathbf{B}$, is in the direction of propagation, z, and there cannot be a component of angular momentum $\mathbf{r}\times(\mathbf{E}\times\mathbf{B})$ in the same direction [10]. However, the fields of the laser modes TEM_{nmq} or TEM_{plq}, unlike those in a coaxial metal waveguide of infinite length, are not strictly transverse [11]. They have small components in the direction of propagation, z. A convenient representation of a linearly polarized laser mode is achieved [12] in the Lorentz gauge using the vector potential

$$\mathbf{A}=\mathbf{x}u(x,y,z)e^{-ikz} \tag{4}$$

where $\mathbf{x}$ is the unit vector in the x direction. The expression $u(x,y,z)$, or $u(r,\phi,z)$, is the complex scalar function describing the distribution of the field amplitude which satisfies the wave equation in the paraxial approximation. In this approximation the second derivatives of E and B fields, and the products of first derivatives, are ignored and $\partial u/\partial z$ taken to be small compared with ku. The cylindrically symmetric solutions $u_{pl}(r,\phi,z)$ which describe Laguerre-Gaussian beams are given by

$$\begin{aligned} u_{pl}(r,\phi,z)=&\frac{C}{(1+z^2/z_R^2)^{1/2}}\left[\frac{r\sqrt{2}}{w(z)}\right]^l L_p^l\left[\frac{2r^2}{w^2(z)}\right] \\ &\times\exp\left[\frac{-r^2}{w^2(z)}\right]\exp\frac{-ikr^2z}{2(z^2+z_R^2)}\exp(-il\phi) \\ &\times\exp\left[i(2p+l+1)\tan^{-1}\frac{z}{z_R}\right] \; , \end{aligned} \tag{5}$$

where z_R is the Rayleigh range, $w(z)$ is the radius of the beam, L_p^l is the associated Laguerre polynomial, C is a constant, and the beam waist is at $z=0$. The Lorentz gauge has the advantage of being readily amenable in all coordinate systems and leads in this case to considerable symmetry in the x and y directions although the results are best expressed in cylindrical coordinates.

Within this description we have shown that the time average of the real part of $\epsilon_0\mathbf{E}\times\mathbf{B}$, which is the linear momentum density, is given by

$$\begin{aligned} \frac{\epsilon_0}{2}(\mathbf{E}^*\times\mathbf{B}+\mathbf{E}\times\mathbf{B}^*)=&i\omega\frac{\epsilon_0}{2}(u^*\nabla u-u\nabla u^*) \\ &+\omega k\epsilon_0|u|^2\mathbf{z} \; , \end{aligned} \tag{6}$$

for a beam of unit amplitude, where $\mathbf{z}$ is the unit vector in the z direction. We may recognize that $u^*\nabla u$ closely echoes the quantum-mechanical expression for the expectation value of linear momentum of a wave function. To achieve this appealing form we have retained the term $\partial u/\partial z$ in the expression for the magnetic field **B**, which could have been ignored.

When applied to the Laguerre-Gaussian distribution given by Eq. (5) for linearly polarized light, the momentum density per unit power is found to be

$$\mathcal{P}=\frac{1}{c}\left[\frac{rz}{(z^2+z_R^2)}|u|^2\mathbf{r}+\frac{l}{kr}|u|^2\boldsymbol{\phi}+|u|^2\mathbf{z}\right] \; , \tag{7}$$

where $\mathbf{r}$ and $\boldsymbol{\phi}$ are unit vectors and $|u|^2\equiv|u(r,\phi,z)|^2$. Here the $\partial u/\partial z$ term has now been neglected. It may be seen that the Poynting vector, given by $c^2\mathcal{P}$, spirals along the direction of propagation; see Fig. 2. The $\mathbf{r}$ component relates to the spread of the beam; the $\boldsymbol{\phi}$ component gives rise to orbital angular momentum in the z direction and the $\mathbf{z}$ component relates to the linear momentum P in the direction of propagation.

Calculation of the time averaged angular momentum density, $\epsilon_0\mathbf{r}\times\langle\mathbf{E}\times\mathbf{B}\rangle$, per unit power yields

$$\mathbf{M}=-\frac{l}{\omega}\frac{z}{r}|u|^2\mathbf{r}+\frac{r}{c}\left[\frac{z^2}{(z^2+z_R^2)}-1\right]|u|^2\boldsymbol{\phi}+\frac{l}{\omega}|u|^2\mathbf{z} \; . \tag{8}$$

The radial and azimuthal components are symmetric about the axis, so that integration over the beam profile leaves only the $\mathbf{z}$ component. The ratio of the flux of angular momentum to that of energy is $L/cP=l/\omega$, while the ratio of angular momentum to linear momentum is now $L/P=l(\lambda/2\pi)$. Our conviction that the Laguerre-Gaussian mode possesses a well-defined orbital angular momentum has thus been justified.

At position (r,ϕ,z) the magnitude of angular momentum density per unit power is given by $M=l/\omega(1+z^2/r^2)^{1/2}|u|^2$, oriented at an angle $\theta=\tan^{-1}z/r$ to the z axis. Locally we have $M_z/\mathcal{P}_z=l(\lambda/2\pi)$ where M_z is the $\mathbf{z}$ component of angular momentum density and $\mathcal{P}_z$ that of the linear momentum density. There is, however, also a local radial component.

We have so far considered linearly polarized light; when the vector potential is generalized to arbitrary polarization we find

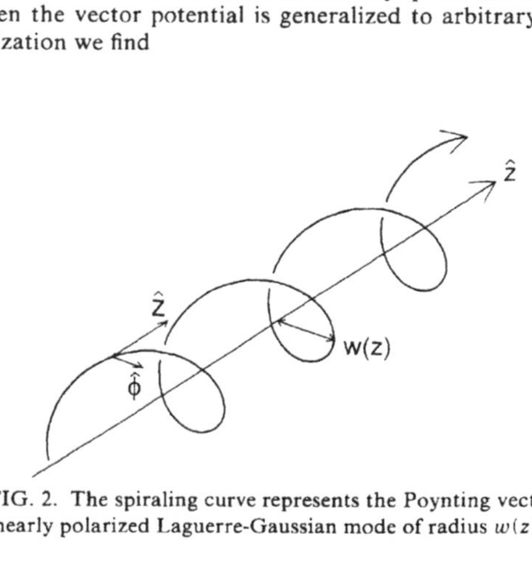

FIG. 2. The spiraling curve represents the Poynting vector of a linearly polarized Laguerre-Gaussian mode of radius $w(z)$.

$$\frac{\epsilon_0}{2}(\mathbf{E}^*\times\mathbf{B}+\mathbf{E}\times\mathbf{B}^*)=i\omega\frac{\epsilon_0}{2}(u^*\nabla u-u\nabla u^*)+\omega k\epsilon_0|u|^2\mathbf{z}+\omega\sigma_z\frac{\epsilon_0}{2}\frac{\partial|u|^2}{\partial r}\mathbf{\Phi}, \tag{9}$$

where the first two terms are polarization independent and relate to orbital angular momentum and the final term is a polarization, or spin, part. These lead to a z component of total angular momentum density per unit power

$$M_z=\frac{l}{\omega}|u|^2+\frac{\sigma_z r}{2\omega}\frac{\partial|u|^2}{\partial r}. \tag{10}$$

The ratio of the angular momentum flux to energy flux now becomes $J/cP=(l+\sigma_z)/\omega$ where $\sigma_z=\mp 1$ for right-handed or left-handed circularly polarized light and $\sigma_z=0$ for linearly polarized light.

We may note that in the paraxial approximation the spin-dependent part of angular momentum density depends upon the gradient of the intensity. Thus at a particular local point the z component of angular momentum flux divided by energy flux does not yield a simple value. However, when the total angular momentum flux is calculated, the integration across the beam profile leads to the simple result of the preceding paragraph.

III. THE DETECTION OF ORBITAL ANGULAR MOMENTUM

In the Beth experiment right-handed circularly polarized light, $-\hbar$, was converted to left-handed circularly polarized light, $+\hbar$, so that $2\hbar$ of spin angular momentum per photon was imparted to the birefringent plate. In the same way the maximum transfer of orbital angular momentum would take place if a Laguerre-Gaussian beam possessing $l\hbar$ angular momentum per photon were converted into one with $-l\hbar$ per photon. The torque arising from the transfer of momentum must then be measured by an appropriate equivalent of the birefringent plate. It is, however, more convenient to discuss the conversion to a beam with zero orbital angular momentum. This is readily possible by transforming a Laguerre-Gaussian to a Hermite-Gaussian distribution, which can be done using a mode convertor with astigmatic optical components.

Abramochkin and Volostnikov [13] have considered the mathematical transformation of laser beams undergoing astigmatism and found an integral transformation of Hermite-Gaussian into Laguerre-Gaussian beams. They recognized that passing the beam through a cylindrical lens can perform the desired conversion. Tamm and Weiss [14] have similarly employed a "mode convertor" involving cylindrical lenses.

Some insight into the mechanism for transformation of modes of arbitrarily high order is readily possible. A Hermite-Gaussian laser beam with its axes along the axes of a cylindrical lens is not changed, apart from a difference in size along the two axes. Therefore the operation of a cylindrical lens on an arbitrary beam pattern is most easily discussed in terms of its Hermite-Gaussian components along the axes of the lens.

The decomposition of a Laguerre-Gaussian mode in terms of Hermite-Gaussian modes can be seen in Eq. (A3) in the Appendix of Abramochkin and Volostnikov's paper [13]. It connects a combination of products of Hermite polynomials to a Laguerre polynomial,

$$\sum_{k=0}^{n+m}(2i)^k P_k^{(n-k,m-k)}(0)H_{n+m-k}(x)H_k(y)=2^{n+m}\times\begin{cases}(-1)^m m!(x+iy)^{n-m}L_m^{n-m}(x^2+y^2) & \text{for } n\geq m\\ (-1)^n n!(x-iy)^{m-n}L_n^{m-n}(x^2+y^2) & \text{for } m>n,\end{cases} \tag{11}$$

where

$$P_k^{(n-k,m-k)}(0)=\frac{(-1)^k}{2^k k!}\frac{d^k}{dt^k}[(1-t)^n(1+t)^m]\bigg|_{t=0}.$$

Also in the appendix of Ref. [13] there is an unnumbered equation which connects the same Hermite polynomials to their 45° transformed value,

$$\sum_{k=0}^{n+m}(-2)^k P_k^{(n-k,m-k)}(0)H_{n+m-k}(x)H_k(y)=(\sqrt{2})^{n+m}H_n\left(\frac{x-y}{\sqrt{2}}\right)H_m\left(\frac{x+y}{\sqrt{2}}\right). \tag{12}$$

It may be seen that the summation in Eq. (11) is the same as that in Eq. (12), save for a factor of $(-i)^k$ where k is the integer associated with the Hermite polynomial in the y direction, $H_k(y)$. But $(-i)^k$, of course, corresponds to an additional change of phase $\pi/2$ for each integer increase in the value of k.

A simple example perhaps makes the result of this clear. From Eq. (12) we may show that

$$(\sqrt{2})^5 H_3\left(\frac{x-y}{\sqrt{2}}\right)H_2\left(\frac{x+y}{\sqrt{2}}\right)=H_5(x)H_0(y)-H_4(x)H_1(y)-2H_3(x)H_2(y)+2H_2(x)H_3(y)+H_1(x)H_4(y)-H_0(x)H_5(y) \tag{13}$$

while from Eq. (11)

$$32re^{i\phi}L_2^1(r^2)=H_5(x)H_0(y)+iH_4(x)H_1(y)+2H_3(x)H_2(y)+2iH_2(x)H_3(y)+H_1(x)H_4(y)+iH_0(x)H_5(y)\ . \qquad (14)$$

We see that a Hermite-Gaussian mode with spatial dependence $H_n(x)H_m(y)$, may, provided the appropriate $\pi/2$ change of phase is achieved, become a single Laguerre-Gaussian mode well defined by L_p^l, or specifically $L_m^{n-m}(r^2)$ for $n \ge m$. We may identify $(n-m)$ as the orbital angular momentum of the photon in units of $\hbar$. Therefore we have converted a Hermite-Gaussian mode with zero orbital angular momentum into a Laguerre-Gaussian mode with $L=l\hbar$. If a change of phase other than $\pi/2$ is introduced, distributions between a Laguerre-Gaussian and a Hermite-Gaussian mode will result with ill-defined orbital angular momentum because such a mode is not an eigenstate of L_z.

As $H_k(y)=(-1)^k H_k(-y)$ we should note that if the left-hand side of Eq. (11) is multiplied inside the summation by $(-1)^k$, the right-hand side then relates to $(x-iy)$ for $n \ge m$ and $(x+iy)$ for $m \ge n$. Therefore the sign of the ϕ dependence of the Laguerre mode becomes changed. In other words, an additional π change of phase for all terms will transform a Laguerre-Gaussian mode with an angular momentum of $l\hbar$ to one with $-l\hbar$.

The equations allow us to recognize the nature of the decomposition of mode patterns into modes along the axes of a cylindrical lens. The presence of the focusing along only one of the axes and not the other, allows the components to propagate with different Gouy phase shifts [15]. In this way the necessary $\pi/2$ phase change to convert a Laguerre-Gaussian mode to a Hermite-Gaussian mode, or vice versa, may be achieved as Tamm and Weiss [14] realized for first-order modes.

Only for the TEM_{10} mode is the intuition that it must be combined with an out of phase version of itself to produce a radially continuous field or intensity distribution, actually true. Using Eqs. (11) and (12) it is easy to show

$$H_1\left(\frac{x-y}{\sqrt{2}}\right)H_0\left(\frac{x+y}{\sqrt{2}}\right)e^{-r^2/w^2}\rightarrow rL_0^1(r^2)e^{-r^2/w^2}e^{i\phi}\ . \qquad (15)$$

It follows from Eq. (14) that to obtain higher-order Laguerre modes it is necessary to combine a number of Hermite-Gaussian distributions: an example is shown in Fig. 3. We have thus established a general recipe for the transformation of Laguerre-Gaussian modes of well-defined angular orbital momentum into Hermite-Gaussian modes, or into a Laguerre-Gaussian mode of opposite angular symmetry. In the experiments currently in progress in our laboratory Hermite-Gaussian modes have been converted into pure, nondegenerate, Laguerre-Gaussian modes by a simple astigmatic optical arrangement which confirms the analysis outlined in the preceding section. The full analysis of the effect of astigmatism on the Gouy phase of laser modes and the detailed design of appropriate mode converters are planned to be published elsewhere [16].

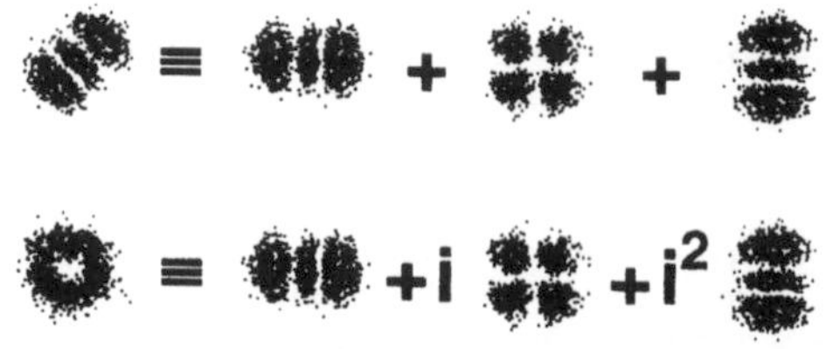

FIG. 3. The decomposition of a TEM_{02} Hermite-Gaussian mode at 45° into a set of Hermite-Gaussian modes and the decomposition of a Laguerre-Gaussian mode into the same, rephased, set.

IV. THE MEASUREMENT OF MECHANICAL TORQUE

Spin angular momentum in the form of circularly polarized light is both produced and detected by birefringence. In the Beth experiment [1] the torque per unit volume is given by $\mathbf{P}\times\mathbf{E}$. As $\mathbf{D}=\kappa\mathbf{E}=\epsilon_0\mathbf{E}+\mathbf{P}$ this may be written as

$$\frac{d\mathbf{J}}{dt}=(\mathbf{D}\times\mathbf{E})\ . \qquad (16)$$

The torque on the birefringent material arises because κ is a tensor; that is, $\mathbf{E}$ is not parallel to $\mathbf{P}$; in fact $\kappa_{xx}=n_x^2$ and $\kappa_{yy}=n_y^2$. The same "tensorial" attributes are required in the measurement of orbital angular momentum.

We have already indicated that a cylindrical lens can introduce phase differences between the Hermite-Gaussian components. Various combinations of cylindrical, or tilted spherical lenses, may be used to yield "retardation plates" of arbitrary thickness for orbital angular momentum. Just as appropriate phasing of the x and y components of the electric field in any light beam is capable of transforming linearly polarized light into circularly, or elliptically, polarized light so the rearrangement of phase introduced by an astigmatic optical element can destroy or create orbital angular momentum. Any measurement of mechanical torque must rely on such a change.

The measurement of a mechanical torque arising from orbital angular momentum should, thus, closely resemble that of spin. A suspended combination of two cylindrical lenses may be made to transform a Laguerre-Gaussian mode to the one with opposite orbital angular momentum, employing the anisotropy of the Gouy phase in the space in between the two lenses [see Fig. 1(b)]. The resultant torque which is predictable in terms of intensity and the value of l will then be compared with that measured from the rotation of the fiber suspension. An appropriate astigmatic birefringent element, or a combination of astigmatic and birefringent parts, would be capable of responding to both orbital and spin contributions simultaneously. This would allow the observation of total angular momentum J. However, a purely astigmatic or birefringent detector will respond only to the uncoupled values of L and S, respectively. It appears that there is no sense in which the spin and orbital contributions are intrinsically coupled except through the mechanical torque applied to such as astigmatic and birefringent device. It may be that perfect decoupling of **L** and **S** is limited to those cases where the paraxial approximation is valid; the decoupling of polarization state and scalar field amplitude is a feature of this approximation.

V. CONCLUSIONS

We have demonstrated that a Laguerre-Gaussian laser mode has a well-defined orbital angular momentum equal to $l\hbar$ per photon, with l the azimuthal mode index. We have outlined how such orbital angular momentum may be removed from the mode and converted into a mechanical torque. We have shown that such a transformation may be achieved by the use of astigmatic optical elements, which may also be used to produce Laguerre-Gaussian modes from the more commonly occurring Hermite-Gaussian modes.

It would appear that all light beams which possess field gradients, and which are not therefore plane waves, will possess a measure of orbital angular momentum. Indeed a badly phased transformation between transverse laser amplitude distributions will in general lead to ill-defined orbital angular momentum. For this reason it is important that stable, nondegenerate, propagating Laguerre-Gaussian polynomial modes are created and entirely transformed. A meaningful measurement of orbital angular momentum will not otherwise result. Although Laguerre-Gaussian modes may be created within the laser [14,19] they are usually degenerate, either simultaneously possessing $\exp(\pm il\phi)$ components or randomly fluctuating between them in time, and so have zero average orbital angular momentum. The reason why Hermite-Gaussian modes are the dominant ones in lasers is because of the presence of intracavity astigmatism [17].

We have been concerned in this paper with the orbital, or the total, angular momentum of the whole light beam. The ratio of the flux of total angular momentum to that of energy only gives $(l+\sigma_z)/\omega$ when we integrate over the whole beam profile. We have not investigated the consequences of the local ratio l/ω inside a linearly polarized Laguerre-Gaussian beam. It is in this regime that interactions with either atoms or small particles will take place; what the effect of orbital momentum of the light will be in this case remains to be investigated.

Other areas of study currently exist which invoke inhomogeneous, transverse, laser fields with field gradients. These include laser cooling, the manipulation of atoms and, expressly, mechanical effects in the interaction of neutral atoms and cylindrically symmetric modes: see [4,18,19]. It is interesting to speculate on whether such investigations might be changed by the existence of orbital angular momentum. It would also be interesting to know the extent to which the language of angular momentum might be a meaningful alternative to that already used in these areas.

ACKNOWLEDGMENTS

It is a pleasure to acknowledge helpful conversations with G. Nienhuis. This work is part of the research program of the Foundation for Fundamental Research on Matter (FOM) and was made possible by financial support from the Netherlands Organization for Scientific Research.

[1] R. A. Beth, Phys. Rev. **50**, 115 (1936).

[2] J. D. Jackson, *Classical Electrodynamics* (Wiley, New York, 1962).

[3] A. Ashkin, Science **210**, 1081 (1980).

[4] A. P. Kazantsev, G. I. Surdutovich, and K. V. Yakovlev, *Mechanical Action of Light on Atoms* (World Scientific, Singapore, 1990).

[5] S. Stenholm, Rev. Mod. Phys. **58**, 699 (1986).

[6] J. H. Poynting, Proc. R. Soc. London, Ser. A **82**, 560 (1909).

[7] D. Marcuse, *Light Transmission Optics* (Van Nostrand, New York, 1972).

[8] A. Messiah, *Quantum Mechanics* (North-Holland, Amsterdam, 1970).

[9] C. Cohen-Tannoudji, J. Dupont-Roc, and G. Grynberg, *Photons and Atoms* (Wiley, New York, 1989), p. 50; J. M. Jauch and F. Rohrlich, *The Theory of Photons and Electrons,* 2nd ed. (Springer-Verlag, Berlin, 1976), p. 34.

[10] The finite dimensions of the light beam in the Beth experiment meant that the field was not constant over the suspended birefringent plate and was not equivalent to a plane wave. A discursive treatment of the role of field gradients in the transfer of angular momentum has been given by J. W. Simmons and M. J. Guttmann, *States, Waves and Photons* (Addison-Wesley, Reading, MA, 1970). Formal discussion of the problem has a long history and an extensive list of references may be found on p. 34 of J. M. Jauch and F. Rohrlich, Ref. [9].

[11] M. Lax, W. H. Louisell, and W. B. McKnight, Phys. Rev. A **11**, 1365 (1975).

[12] H. A. Haus, *Waves and Fields in Optoelectronics* (Prentice-Hall, Englewood Cliffs, NJ, 1984).

[13] E. Abramochkin and V. Volostnikov, Opt. Commun. **83**, 123 (1991).

[14] C. Tamm and C. O. Weiss, J. Opt. Soc. Am. B **7**, 1034 (1990).

[15] A. E. Siegman, *Lasers* (University Science Books, Mill Valley, CA, 1986).

[16] M. W. Beijersbergen, L. Allen, and J. P. Woerdman (unpublished).

[17] S. L. Chao and J. M. Forsyth, J. Opt. Soc. Am. **65**, 867 (1975).

[18] P. Colet, M. San Miguel, M. Bambilla, and L. A. Lugiato, Phys. Rev. A **43**, 3862 (1991).

[19] M. Brambilla, L. A. Lugiato, V. Penna, F. Prati, C. Tamm, and C. O. Weiss, Phys. Rev. A **43**, 5090 (1991); **43**, 5114 (1991).

PAPER 2.2

Eigenfunction description of laser beams and orbital angular momentum of light

S.J. van Enk and G. Nienhuis
Huygens Laboratorium, Rijksuniversiteit Leiden, Postbus 9504, 2300 RA Leiden, The Netherlands

Received 9 June 1992

The propagation of light beams through astigmatic lens systems is accompanied by a transfer of orbital angular momentum. We develop a method to describe propagating light beams by operators of which the field is an eigenfunction. This method is applied to determine when an astigmatic lens system transforms gaussian beams into other gaussian beams and where in the system angular momentum is transferred. We show that the Gouy phase is equal to the dynamical phase of a quantummechanical harmonic oscillator with time-dependent energy.

1. Introduction

With the transverse intensity gradient of a beam of light corresponds a density of transverse momentum. This in turn may lead to a nonvanishing orbital angular momentum along the beam axis. Recently, it was demonstrated that astigmatic lens systems may change this angular momentum [1]. Since lenses normally do not change the polarization of the light, the spin is conserved. In an experiment [2], Hermite–gaussian beams which possess no orbital angular momentum were transformed into Laguerre–gaussian beams, which have a well-defined orbital angular momentum per photon. Since the total angular momentum of matter and light is conserved, this transformation is accompanied by a small torque on the lens system. The transformation of laser beams using astigmatic lenses has recently been investigated experimentally by Tamm and Weiss [3].

In this paper we present a novel theoretical description of light beams, and we apply it to this type of experiments. The method exploits some well-known similarities between quantum mechanics and paraxial wave optics, and makes use of a state-vector representation developed by Stoler [4]. Instead of giving analytical expressions for the electromagnetic field, we specify the field by (hermitian) operators of which the field is an eigenfunction, and by the corresponding eigenvalues. Two operators suffice to determine the spatial dependence of a monochromatic beam and one operator determines its polarization. The propagation of laser light through optical systems can then be described by the evolution of the 'eigenoperators' of the light field. In this way, we derive conditions under which gaussian beams are transformed into other gaussian beams. A case of special interest occurs when Hermite–gaussians are converted into Laguerre–gaussian modes. Analogously, the change of angular momentum of a field by lenses and lens systems is described by the evolution of the corresponding operator $\hat{L}_z$, in the same manner as in the Heisenberg picture in quantum mechanics. We also demonstrate that the Gouy phase is related to the dynamical phase of a time-dependent harmonic oscillator.

2. Linear and angular momentum of classical electromagnetic fields

The local densities of linear momentum and angular momentum of the source-free classical electromagnetic (em) field in vacuum are given by [5,6]

$$p=\epsilon_0 \boldsymbol{E}\times\boldsymbol{B} ,$$

$$j=\epsilon_0 \boldsymbol{r}\times(\boldsymbol{E}\times\boldsymbol{B}) , \qquad (1)$$

in terms of the electric field $\boldsymbol{E}$ and the magnetic field $\boldsymbol{B}$. These densities correspond to the Noether currents associated with the invariance of the free Maxwell equations under spatial translations and rotations, respectively [6,7]. The total linear and angular momentum of the em field, defined by

$$\boldsymbol{P}=\int \mathrm{d}\boldsymbol{r}\,\boldsymbol{p}\,, \quad \boldsymbol{J}=\int \mathrm{d}\boldsymbol{r}\,\boldsymbol{j}, \tag{2}$$

are therefore conserved quantities in vacuum.

We now consider monochromatic fields with frequency ω, and we use the complex notation

$$\boldsymbol{E}=(\boldsymbol{E}\,\mathrm{e}^{-\mathrm{i}\omega t}+\boldsymbol{E}^*\,\mathrm{e}^{\mathrm{i}\omega t})/2\,,$$
$$\boldsymbol{B}=(\boldsymbol{B}\,\mathrm{e}^{-\mathrm{i}\omega t}+\boldsymbol{B}^*\,\mathrm{e}^{\mathrm{i}\omega t})/2\,, \tag{3}$$

where the asterisk denotes complex conjugation. Then we can eliminate the magnetic field from eqs. (2) by using the Maxwell equation

$$\mathrm{i}\omega\boldsymbol{B}=\nabla\times\boldsymbol{E}\,. \tag{4}$$

For fields that vanish sufficiently fast for $|\boldsymbol{r}|\to\infty$, partial integration in eqs. (2) leads to the following expression for the total linear momentum

$$\boldsymbol{P}=\frac{\epsilon_0}{2\mathrm{i}\omega}\int \mathrm{d}\boldsymbol{r}\sum_{j=x,y,z} E_j^*\nabla E_j\,, \tag{5}$$

where we used that $\nabla\cdot\boldsymbol{E}=0$. The total momentum is manifestly independent of time. The expression (5) takes a form similar to the quantum-mechanical expression for the expectation value of the linear momentum of a particle. This has been noted before in a reciprocal (Fourier) representation [6–8]. In a similar way the total angular momentum can be written as

$$\boldsymbol{J}=\frac{\epsilon_0}{2\mathrm{i}\omega}\int \mathrm{d}\boldsymbol{r}\sum_{j=x,y,z} E_j^*(\boldsymbol{r}\times\nabla)E_j+\frac{\epsilon_0}{2\mathrm{i}\omega}\int \mathrm{d}\boldsymbol{r}\,\boldsymbol{E}^*\times\boldsymbol{E}$$
$$\equiv\boldsymbol{L}+\boldsymbol{S}\,. \tag{6}$$

The interpretation of the first and second term on the right hand side of eq. (6) as orbital angular momentum and spin is, although seemingly obvious, not without fundamental difficulties [6–8].

3. The paraxial approximation

Paraxial wave optics [9–11] describes the propagation of light beams whose transverse dimensions are much smaller than the typical longitudinal distance over which the field changes in magnitude. We start by recalling some results from ref. [9]. The transverse dimensions of the beam have the order of magnitude of the beam waist w_0, which is assumed to be much smaller than the diffraction length $l=kw_0^2$, with $k=\omega/c$ the wave number. One writes for the electric-field component of a beam propagating in the z direction

$$\boldsymbol{E}=\mathrm{e}^{\mathrm{i}kz}\boldsymbol{F}\,. \tag{7}$$

The small parameter w_0/l is used as an expansion parameter, which means that derivatives of $\boldsymbol{F}$ with respect to z can be neglected compared to the transverse derivatives. The field $\boldsymbol{F}$ satisfies to lowest order in w_0/l the paraxial wave equation

$$2\mathrm{i}k\frac{\partial}{\partial z}\boldsymbol{F}=-\left(\frac{\partial^2}{\partial x^2}+\frac{\partial^2}{\partial y^2}\right)\boldsymbol{F}\,, \tag{8}$$

which determines the propagation of the field in the z direction for a given field distribution in a plane $z=z_0$. Furthermore, the z component of $\boldsymbol{F}$ is smaller than the transverse components by a factor w_0/l.

Within the paraxial approximation we find it convenient to consider quantities of the em field that are defined as averages over planes with z=constant. Thus, the linear and angular momentum per unit of length in a plane $z=z_0$ are given by

$$\mathscr{P}(z_0)=\iint \mathrm{d}x\,\mathrm{d}y\,\boldsymbol{p}(x,y,z_0)\,,$$
$$\mathscr{J}(z_0)=\iint \mathrm{d}x\,\mathrm{d}y\,\boldsymbol{j}(x,y,z_0)\,. \tag{9}$$

Since the integrations in these definitions do not extend over all space, the steps leading to the appealing forms (5) and (6) are not necessarily valid here. For instance, the quantities (9) are not independent of time. Therefore, we will always average such quantities over a period $2\pi/\omega$. Then, in the paraxial approximation the (time-averaged) transverse components of $\mathscr{P}$, and the z component of $\mathscr{J}$ can be rewritten to lowest order in w_0/l as

$$\mathscr{P}_x = \sum_{j=x,y} \frac{\epsilon_0}{2i\omega} \iint dx\, dy\, F_j^* \frac{\partial}{\partial x} F_j ,$$

$$\mathscr{P}_y = \sum_{j=x,y} \frac{\epsilon_0}{2i\omega} \iint dx\, dy\, F_j^* \frac{\partial}{\partial y} F_j ,$$

$$\mathscr{J}_z = \sum_{j=x,y} \frac{\epsilon_0}{2i\omega} \iint dx\, dy\, F_j^* \left(x\frac{\partial}{\partial y} - y\frac{\partial}{\partial x} \right) F_j$$
$$+ \frac{\epsilon_0}{2i\omega} \iint dx\, dy\, (F_x^* F_y - F_y^* F_x)$$
$$\equiv \mathscr{L}_z + \mathscr{S}_z . \qquad (10)$$

Here we explicitly neglected derivatives of $\boldsymbol{F}$ with respect to z compared with transverse derivatives, and used that the z component of $\boldsymbol{F}$ is small [9]. The z component of the linear momentum per unit length is to lowest order in w_0/l given by

$$\mathscr{P}_z = \sum_{j=x,y} \frac{\epsilon_0}{2\omega} \iint dx\, dy\, F_j^* k F_j . \qquad (11)$$

From the well-known expression for the local energy density of the em field [5,6],

$$u = \frac{\epsilon_0}{2} (\boldsymbol{E}^2 + c^2 \boldsymbol{B}^2) , \qquad (12)$$

we obtain the time-averaged field energy per unit of length $\mathscr{E}$,

$$\mathscr{E} = \mathscr{N} \hbar\omega = \sum_{j=x,y} \frac{\epsilon_0}{2} \iint dx\, dy\, F_j^* F_j , \qquad (13)$$

where $\mathscr{N}$ is the number of photons per unit of length. Division of quantities like eqs. (10) and (11) by $\mathscr{N}$ yields the value of that quantity per photon. For instance, the momentum in the z direction is, obviously, equal to $\hbar k$ per photon.

4. Operator formalism and notation

4.1. Schrödinger picture

The propagation of light beams can also be described in an operator formalism which is very similar to the operator description of the hamiltonian evolution of quantum-mechanical states. We use here the formalism developed by Stoler [4]. Moreover, the analogy between the expressions (10)–(13) for time-averaged quantities of the classical em field, and the quantum-mechanical expectation values of the same quantities for particles, allows us to extend that formalism (see also ref. [11]). We use the following notations and conventions:

- the field $\boldsymbol{F}(x, y, z)$ is represented by the ket vector $|\boldsymbol{F}(z)\rangle$;
- operators acting upon the field $\boldsymbol{F}$ are represented by corresponding operators, denoted by a caret, acting upon the ket $|\boldsymbol{F}\rangle$;
- the evolution of a field propagating from $z=z_0$ to $z=z_1$ through a given lossless optical system is determined by a unitary operator $\hat{h}$, according to

$$|\boldsymbol{F}(z_1)\rangle = \hat{h} |\boldsymbol{F}(z_0)\rangle ; \qquad (14)$$

the coordinate z plays the same role as time in ordinary quantum mechanics;

- the scalar product $\langle \boldsymbol{F}(z) \| \boldsymbol{G}(z) \rangle$ is defined as

$$\langle \boldsymbol{F}(z) \| \boldsymbol{G}(z) \rangle$$
$$= \frac{\epsilon_0}{2\omega} \iint dx\, dy\, \boldsymbol{F}^*(x, y, z) \cdot \boldsymbol{G}(x, y, z)$$
$$= \sum_{j=x,y} \frac{\epsilon_0}{2\omega} \iint dx\, dy\, F_j^*(x, y, z) G_j(x, y, z) , \qquad (15)$$

to lowest order in w_0/l.

When $\hat{Q}$ is an hermitian operator, the quantity $\mathscr{Q}(z)$ defined by

$$\mathscr{Q}(z) = \langle \boldsymbol{F}(z) | \hat{Q} | \boldsymbol{F}(z) \rangle \qquad (16)$$

is real. For example, when we introduce the operators for transverse momentum

$$\hat{p}_x = \frac{1}{i} \frac{\partial}{\partial x} , \qquad \hat{p}_y = \frac{1}{i} \frac{\partial}{\partial y} , \qquad (17)$$

then it follows from eqs. (10) that

$$\mathscr{P}_x = \langle \boldsymbol{F} | \hat{p}_x | \boldsymbol{F} \rangle , \qquad \mathscr{P}_y = \langle \boldsymbol{F} | \hat{p}_y | \boldsymbol{F} \rangle . \qquad (18)$$

The momentum operators obey the usual commutation relations with the position operators $\hat{x}$ and $\hat{y}$,

$$[\hat{x}, \hat{p}_x] = [\hat{y}, \hat{p}_y] = i . \qquad (19)$$

As another example, we introduce the operator $\hat{L}_z$ for the z component of the orbital angular momentum acting upon the external degrees of freedom of the field, by

$$\hat{L}_z = \hat{x}\hat{p}_y - \hat{y}\hat{p}_x . \qquad (20)$$

The spin operator $\hat{S}_z$ on the other hand acts upon the vectorial indices of the field,

$$\hat{S}_z | F_j \rangle = \sum_{k=x,y} \mathrm{i}\epsilon_{jkz} | F_k \rangle , \qquad (21)$$

with ϵ_{jkl} the completely anti-symmetrical Levi–Civita tensor. A matrix representation of the spin operator is therefore

$$\hat{S}_z = \begin{pmatrix} 0 & \mathrm{i} \\ -\mathrm{i} & 0 \end{pmatrix} . \qquad (22)$$

The eigenvalues of $\hat{S}_z$ are, of course, equal to ± 1. With these operators substituted for $\hat{Q}$ we can write eqs. (10) for $\mathscr{L}_z$ and $\mathscr{S}_z$ in the form (16).

The evolution of a light beam propagating in vacuum is found from the formal solution of eq. (8) [4]. It can be written in the form (14), with $\hat{h}$ the free propagation operator describing evolution,

$$\hat{h} = \exp\left(- \frac{\mathrm{i}(z_1 - z_0)}{2k} \hat{p}^2 \right) , \qquad (23)$$

where we defined $\hat{p}^2 = \hat{p}_x^2 + \hat{p}_y^2$.

4.2. Heisenberg picture

When a field evolves according to eq. (14), the evolution of a quantity $\mathscr{Q}(z)$, defined in eq. (16), is given by

$$\mathscr{Q}(z_1) = \langle \boldsymbol{F}(z_0) | \hat{h}^\dagger \hat{Q} \hat{h} | \boldsymbol{F}(z_0) \rangle , \qquad (24)$$

with $\hat{h}^\dagger$ the Hermitian conjugate of $\hat{h}$. As in the Heisenberg picture in quantum mechanics we can interpret the operator $\hat{h}^\dagger \hat{Q} \hat{h}$ as giving the evolution of the quantity corresponding to the operator $\hat{Q}$, where now the field is described by the state vector $|\boldsymbol{F}(z_0)\rangle$ at a single value z_0 of z. This picture is very useful when examining certain properties of physical quantities of the form (16) that are independent of the fields. For instance, if the operator $\hat{Q}$ commutes with the operator $\hat{p}^2$, it also commutes with the free propagation operator (23). This implies that

$$\hat{h}^\dagger \hat{Q} \hat{h} = \hat{Q} , \qquad (25)$$

which in turn implies that the quantity $\mathscr{Q}(z)$ is conserved under free propagation, i.e.

$$\frac{\mathrm{d}\mathscr{Q}}{\mathrm{d}z} = 0 . \qquad (26)$$

Thus the z components of both orbital angular momentum and spin are conserved during free propagation, since the operators $\hat{L}_z$ and $\hat{S}_z$ both commute with $\hat{p}^2$. This means that their value per unit length is uniform along z. Also the energy per unit length (13) is a conserved quantity, as it is represented by the identity operator. Note that 'conserved' here does not refer to conservation in time, but rather to conservation with the position z.

5. Ideal lenses

A lens is called ideal when it is sufficiently thin, so that propagation within the lens is negligible. An ideal lens modifies an incoming field distribution $\boldsymbol{F}(x, y, z)$ by adding a local phase $\psi(x, y)$ to the field, where ψ is a real function of (x, y) [12]. If the lens has a constant refractive index n, the phase change ψ is due to the local change of the optical path, which is proportional to the local thickness $b(x, y)$ of the lens,

$$\psi(x, y) = kb(x, y)(n-1) . \qquad (27)$$

In the operator formalism of the preceding section, the ideal lens is represented by the operator

$$\hat{T} = \exp[\mathrm{i}\psi(\hat{x}, \hat{y})] . \qquad (28)$$

In particular, a cylindrical lens with its axis oriented at an angle γ with the positive x axis is described by the phase

$$\psi(x, y) = - \frac{k}{2f}(x \cos\gamma + y \sin\gamma)^2 , \qquad (29)$$

with f the focal length of the lens along its axis. The focal length in the orthogonal direction is infinite. Within the Heisenberg picture it is easy to prove that the z component of the spin, $\mathscr{S}_z$, cannot be changed by a lens. Since the operator $\hat{S}_z$ does not act upon the coordinates of the field,

$$\hat{T}^\dagger \hat{S}_z \hat{T} = \hat{S}_z , \qquad (30)$$

for arbitrary $\hat{T}$ defined in eq. (28). The orbital part of the angular momentum does change in general, since

$$\hat{T}^{\dagger}\hat{L}_z\hat{T}=\hat{L}_z+\hat{x}\frac{\partial\psi}{\partial\hat{y}}-\hat{y}\frac{\partial\psi}{\partial\hat{x}}$$
$$\equiv\hat{L}_z+\delta\hat{L}_z\,. \qquad (31)$$

For a general field $|\boldsymbol{F}_{\text{in}}\rangle$ incident on the lens, the difference in the value of $\mathscr{L}_z$ in the output field and its value in the input field is

$$\delta\mathscr{L}_z=\langle\boldsymbol{F}_{\text{in}}\,|\,\delta\hat{L}_z\,|\,\boldsymbol{F}_{\text{in}}\rangle\,. \qquad (32)$$

Therefore the angular momentum absorbed by the lens per unit time is $-c\delta\mathscr{L}_z$. This is the total torque on the lens. In terms of the output field $|\boldsymbol{F}_{\text{out}}\rangle$ the relation (32) becomes

$$\delta\mathscr{L}_z=-\langle\boldsymbol{F}_{\text{out}}\,|\,\delta\hat{L}_z\,|\,\boldsymbol{F}_{\text{out}}\rangle\,, \qquad (33)$$

which follows from the relation

$$\hat{T}\hat{L}_z\hat{T}^{\dagger}=\hat{L}_z-\delta\hat{L}_z\,. \qquad (34)$$

For a spherical lens the operator $\delta\hat{L}_z$ vanishes. Hence, such a lens can never absorb angular momentum in the z direction from an em field. For a cylindrical lens satisfying eq. (29) one finds

$$\delta\hat{L}_z(\gamma)=\frac{k}{2f}[2\hat{x}\hat{y}\cos 2\gamma-(\hat{x}^2-\hat{y}^2)\sin 2\gamma]\,. \qquad (35)$$

From eq. (35) it follows that

$$\delta\hat{L}_z(\gamma\pm\pi/2)=-\delta\hat{L}_z(\gamma)\,. \qquad (36)$$

Thus, when a cylindrical lens is rotated over $\pi/2$ about the z axis, the amount of orbital angular momentum absorbed by the lens changes sign. This implies, that for any incoming field there is always an orientation γ_0 of the lens for which $\mathscr{L}_z$ of that field is not changed by the lens. Then it follows from eq. (35) again that the maximum change in the value of $\mathscr{L}_z$, $\delta\mathscr{L}_{\text{max}}$, is attained for an orientation angle $\gamma=\gamma_0\pm\pi/4$. In fact, the change in orbital angular momentum of an arbitrary field due to a cylindrical lens can always be written as

$$\delta\mathscr{L}_z=\pm\delta\mathscr{L}_{\text{max}}\sin(2\gamma-2\gamma_0)\,. \qquad (37)$$

6. Applications

6.1. Circular fields

As is well known, the operator $\hat{L}_z$ takes the simple form

$$\hat{L}_z=-\mathrm{i}\frac{\partial}{\partial\phi} \qquad (38)$$

in cylindrical coordinates

$$x=r\cos\phi;\quad y=r\sin\phi;\quad z=z\,. \qquad (39)$$

We define a circular field as an eigenfunction of $\hat{L}_z$. These fields have an azimuthal dependence given by $\exp(\mathrm{i}l\phi)$. Since ϕ is a periodic variable, l must be an integer. In a circular field with eigenvalue l the orbital angular momentum per photon amounts to $l\hbar$ [1], and the intensity is independent of ϕ. Moreover, since $\hat{L}_z$ commutes with the free propagation operator, circular fields remain circular during free propagation. We shall denote these fields by kets $|l\rangle$. Examples of circular fields are the Laguerre–gaussian modes [10].

We can now calculate the effect of an arbitrary ideal lens on the orbital angular momentum of a circular field. The change $\delta\mathscr{L}_z$ in orbital angular momentum of an arbitrary input field $|\boldsymbol{F}\rangle$, eq. (32), can be rewritten in cylindrical coordinates:

$$\delta\mathscr{L}_z=\langle\boldsymbol{F}|\,\partial\psi/\partial\phi\,|\boldsymbol{F}\rangle$$
$$=\frac{\epsilon_0}{2\omega}\int_0^{\infty}\mathrm{d}r\int_0^{2\pi}r\,\mathrm{d}\phi\,\boldsymbol{F}^*\cdot\boldsymbol{F}\frac{\partial\psi}{\partial\phi}$$
$$=-\frac{\epsilon_0}{2\omega}\int_0^{\infty}\mathrm{d}r\int_0^{2\pi}r\,\mathrm{d}\phi\,\psi\frac{\partial}{\partial\phi}(\boldsymbol{F}^*\cdot\boldsymbol{F})\,, \qquad (40)$$

where we used partial integration to obtain the last equality. From this equality it is clear that no ideal lens can change the orbital angular momentum of a circular field. Hence

$$\delta\mathscr{L}_z=\langle l|\,\delta\hat{L}_z\,|l\rangle=0\,. \qquad (41)$$

With eq. (33), this also shows that the lens cannot have changed the orbital angular momentum of a field when the *out*coming beam is circular.

6.2. A cylindrical lens system

In this subsection we consider a particular optical system, which has been used in the experiments mentioned in the Introduction [2]. The system consists of two identical cylindrical lenses, oriented along the x axis, with focal length f, and separated from each other by a distance $2d$ (fig. 1). In ref. [3] a slightly different configuration was used, in that the two lenses had different focal lengths. Our results can easily be generalized to include this case. However, an arbitrary conversion of a gaussian beam can already be realized with lenses with equal focal lengths (see sect. 7). The propagation operator $\hbar$ for this system is

$$\hbar = \exp\left(\frac{-ik\hat{x}^2}{2f}\right)\exp\left(\frac{-id\hat{p}^2}{k}\right)\exp\left(\frac{-ik\hat{x}^2}{2f}\right). \tag{42}$$

The action of the system can be specified by giving the evolution of the coordinate operators $\hat{x}$ and $\hat{y}$, and of the momentum operators $\hat{p}_x$ and $\hat{p}_y$ in the Heisenberg picture,

$$\hbar^\dagger \hat{x}\hbar = \hat{x}\left(1-\frac{2d}{f}\right)+\frac{2d}{k}\hat{p}_x ,$$

$$\hbar^\dagger \hat{y}\hbar = \hat{y}+\frac{2d}{k}\hat{p}_y ,$$

$$\hbar^\dagger \hat{p}_x\hbar = \hat{p}_x\left(1-\frac{2d}{f}\right)-\frac{2k}{f}\left(1-\frac{d}{f}\right)\hat{x} ,$$

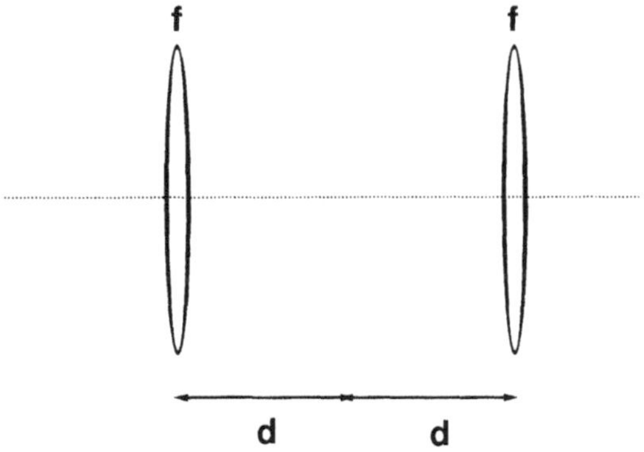

Fig. 1. Configuration of two parallel cylindrical lenses with focal length f at mutual distance $2d$.

$$\hbar^\dagger \hat{p}_y\hbar = \hat{p}_y , \tag{43}$$

obtained by means of the Baker–Campbell–Hausdorff formula [13]

$$\exp(B)A\exp(-B)$$
$$= A + [B, A] + \frac{1}{2!}[B, [B, A]] + \cdots . \tag{44}$$

From eqs. (43) one finds that the orbital angular momentum changes according to

$$\hbar^\dagger \hat{L}_z \hbar = \hat{L}_z\left(1-\frac{2d^2}{f^2}\right)+\frac{2k}{f}\left(1-\frac{d}{f}\right)\hat{x}\hat{y}+\frac{4d^2}{kf}\hat{p}_x\hat{p}_y$$
$$+(\hat{x}\hat{p}_y+\hat{y}\hat{p}_x)\left(\frac{2d}{f}-\frac{2d^2}{f^2}\right). \tag{45}$$

In particular, for incoming circular fields $|l\rangle$ only the first term on the right hand side of eq. (45) contributes, and one finds for the change in $\mathscr{L}_z$

$$\delta\mathscr{L}_z = -\frac{2d^2}{f^2}\langle l|\hat{L}_z|l\rangle , \tag{46}$$

so that the lens system absorbs an amount $2l\hbar d^2/f^2$ of angular momentum per photon. However, the output field cannot be circular. In fact, when both in- and outcoming fields would be circular, then neither of the two lenses could have changed $\mathscr{L}_z$, as shown in section 6.1; furthermore, during the free propagation between the lenses $\mathscr{L}_z$ is conserved (section 4.2), so that the orbital angular momentum of in- and outcoming fields would have to be the same. This is in contradiction with eq. (46), since the multiplication factor is always negative.

From similar arguments is clear that the change in angular momentum of an incoming circular field takes place exclusively at the second lens.

Moreover, eq. (46) shows that we must have $d=f/\sqrt{2}$ in order that the lens system changes a circular field $|l\rangle$ with $l\neq 0$ into a field with zero angular momentum. Conversely, the same condition is required to produce a circular field $|l\rangle$ with $l\neq 0$ from an input field that carries no angular momentum (cf. ref. [2]).

6.3. A paradox

Geometrical optics suggests a simple way to convert a circular field $|l\rangle$ into the opposite state $|-l\rangle$.

This is achieved when the system inverts the x direction, without changing the y direction. For a well-collimated beam this simple reflection is produced by the configuration of fig. 2, with $d=f$. This configuration has been termed a π convertor [2]. The picture of x inversion is confirmed by eqs. (43), provided that the last terms of the first two equations are negligible. This requires that

$$\delta \equiv d/kw^2 \ll 1 , \tag{47}$$

where w is the transverse dimension of the incoming beam. Then the lens system transforms a field $|\boldsymbol{F}\rangle$ according to

$$\boldsymbol{F}(x, y) \to \boldsymbol{F}(-x, y) , \tag{48}$$

apart from a phase factor, and a small $\mathcal{O}(\delta)$ correction. When applied to a circular field $|l\rangle$, this transformation (48) yields another circular field $|-l\rangle$. This simple geometrical picture is confirmed by eq. (46), which shows that for $d=f$ the momentum transfer is exactly $-2\hbar l$ per photon. However, this transformation is in contradiction with the conclusion of section 6.2, namely that for a circular input field the output field cannot be circular. In fact, the conversion is only exact in the geometrical optics limit $k \to \infty$ (so that $\delta \to 0$). Hence, when the input field is exactly the circular state $|l\rangle$, then the output field must be written as $|-l\rangle + \delta|s\rangle$, where $|s\rangle$ stands for some superposition of circular fields. Then the input field has an angular momentum per photon equal to $l\hbar$, and for the output field this value is exactly $-l\hbar$ (when $d=f$). Furthermore, the transfer of angular momentum must occur at the second lens, as we showed in section 6.2.

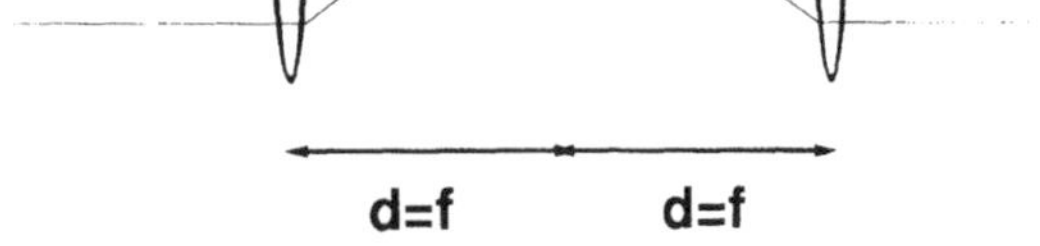

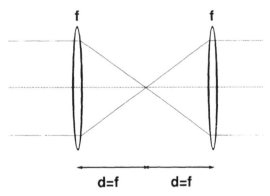

Fig. 2. In the geometrical optics limit the lens system of fig. 1. inverts the image in one direction when $d=f$.

This raises another problem: on the one hand the second lens transfers $2l\hbar$ per photon when δ is small but finite, while on the other hand there can be no transfer when the output field is circular (i.e. for $\delta=0$). This discontinuity can be explained from eq. (29). The phase that the lens adds to the field is of the order of $1/\delta$. Therefore the effect of the second lens on the orbital angular momentum can be arbitrarily large for $\delta \to 0$. Thus, for an *exact* circular output field there is no transfer of $\mathcal{L}_z$, but small $\mathcal{O}(\delta)$ corrections to a pure circular field give in general rise to a finite transfer of $\mathcal{L}_z$ of order unity in units of $\hbar$ per photon. More precisely, since the output field is given by $|-l\rangle + \delta|s\rangle$, one finds from eqs. (33) and (35) the finite change in $\mathcal{L}_z$ by the second cylindrical lens in the limit $\delta \to 0$,

$$\delta\mathcal{L}_z = -2\,\mathrm{Re}\langle s|\hat{x}\hat{y}/w^2|-l\rangle . \tag{49}$$

A similar argument is valid when the *input* field is not exactly circular.

We conclude that the total transfer of angular momentum is close to $-2\hbar l$ per photon when the input field is nearly circular, and when d is close to f. However, the location where the angular momentum is transferred is very sensitive to the input state: a small admixture of other states can give a drastic variation in the distribution of the transfer over the two lenses.

For noncircular input states, eq. (45) shows that the total amount of angular momentum transfer depends in a very sensitive way on d (or f). This is due to the second term in eq. (45), which is proportional to $1/\delta$. When the average of xy vanishes, which is the case for circular fields, this term does not contribute, and the sensitivity disappears.

7. Eigenoperators of fields

We earlier defined circular fields by requiring them to be eigenfunctions of the operator $\hat{L}_z$. In a similar way we can specify a field $|\boldsymbol{F}(z)\rangle$, for given wave number k, by giving three commuting hermitian operators of which the field is an eigenmode, and the corresponding eigenvalues. One of the operators, $\hat{S}(z)$, refers to the vector character (polarization or spin) of the field. The other two operators $\hat{K}_i(z)$, for

$i=1, 2$, act upon the translational degrees of freedom of the field, and determine its (x,y) dependence. Thus, the field $|\boldsymbol{F}(z)\rangle$ is uniquely determined, up to a normalization factor and an arbitrary phase, by

$$\hat{S}(z)|\boldsymbol{F}(z)\rangle = s|\boldsymbol{F}(z)\rangle ,$$
$$\hat{K}_i(z)|\boldsymbol{F}(z)\rangle = \mu_i|\boldsymbol{F}(z)\rangle . \qquad (50)$$

The operators $\hat{S}$ and $\hat{K}_i$ may be called eigenoperators of the field. The eigenvalue of the polarization operator $\hat{S}$ equals $s=\pm 1$, and the eigenvalues μ_i are real, and independent of z (see below). When a field propagates through a given optical system, it evolves according to eq. (14), and it will in general no longer satisfy eqs. (50). Instead, the field $\hat{h}|\boldsymbol{F}\rangle$ is an eigenfunction of the operators

$$\hat{S}' = \hat{h}\hat{S}\hat{h}^\dagger ,$$
$$\hat{K}_i' = \hat{h}\hat{K}_i\hat{h}^\dagger , \qquad (51)$$

with the same eigenvalues as in eqs. (50). The evolution of an em field is thus also described (up to a phase factor) by the evolution (eq. (51)) of its eigenoperators. This description is sometimes more convenient than dealing with the explicit spatial dependence of the fields. Notice that the operators $\hat{K}_i(z)$ as a function of z do not obey the Heisenberg equation, but rather the Liouville–Von Neumann equation.

Since the optical systems we consider in this paper do not contain polarization-changing elements, polarization is conserved, and will henceforth be left out of consideration. From now on we only discuss scalar fields.

7.1. *Hermite- and Laguerre-gaussian modes*

We apply the eigenoperator description to Hermite–gaussian (HG) and Laguerre–gaussian (LG) modes, which are commonly used in laser physics [10]. We first give the eigenoperators of these fields in their focal plane, which is the plane where the transverse dimension of the beam has a minimum. The focal plane is chosen to be the plane $z=0$. In an arbitrary plane $z=z_0$ the fields can then be constructed formally by letting the fields propagate freely from $z=0$ to $z=z_0$. In the next section we will show that the modes defined in this way are indeed the HG and LG modes.

We define a HG field in its focal plane by the eigenvalue equations

$$\frac{1}{2}\left(\frac{k}{\beta}\hat{x}^2 + \frac{\beta}{k}\hat{p}_x^2\right)|H^\beta_{nm}(0)\rangle = (n+\tfrac{1}{2})|H^\beta_{nm}(0)\rangle ,$$
$$\frac{1}{2}\left(\frac{k}{\beta}\hat{y}^2 + \frac{\beta}{k}\hat{p}_y^2\right)|H^\beta_{nm}(0)\rangle = (m+\tfrac{1}{2})|H^\beta_{nm}(0)\rangle . \qquad (52)$$

Thus a HG field $|H^\beta_{nm}(0)\rangle$ is an eigenfunction of a 'Hamiltonian' of two identical harmonic oscillators in the x and y direction, with 'energy' $n+\frac{1}{2}$ and $m+\frac{1}{2}$, respectively. The parameter β can be identified with the Rayleigh range or half the confocal parameter of the beam [10]. Analogously, LG fields with the same Rayleigh range β can be defined by

$$\hat{L}_z|L^\beta_{lN}(0)\rangle = l|L^\beta_{lN}(0)\rangle ,$$
$$\frac{1}{2}\left(\frac{k}{\beta}\hat{r}^2 + \frac{\beta}{k}\hat{p}^2\right)|L^\beta_{lN}(0)\rangle = (N+1)|L^\beta_{lN}(0)\rangle , \qquad (53)$$

where $\hat{r}^2 = \hat{x}^2 + \hat{y}^2$. Thus a LG field $|L^\beta_{lN}(0)\rangle$ is an eigenfunction of $\hat{L}_z$, with eigenvalue l, and of a 2-D degenerate harmonic oscillator, with 'energy' $N+1$. The latter operator is just the sum of the two harmonic oscillators in eqs. (52). From quantum mechanics it is known that one obtains solutions only for $|l| \le N$. The two eigenvalue equations (53) determine the ϕ and r dependence of LG fields. By applying the free propagation operator (23) to the fields at $z=0$ we obtain the HG and LG modes for arbitrary z,

$$|H^\beta_{nm}(z)\rangle = \exp\left(\frac{-iz}{2k}\hat{p}^2\right)|H^\beta_{nm}(0)\rangle ,$$
$$|L^\beta_{lN}(z)\rangle = \exp\left(\frac{-iz}{2k}\hat{p}^2\right)|L^\beta_{lN}(0)\rangle . \qquad (54)$$

For $z\neq 0$ these fields are no longer eigenfunctions of the same, or of any other, harmonic oscillator. Namely, we find that the eigenoperators of the HG mode evolve during free propagation as

$$\hat{K}_1(z) = \frac{1}{2}\left[\frac{k}{\beta}\hat{x}^2 + \left(\frac{\beta}{k} + \frac{z^2}{k\beta}\right)\hat{p}_x^2 - \frac{z}{\beta}\hat{t}_x\right],$$
$$\hat{K}_2(z) = \frac{1}{2}\left[\frac{k}{\beta}\hat{y}^2 + \left(\frac{\beta}{k} + \frac{z^2}{k\beta}\right)\hat{p}_y^2 - \frac{z}{\beta}\hat{t}_y\right], \qquad (55)$$

and the eigenvalues remain $n+\frac{1}{2}$ and $m+\frac{1}{2}$. Relation (55) follows from eqs. (51) with eq. (23) substituted for $\hat{h}$ while using the BCH formula (44). We defined here the operators $\hat{t}_x$ and $\hat{t}_y$ and their sum $\hat{t}$ by

$$\hat{t}_x = \hat{x}\hat{p}_x + \hat{p}_x\hat{x} ,$$
$$\hat{t}_y = \hat{y}\hat{p}_y + \hat{p}_y\hat{y} ,$$
$$\hat{t} = \hat{t}_x + \hat{t}_y . \quad (56)$$

The LG mode $|L^{\beta}_{lN}(z)\rangle$ is an eigenmode of $\hat{L}_z$ with eigenvalue l for all z, since $\hat{L}_z$ commutes with $\hat{p}^2$. Its second eigenoperator is given by $\hat{K}(z) = \hat{K}_r(z) + \hat{K}_2(z)$, and the eigenvalue remains $N+1$.

7.2. Gaussian modes

We have shown that HG and LG modes have one eigenoperator in common, of the general form

$$\hat{K}(\boldsymbol{a}) = a_1\hat{r}^2 + a_2\hat{p}^2 + a_3\hat{t} , \quad (57)$$

which is invariant under rotations about the z axis. In their focal plane both fields are eigenfunctions of a 2D degenerate harmonic oscillator (DHO), since there a_3 vanishes. Fields with an eigenoperator of the form (57) with arbitrary values a_1, a_2 and a_3 will be called gaussian fields. This is in accordance with textbook nomenclature [10], where gaussian fields are written as an eigenfunction of a DHO multiplied by a position-dependent phase factor (see next section). Such fields remain gaussian during their free propagation. Furthermore, it is easily proved that any gaussian field can be written as a linear superposition of HG modes $|H^{\beta}_{nm}(z)\rangle$ with $n+m$=constant, and with a properly selected value of β and of the position of the focal plane. The converse statement is obviously also true. Thus, any HG mode $|H^{\beta}_{nm}(z)\rangle$ that is rotated about the z axis can be written as such a superposition. The same is true for the LG mode $|L^{\beta}_{lN}(z)\rangle$, where $n+m$ must equal N.

7.3. Mode convertors

The eigenoperator (57) of a gaussian field will be transformed by a cylindrical lens into an operator that is no longer invariant under rotations about the z axis. Then the output field is not gaussian. However, in special cases, a lens system may convert a gaussian field into another gaussian field. Then we call the lens system a mode convertor [1,2]. The condition for a mode convertor is, that the operator

$$\hat{K}' = \hat{h}\hat{K}(\boldsymbol{a})\hat{h}^{\dagger} \quad (58)$$

be of the form (57) again. We substitute for $\hat{h}$ the evolution operator (42) of the lens system from section 6.2. For a given incoming field, this yields the conditions under which the lens system is a mode convertor for this field. We obtain

$$d = k\frac{a_3}{a_1} ,$$

$$f = k\frac{a_3 a_2}{a_2 a_1 - 2a_3^2} , \quad (59)$$

in terms of the coefficients a_i of the eigenoperator (57) of the incoming field. For incoming HG modes with eigenoperators (55) we can substitute $a_1 = k/\beta$ and $a_3 = -z/\beta$ in the conditions (59), which then become

$$z = -d ,$$

$$f = d\frac{\beta^2 + d^2}{\beta^2 - d^2} . \quad (60)$$

The first condition means that for a mode convertor the focal plane of the incoming HG mode must be halfway between the two lenses [2].

7.4. Production of circular fields

We now wish to investigate when it is possible to convert a HG mode $|H^{\beta}_{nm}(z)\rangle$ into a field that has the orbital angular momentum operator $\hat{L}_z$ as an eigenoperator, that is, into a circular field.

The two eigenoperators of the HG mode were given in eqs. (55). Eigenoperators of this form will keep the same form during their propagation through the cylindrical lens system of section 6.2. The reason is (i) that the evolution operator (42) of the lens system consists of exponentials of the operators occurring in (55) and (ii) that these operators form a closed (Lie) algebra. In fact, they generate the direct product SU(1,1)×SU(1,1) [13]. The operators (55) will therefore evolve, according to eqs. (51), within that algebra during the entire propagation. This algebra does not contain $\hat{L}_z$. Therefore, the lens system satisfying eq. (42) alone is not sufficient to produce an eigenfunction of $\hat{L}_z$ from a HG mode $|H^{\beta}_{nm}(z)\rangle$. Both eigenoperators of this field are and remain of the form (55).

However, application of a rotation about the z axis to a HG mode before it enters the lens system will transform eigenoperators in such a way that they will belong to a larger algebra which does contain $\hat{L}_z$. The reason is, that rotations around the z axis are generated by $\hat{L}_z$: the rotation operator $\hat{R}_\gamma$ describing a rotation around the z axis over an angle γ is given by

$$\hat{R}_\gamma = \exp(i\gamma\hat{L}_z) . \tag{61}$$

If the HG mode $|H^{\beta}_{nm}(z)\rangle$, rotated over an angle γ and having passed the lens system, is to be an eigenfunction of $\hat{L}_z$, we must require $\hat{L}_z$ to be a linear combination of the two eigenoperators of the output field

$$\hat{K}'_i = \hat{h}\hat{R}_\gamma\hat{K}_i\hat{R}_\gamma^{-1}\hat{h}^\dagger , \tag{62}$$

for $i=1, 2$, with $\hat{K}_1$ and $\hat{K}_2$ the eigenoperators of the incoming field, defined in eqs. (55). Conversely, the operator

$$\hat{L}'_z = \hat{R}_\gamma^{-1}\hat{h}^\dagger\hat{L}_z\hat{h}\hat{R}_\gamma \tag{63}$$

must be a linear combination of the operators $\hat{K}_1$ and $\hat{K}_2$. Using eq. (45) one finds

$$\begin{aligned}\hat{R}_\gamma^{-1}\hat{h}^\dagger\hat{L}_z\hat{h}\hat{R}_\gamma = {} & \hat{L}_z\left(1-\frac{2d^2}{f^2}\right) \\ & + \left(\frac{2d}{f}-\frac{2d^2}{f^2}\right)[(\hat{x}\hat{p}_y+\hat{y}\hat{p}_x)\cos 2\gamma \\ & +\sin 2\gamma(\hat{y}\hat{p}_y-\hat{x}\hat{p}_x)] \\ & + \frac{4d^2}{kf}[\hat{p}_x\hat{p}_y\cos 2\gamma+\tfrac{1}{2}\sin 2\gamma(p_y^2-\hat{p}_x^2)] \\ & + \frac{2k}{f}\left(1-\frac{d}{f}\right)[\hat{x}\hat{y}\cos 2\gamma+\tfrac{1}{2}\sin 2\gamma(\hat{y}^2-\hat{x}^2)] .\end{aligned} \tag{64}$$

Inspection of eq. (64) shows that this expression is indeed a linear combination of $\hat{K}_1$ and $\hat{K}_2$:

$$\hat{R}_\gamma^{-1}\hat{h}^\dagger\hat{L}_z\hat{h}\hat{R}_\gamma = \pm(\hat{K}_2-\hat{K}_1) , \tag{65}$$

if and only if

$$\begin{aligned}&\gamma=\pm\pi/4 , \quad f=d\sqrt{2} , \\ &z=-d , \quad \beta=d(1+\sqrt{2}) .\end{aligned} \tag{66}$$

These restrictions are the same as those given for the '$\pi/2$ convertor' [2]. Here we have proved that the rotation angle γ must equal $\pm\pi/4$, and that the focal plane of the HG mode must be half way between the two lenses. For the eigenvalue l of $\hat{L}_z$ one finds, from eq. (65),

$$l=\pm(m-n) , \tag{67}$$

because the eigenvalues corresponding to $\hat{K}_2$ and $\hat{K}_1$ are $m+\frac{1}{2}$ and $n+\frac{1}{2}$, respectively. Since the relations (66) are consistent with eqs. (60), the lens system acts in this case also as a mode convertor. That is, the outcoming mode is not just an eigenfunction of $\hat{L}_z$, it is in fact the LG field $|L^{\beta}_{l,n+m}(d)\rangle$.

8. Properties of gaussian modes

We show here that the eigenoperators provide a simple way to explain that a freely propagating field, which is an eigenfunction of a 2D degenerate harmonic oscillator (DHO) in one plane, can be written as the product of a local phase factor and an eigenfunction of another DHO, in any other plane z=constant. Subsequently, we show that the Gouy phase of gaussian beams [10] can be identified with the dynamical phase of a quantum-mechanical DHO with time-dependent energy.

We consider a field $|F_N(0)\rangle$ that is an eigenfunction of the DHO

$$\hat{K}(0) = \frac{1}{2}\left[\frac{\hat{r}^2}{\alpha_0}+\alpha_0\hat{p}^2\right], \tag{68}$$

with eigenvalue $N+1$. When the field propagates through the vacuum, it becomes an eigenfunction of the operator

$$\hat{K}(z) \equiv \exp\left(\frac{-iz}{2k}\hat{p}^2\right)\frac{1}{2}\left[\frac{\hat{r}^2}{\alpha_0}+\alpha_0\hat{p}^2\right]\exp\left(\frac{iz}{2k}\hat{p}^2\right), \tag{69}$$

at position z. We now search for parameters R and α such that this eigenoperator $\hat{K}(z)$ is again a DHO up to a position-dependent phase factor. More precisely, we require

$$\hat{K}(z) = \exp\left(\frac{ik}{2R}\hat{r}^2\right)\frac{1}{2}\left[\frac{\hat{r}^2}{\alpha}+\alpha\hat{p}^2\right]\exp\left(\frac{-ik}{2R}\hat{r}^2\right). \tag{70}$$

This relation implies that the field $|F(z)\rangle$ at position z is the product of the phase factor $\exp(ik\hat{r}^2/2R)$ and an eigenfunction of a DHO of the form (68) with α_0 replaced by $\alpha(z)$. Its eigenvalue is $N+1$. Using the BCH formula (44) we find the unique solution of eq. (70) to be

$$R(z)=\frac{z^2+\alpha_0^2k^2}{z},$$

$$\alpha(z)=\frac{z^2+\alpha_0^2k^2}{k^2\alpha_0}. \tag{71}$$

For HG and LG modes we may substitute the value $\alpha_0=\beta/k$, according to eqs. (52) and (53). Then we find

$$R(z)=\frac{z^2+\beta^2}{z},$$

$$\alpha(z)=\frac{z^2+\beta^2}{k\beta}. \tag{72}$$

These parameters are, indeed, equal to the local radius of curvature of these modes, and to the square of the local beam radius, respectively [10].

8.1. Gouy phase

As just shown we can rewrite the field $|F_N(z)\rangle$ as

$$|F_N(z)\rangle=\exp\left(\frac{ik\hat{r}^2}{2R(z)}\right)|H_N(\alpha,z)\rangle, \tag{73}$$

where $|H_N(\alpha,z)\rangle$ is an eigenfunction of a DHO with eigenvalue $N+1$. Using the fact that $|F_N(z)\rangle$ satisfies the paraxial wave equation, we find, after substituting eq. (73),

$$2ik\frac{\mathrm{d}}{\mathrm{d}z}|H_N(\alpha,z)\rangle=\left[k^2\frac{1}{R^2}\frac{\partial R}{\partial z}\hat{r}^2\right.$$
$$\left.+\exp\left(\frac{-ik}{2R}\hat{r}^2\right)\hat{p}^2\exp\left(\frac{+ik}{2R}\hat{r}^2\right)\right]|H_N(\alpha,z)\rangle$$
$$=\left[\frac{\hat{r}^2}{\alpha^2}+\hat{p}^2+\frac{k}{R}\hat{t}\right]|H_N(\alpha,z)\rangle$$
$$=\left[\frac{\hat{r}^2}{\alpha^2}+\hat{p}^2+\frac{k}{2\alpha}\frac{\partial\alpha}{\partial z}\hat{t}\right]|H_N(\alpha,z)\rangle, \tag{74}$$

with $\hat{t}$ defined in eq. (56). The field distribution corresponding to $|H_N(\alpha,z)\rangle$ can be written as a real function of x and y, represented by the ket $|G_N(\alpha)\rangle$, multiplied by an overall z-dependent phase factor,

$$|H_N(\alpha,z)\rangle=\exp(-i\phi_N(z))|G_N(\alpha)\rangle. \tag{75}$$

The last term on the right hand side of eq. (74) being proportional to $\mathrm{d}\alpha/\mathrm{d}z$ results from the z dependence of the real part $|G_N(\alpha)\rangle$ in eq. (75) through $\alpha(z)$. Indeed one can show that

$$\hat{t}|G_N(\alpha)\rangle=4i\alpha\frac{\partial}{\partial\alpha}|G_N(\alpha)\rangle. \tag{76}$$

The remaining two terms on the right hand side of eq. (74) therefore determine the explicit z dependence of $|H_N(\alpha,z)\rangle$, and hence the phase $\phi_N(z)$. This remaining part has the form of a Hamiltonian for an harmonic oscillator with constant mass and position-dependent frequency. The ket $|H_N(\alpha,z)\rangle$ is an eigenfunction of this Hamiltonian with position-dependent eigenvalue ('energy') $2(N+1)/\alpha(z)$. Therefore, the phase $\phi_N(z)$ is given by

$$\phi_N(z)=\int_0^z \mathrm{d}s\,\frac{2(N+1)}{2k\alpha(s)}. \tag{77}$$

For HG and LG modes the substitution of $\alpha_0=\beta/k$ leads to the well-known expression for the Gouy phase [10]:

$$\phi(z)=\tan^{-1}(z/\beta), \tag{78}$$

and $\phi_N=(N+1)\phi$. Thus HG and LG modes evolve, apart form the phase factor in eq. (73) in the same way as the state of a quantum-mechanical harmonic oscillator with time-dependent energy in the adiabatic regime, when it starts in an eigenstate. Namely, in the adiabatic approximation one neglects the time derivative of the hamiltonian. Here, on the other hand, the presence of the third term in eq. (74) proportional to the derivative of $\alpha(z)$ ensures that the ket $|H_N(\alpha,z)\rangle$ remains an *exact* eigenfunction of the Hamiltonian.

9. Conclusions

We discussed the change in orbital angular momentum of light beams during their passage through systems of cylindrical lenses. We presented a formalism which makes use of some formal analogies between quantum mechanics and paraxial wave optics. For instance, a light field can be represented by a state vector, with the coordinate z as a propagation variable, which plays the same role as time in quantum mechanics [4]. We showed that physical quantities such as the linear and angular momentum per unit beam length can be given in the form of expectation values of corresponding operators in the state associated with the field.

In the 'Schrödinger picture', the propagation of fields along the z axis is described by a unitary operator. In the 'Heisenberg picture' the evolution of physical quantities rather than of the fields is given. By using the Heisenberg picture, we proved that, e.g., the z component of orbital angular momentum is uniform along z for arbitrary freely propagating light beams. Furthermore, we calculated the amount of angular momentum absorbed by ideal lenses, and described effects of cylindrical lenses for arbitrary incident fields.

We then discussed the action of cylindrical lens systems on the linear and orbital angular momentum of light beams, and determined exactly where and how much angular momentum is transferred. We also found a peculiar paradox for a particular lens system in the geometrical optics limit. It turns out that this limit displays a singularity for lens systems.

We argued that a light field is determined at each position z by three eigenoperators which have the field as an eigenfunction, and by its eigenvalues. Propagation of a field can alternatively be described by giving the evolution of its eigenoperators. In particular, we applied this formalism to define Hermite–gaussian (HG) and Laguerre–gaussian (LG) modes [10], and the definition for more general gaussian modes arises naturally. Using this formalism we were able to find in a straightforward way when an astigmatic lens system converts one gaussain mode into another. We also derived the necessary and sufficient conditions under which HG modes can be converted into LG modes and conversely. This possibility was derived in a completely different way in ref. [2].

Acknowledgements

We acknowledge stimulating discussions with L. Allen, M.W. Beijersbergen and J.P. Woerdman. This work is part of the research program of the Stichting voor Fundamenteel Onderzoek der Materie (FOM) which is financially supported by the Nederlandse Organisatie voor Wetenschappelijk Onderzoek (NWO).

References

[1] L. Allen, M.W. Beijersbergen, R.J.C. Spreeuw and J.P. Woerdman, Phys. Rev. A 45 (1992) 8185.
[2] M.W. Beijersbergen, L. Allen, H.E.L.O. van der Veen and J.P. Woerdman, to be published.
[3] C. Tamm and C.O. Weiss, J. Opt. Soc. Am. B 7 (1990) 1034.
[4] D. Stoler, J. Opt. Soc. Am. B 71 (1981) 334.
[5] J.D. Jackson, Classical Electrodynamics (Wiley, New York, 1962).
[6] C. Cohen-Tannoudji, J. Dupont-Roc and G. Grynberg, Photons and Atoms (Wiley, New York, 1989).
[7] J.M. Jauch and F. Rohrlich, The Theory of Photons and Electrons (Springer-Verlag, Berlin, 1976).
[8] J.W. Simmons and M.J. Guttmann, States, Waves and Photons (Addison-Wesley, Reading, MA, 1970).
[9] M. Lax, W.H. Louisell and W.B. McKnight, Phys. Rev. A 11 (1975) 1365.
[10] A.E. Siegman, Lasers (University Science Book, Mill Valley, 1986).
[11] D. Marcuse, Light Transmission Optics (Van Nostrand, New York, 1972).
[12] J.W. Goodman, Introduction to Fourier Optics (McGraw-Hill, New York, 1968).
[13] R. Gilmore, Lie Groups, Lie Algebras and Some of Their Applications (Wiley, New York, 1974).

PAPER 2.3

Paraxial wave optics and harmonic oscillators

G. Nienhuis and L. Allen*

Huygens Laboratorium, Rijksuniversiteit Leiden, Postbus 9504, 2300 RA Leiden, The Netherlands

(Received 23 November 1992)

The operator algebra of the quantum harmonic oscillator is applied to the description of Gaussian modes of a laser beam. Higher-order modes of the Hermite-Gaussian or the Laguerre-Gaussian form are generated from the fundamental mode by ladder operators. This approach allows the description of both free propagation and refraction by ideal astigmatic lenses. The paraxial optics analog of a coherent state is shown to be a light beam with a displaced beam axis which is refracted by lenses according to geometric optics. The expectation value of the orbital angular momentum of a paraxial beam of light is found to be expressible in terms of a contribution analogous to the angular momentum of the oscillator plus contributions which arise from the ellipticity of the wave fronts and of the light spot. This clarifies the process by which a transfer of orbital angular momentum between a light beam and astigmatic lenses or diaphragms occurs.

PACS number(s): 42.50.Vk, 42.25.Bs

I. INTRODUCTION

It is well known that the analytical form of Gaussian modes of a laser beam resembles the wave functions of the stationary states of a two-dimensional quantum-mechanical harmonic oscillator [1,2]. This suggests that algebraic treatments of the harmonic oscillator may be fruitfully applied to wave optics. Such a treatment can give insight into the structure and the properties of paraxial modes, and the connection between modes of different order. Moreover, algebraic methods often greatly simplify explicit calculations of physical quantities. Recently it has been pointed out that a Laguerre-Gaussian beam carries an orbital angular momentum along its propagation direction [3]. This angular momentum arises from the transverse momentum density of the field. It might be expected that the orbital angular momentum of a beam of light be analogous to the angular momentum of the harmonic oscillator.

In this paper, operator algebra is applied to describe Gaussian modes of a laser beam in the presence of ideal, but possibly, astigmatic lenses. Raising and lowering operators are introduced, which generate all higher-order modes from the fundamental one. These operators depend in a simple way on the coordinate in the propagation direction, and they can be expressed as a unitary transformation of the real harmonic-oscillator ladder operators. The ladder operators are characterized by three z-dependent beam parameter for each transverse dimension. These parameters are the radius of curvature, the spot size, and the phase. The algebraic connection between modes with different mode indices is the same as that for the number states of the harmonic oscillator. The fundamental mode is the eigenvector of the lowering operator with eigenvalue zero. The connection between modes of different order, and between Laguerre-Gaussian and Hermite-Gaussian modes, follows directly. Eigenvectors with different eigenvalues, which are analogous to the coherent states, represent beams with a displaced axis. The axis of the beam simply obeys the rules of geometric optics while leaving unaffected the three beam parameters, which reflect the wave-optical nature of the beam. This separation between geometric-optical and wave-optical aspects of the beam during its propagating through lens systems follows in a direct fashion from the algebraic treatment. Finally, we apply the operator algebra to the description of the orbital angular momentum of a Gaussian beam, the expression for which is found to separate into a term that is analogous to the orbital angular momentum of the isotropic two-dimensional harmonic oscillator, a contribution expressing the ellipticity of the spot size, and a term arising from the ellipticity of the wave fronts. Only this final term is modified when the beam passes an astigmatic lens. The torque exerted on the lens vanishes when the intensity distribution of the beam is symmetric with respect to one of the axes of the lens. This explains why a nonastigmatic Laguerre-Gaussian beam cannot transfer angular momentum to an astigmatic lens [4]. However, we demonstrate that such a beam transfers a significant amount of angular momentum to the lens if one applies angular aperturing of the beam just in front of the lens. Such a transfer depends only on the radial mode index.

II. OPERATOR ALGEBRA FOR FREE PARAXIAL MODE

As is well known, a complete set of solutions of the wave equation of a beam of light in the paraxial approximation consists of the products of a Gaussian with a Hermite polynomial H_n [1,2]. For a light beam propagating in the z direction, and a single transverse dimension x, these Hermite-Gaussian modes can be expressed as the normalized functions [2]

$$u_n(x,z)=\exp\left[\frac{ikx^2}{2s(z)}-i\chi(z)(n+\tfrac{1}{2})\right]\times\frac{1}{\sqrt{\gamma(z)}}\psi_n\left[\frac{x}{\gamma(z)}\right]. \tag{2.1}$$

Here the functions ψ_n represent the real normalized eigenfunctions of the harmonic oscillator, defined by the Hamiltonian

$$H=\frac{1}{2}\left[-\frac{\partial^2}{\partial\xi^2}+\xi^2\right] . \tag{2.2}$$

Their explicit expressions are

$$\psi_n(\xi)=[2^n n!\sqrt{\pi}]^{-1/2}\exp(-\xi^2/2)H_n(\xi) . \tag{2.3}$$

The formal analogy between the Hermite-Gaussian modes and the harmonic-oscillator eigenstates offers the possibility of applying the operator algebra of the harmonic oscillator to paraxial beam optics.

The Hermite-Gaussian mode functions (2.1) depend on three z-dependent mode parameters γ, s, and χ. The expressions for the spot size γ and radius of curvature of the wave front s can be combined in the single complex equality

$$\frac{1}{\gamma^2}-\frac{ik}{s}=\frac{k}{b+iz} , \tag{2.4}$$

with k the wave number of the light. The parameter b is the Rayleigh range of the beam, which determines the size of the focal region. Finally, the phase factor χ is given by

$$\tan\chi=z/b . \tag{2.5}$$

In the present case of Gaussian beams, this factor describes the phase jump of π that occurs over the focal region of any spherical converging wave, which was first recognized by Gouy [5,2]. The basis set (2.1) depends on the value of the Rayleigh range b and on the location of the focal plane, which we have chosen at $z=0$. The spot size at focus is equal to $\sqrt{b/k}$, and it increases as $z/\sqrt{bk}$ for $z\gg b$. Hence a small spot size at focus implies a large divergence angle of the beam.

Equation (2.1) displays both the analogy and the difference between the propagation of a Hermite-Gaussian mode of light and the evolution of a harmonic oscillator. The diffraction of the beam is expressed by a variation of the spot size γ and the radius of curvature s with z. This is absent for the oscillator. As may be noted in (2.1), the variation of the Gouy phase χ during propagation, which multiplies the level energy $n+\frac{1}{2}$, is analogous to the phase Ωt of the oscillator with frequency Ω during its evolution. In this sense, the free propagation of a Gaussian beam from $-\infty$ to ∞ corresponds to half a cycle of the oscillator.

The Hermite-Gaussian modes (2.1) are solutions of the paraxial wave equation [1,2,6]

$$\frac{\partial^2}{\partial x^2}u(x,z)=-2ik\frac{\partial}{\partial z}u(x,z) . \tag{2.6}$$

When the variations of u in the transverse direction x are small over a wavelength, a solution u of (2.6) determines, to a good approximation, the electric field $\mathbf{E}$ of a monochromatic light wave with frequency $\omega=ck$, of the form

The constant transverse vector $\mathbf{E}_0$ determines the polarization and the amplitude of the paraxial beam. Equation (2.6) has the same form as Schrödinger's equation for a free particle in one dimension, with the z coordinate replacing the time variable [7]. The formalism of quantum mechanics in terms of a Hilbert state space, and with linear operators representing observables, can be carried over directly to a classical light beam in the paraxial approximation, as has been done by Stoler [8]. In the same spirit, we introduce state vectors $|u(z)\rangle$, which have $u(x,z)$ for the mode function. Furthermore, we define the coordinate operator X, and the momentum operator P, according to the equalities

$$Xu(x,z)=xu(x,z),\quad Pu(x,z)=\frac{1}{i}\frac{\partial}{\partial x}u(x,z) . \tag{2.8}$$

(Operators acting on states vectors are indicated by roman capitals throughout this paper.) Then the propagation equation (2.6) can be put in the form

$$\frac{d}{dz}|u(z)\rangle=-\frac{i}{2k}P^2|u(z)\rangle . \tag{2.9}$$

Equation (2.9) has the formal solution

$$|u(z)\rangle=U(z)|u(0)\rangle , \tag{2.10}$$

with the propagation operator U defined by

$$U(z)=\exp\left[-\frac{i}{2k}P^2z\right] . \tag{2.11}$$

To complete the analogy with quantum mechanics, we assume normalized state vectors obeying the identity

$$\langle u(z)|u(z)\rangle\equiv\int dx\, u^*(x,z)u(x,z)=1 . \tag{2.12}$$

When $|u\rangle$ is normalized for one position z, the normalization for all other z values is automatic. For an inner product defined as in (2.12), the operators X and P are Hermitian. The expectation value of operators is defined in the standard way as

$$\langle P(z)\rangle\equiv\langle u(z)|P|u(z)\rangle=\int dx\, u^*(x,z)Pu(x,z) . \tag{2.13}$$

It may be shown that with this normalization (2.12) $\hbar\langle P(z)\rangle$ is equal to the transverse momentum per photon in the beam [4].

A. Ladder operators

The expressions (2.1), which are solutions of the paraxial wave equation (2.6), can be obtained by operator algebra techniques. Our starting point is the introduction of z-dependent field operators, which allow a series of solutions to be created from one. For $z=0$, we define the operators

$$A(0)=\frac{1}{\sqrt{2bk}}[kX+ibP],\quad A^\dagger(0)=\frac{1}{\sqrt{2bk}}[kX-ibP] , \tag{2.14}$$

real: they transform a real function into another real function. The z dependence of these operators is defined by the requirement that, for any solution $|u(z)\rangle$ of (2.9), $A^{\dagger}(z)|u(z)\rangle$ is also a solution. This implies that

$$A(z)=U(z)A(0)U^{\dagger}(z),\quad A^{\dagger}(z)=U(z)A^{\dagger}(0)U^{\dagger}(z)\ . \tag{2.15}$$

For free propagation, explicit expressions for these z-dependent operators follow from the operator identity

$$\exp\left[-\frac{i}{2k}P^2z\right]X\exp\left[\frac{i}{2k}P^2z\right]=X-\frac{z}{k}P\ . \tag{2.16}$$

Combining (2.15) and (2.16) gives the simple z dependence

$$A(z)=\frac{1}{\sqrt{2bk}}[kX+i(b+iz)P]\ ,$$
$$A^{\dagger}(z)=\frac{1}{\sqrt{2bk}}[kX-i(b-iz)P]\ . \tag{2.17}$$

These operators obey the standard commutation rule

$$[A(z),A^{\dagger}(z)]=1 \tag{2.18}$$

for boson raising and lowering operators and, as shown in every quantum mechanics textbook, this commutation rule is sufficient to prove that the number operator

$$N(z)=A^{\dagger}(z)A(z) \tag{2.19}$$

has the natural numbers as its eigenvalue spectrum. If we indicate the eigenstates for $z=0$ as $|u_n(0)\rangle$, and define their z dependence by

$$|u_n(z)\rangle=U(z)|u_n(0)\rangle\ , \tag{2.20}$$

then these states $|u_n(z)\rangle$ are solutions of (2.9), obeying the eigenvalue relation

$$N(z)|u_n(z)\rangle=n|u_n(z)\rangle\ , \tag{2.21}$$

for $n=0,1,2,\ldots$. Moreover, the phases of these states can be chosen so that

$$A^{\dagger}(z)|u_n(z)\rangle=\sqrt{n+1}|u_{n+1}(z)\rangle\ ,$$
$$A(z)|u_n(z)\rangle=\sqrt{n}\,|u_{n-1}(z)\rangle\ . \tag{2.22}$$

Hence, as for the harmonic oscillator, the higher-order modes can be obtained from the fundamental mode by applying the raising operator, according to

$$|u_n(z)\rangle=\frac{1}{\sqrt{n!}}[A^{\dagger}(z)]^n|u_0(z)\rangle\ . \tag{2.23}$$

We have thus proved the existence of a set of normalized solutions $|u_n(z)\rangle$ of (2.9), which are coupled by the operators $A^{\dagger}(z)$ and $A(z)$ according to the usual relations for the ladder operators of the harmonic oscillator.

These results allow the analytic form of the Hermite-Gaussian modes (2.1) to be explained in terms of the harmonic-oscillator eigenfunctions. In order to show this, we rewrite the operators (2.17) as a transformation of real operators, by using the transformation

$$\exp\left[\frac{ik}{2s}X^2\right]P\exp\left[-\frac{ik}{2s}X^2\right]=P-\frac{k}{s}X\ , \tag{2.24}$$

which is fully analogous to (2.16). This allows us to write (2.17) in the alternative form

$$A(z)=\exp\left[\frac{ik}{2s}X^2\right]B(z)\exp\left[-\frac{ik}{2s}X^2\right]\exp(i\chi)\ , \tag{2.25}$$

with $B(z)$ the real operator

$$B(z)=\frac{1}{\sqrt{2}}\left[\frac{X}{\gamma}+i\gamma P\right]\ . \tag{2.26}$$

The conjugate expressions hold for $A^{\dagger}$ and $B^{\dagger}$. As is obvious from (2.1), when A operates on a mode $u_n(x,z)$, the real operator B acts on the harmonic-oscillator state ψ_n. The z-dependent quantities s, γ, and χ are given in (2.4)–(2.6). As the higher-order modes $|u_n(z)\rangle$ can be generated from the fundamental one by repeated action of $A^{\dagger}$, it is sufficient to find the analytical form of the mode function $u_0(x,z)$. This is not difficult if one realizes that the lowering operators $A(z)$ must give zero when operating on the fundamental mode. For the normalized mode function, this gives as a solution of (2.6)

$$u_0(x,z)=\left[\frac{bk}{\pi}\right]^{1/4}\frac{1}{\sqrt{b+iz}}\exp\left[-\frac{kx^2}{2(b+iz)}\right]\ , \tag{2.27}$$

which may be rewritten in terms of the ground state of the harmonic oscillator as

$$u_0(x,z)=\exp\left[\frac{ik}{2s}x^2\right]\exp\left[-\frac{i\chi}{2}\right]\frac{1}{\sqrt{\gamma}}\psi_0\left[\frac{x}{\gamma}\right]\ . \tag{2.28}$$

The analytical form (2.1) of the higher-order modes follows directly after applying Eq. (2.23) and the recognition that the operator $B^{\dagger}$ is the raising operator for the harmonic-oscillator eigenfunctions with Gaussian width γ.

B. Coherent states and the Gouy phase

In view of the analogy of the complete set of Hermite-Gaussian modes with a harmonic oscillator, or equivalently with a single quantized field mode, it is natural to consider the analog of a coherent state [9]. Such a state arises from the ground state after the application of a displacement in phase space. The displacement operator

$$D(0)=\exp[iq_0X-ia_0P] \tag{2.29}$$

displaces position over a_0, and momentum over q_0, according to the relations [10]

$$D^{\dagger}(0)XD(0)=X+a_0,\quad D^{\dagger}(0)PD(0)=P+q_0\ . \tag{2.30}$$

We consider the mode which, for $z=0$, is equivalent to the displaced ground state

$$|u(0)\rangle=D(0)|u_0(0)\rangle\ . \tag{2.31}$$

This state is an eigenstate of the lowering operator $A(0)$ with complex eigenvalue

$$\alpha_0=\frac{1}{\sqrt{2bk}}[ka_0+ibq_0] . \quad (2.32)$$

The solution of (2.9), which is given by the coherent state (2.31) for $z=0$, can be expressed as

$$|u(z)\rangle=D(z)|u_0(z)\rangle , \quad (2.33)$$

with

$$D(z)=U(z)D(0)U^{\dagger}(z) . \quad (2.34)$$

Obviously, $|u(z)\rangle$ is an eigenvector of $A(z)$ with eigenvalue α_0. Application of (2.16) shows that $D(z)$ is again a displacement operator, and takes the form

$$D(z)=\exp[iq(z)X-ia(z)P] , \quad (2.35)$$

with the z-dependent displacements of position and momentum

$$a(z)=a_0+\frac{z}{k}q_0, \quad q(z)=q_0 . \quad (2.36)$$

Hence the state $|u(z)\rangle$, which is the analog of a coherent state, is simply the fundamental mode $|u_0(z)\rangle$, with the transverse momentum displaced by the constant amount q_0, and the position by $a(z)$, as given by (2.36). This equation simply describes a ray of light, tilted with respect to the z axis at an angle q_0/k.

In view of the analogy between the Gouy phase χ and the phase of the oscillator, we may expect that the variation of the displacements (2.36) with z corresponds to a variation in time of the displacement of a coherent state of the oscillator during its evolution. This may be illustrated by rewriting the displacement operator in the form

$$D(z)=\exp[\alpha_0A^{\dagger}(z)-\alpha_0^*A(z)] , \quad (2.37)$$

with α_0 given in (2.32). If we substitute the alternative form (2.25) of A and $A^{\dagger}$ into (2.37), we obtain an expression for D in terms of the real field operators

$$D(z)=\exp\left[\frac{ik}{2s}X^2\right]\exp[\alpha(z)B^{\dagger}(z)-\alpha^*(z)B(z)] \times\exp\left[-\frac{ik}{2s}X^2\right] , \quad (2.38)$$

with

$$\alpha(z)=\alpha_0e^{-i\chi} . \quad (2.39)$$

In Eq. (2.39), the z-dependent displacement is described by the dimensionless quantity α, which indicates the relative displacement in units of γ. In contrast, the displacements in position and momentum in Eq. (2.35) are indicated by a and q on an absolute scale. Hence the Gouy phase jump over π near a focus, which corresponds to an oscillation over half a cycle of the relative displacement, corresponds to the rectilinear motion of the light on an absolute scale.

III. IDEAL LENSES

For the description of experiment, it is necessary to extend the field-operator description by including optical elements, such as lenses and diaphragms. An ideal lens is sufficiently thin so that no propagation occurs in the lens. Its only effect is to add an x-dependent phase factor to the field. When a lens with focal length f with its center on the z axis is located at the position z_1, the relation between the field incident on the lens and the outgoing field is given [8] by

$$u(x,z_+)=\exp\left[-\frac{ik}{2f}x^2\right]u(x,z_-) , \quad (3.1)$$

with z_+ (z_-) a position immediately behind (before) the lens. This local phase jump can be described by writing for the propagation operator

$$U(z_+)=\exp\left[-\frac{ik}{2f}X^2\right]U(z_-) . \quad (3.2)$$

When the lens at position z_1 is the only one in the interval $[0,z]$, the propagation operator is equal to

$$U(z)=\exp\left[-\frac{i}{2k}P^2(z-z_1)\right]\exp\left[-\frac{ik}{2f}X^2\right] \times\exp\left[-\frac{i}{2k}P^2z_1\right] , \quad (3.3)$$

for $z\geq z_1$. In this way, we can compose the propagation operator for an optical axis with an arbitrary set of lenses, with free propagation in between lenses.

A. Ladder operators and fundamental mode

If we start with a definition of a lowering operator $A(0)$, as in Eq. (2.14), then the propagation operator U defines the field operators for all values of z, according to (2.15). As the commutation relation (2.18) is unaffected by the presence of lenses, we still have the fundamental mode as an eigenstate of $A(z)$ with eigenvalue zero, together with the higher-order modes, which are related by (2.22). The analytical expressions are modified to account for the presence of lenses, but the operator algebra remains the same.

It is important to recognize that the explicit analytical expressions for all Hermite-Gaussian modes are fully determined by the radius of curvature s, the spot size γ, and the phase χ. The z dependence of these three parameters, which fully describe the fundamental mode for a given configuration of lenses, defines the framework of the operator algebra. To understand this, it is sufficient to note that a field operator A, defined as an arbitrary linear combination of the operators X and P for a single z value, keeps a similar form for all positions. This is obvious from the transformation rules (2.16) and (2.24). Physically, this reflects the fact that a Gaussian beam remains Gaussian during free propagation and in passage through (ideal) lenses.

If an optical axis with an arbitrary number of lenses of arbitrary focal length is considered, we can define the lowering operator A at one position as a linear combination of X and P, and define the z dependence of $A(z)$ in terms of the propagation operator, as in (2.15). Then A retains a similar form for all z, and we may write

$$A(z)=\frac{1}{\sqrt{2}}[\kappa(z)X+i\beta(z)P] , \tag{3.4}$$

with κ and β complex-valued functions of z. In order that A and $A^\dagger$ obey the commutation rule (2.18), κ and β must be related by the equality

$$\mathrm{Re}\kappa\beta^*=1 . \tag{3.5}$$

Upon inversion, Eq. (3.4) takes the form

$$X=\frac{1}{\sqrt{2}}[\beta^* A+\beta A^\dagger], \quad P=\frac{1}{i\sqrt{2}}[\kappa^* A-\kappa A^\dagger] . \tag{3.6}$$

In an interval without lenses, where free propagation occurs, the z dependence of κ and β may be determined using the transformation (2.16). We find that κ is constant in such an interval, whereas β varies linearly with z. The derivatives are

$$\frac{d\beta}{dz}=\frac{i\kappa}{k}, \quad \frac{d\kappa}{dz}=0 . \tag{3.7}$$

On the other hand, application of (2.24) shows that a lens with focal length f does not change β, but it does cause a jump in κ. The result is

$$\beta(z_+)=\beta(z_-), \quad \kappa(z_+)=\kappa(z_-)+\frac{ik\beta}{f} . \tag{3.8}$$

Hence, β is a continuous function that varies linearly between lenses with a slope proportional to κ. Conversely, κ makes jumps at the lens positions with a jump size that depends on the local value of β. In the absence of any lens, we again find the ladder operators to be given by (2.17).

The z-dependent lowering operator A has an eigenvector $|u_0(z)\rangle$ with eigenvalue zero, which is a solution of (2.7) with a propagation operator that is modified by the lenses. This light beam is the fundamental Gaussian mode for this arbitrary lens configuration, and its normalized wave function is equal to

$$u_0(x,z)=[\beta\sqrt{\pi}]^{-1/2}\exp\left[-\frac{\kappa x^2}{2\beta}\right] . \tag{3.9}$$

It is important to note that the intensity distribution has width $|\beta|$, whereas the momentum distribution, which is determined by the Fourier transform of (3.9), has width $|\kappa|$. When κ and β vary according to the rules (3.6) and (3.7), this expression obeys the evolution equation (2.6) between two lens positions and makes the phase jump (3.1) at the lens positions. This result generalizes that for which no lenses are present given in (2.27).

B. Higher-order modes

In order to obtain the analytic form for the higher-order modes, it is convenient to express the lowering operator (3.4) and the fundamental mode function (3.9) in terms of the radius of curvature s, the spot size γ, and the phase χ, as in (2.25) and (2.28). The quantities β and κ, with the normalization (3.5), uniquely determine the values of the fundamental mode parameters s, γ, and χ, by the relations

$$s=-k\left[\mathrm{Im}\frac{\kappa}{\beta}\right]^{-1}, \quad \gamma=|\beta|, \quad \chi=\arg\beta . \tag{3.10}$$

These results generalize (2.4) and (2.5). In the intervals between lenses, the variation of s, γ, and χ is determined by the linear variation of β with z, as expressed in (3.7). In the region of negative values of s, the beam converges and γ decreases, while the beam diverges where s is positive. A focus occurs where κ has the same argument as β, so that $s=0$. The phase angle χ can only increase, and it varies most rapidly at a focus. Since a lens does not change β, both γ and χ remain unmodified by the lens. According to (3.8), κ/β changes by ik/f, and the only effect of the lens is a change of the radius of curvature,

$$\frac{1}{s(z_+)}=\frac{1}{s(z_-)}-\frac{1}{f} , \tag{3.11}$$

in accordance with (3.1). This is just the lens formula of geometric optics. Since the expressions (2.25)–(2.27) remain valid, we can apply the raising operator to the fundamental mode repeatedly to obtain the higher-order modes. We conclude that they are given by the same expression (2.1) as in the case of free propagation, but with the behavior of the fundamental mode parameters as determined in this section. In the presence of lenses, the spot size γ can pass through several foci. Correspondingly, the Gouy phase χ can execute a series of π jumps. This is equivalent to the harmonic oscillator, which is the analog of the light beam in the presence of lenses, performing a number of cycles. The algebraic connection between the modes of various orders is completely unaffected by these changes in the three fundamental mode parameters.

C. Coherent states and ray optics

The discussion of Sec. II B on displaced fundamental modes in analogy to coherent states remains largely valid in the presence of lenses. The displacement of the spot at $z=0$, as described by $D(0)$ defined in (2.29), leads to a displacement for arbitrary z, where the operator $D(z)$ is still defined in (2.34). The only difference is that now the propagation operator $U(z)$ is changed. Likewise, the expression (2.37) for $D(z)$ in terms of the z-dependent field operators A and $A^\dagger$ remains valid, and, moreover, these operators still obey Eqs. (2.25) and (2.26), and (2.38) and (2.39) still hold. Hence the expression for the displacement on the relative scale in units of γ remains the same. The effect of the lenses is completely hidden in the modified behavior of the fundamental mode parameters s, γ, and χ.

The refraction by the lenses becomes directly obvious when we express the displacement on the absolute scale, by rewriting $D(z)$ in the form (2.35) in terms of a momentum displacement $q(z)$ and a position displacement $a(z)$. In an interval between two lenses, where free propagation occurs, the z dependence of q and a is determined by the transformation (2.16). One finds that a varies linearly with z, whereas q is constant. The derivatives are

$$\frac{da}{dz}=\frac{q}{k}, \quad \frac{dq}{dz}=0 . \tag{3.12}$$

Applying (3.11) shows that the effect of a lens with focal length f is described by the transformation

$$a(z_+)=a(z_-), \quad q(z_+)=q(z_-)-\frac{ka}{f} . \tag{3.13}$$

One should recall that q/k has the significance of the angle of a light ray with the z axis, whereas a is the distance of the ray from the axis at the position z. Hence Eq. (3.12) describes the rectilinear motion of a ray of light in free space, while (3.13) describes the refraction of the ray by a lens with focal length f.

Hence, the behavior of the axis of a beam in a displaced fundamental mode (the analog of a coherent state) simply obeys the laws of geometric optics. This behavior is completely independent of the variation of the fundamental beam parameters s, γ, and χ, which are a reflection of the wave-optical nature of the beam. The variation of these beam parameters is governed by the transformation (3.7) and (3.8) for β and κ, which strongly resemble the geometric-optical rules (3.12) and (3.13). This is no accident, as both pairs of rules follow from the transformations (2.16) and (2.24). In summary, a Gaussian beam is completely specified by its first and second moments. The first moments give the average position and momentum: namely a and q. They determine the geometric-optical properties of the beam. The second moments, which are specified by the complex quantities β and κ, determine the position and momentum widths of the fundamental mode, and describe the wave aspect of the beam.

IV. ORBITAL ANGULAR MOMENTUM

In this section, we apply the operator formalism to describe the orbital angular momentum of a paraxial beam of light. This requires the explicit consideration of the two transverse dimensions. Then we have two momentum operators P_x and P_y for the x and the y direction, and two position operators X and Y. We introduce the operator

$$L=XP_y-YP_x , \tag{4.1}$$

in analogy to the quantum operator of orbital angular momentum. In cylindrical coordinates r and ϕ, with

$$x=r\cos\phi, \quad y=r\sin\phi , \tag{4.2}$$

L takes the well-known form

$$L=\frac{1}{i}\frac{\partial}{\partial\phi} . \tag{4.3}$$

The total angular momentum of a radiation field is given by Maxwell's theory in the form [11]

$$\mathbf{J}=\varepsilon_0\int d\mathbf{r}\{\mathbf{r}\times[\mathbf{E}(\mathbf{r})\times\mathbf{B}(\mathbf{r})]\} . \tag{4.4}$$

This angular momentum can be separated into a term resulting from the derivatives of the amplitudes, and a term that derives from the polarization of the field. With some caution, this may be viewed as a separation into orbital angular momentum and spin of the field [12]. For a monochromatic field in the paraxial approximation (2.7), the z component J_z is separated as [4]

$$J_z=\frac{\varepsilon_0}{2\omega}\int dx\,dy\,dz\left[\mathbf{E}_0^*\cdot\mathbf{E}_0u^*Lu+\frac{1}{i}u^*u(E_{0x}^*E_{0y}-E_{0y}^*E_{0x})\right] . \tag{4.5}$$

This implies that for an arbitrary normalized mode $|u(z)\rangle$, the expectation value $\hbar\langle L\rangle$ is equal to that of the orbital angular momentum in the z direction per photon [4].

For each transverse dimension, we have independent values for the fundamental mode parameters s, γ, and χ, which we distinguish with suffices x and y. The corresponding lowering operators A_x and A_y are determined by these parameters, as in (2.25), in terms of the real operators B_x and B_y. In the presence of cylindrical lenses, the operators for the two dimensions can be treated independently only when the lens axes coincide with the x and the y axis, which is what will be assumed from now on. Hence the spot size and the wave front will, in general, be elliptical and with coinciding axes. We exclude for the moment the general astigmatism that arises by using cylindrical lenses oriented at oblique angles to each other [13]. Then the Hermite-Gaussian modes $|u_{nn'}(z)\rangle$ have mode functions $u_{nn'}(\mathbf{r})=u_n(x,z)u_{n'}(y,z)$, which are products of one-dimensional modes. All higher-order modes can then be generated by repeated action of ladder operators on the fundamental mode $|u_{00}(z)\rangle$.

Laguerre-Gaussian modes are the laser mode analog of the angular-momentum eigenstates of the isotropic two-dimensional harmonic oscillator. It has been shown that a higher-order Hermite-Gaussian beam can be converted into a Laguerre-Gaussian beam by using astigmatic lenses [14–16]. We wish to demonstrate that this and similar conversions are conveniently understood by operator algebra. First, we give the connection between these modes in an operator form.

A. Hermite-Gaussian, Laguerre-Gaussian, and elliptical Gaussian modes

Generalizing the algebraic treatment of the two-dimensional harmonic oscillator [17], we introduce mixed ladder operators by the definition

$$A_\pm=\frac{1}{\sqrt{2}}[A_x\mp iA_ye^{i\theta}] , \tag{4.6}$$

in terms of a phase θ that is independent of z. These operators obey the commutation rules

$$[A_\pm(z),A_\pm^\dagger(z)]=1, \quad [A_\pm(z),A_\mp^\dagger(z)]=0 . \tag{4.7}$$

The operators for the sum and the difference of the number operators are

$$N(z)=N_+(z)+N_-(z), \quad M(z)=N_+(z)-N_-(z) , \tag{4.8}$$

with

$$N_\pm(z)=A_\pm^\dagger(z)A_\pm(z) . \tag{4.9}$$

In terms of the field operators A_x and A_y, the operators N and M are

$$N=A_x^\dagger A_x+A_y^\dagger A_y, \quad M=iA_y^\dagger A_x e^{-i\theta}-iA_x^\dagger A_y e^{i\theta} . \tag{4.10}$$

Because N and M commute, they have a common basis of eigenstates $|u_{nm}(z)\rangle$, with eigenvalues n and m. As A_+ or A_- decrease the eigenvalue of N_+ and N_- by one unit, the lowering operators $A_\pm$ both decrease the total excitation number n, whereas A_+ decreases m and A_- increases m by one unit. This is apparent from the commutation rules

$$[N(z), A_\pm(z)]=-A_\pm(z), \quad [M(z), A_\pm(z)]=\mp A_\pm(z) . \tag{4.11}$$

With the proper phase convention of the eigenfunctions, the explicit action of the ladder operators may be given by

$$A_\pm(z)|u_{nm}(z)\rangle=\left[\frac{n\pm m}{2}\right]^{1/2}|u_{n-1,m\mp 1}(z)\rangle ,$$
$$A_\pm^\dagger(z)|u_{nm}(z)\rangle=\left[\frac{n\pm m+2}{2}\right]^{1/2}|u_{n+1,m\pm 1}(z)\rangle . \tag{4.12}$$

The total mode number n can take all natural values $0,1,2,\ldots$, and for each value of n, m can take the $n+1$ values $-n,-n+2,\ldots,n-2,n$. The mode $|u_{nm}\rangle$ is created from the fundamental mode by applying the raising operators $A_+^\dagger$ and $A_-^\dagger$, and one finds

$$|u_{nm}(z)\rangle=\left[\frac{1}{p!q!}\right]^{1/2}[A_-^\dagger(z)]^q[A_+^\dagger(z)]^p|u_{00}(z)\rangle , \tag{4.13}$$

where we introduced the integers p and q, defined by

$$p=\frac{n+m}{2}, \quad q=\frac{n-m}{2} \tag{4.14}$$

in terms of n and m. This result allows the expansion of the eigenfunctions $u_{nm}(\mathbf{r})$ of N and M in terms of the Hermite-Gaussian modes. If we substitute the definitions (4.6) into (4.13), the brackets can be worked out in terms of the coefficients g_s, which are defined by the expansion

$$(1-t)^q(1+t)^p=\sum_{s=0}^{n} g_s t^s . \tag{4.15}$$

As the fundamental mode u_{00} is the product of two one-dimensional fundamental modes, we may directly apply (2.23) for each transverse dimension, with the result

$$u_{nm}(\mathbf{r})=\sum_{s=0}^{p+q} g_s i^s e^{-is\theta}\left[\frac{(p+q-s)!s!}{2^{p+q}p!q!}\right]^{1/2} \times u_{p+q-s}(x,z)u_s(y,z) . \tag{4.16}$$

The equalities (4.10)–(4.13) and (4.16) hold for any beam specified by its fundamental mode parameters. In the special case of a nonastigmatic region of the beam, where $\gamma_x=\gamma_y=\gamma$ and $s_x=s_y=s$, the phases χ_x and χ_y can only differ by a constant. Then the expression for $A_\pm$ in terms of the real ladder operators B_x and B_y takes the form

$$A_\pm=e^{i\chi_x}\exp\left[\frac{ik}{2s}(X^2+Y^2)\right]B_\pm\exp\left[-\frac{ik}{2s}(X^2+Y^2)\right] . \tag{4.17}$$

Here we denote

$$B_\pm=\frac{1}{\sqrt{2}}(B_x\mp ie^{i\zeta}B_y) , \tag{4.18}$$

where B_x and B_y are determined by the spot sizes $\gamma_x=\gamma_y=\gamma$, in analogy to (2.26), and where

$$\zeta=\chi_y-\chi_x+\theta . \tag{4.19}$$

As the fundamental mode function takes the form

$$u_{00}(\mathbf{r})=\exp\left[\frac{ikr^2}{2s}\right]\exp\left[-\frac{i}{2}(\chi_x+\chi_y)\right] \times\frac{1}{\gamma}\psi_0\left[\frac{x}{\gamma}\right]\psi_0\left[\frac{y}{\gamma}\right] , \tag{4.20}$$

operating on u_{00} with $A_\pm$ is equivalent to operating on the harmonic-oscillator ground state with $B_\pm$. This demonstrates that it is the value of ζ that determines the nature of the mode $|u_{nm}\rangle$ in a nonastigmatic region.

When $\zeta=\pi/2$, the operators $B_\pm$ are equal to

$$B_\pm=\frac{1}{\sqrt{2}}[B_x\pm B_y] . \tag{4.21}$$

Substituting (4.17) and (4.20) in (4.13) then gives for the higher-order mode functions

$$u_{nm}(\mathbf{r})=\exp\left[\frac{ik}{2s}r^2-\frac{i}{2}(\chi_x+\chi_y)-i\chi_x(p+q)\right] \times\frac{1}{\gamma}\psi_p\left[\frac{x+y}{\gamma\sqrt{2}}\right]\psi_q\left[\frac{x-y}{\gamma\sqrt{2}}\right] , \tag{4.22}$$

with p and q defined by (4.14). The functions (4.22) are normalized Hermite-Gaussian modes with their symmetry axis rotated over 45°. If we combine (4.22) with (4.16), we obtain the transformation of these rotated Hermite-Gaussian modes in the nonrotated Hermite-Gaussian modes. In the special case that $\chi_x=\chi_y$, so that $\theta=\zeta=\pi/2$, this is equivalent to an analytical expansion of products of Hermite polynomials in $(x\pm y)/\sqrt{2}$, as given by Abramochkin and Volostnikov [14].

When $\zeta=0$, we find that

$$B_{\pm}=\frac{1}{\sqrt{2}}[B_x \mp iB_y] , \qquad (4.23)$$

which are circular ladder operators. It is easy to verify that, in this case, the operator M is equal to the angular momentum L, and the quantum number $m=l$ is the corresponding eigenvalue. The beam carries an orbital angular momentum $\hbar l$ per photon [3]. For the mode functions, we obtain, after using (4.15),

$$u_{nl}(\mathbf{r})=\exp\left[\frac{ik}{2s}r^2-i\chi(n+1)\right]\frac{1}{\gamma}\psi_{nl}\left[\frac{x}{\gamma},\frac{y}{\gamma}\right] , \qquad (4.24)$$

with ψ_{nl} the wave functions of the isotropic harmonic oscillator that are eigenfunctions of both energy and angular momentum, with quantum numbers n and l. The azimuthal dependence of ψ_{nl} is given by $\exp(il\phi)$, and its radial part has the form [17,2]

$$\exp(-r^2/2\gamma^2)r^{|l|}L_p^{|l|}(r^2/\gamma^2) ,$$

with $p=(n-|l|)/2$, and $L_p^{|l|}$ the generalized Laguerre polynomial. As both the Laguerre-Gaussian and the Hermite-Gaussian modes form a complete set, a basis transformation must exist. The analytical derivation of this transformation is painstaking and not very transparent [15]. The transformation, obtained by operator algebra, is given by (4.16) in the special case that $\theta=0$. A similar result has recently been found by group-theoretic methods by Danakas and Aravind [18].

It is natural to compare the Laguerre-Gaussian modes to circular polarization and the Hermite-Gaussian modes to linear polarization. We have seen that they arise as special cases of the modes $|u_{nm}\rangle$ defined by (4.13) with (4.17) and (4.18), for $\zeta=0$ and $\zeta=\pi/2$, respectively. For intermediate values of ζ, the modes $|u_{nm}\rangle$ have an elliptical nature. This is obvious when we rewrite (4.18) in the form

$$B_{\pm}=\left[\frac{1}{\sqrt{2}}(B_x \pm iB_y)\cos\left[\frac{\zeta}{2}-\frac{\pi}{4}\right] - \frac{i}{\sqrt{2}}(B_x \mp iB_y)\sin\left[\frac{\zeta}{2}-\frac{\pi}{4}\right]\right] \times\exp\left[i\left[\frac{\zeta}{2}-\frac{\pi}{4}\right]\right] . \qquad (4.25)$$

The corresponding creation operators $A_{\pm}^{\dagger}$ acting on the fundamental mode generate elliptical patterns of the transverse momentum density, with the axes oriented in the xy plane along the lines $x=\pm y$. This ellipticity should not be confused with the elliptic intensity distribution in the case of an astigmatic beam.

B. Mode conversion

In general, a mode converter is defined as a configuration of lenses that transforms a nonastigmatic input beam into a nonastigmatic output beam with an astigmatic region between them. Let us consider a configuration of astigmatic lenses with their elliptical axes oriented along the x and the y axis. The lenses are located between z_1 and z_2. The incoming beam for $z \le z_1$ is supposed to be nonastigmatic, so that $\gamma_x=\gamma_y=\gamma$ and $s_x=s_y=s$. The lens configuration determines the values of the fundamental mode parameters s_x, γ_x, χ_x and s_y, γ_y, and χ_y for all values of z. The output beam for $z \ge z_2$ is also nonastigmatic, provided that the last lens is located at a position where $\gamma_x=\gamma_y$, and that it has different focal lengths f_x and f_y along the two axes that make up for a possible difference in s_x and s_y, according to

$$\frac{1}{s_x(z_-)}-\frac{1}{f_x}=\frac{1}{s_y(z_-)}-\frac{1}{f_y} . \qquad (4.26)$$

As the characteristics of the mode $|u_{nm}\rangle$ in a nonastigmatic region are determined by the value of the parameter ζ, the properties of a converter are fully specified by the difference between the ζ values in the output and the input beam, which is equal to

$$\Delta\zeta=(\zeta)_{\text{out}}-(\zeta)_{\text{in}}=(\chi_x-\chi_y)_{\text{out}}-(\chi_x-\chi_y)_{\text{in}} . \qquad (4.27)$$

Consider a Hermite-Gaussian input beam with its axes oriented at 45° with respect to the axes of the lenses, so that

$$u_{\text{in}}(z)=u_p\left[\frac{x+y}{\sqrt{2}},z\right]u_q\left[\frac{x-y}{\sqrt{2}},z\right] , \qquad (4.28)$$

with u_p and u_q the one-dimensional Hermite-Gaussian mode functions (2.1). According to (4.22), this input mode is identical to the mode $|u_{nm}\rangle$ with $\zeta=\pi/2$, and with $n=p+q$, and $m=p-q$. In order that the output beam be in the Laguerre-Gaussian mode (4.24), we must have $(\zeta)_{\text{out}}=0$, or $\Delta\zeta=-\pi/2$. The angular-momentum quantum number is $l=p-q$. Such a $\pi/2$ converter has recently been realized with a system consisting of two identical cylindrical lenses [16]. Then the Rayleigh range is the same at input and output. As the Gouy phases χ_x and χ_y cannot increase by more than π in a region of free propagation, the value of $\Delta\zeta$ for a system of only two lenses must obey the inequality $-\pi<\Delta\zeta<\pi$. For a system consisting of more lenses, $\Delta\zeta$ can be outside this range and the Rayleigh range can be different at input and output. Inspection of (4.17) and (4.18) shows that for $\zeta=\pi$, the operator M is equal to $-L$. Hence a lens system with $\Delta\zeta=\pi$ inverts the azimuthal mode index l of a Laguerre-Gaussian beam. It cannot, however, be realized with less than three lenses [4].

Our treatment demonstrates that the conversion properties of an arbitrary astigmatic lens system with a single axis with a nonastigmatic input and output is determined by the single parameter $\Delta\zeta$, which is the difference of the change in Gouy phase for the two axes of the lens system. As indicated in (4.18), this phase determines the change in the ladder operator $B_{\pm}$ from the input to the output region. An elliptical Gaussian beam specified by a value of ζ is realized by a converter with an appropriately selected value of $\Delta\zeta$ when the input beam is Hermite-Gaussian or Laguerre-Gaussian. The action of the converter results from the modification of the eigenoperator M, while the eigenvalues n and m remain unchanged. The propagation of the beam through the lens system is fully analogous to the evolution of a two-dimensional harmonic oscillator, with a modification of the oscillation frequencies Ω_x and Ω_y. The change in the phase χ over a propagation distance plays the role of the oscillator phase $\int dt\, \Omega(t)$ over a time interval. In each case, it is the total phase difference between the two directions that determines the state change for an arbitrary initial state. When the mode function is simply the product of one-dimensional functions for the x and the y direction, this phase difference does not affect the nature of the mode. It only plays a role for an entangled state, which is a linear combination of different products of one-dimensional eigenstates for the two axes. This explains why an input Hermite-Gaussian mode must have axes that do not coincide with the symmetry axes of the lenses, in order for mode conversion to take place.

C. Angular-momentum transfer

It is informative to derive expressions for the transfer of orbital angular momentum when a cylindrical lens is traversed by a possibly astigmatic Gaussian beam that is an eigenfunction of N. Again, clearly, it is necessary to consider both transverse directions. The transverse field amplitude for a given position z of the beam is generally described by the function $u(x,y)$, that can be separated as

$$u(\mathbf{r})=\exp\left[\frac{ik}{2}\left[\frac{x^2}{s_x}+\frac{y^2}{s_y}\right]\right]\frac{1}{\sqrt{\gamma_x\gamma_y}}\Psi(\xi,\eta)\ . \qquad (4.29)$$

We introduced the scaled transverse coordinates and momenta

$$\xi=x/\gamma_x,\quad \eta=y/\gamma_y,\quad P_\xi=\frac{1}{i}\frac{\partial}{\partial\xi},\quad P_\eta=\frac{1}{i}\frac{\partial}{\partial\eta}\ , \qquad (4.30)$$

so that Ψ is the normalized wave function of an energy eigenstate of a two-dimensional isotropic harmonic oscillator. As we are interested only in the distribution at a given position z, we suppressed the z dependence in (4.29). The Gouy phases χ_x and χ_y are absorbed in Ψ. When γ_x and γ_y are different, the Gaussian part of u has unequal widths along the two axes, whereas the function Ψ is unsqueezed. If we evaluate the orbital angular momentum per photon $\hbar\langle L\rangle=\hbar\langle u|L|u\rangle$ for the beam (4.29) while using (2.24), we can express $\langle L\rangle$ as an expectation value over the wave function Ψ in terms of the scaled coordinates, with the result

$$\begin{aligned}\langle L\rangle=&k\gamma_x\gamma_y\left[\frac{1}{s_y}-\frac{1}{s_x}\right]\langle\Psi|\xi\eta|\Psi\rangle\\&+\frac{1}{2}\left[\frac{\gamma_x}{\gamma_y}-\frac{\gamma_y}{\gamma_x}\right]\langle\Psi|(\xi P_\eta+\eta P_\xi)|\Psi\rangle\\&\times\frac{1}{2}\left[\frac{\gamma_x}{\gamma_y}+\frac{\gamma_y}{\gamma_x}\right]\langle\Psi|(\xi P_\eta-\eta P_\xi)|\Psi\rangle\ .\end{aligned} \qquad (4.31)$$

The first term on the right-hand side of (4.31) results from the difference in the two radii of curvature. This reflects the ellipticity of the wave front. The second term is due to the difference in spot size and results from the elliptical shape of the light spot. The last term contains the angular momentum in scaled coordinates. It is the only surviving term in a nonastigmatic beam. Note that for a freely propagating astigmatic beam, all three terms in (4.31) vary with z, whereas their sum remains constant.

If the beam described by u passes a cylindrical lens, with focal length f in the x direction, the field leaving the lens is given by $u(\mathbf{r})\exp(-ikx^2/2f)$. Hence the last two terms in (4.31) are unaffected and only the first one changes, due to the change in s_x. The net increase of angular momentum per photon passing the lens is

$$\hbar\langle\delta L\rangle=\frac{\hbar k}{f}\gamma_x\gamma_y\langle\Psi|\xi\eta|\Psi\rangle=\frac{\hbar k}{f}\langle u|xy|u\rangle\ . \qquad (4.32)$$

This term determines the torque that the field exerts on the lens. It vanishes when the intensity distribution $|u|^2$ has the x or the y axis as symmetry axis. In particular, the torque on the lens vanishes when the input beam is an astigmatic Hermite-Gaussian mode $u_n(x,z)u_{n'}(y,z)$. This is understandable, as the lens orientation is at an extremum of energy. Furthermore, the angular-momentum transfer (4.32) vanishes for a Laguerre-Gaussian input mode, and for symmetry reasons [4] it will also vanish when the beam leaving the lens is Laguerre-Gaussian. Notice that this angular-momentum transfer is proportional to the wave number k. Therefore, when this matrix element is nonvanishing, it can be as large as many units $\hbar$, even for modes of moderate (nonzero) order.

D. Angular aperturing and angular momentum

In Sec. IV C, it was argued that a Laguerre-Gaussian beam cannot transfer angular momentum to an astigmatic lens. However, when angular aperturing is applied just in front of the lens, an appreciable transfer of angular momentum can occur. Suppose that the Laguerre-Gaussian mode (4.24) passes a diaphragm with a transmittance that depends only on the azimuthal angle ϕ, and not on the radial coordinate r. Then the mode function leaving the diaphragm can be given in polar coordinates as

$$u(\mathbf{r})=\exp\left[\frac{ik}{2s}r^2-i\chi(n+1)\right]\frac{1}{\gamma\sqrt{2\pi}}R_{nl}(r/\gamma)\times e^{il\phi}A(\phi)\ , \qquad (4.33)$$

with R_{nl} the normalized radial-wave function of the angular-momentum eigenstates ψ_{nl} of the two-dimensional harmonic oscillator. For an ideal diaphragm, the transmittance function $A(\phi)$ attains only the value zero or 1. When a cylindrical lens with focal length f in the x direction is placed immediately after the diaphragm, so that no propagation occurs between them, the transfer of angular momentum to the lens is found after substitution of (4.33) in (4.32). The result is the product of a radial and an angular average. By using the well-known fact that the average potential and kinetic energy are equal in an energy eigenstate of the harmonic oscillator, we obtain for the transfer of angular momentum per incident photon the expression

$$\hbar\langle \delta L \rangle = \frac{\hbar k}{2f}\gamma^2(n+1)\frac{1}{2\pi}\int_0^{2\pi} d\phi |A(\phi)|^2 \sin(2\phi) \ . \quad (4.34)$$

This result is independent of the azimuthal mode index l, and it is proportional to a Fourier coefficient of the ϕ-dependent transmission function $|A(\phi)|^2$. The angular momentum transferred to the lens can be much larger than the angular momentum per photon $\hbar l$ in the incident beam. The transfer is maximal when $|A(\phi)|^2$ is 1 for $0<\phi<\pi/2$, and $\pi<\phi<3\pi/2$, and zero elsewhere.

V. CONCLUSIONS

The propagation of Gaussian laser beams in free space, and their refraction by ideal astigmatic lenses, are described by harmonic-oscillator operator algebra. A mode is characterized by z-dependent eigenoperators and constant eigenvalues. This explains why the Hermite-Gaussian form of a beam is conserved during its free propagation and at refraction by lens systems. Moreover, it shows that the displacement of the axis of a slightly misaligned beam follows the rules of geometric optics, which does not affect the beam diffraction. Such a displaced beam is the analog of a coherent state of the harmonic oscillator. The relationship between Laguerre-Gaussian and Hermite-Gaussian modes takes a simple algebraic form. Conversion from one mode into another by astigmatic lenses is directly characterized by a single phase difference. The orbital angular momentum of a light beam in its propagation direction is separated into a contribution that resembles the angular momentum of a harmonic oscillator and terms that originate from the astigmatism of the beam. These latter terms can be considerably larger than the first. The torque of a beam exerted on an astigmatic lens is expressed in terms of a single expectation value. When a Laguerre-Gaussian beam, with angular momentum $\hbar l$ per photon, traverses an astigmatic lens after passing an angular aperture, it may transfer more angular momentum to the lens than $\hbar l$ per photon. Apart from the physical interest of these results, they demonstrate the advantage of the algebraic method in terms of ladder operators for paraxial optics.

ACKNOWLEDGMENT

This work is part of the research program of the Stichting voor Fundamenteel Onderzoek der Materie (FOM), which is financially supported by the Nederlandse Organisatie voor Wetenschappelijk Onderzoek (NWO).

*Present address: Physics Department, University of Essex, England.

[1] H. A. Haus, *Waves and Fields in Optoelectronics* (Prentice-Hall, Englewood Cliffs, NJ, 1984).
[2] A. E. Siegman, *Lasers* (University Science Books, Mill Valley, CA, 1986).
[3] L. Allen, M. W. Beijersbergen, R. J. C. Spreeuw, and J. P. Woerdman, Phys. Rev. A **45**, 8185 (1992).
[4] S. J. van Enk and G. Nienhuis, Opt. Commun. **94**, 147 (1992).
[5] L. G. Gouy, C. R. Acad. Sci. **110**, 1251 (1890).
[6] M. Lax, W. H. Louisell, and W. B. McKnight, Phys. Rev. A **11**, 1365 (1975).
[7] D. Gloge and D. Marcuse, J. Opt. Soc. Am. **59**, 1629 (1969).
[8] D. Stoler, J. Opt. Soc. Am. **71**, 334 (1981).
[9] R. Loudon, *The Quantum Theory of Light* (Clarendon, Oxford, 1981).
[10] P. Meystre and M. Sargent III, *Elements of Quantum Optics* (Springer, Berlin, 1990).
[11] J. D. Jackson, *Classical Electrodynamics* (Wiley, New York, 1962).
[12] J. W. Simmons and M. J. Guttmann, *States, Waves and Photons* (Addison-Wesley, Reading, MA, 1970).
[13] J. A. Arnaud and H. Kogelnik, Appl. Opt. **8**, 1687 (1969).
[14] E. Abramochkin and V. Volostnikov, Opt. Commun. **83**, 123 (1991).
[15] C. Tamm and C. O. Weiss, J. Opt. Soc. Am. B **7**, 1034 (1990).
[16] M. W. Beijersbergen, L. Allen, H. E. L. O. van der Veen, and J. P. Woerdman, Opt. Commun. **96**, 123 (1993).
[17] A. Messiah, *Mécanique Quantique* (Dunod, Paris, 1964).
[18] S. Danakas and P. K. Aravind, Phys. Rev. A **45**, 1973 (1992).

PAPER 2.4

Matrix formulation for the propagation of light beams with orbital and spin angular momenta

L. Allen
School of Physics and Astronomy, University of St. Andrews, Fife KY16 9SS, Scotland

J. Courtial and M. J. Padgett
Department of Physics and Astronomy, Kelvin Building, University of Glasgow, Glasgow G12 8QQ, Scotland
(Received 26 May 1999)

Jones matrices describe the polarization, or spin angular momentum, of a light beam as it passes through an optical system. We devise an equivalent of the Jones matrix formulation for light possessing orbital angular momentum. The matrices are then developed to account for light that has both spin and orbital angular momentum. [S1063-651X(99)07412-7]

PACS number(s): 42.25.Bs

The behavior of polarized light as it passes through birefringent optical components is well described by an approach devised by Jones [1]. The beam is represented in terms of a column vector, the Jones vector, the elements of which are the complex amplitudes of the orthogonally polarized components. Optical components such as linear and circular polarizers, phase retarders, etc., are represented by 2×2 matrices called Jones matrices: matrix multiplication accounts for the passage of the light through a sequence of optical elements. An alternative formulation derived by Mueller [2], based on the Stokes parameters, is not able to deal fully with the superposition of coherent beams, but has the advantage of being applicable to partially as well as fully polarized light. A full account of both the Jones and Mueller matrices and their application, has been given by Gerrard and Burch [3].

Polarized light is associated with spin angular momentum. It is now well established [4–10] that light beams may possess well defined orbital angular momentum. Beams with an azimuthal phase dependence of $\exp(il\phi)$, such as Laguerre Gaussian beams, have an orbital angular momentum of $l\hbar$ per photon [4]. Such modes may be readily created in the laboratory by passing the Hermite Gaussian modes usually emitted by lasers, through a mode converter [5] which consists of two canonically disposed cylindrical lenses oriented at 45° to the axes of the mode. The indices (n, m) characterizing the Hermite Gaussian modes give the indices (l, p) of the Laguerre Gaussian modes, where $l=|n-m|$ and $p=\min(m,n)$ [4].

Hermite Gaussian and Laguerre Gaussian modes both form complete, orthogonal, basis sets from which any arbitrary field distribution may be described. The order of a mode is defined [5] by $N=n+m=2p+|l|$ and does not change as the mode is converted or rotated. There is an exact analogy between a waveplate for polarized light and a mode converter; see Fig. 1. Just as a phase shift is introduced between orthogonal polarization components by birefringent waveplates, so mode converters based on cylindrical lenses introduce a phase shift between orthogonal modes of the same order. This equivalence allows polarization and mode structure to be treated in similar ways and leads to a formulation, analogous to the Jones matrix approach, for modes of order N.

The Poincaré sphere is an equivalent representation to the Jones matrix formulation for polarized light. Any state of polarization may be represented by a point on the sphere; waveplates and other polarizing elements then move the polarization state to another position on the surface of the sphere. It has been shown [11], that there exists an orbital angular momentum equivalent to the Poincaré sphere for modes of order $N=1$. A Poincaré sphere approach to arbi-

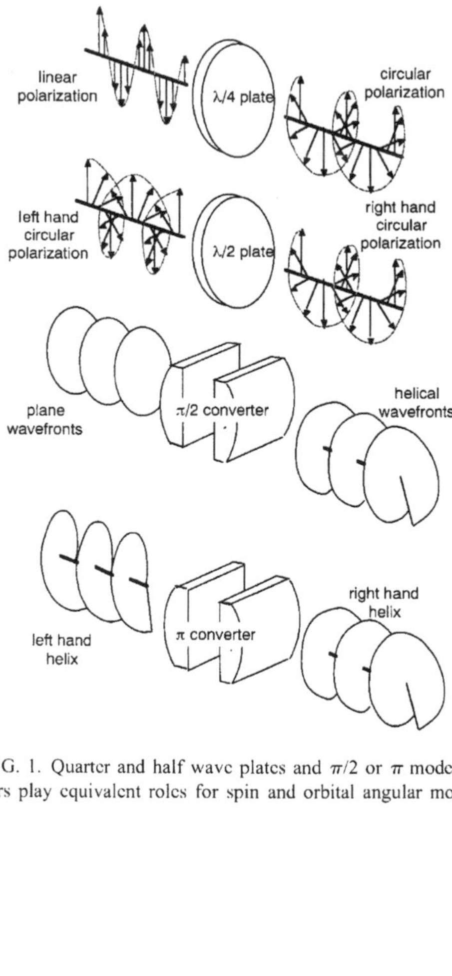

FIG. 1. Quarter and half wave plates and $\pi/2$ or π mode converters play equivalent roles for spin and orbital angular momentum.

TABLE I. $[(N+1)\times(N+1)]$ matrices for modes of order N.

$\pi/2$ converter

$$[C(\pi/2)]=\begin{bmatrix} 1 & 0 & 0 & 0 & 0 & 0 & \cdots & \cdots \\ 0 & -i & 0 & 0 & 0 & 0 & \cdots & \cdots \\ 0 & 0 & -1 & 0 & 0 & 0 & \cdots & \cdots \\ 0 & 0 & 0 & i & 0 & 0 & \cdots & \cdots \\ 0 & 0 & 0 & 0 & 1 & 0 & \cdots & \cdots \\ 0 & 0 & 0 & 0 & 0 & -i & \cdots & \cdots \\ \cdots & \cdots & \cdots & \cdots & \cdots & \cdots & \cdots & \cdots \\ \cdots & \cdots & \cdots & \cdots & \cdots & \cdots & \cdots & \cdots \end{bmatrix}$$

π converter

$$[C(\pi)]=\begin{bmatrix} 1 & 0 & 0 & 0 & 0 & 0 & \cdots & \cdots \\ 0 & -1 & 0 & 0 & 0 & 0 & \cdots & \cdots \\ 0 & 0 & 1 & 0 & 0 & 0 & \cdots & \cdots \\ 0 & 0 & 0 & -1 & 0 & 0 & \cdots & \cdots \\ 0 & 0 & 0 & 0 & 1 & 0 & \cdots & \cdots \\ 0 & 0 & 0 & 0 & 0 & -1 & \cdots & \cdots \\ \cdots & \cdots & \cdots & \cdots & \cdots & \cdots & \cdots & \cdots \\ \cdots & \cdots & \cdots & \cdots & \cdots & \cdots & \cdots & \cdots \end{bmatrix}$$

Mode filter (e.g. $HG_{N-2,2}$)

$$[F(N-2,2)]=\begin{bmatrix} 0 & 0 & 0 & 0 & 0 & 0 & \cdots & \cdots \\ 0 & 0 & 0 & 0 & 0 & 0 & \cdots & \cdots \\ 0 & 0 & 1 & 0 & 0 & 0 & \cdots & \cdots \\ 0 & 0 & 0 & 0 & 0 & 0 & \cdots & \cdots \\ 0 & 0 & 0 & 0 & 0 & 0 & \cdots & \cdots \\ 0 & 0 & 0 & 0 & 0 & 0 & \cdots & \cdots \\ \cdots & \cdots & \cdots & \cdots & \cdots & \cdots & \cdots & \cdots \\ \cdots & \cdots & \cdots & \cdots & \cdots & \cdots & \cdots & \cdots \end{bmatrix}$$

Mode rotation matrix for $N=1$

$$[\mathrm{rot}(\phi)]=\begin{bmatrix} \cos(\phi) & \sin(\phi) \\ -\sin(\phi) & \cos(\phi) \end{bmatrix}$$

Mode rotation matrix for $N=2$

$$[\mathrm{rot}(\phi)]=\begin{bmatrix} \cos^2\phi & \dfrac{\sin 2\phi}{\sqrt{2}} & \sin^2\phi \\ \dfrac{-\sin 2\phi}{\sqrt{2}} & \cos 2\phi & \dfrac{\sin 2\phi}{\sqrt{2}} \\ \sin^2\phi & \dfrac{-\sin 2\phi}{\sqrt{2}} & \cos^2\phi \end{bmatrix}$$

trary order of N is, however, not easy to devise as higher order geometries are required. This is not a difficulty for a matrix formulation.

In this paper we show that just as there is a two element Jones column vector and a set of 2×2 matrices for polarized light, so there is an $(N+1)$-element column vector and a set of $[(N+1)\times(N+1)]$ matrices to describe the passage of modes of order N, which may possess orbital angular momentum, through a series of optical components such as mode converters and beam rotators. For these matrices to apply, N must be conserved and the modes must have the same Rayleigh range and beam waist. Most imaging systems

TABLE I (*continued*).

Mode rotation matrix for $N=3$

$$[\text{rot}(\phi)]=\begin{bmatrix} \cos^3\phi & \sqrt{3}\cos^2\phi\sin\phi & \sqrt{3}\cos\phi\sin^2\phi & \sin^3\phi \\ -\sqrt{3}\cos^2\phi\sin\phi & \dfrac{\cos\phi+3\cos 3\phi}{4} & \dfrac{-\sin\phi+3\sin 3\phi}{4} & \sqrt{3}\cos\phi\sin^2\phi \\ \sqrt{3}\cos\phi\sin^2\phi & \dfrac{\sin\phi-3\sin 3\phi}{4} & \dfrac{\cos\phi+3\cos 3\phi}{4} & \sqrt{3}\cos^2\phi\sin\phi \\ -\sin^3\phi & \sqrt{3}\cos\phi\sin^2\phi & -\sqrt{3}\cos^2\phi\sin\phi & \cos^3\phi \end{bmatrix}$$

satisfy this requirement, but those with optical components which may result in energy exchange between modes such as optical fibres, holograms apertures and nonlinear crystals [9], will not.

Any mode of order N can be expanded as the sum of $(N+1)$ Hermite Gaussian, or Laguerre Gaussian, modes of the same order [4,5]. It follows that such a mode may be represented by a column vector with $(N+1)$ elements

$$\begin{bmatrix} a_{N,0} \\ a_{N-1,0} \\ \cdots \\ \cdots \\ a_{1,N-1} \\ a_{0,N} \end{bmatrix},$$

where $a_{n,m}$ is the complex amplitude coefficient of the (n, m) Hermite Gaussian mode. For example, a Hermite Gaussian (2, 0) mode oriented at 45° to the laboratory frame is represented by a three element column vector containing the complex amplitude weightings of the (2,0), (1,1), (0,2) Hermite Gaussian modes, respectively. This rotated Hermite Gaussian mode is transformed to a Laguerre Gaussian, by passing the beam through a $\pi/2$ converter, which introduces phase shifts between the constituent modes of $(m-n)\pi/4$.

The matrices were developed by consideration of the role of each optical component. The way in which a $\pi/2$ converter transforms a (n,m) Hermite Gaussian mode to a (p,l) Laguerre Gaussian mode is well understood [5]. This converts a mode with no orbital angular momentum into one possessing it. The relative phases of the initial and final component modes are known and the nature of the required matrix follows easily enough. Likewise the matrix for a π converter, which converts a mode with orbital angular momentum l to one with $-l$, follows readily. It is rather more difficult to determine the form of the matrix for the rotation of a mode. Each element of the matrix can be deduced by considering the form of the decomposition of each of the individual modes as it is rotated through an angle ϕ. We may note that components of the higher orders of the rotation matrix follow from the transformation of angular momentum eigenfunctions under finite rotation, given by expressions (4.1.15) or (4.1.23), in the book by Edmonds [12] on angular momentum.

The optical component analogous to a polarizer is a mode filter which selects one constituent mode from those in the column vector. Such filters have been developed and employed to preferentially transmit specific modes [13]. The matrices for $\pi/2$ converters, π converters, mode filters and mode rotations are given in Table I.

To show how the matrices may be used, consider again the (2,0) Hermite Gaussian mode. The column vector representing such an input mode is that for mode order $N=n+m=2$. The vector, therefore, has three components. The first term in the column vector indicates that the input mode is indeed the (2,0) mode and the subsequent zeros show that there is no admixture of (1,1) or (0,2) modes. When the mode is oriented at 45° and passed through a $\pi/2$ converter, the column vector describing the output beam is given by

$$[C(\pi/2)]\times[\text{rot}(45°)]\times[HG_{2,0}]$$

$$\begin{bmatrix} 1 & 0 & 0 \\ 0 & -i & 0 \\ 0 & 0 & -1 \end{bmatrix}\times\begin{bmatrix} \dfrac{1}{2} & \dfrac{1}{\sqrt{2}} & \dfrac{1}{2} \\ \dfrac{-1}{\sqrt{2}} & 0 & \dfrac{1}{\sqrt{2}} \\ \dfrac{1}{2} & \dfrac{-1}{\sqrt{2}} & \dfrac{1}{2} \end{bmatrix}\times\begin{bmatrix} 1 \\ 0 \\ 0 \end{bmatrix}=\begin{bmatrix} \dfrac{1}{2} \\ \dfrac{i}{\sqrt{2}} \\ \dfrac{-1}{2} \end{bmatrix}.$$

From Beijersbergen *et al.* [5], we recognize the Hermite Gaussian coefficients of this output beam as a Laguerre Gaussian mode with indices $l=2$ and $p=0$.

The coefficients of the column vectors in the Jones matrix formulation correspond, respectively, to vertically and horizontally polarized light. It would be possible to formulate an equivalent representation in which the coefficients were for right and left handed circular polarizations. In the same way, our beam description in terms of Hermite Gaussian modes could be replaced with a description in terms of Laguerre Gaussian modes. Conceptually there is no difference between the two approaches, except that in our formulation the matrices for optical components are simple and the rotation matrix relatively complicated while for a column vector based on Laguerre Gaussian modes, the inverse would be the case.

TABLE II. $[2(N+1)\times 2(N+1)]$ matrices for polarized modes of order N.

$\pi/2$ converter

$$[C(\pi/2)]=\begin{bmatrix} 1&0&0&0&0&0&0&0&0&0&\cdots&\cdots\\ 0&1&0&0&0&0&0&0&0&0&\cdots&\cdots\\ 0&0&-i&0&0&0&0&0&0&0&\cdots&\cdots\\ 0&0&0&-i&0&0&0&0&0&0&\cdots&\cdots\\ 0&0&0&0&-1&0&0&0&0&0&\cdots&\cdots\\ 0&0&0&0&0&-1&0&0&0&0&\cdots&\cdots\\ 0&0&0&0&0&0&i&0&0&0&\cdots&\cdots\\ 0&0&0&0&0&0&0&i&0&0&\cdots&\cdots\\ 0&0&0&0&0&0&0&0&1&0&\cdots&\cdots\\ 0&0&0&0&0&0&0&0&0&1&\cdots&\cdots\\ \cdots&\cdots&\cdots&\cdots&\cdots&\cdots&\cdots&\cdots&\cdots&\cdots&\cdots&\cdots\\ \cdots&\cdots&\cdots&\cdots&\cdots&\cdots&\cdots&\cdots&\cdots&\cdots&\cdots&\cdots \end{bmatrix}$$

π converter

$$[C(\pi)]=\begin{bmatrix} 1&0&0&0&0&0&0&0&0&0&\cdots&\cdots\\ 0&1&0&0&0&0&0&0&0&0&\cdots&\cdots\\ 0&0&-1&0&0&0&0&0&0&0&\cdots&\cdots\\ 0&0&0&-1&0&0&0&0&0&0&\cdots&\cdots\\ 0&0&0&0&1&0&0&0&0&0&\cdots&\cdots\\ 0&0&0&0&0&1&0&0&0&0&\cdots&\cdots\\ 0&0&0&0&0&0&-1&0&0&0&\cdots&\cdots\\ 0&0&0&0&0&0&0&-1&0&0&\cdots&\cdots\\ 0&0&0&0&0&0&0&0&1&0&\cdots&\cdots\\ 0&0&0&0&0&0&0&0&0&1&\cdots&\cdots\\ \cdots&\cdots&\cdots&\cdots&\cdots&\cdots&\cdots&\cdots&\cdots&\cdots&\cdots&\cdots\\ \cdots&\cdots&\cdots&\cdots&\cdots&\cdots&\cdots&\cdots&\cdots&\cdots&\cdots&\cdots \end{bmatrix}$$

$\lambda/4$ waveplate

$$[W(\lambda/4)]=\begin{bmatrix} 1&0&0&0&0&0&0&0&0&0&\cdots&\cdots\\ 0&i&0&0&0&0&0&0&0&0&\cdots&\cdots\\ 0&0&1&0&0&0&0&0&0&0&\cdots&\cdots\\ 0&0&0&i&0&0&0&0&0&0&\cdots&\cdots\\ 0&0&0&0&1&0&0&0&0&0&\cdots&\cdots\\ 0&0&0&0&0&i&0&0&0&0&\cdots&\cdots\\ 0&0&0&0&0&0&1&0&0&0&\cdots&\cdots\\ 0&0&0&0&0&0&0&i&0&0&\cdots&\cdots\\ 0&0&0&0&0&0&0&0&1&0&\cdots&\cdots\\ 0&0&0&0&0&0&0&0&0&i&\cdots&\cdots\\ \cdots&\cdots&\cdots&\cdots&\cdots&\cdots&\cdots&\cdots&\cdots&\cdots&\cdots&\cdots\\ \cdots&\cdots&\cdots&\cdots&\cdots&\cdots&\cdots&\cdots&\cdots&\cdots&\cdots&\cdots \end{bmatrix}$$

Light with a particular mode structure, and so a defined orbital angular momentum may also be polarized and have a defined spin angular momentum. In many instances the two attributes of the beam will remain totally uncoupled and the resultant beam may be determined by two independent calculations: one with the Jones matrices and the other with the matrices developed for orbital angular momentum. However, this is not always the case; for example in the calculation of the eigenmodes of beams within laser resonators "out-of-plane" ring geometries and birefringent optical components

TABLE II (*continued*).

λ/2 waveplate

$$[W(\lambda/2)]=\begin{bmatrix}
1 & 0 & 0 & 0 & 0 & 0 & 0 & 0 & 0 & 0 & \cdots & \cdots \\
0 & -1 & 0 & 0 & 0 & 0 & 0 & 0 & 0 & 0 & \cdots & \cdots \\
0 & 0 & 1 & 0 & 0 & 0 & 0 & 0 & 0 & 0 & \cdots & \cdots \\
0 & 0 & 0 & -1 & 0 & 0 & 0 & 0 & 0 & 0 & \cdots & \cdots \\
0 & 0 & 0 & 0 & 1 & 0 & 0 & 0 & 0 & 0 & \cdots & \cdots \\
0 & 0 & 0 & 0 & 0 & -1 & 0 & 0 & 0 & 0 & \cdots & \cdots \\
0 & 0 & 0 & 0 & 0 & 0 & 1 & 0 & 0 & 0 & \cdots & \cdots \\
0 & 0 & 0 & 0 & 0 & 0 & 0 & -1 & 0 & 0 & \cdots & \cdots \\
0 & 0 & 0 & 0 & 0 & 0 & 0 & 0 & 1 & 0 & \cdots & \cdots \\
0 & 0 & 0 & 0 & 0 & 0 & 0 & 0 & 0 & -1 & \cdots & \cdots \\
\cdots & \cdots & \cdots & \cdots & \cdots & \cdots & \cdots & \cdots & \cdots & \cdots & \cdots & \cdots \\
\cdots & \cdots & \cdots & \cdots & \cdots & \cdots & \cdots & \cdots & \cdots & \cdots & \cdots & \cdots
\end{bmatrix}$$

Mode filter (e.g., $HG_{N-2,2}$)

$$[F(N-2,2)]=\begin{bmatrix}
0 & 0 & 0 & 0 & 0 & 0 & 0 & 0 & 0 & 0 & \cdots & \cdots \\
0 & 0 & 0 & 0 & 0 & 0 & 0 & 0 & 0 & 0 & \cdots & \cdots \\
0 & 0 & 0 & 0 & 0 & 0 & 0 & 0 & 0 & 0 & \cdots & \cdots \\
0 & 0 & 0 & 0 & 0 & 0 & 0 & 0 & 0 & 0 & \cdots & \cdots \\
0 & 0 & 0 & 0 & 1 & 0 & 0 & 0 & 0 & 0 & \cdots & \cdots \\
0 & 0 & 0 & 0 & 0 & 1 & 0 & 0 & 0 & 0 & \cdots & \cdots \\
0 & 0 & 0 & 0 & 0 & 0 & 0 & 0 & 0 & 0 & \cdots & \cdots \\
0 & 0 & 0 & 0 & 0 & 0 & 0 & 0 & 0 & 0 & \cdots & \cdots \\
0 & 0 & 0 & 0 & 0 & 0 & 0 & 0 & 0 & 0 & \cdots & \cdots \\
0 & 0 & 0 & 0 & 0 & 0 & 0 & 0 & 0 & 0 & \cdots & \cdots \\
\cdots & \cdots & \cdots & \cdots & \cdots & \cdots & \cdots & \cdots & \cdots & \cdots & \cdots & \cdots \\
\cdots & \cdots & \cdots & \cdots & \cdots & \cdots & \cdots & \cdots & \cdots & \cdots & \cdots & \cdots
\end{bmatrix}$$

Polarizer: for vertical polarization

$$[P(\updownarrow)]=\begin{bmatrix}
0 & 0 & 0 & 0 & 0 & 0 & 0 & 0 & 0 & 0 & \cdots & \cdots \\
0 & 1 & 0 & 0 & 0 & 0 & 0 & 0 & 0 & 0 & \cdots & \cdots \\
0 & 0 & 0 & 0 & 0 & 0 & 0 & 0 & 0 & 0 & \cdots & \cdots \\
0 & 0 & 0 & 1 & 0 & 0 & 0 & 0 & 0 & 0 & \cdots & \cdots \\
0 & 0 & 0 & 0 & 0 & 0 & 0 & 0 & 0 & 0 & \cdots & \cdots \\
0 & 0 & 0 & 0 & 0 & 1 & 0 & 0 & 0 & 0 & \cdots & \cdots \\
0 & 0 & 0 & 0 & 0 & 0 & 0 & 0 & 0 & 0 & \cdots & \cdots \\
0 & 0 & 0 & 0 & 0 & 0 & 0 & 1 & 0 & 0 & \cdots & \cdots \\
0 & 0 & 0 & 0 & 0 & 0 & 0 & 0 & 0 & 0 & \cdots & \cdots \\
0 & 0 & 0 & 0 & 0 & 0 & 0 & 0 & 0 & 1 & \cdots & \cdots \\
\cdots & \cdots & \cdots & \cdots & \cdots & \cdots & \cdots & \cdots & \cdots & \cdots & \cdots & \cdots \\
\cdots & \cdots & \cdots & \cdots & \cdots & \cdots & \cdots & \cdots & \cdots & \cdots & \cdots & \cdots
\end{bmatrix}$$

give rise to rotations of the mode profile and state of polarization, respectively [14]. The corresponding eigenvalues relate to the round trip loss and it is unclear how they could be determined from independent calculation of the orbital and spin eigenmodes. Another example, which we analyze below, is the recently studied rotational frequency shift [10] where the orbital and spin angular momentum interact in such a way that the total angular momentum is the important parameter, not the individual contributions.

More generally, the study of the orbital angular momen-

TABLE II (*continued*).

Mode rotation matrix for $N=1$

$$[\mathrm{rot}_{\mathrm{mode}}(\phi)]=\begin{bmatrix}\cos(\phi) & 0 & \sin(\phi) & 0\\ 0 & \cos(\phi) & 0 & \sin(\phi)\\ -\sin(\phi) & 0 & \cos(\phi) & 0\\ 0 & -\sin(\phi) & 0 & \cos(\phi)\end{bmatrix}$$

Mode rotation matrix for $N=2$

$$[\mathrm{rot}_{\mathrm{mode}}(\phi)]=\begin{bmatrix}\cos^2\phi & 0 & \frac{\sin 2\phi}{\sqrt{2}} & 0 & \sin^2\phi & 0\\ 0 & \cos^2\phi & 0 & \frac{\sin 2\phi}{\sqrt{2}} & 0 & \sin^2\phi\\ \frac{-\sin 2\phi}{\sqrt{2}} & 0 & \cos 2\phi & 0 & \frac{\sin 2\phi}{\sqrt{2}} & 0\\ 0 & \frac{-\sin 2\phi}{\sqrt{2}} & 0 & \cos 2\phi & 0 & \frac{\sin 2\phi}{\sqrt{2}}\\ \sin^2\phi & 0 & \frac{-\sin 2\phi}{\sqrt{2}} & 0 & \cos^2\phi & 0\\ 0 & \sin^2\phi & 0 & \frac{-\sin 2\phi}{\sqrt{2}} & 0 & \cos^2\phi\end{bmatrix}$$

Polarization rotation matrix for $N=1$

$$[\mathrm{rot}_{\mathrm{mode}}(\phi)]=\begin{bmatrix}\cos(\phi) & \sin(\phi) & 0 & 0\\ -\sin(\phi) & \cos(\phi) & 0 & 0\\ 0 & 0 & \cos(\phi) & \sin(\phi)\\ 0 & 0 & -\sin(\phi) & \cos(\phi)\end{bmatrix}$$

Polarization rotation matrix for $N=2$

$$[\mathrm{rot}_{\mathrm{mode}}(\phi)]=\begin{bmatrix}\cos\phi & \sin\phi & 0 & 0 & 0 & 0\\ -\sin\phi & \cos\phi & 0 & 0 & 0 & 0\\ 0 & 0 & \cos\phi & \sin\phi & 0 & 0\\ 0 & 0 & -\sin\phi & \cos\phi & 0 & 0\\ 0 & 0 & 0 & 0 & \cos\phi & \sin\phi\\ 0 & 0 & 0 & 0 & -\sin\phi & \cos\phi\end{bmatrix}$$

tum of light beams is a rapidly expanding field and it is not yet clear where it may lead. It is possible that there will be other applications where the interaction of polarization and orbital angular momentum is best examined within a single formalism. It is therefore of considerable interest to extend the matrices describing the mode composition to include simultaneously the polarization behavior previously described by Jones matrices. Such matrices would then be able to model the propagation of both spin and orbital angular momentum, through an optical system comprising both polarizing and mode transforming components.

Any combination of polarized modes of order N can be expressed as a $2(N+1)$ element column vector containing the complex weightings of orthogonally polarized Hermite Gaussian modes of the same order, as with

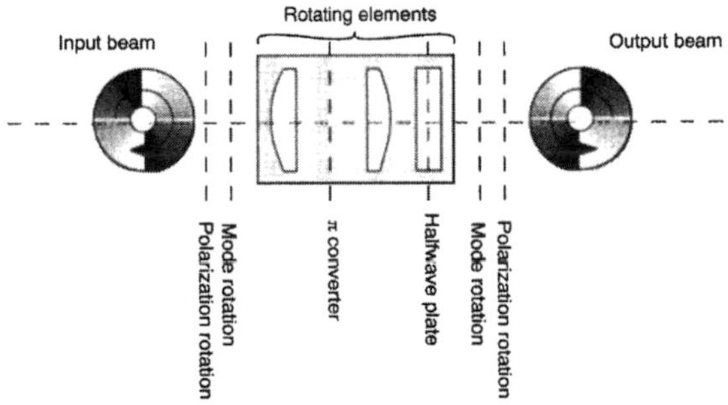

FIG. 2. Experimental configuration for the observation of the rotational frequency shift. (Arrows and shading on the mode distributions indicate the polarization and the sense of the azimuthal phase, respectively.)

$$\begin{bmatrix} a_{N,0}(\leftrightarrow) \\ a_{N,0}(\updownarrow) \\ a_{N-1,0}(\leftrightarrow) \\ a_{N-1,0}(\updownarrow) \\ \cdots \\ \cdots \\ \cdots \\ \cdots \\ a_{1,N-1}(\leftrightarrow) \\ a_{1,N-1}(\updownarrow) \\ a_{0,N}(\leftrightarrow) \\ a_{0,N}(\updownarrow) \end{bmatrix},$$

where the arrows in brackets denote the orientation of the polarization. The intensity is simply given by the square of the modulus of the vector representing the beam.

The $[2(N+1)\times 2(N+1)]$ matrices for the mode converters are based on those obtained previously, but each term must be repeated along the diagonal so that both polarizations are rephased. The $[2(N+1)\times 2(N+1)]$ matrices for waveplates are based on the corresponding Jones matrices, but the whole of each matrix must be repeated along the diagonal to act on each mode independently.

The rotation matrices are more complicated. The mode may be rotated with an image rotator and the polarization independently rotated by the use of a waveplate. The mode rotation matrix includes the same terms as our N dependent matrix given in Table I, but they must be diagonally repeated to rotate modes of both polarizations and interspaced with zeros to prevent their mixing. Similarly, the matrix for the rotation of the polarization is identical to that used within the Jones matrix formulation, but is diagonally repeated to rotate the polarization state of each of the individual Hermite Gaussian modes. The matrices for mode converters, waveplates, mode filters, polarizers and rotations are given in Table II.

To confirm the effectiveness of the matrix formulation, we modelled our rotational frequency shift experiment [10] for a circularly polarized, $\sigma=\pm 1$, Laguerre Gaussian mode. This enables us to find the frequency shift which results from the rotation of a beam containing both spin and orbital angular momentum. In the experiment, beam rotation was introduced by simultaneously rotating a π converter and half-wave plate; see Fig. 2. The matrix description of the output beam becomes

$$[\mathrm{rot}_{\mathrm{pol}}(\phi)]\times[\mathrm{rot}_{\mathrm{mode}}(\phi)]\times[W(\lambda/2)]\times[C(\pi)]$$
$$\times[\mathrm{rot}_{\mathrm{mode}}(-\phi)]\times[\mathrm{rot}_{\mathrm{pol}}(-\phi)]\times[LG_{l,p}(\sigma=\pm 1)].$$

A change in the handedness of the circularly polarized light, such that the spin angular momentum component is either additive or subtractive from the orbital angular momentum, gives a total angular momentum of the beam of $(l\pm 1)\hbar$ per photon. The frequency shift may be correctly deduced by comparison with its nonrotated equivalent to be $(l\pm 1)\Omega$, where Ω is the angular frequency of the beam. This offers powerful evidence of the utility of the matrices.

We have shown that there is a set of matrices equivalent to the Jones matrices which can express the behavior of light possessing orbital angular momentum and have developed them to account simultaneously for polarization. These matrices can be used in any optical system which conserves mode order. The equivalent of the Mueller 4×4 matrices, which depend upon the four Stokes parameters, for orbital angular momentum is currently the basis of further investigation.

It is a pleasure to acknowledge the contribution of Professor S. M. Barnett who drew to our attention to the rotation matrix in Ref. [12].

[1] R. C. Jones, J. Opt. Soc. Am. **46**, 126 (1956), and references therein.

[2] H. Mueller, J. Opt. Soc. Am. **38**, 661 (1948).

[3] A. Gerrard and J. M. Burch, *Introduction to Matrix Methods in Optics* (Wiley, New York, 1975).

[4] L. Allen, M. W. Beijersbergen, R. J. C. Spreeuw, and J. P. Woerdman, Phys. Rev. A **45**, 8185 (1992).

[5] M. W. Beijersbergen, L. Allen, H. E. L. O. van der Veen, and J. P. Woerdman, Opt. Commun. **96**, 123 (1993).

[6] S. J. van Enk and G. Nienhuis, Opt. Commun. **94**, 147 (1992).

[7] S. M. Barnett and L. Allen, Opt. Commun. **110**, 670 (1994).

[8] N. B. Simpson, K. Dholakia, L. Allen, and M. J. Padgett, Opt. Lett. **22**, 52 (1997).

[9] K. Dholakia, N. B. Simpson, M. J. Padgett, and L. Allen, Phys. Rev. A **54**, R3742 (1996).

[10] J. Courtial, K. Dholakia, D. A. Robertson, L. Allen, and M. J. Padgett, Phys. Rev. Lett. **80**, 3217 (1998).

[11] M. J. Padgett and J. Courtial, Opt. Lett. **24**, 430 (1998).

[12] A. R. Edmonds, *Angular Momentum in Quantum Mechanics* (Princeton University Press, Princeton, 1957).

[13] N. R. Heckenberg, R. McDuff, C. P. Smith, H. Rubinsztein-Dunlop, and M. J. Wegener Opt. Quantum Electron. **24**, S951 (1992).

[14] C. Yelland, J. Hong, M. J. Padgett, M. H. Dunn, and W. Sibbett, Opt. Commun. **109**, 451 (1994).

PAPER 2.5

Paraxial beams of spinning light

Michael Berry

H H Wills Physics Laboratory, Tyndall Avenue, Bristol BS8 1TL, United Kingdom

Abstract

A simple derivation is given for the known result that the component of total angular momentum along the propagation direction of a general paraxial beam can be separated into orbital and spin parts. Comments are made about orbital angular momentum and wave dislocations, and about how the orbital and spin angular momenta can be changed by propagation in a refracting medium.

(Published in *Singular optics* (Eds. M S Soskin and M V Vasnetsov SPIE **3487** (1998) pp6-11)

1. Angular momentum decomposition

It is now well established by theory and experiment that beams of light can carry both orbital and spin angular momentum [1-4], at least in the paraxial approximation [5]. In spite of this widespread understanding, it seems worthwhile to present the argument in its most general form, and that is my main purpose here - without claiming originality for the central ideas. The aims are to represent the light in terms of general transverse fields (i.e. not necessarily as a superosition of Gauss-Laguerre beams), to allow the polarization to vary with position, to introduce paraxiality in the most direct way, and to write the result in a way that resembles quantum mechanics, so that the separation into the two kinds of angular momentum is compelling. In addition, I will make some remarks about the significance of the two components, and how they can be changed by the medium through which the light is propagating.

Let monochromatic light with frequency ω (and wavenumber $k=\omega/c$) travel in vacuum in the paraxial direction z (unit vector $\mathbf{e}_z$). A precise specification of z will be given later. Let positions in the transverse plane be denoted by **r**, let the transverse part of a general vector be denoted by **V**, and let the full three-dimensional counterparts of these quantities be denoted by the suffix "tot". Thus

$$\mathbf{r} = (x, y), \quad \mathbf{V} = \left(A_x, A_y\right), \quad \mathbf{r}_{\text{tot}} = (x, y, z), \quad \mathbf{V}_{\text{tot}} = \left(A_x, A_y, A_z\right) \tag{1}$$

We will calculate the z component of the total angular momentum per photon in unit length of a transverse slice of the beam about the axis **r**=0, and denote this quantity by J_z. We use standard notation for the complex electromagnetic fields, and the well known results [6] that the angular momentum density is

$$\operatorname{Re} \mathbf{r} \times \left(\mathbf{D}_{\text{tot}}^* \times \mathbf{B}_{\text{tot}}\right) \tag{2}$$

and the energy density is

$$\operatorname{Re} \tfrac{1}{2} \left(\mathbf{E}_{\text{tot}}^* \cdot \mathbf{D}_{\text{tot}} + \mathbf{B}_{\text{tot}}^* \cdot \mathbf{H}_{\text{tot}}\right) \tag{3}$$

Then

$$J_z = \hbar\omega \frac{\operatorname{Re} \iint \mathrm{d}x\mathrm{d}y\, \mathbf{r} \times \left(\mathbf{D}_{\text{tot}}^* \times \mathbf{B}_{\text{tot}}\right) \cdot \mathbf{e}_z}{\operatorname{Re} \frac{1}{2} \iint \mathrm{d}x\mathrm{d}y \left(\mathbf{E}_{\text{tot}}^* \cdot \mathbf{D}_{\text{tot}} + \mathbf{B}_{\text{tot}}^* \cdot \mathbf{H}_{\text{tot}}\right)} \tag{4}$$

Maxwell's equations and the constitutive relations can be used to eliminate all fields except $\mathbf{E}_{\text{tot}}$, with the result

$$J_z = \hbar \frac{\operatorname{Re}(-\mathrm{i}) \iint \mathrm{d}x\mathrm{d}y\, \mathbf{r} \times \left(\mathbf{E}_{\text{tot}}^* \times (\nabla \times \mathbf{E}_{\text{tot}})\right) \cdot \mathbf{e}_z}{\operatorname{Re} \frac{1}{2} \iint \mathrm{d}x\mathrm{d}y \left(\mathbf{E}_{\text{tot}}^* \cdot \mathbf{E}_{\text{tot}} + \frac{1}{k^2} \left(\nabla \times \mathbf{E}_{\text{tot}}^*\right) \cdot \left(\nabla \times \mathbf{E}_{\text{tot}}\right)\right)} \tag{5}$$

In this exact formula, all components of $\mathbf{E}_{\text{tot}}$ can vary with x,y and z.

Now we invoke the paraxial approximation and then $\nabla \cdot \mathbf{E}_{\text{tot}} = 0$ to eliminate all z derivatives and z components to order $1/k$:

$$\partial_z \mathbf{E}_{\text{tot}} \approx \mathrm{i}k \mathbf{E}_{\text{tot}}, \quad E_z \approx \frac{\mathrm{i}}{k} \nabla \cdot \mathbf{E} \tag{6}$$

Thus, in the denominator in (5),

$$\mathbf{E}_{\text{tot}}^{*}\cdot\mathbf{E}_{\text{tot}}\approx\frac{1}{k^{2}}\left(\nabla\times\mathbf{E}_{\text{tot}}^{*}\right)\cdot\left(\nabla\times\mathbf{E}_{\text{tot}}\right)\approx\mathbf{E}^{*}\cdot\mathbf{E} \tag{7}$$

The numerator can be similarly simplified using (6) and also integration by parts with respect to x and y, leading after a little algebra to

$$J_{z}\approx\frac{\operatorname{Re}(-\mathrm{i}\hbar)\iint\mathrm{d}x\mathrm{d}y\left\{\mathbf{E}^{*}\cdot\left[\mathbf{e}_{z}\cdot\left(\mathbf{r}\times(-\mathrm{i}\nabla)\right)\right]\mathbf{E}+\mathbf{e}_{z}\cdot\mathbf{E}^{*}\times\mathbf{E}\right\}}{\operatorname{Re}\iint\mathrm{d}x\mathrm{d}y\,\mathbf{E}^{*}\cdot\mathbf{E}} \tag{8}$$

It is well known [5] that a similar expression, with all vectors replaced by their three-dimensional counterparts, can be obtained exactly - that is, without the paraxial approximation - if the quantity being calculated is the total angular momentum in the whole three-dimensional beam; but this depends on integration with respect to z, an option unavailable in our calculation for a transverse slice.

In the final step, we introduce a basis of circular polarizations, replacing **E** by the column vector

$$|\psi\rangle\equiv\begin{pmatrix}\psi_{+}\\ \psi_{-}\end{pmatrix}\equiv\frac{1}{\sqrt{2}}\begin{pmatrix}E_{x}-\mathrm{i}E_{y}\\ E_{x}+\mathrm{i}E_{y}\end{pmatrix} \tag{9}$$

Then we define scalar products as

$$\langle\phi|\psi\rangle\equiv\iint\mathrm{d}x\mathrm{d}y\begin{pmatrix}\phi_{+}^{*}(\mathbf{r}) & \phi_{-}^{*}(\mathbf{r})\end{pmatrix}\begin{pmatrix}\psi_{+}(\mathbf{r})\\ \psi_{-}(\mathbf{r})\end{pmatrix} \tag{10}$$

and recognize

$$\mathbf{p}=-\mathrm{i}\hbar\nabla \tag{11}$$

as the momentum operator, and

$$l_{z}=\mathbf{r}\times\mathbf{p}\cdot\mathbf{e}_{z},\quad s_{z}=\hbar\begin{pmatrix}1 & 0\\ 0 & -1\end{pmatrix} \tag{12}$$

as the operators for the z components of the orbital angular momentum, and the spin angular momentum of a spin-1 particle. Thus (8) becomes the well known expression

$$J_{z}\approx\frac{\langle\psi|l_{z}|\psi\rangle+\langle\psi|s_{z}|\psi\rangle}{\langle\psi|\psi\rangle}\equiv L_{z}+S_{z} \tag{13}$$

This is valid in the paraxial approximation, that is when quantities of order $1/k$ are neglected. Henceforth we assume without loss of generality that the beam is normalized with $\langle\psi|\psi\rangle=1$.

2. Remarks on spin and orbital parts of the angular momentum

It might be thought that the difference between the spin and orbital contributions S_z and L_z in (13) is that S_z is independent of the axis $\mathbf{r}=0$ about which J_z is calculated, whereas L_z depends on this axis. But this can be wrong, as the following argument shows. Under a shift of origin, $\mathbf{r}\to\mathbf{r}_0+\mathbf{r}$, the orbital angular momentum shifts too:

$$L_z \to L_z + \mathbf{e}_z \cdot \mathbf{r}_0 \times \mathbf{P}, \quad \text{where } \mathbf{P} \equiv \langle\psi|\mathbf{p}|\psi\rangle \tag{14}$$

Here $\mathbf{P}$ is the transverse current in the beam. If $\mathbf{P}$ vanishes, the shift $\mathbf{r}\to\mathbf{r}_0+\mathbf{r}$ leaves L_z unaltered. But for a paraxial beam $\mathbf{P}$ will always be small compared with the current along the paraxial direction z. Therefore it is possible and indeed reasonable to stipulate z as the direction for which $\mathbf{P}$ is *exactly* zero. With paraxiality thus defined, *both* components of J_z are invariant under a shift of axis, so this invariance cannot be used to define S.

The true distinction between orbital and spin angular momentum is of course that L_z depends on the spatial structure of the beam irrespective of polarization, whereas S_z depends on the state of polarization of the beam. There are subtleties concealed here too. To appreciate these, note that, from (9), (10, (12) and (13),

$$S_z = \hbar\langle\psi|\begin{pmatrix}1 & 0\\ 0 & -1\end{pmatrix}|\psi\rangle = \hbar\iint \mathrm{d}x\mathrm{d}y\left(|\psi_+(\mathbf{r})|^2 - |\psi_-(\mathbf{r})|^2\right) \tag{15}$$

so that $S_z=+\hbar$ for right-circularly polarized light ($\psi_-=0$), and $S_z=-\hbar$ for left-circularly polarized light ($\psi_+=0$), as expected for spin-1 particles. For linearly polarized light ($|\psi_+|=|\psi_-|$), $S_z=0$. A complete description of the polarization state involves the other two Pauli matrices, and suggests defining

$$S_x = \hbar\langle\psi|\begin{pmatrix}0 & 1\\ 1 & 0\end{pmatrix}|\psi\rangle, \quad S_y = \hbar\langle\psi|\begin{pmatrix}0 & -\mathrm{i}\\ \mathrm{i} & 0\end{pmatrix}|\psi\rangle \tag{16}$$

But S_x, S_y and S_z must not be regarded as the three components of a spin angular momentum vector of the light, although they give a complete description of the state of polarization of the beam. One reason is that they have the wrong commutation relations, e.g. $[S_x, S_y]=2i\hbar S_z$ (rather than $i\hbar S_z$ as would be appropriate to a spin-1 particle); another is that $S_x^2+S_y^2+S_y^2=3\hbar^2$(rather than $2\hbar^2$ as would be appropriate to a spin-1 particle).

One of several correct representations of spin-1 matrices is that implied by the three-dimensional version of the last term in (8), namely

$$\mathrm{Re}\left(-i\hbar\mathbf{E}^*_{\mathrm{tot}}\times\mathbf{E}_{\mathrm{tot}}\right)=\mathbf{E}^*_{\mathrm{tot}}\cdot\mathbf{S}_{\mathrm{tot}}\cdot\mathbf{E}_{\mathrm{tot}} \tag{17}$$

where

$$\mathbf{S}_{\mathrm{tot}}=-i\hbar\left\{\begin{pmatrix}0&0&0\\0&0&1\\0&-1&0\end{pmatrix},\begin{pmatrix}0&0&-1\\0&0&0\\1&0&0\end{pmatrix},\begin{pmatrix}0&1&0\\-1&0&0\\0&0&0\end{pmatrix}\right\} \tag{18}$$

Strangely, however, these spin-1 matrices, unlike those in (15) and (16), do not give a complete description of the state of polarization of the beam: for any transverse wave, whatever its polarization, the transverse components of the expectation value of $\mathbf{S}_{\mathrm{tot}}$ vanish, leaving only the single (real) z component nonvanishing, and unable to discriminate, for example, between the different possible states of linear polarization. There is another well known 'paradox' associated with the full three-dimensional angular momentum of a transverse light wave (e.g. a plane wave): the exact equation (2) implies that all components of total angular momentum vanish for such a wave, whereas (13) predicts a nonvanishing value for J_z for a circularly polarized transverse wave. The resolution lies in the fact that (13) applies to a beam whose lateral extent is effectively finite, which can therefore not be exactly transverse; in the derivation of (13), the longitudinal component (approximated by (6)) contributed through the integration by parts.

Turning now to the orbital angular momentum L_z, we note the common opinion [7] that this quantity is associated with optical dislocation lines (phase singularities) [8-11]. In general, however, there is no such association, as the following argument shows.

Since L_z is associated with the spatial variation of the beam, it suffices to take for $|\psi\rangle$ a scalar wavefunction $\psi(\mathbf{r},z)$, satisfying the paraxial wave equation. We let ψ be a 'spiral beam' [12, 13] with waist size w; In the focal plane z=0 this has the form

$$\psi(\mathbf{r},0) = f(\zeta)\exp\left\{-r^2/2w^2\right\} \quad \left(\zeta \equiv x + iy,\ \ r \equiv \sqrt{\left(x^2+y^2\right)}\right) \tag{19}$$

Away from the focal plane, ψ preserves this form while expanding and rotating. A short calculation gives

$$L_z = \hbar\frac{\mathrm{Re}\iint \mathrm{d}x\mathrm{d}y\exp\left(-r^2/w^2\right)f(\zeta)^{*}\zeta\partial_\zeta f(\zeta)}{\mathrm{Re}\iint \mathrm{d}x\mathrm{d}y\exp\left(-r^2/w^2\right)\left|f(\zeta)\right|^2} \tag{20}$$

If $f(\zeta)$ is an Mth order polynomial, that is

$$f(\zeta) = \sum_{m=0}^{M} f_m\zeta^m \tag{21}$$

then

$$L_z = \hbar\frac{\sum_{m=1}^{M} m\,m!\left|f_m\right|^2 w^{2m}}{\sum_{m=0}^{M} m!\left|f_m\right|^2 w^{2m}} \equiv \langle m\rangle\hbar \tag{22}$$

Now, dislocations are the zeros of ψ, that is the zeros of $f(\zeta)$, and by the fundamental theorem of algebra there are exactly M of them for the spiral beam with f given by (21). Moreover, they all have the same sign, so the total dislocation strength [11] is also M. In the special case of a pure Gaussian beam, where only the term m=M in the series (21) is nonzero, $\langle m\rangle = M$ and there is a sense in which it is correct to ascribe the (quantized) orbital angular momentum to the Mth order (screw) dislocation. But in all other cases, $\langle m\rangle \neq M$ and the association between L_z and dislocations is false. The easy way to see that this must be so is to note that since dislocations are zeros they have zero weight in the integral (20) for L_z.

To avoid confusion, I should remark that there is an association between a dislocation line and the torque exerted on a small particle [2-4], arising from the vortex structure close to the line. This very interesting effect concerns the angular momentum density in the vicinity of the particle, rather than the integrated quantity L_z considered here.

3. Beam propagation in a refracting medium

If the beam propagates paraxially in a homogenous and isotropic medium, no torque can be exerted on it, and L_z and S_z are separately conserved. However, an inhomogenous medium can exert a torque that changes L_z, and an anisotropic medium can exert a torque that changes S_z. To quantify these effects, let the medium have dielectric tensor

$$\mathsf{e}(\mathbf{r}) = \varepsilon_0\left(1 + 2\mathsf{n}(\mathbf{r})\right) \tag{23}$$

Here n is a transverse refractive index tensor, hermitean because the medium is assumed transparent, whose components

$$\mathsf{n}(\mathbf{r}) = \begin{pmatrix} n_{xx}(\mathbf{r}) & n_{xy}(\mathbf{r}) \\ n_{xy}(\mathbf{r}) & n_{zz}(\mathbf{r}) \end{pmatrix} \tag{24}$$

are all much less than unity since the propagation is paraxial. In this case contributions from n to J_z are of higher than paraxial order, and the argument leading from (2) and (3) to the final formula (13) is unaffected by the anisotropy and variations of the medium.

However, propagation of the angular momentum is affected, in a manner determined by the paraxial wave equation for **E**. From Maxwell's equations, this is

$$ik\partial_z\mathbf{E} = -\tfrac{1}{2}\left(\partial_x^2 + \partial_x^2\right)\mathbf{E} - k^2\mathsf{n}(\mathbf{r})\cdot\mathbf{E} \tag{25}$$

Transforming to the basis (9) of circular polarizations gives the 'Schrödinger lookalike' equation

$$ik\partial_z|\psi\rangle = \mathsf{H}|\psi\rangle \tag{26}$$

Here the 'hamiltonian' H (now setting $\hbar$=1 for simplicity of notation) is

$$\mathsf{H} = \tfrac{1}{2}\mathbf{p}\cdot\mathbf{p} - k^2\left\{W_0(\mathbf{r}) + \mathbf{W}(\mathbf{r})\cdot\sigma\right\} \tag{27}$$

in which σ is the (three-dimensional) vector of Pauli matrices and $W_0(\mathbf{r})$ and the three-dimensional vector $\mathbf{W}(\mathbf{r})$ are (invoking the hermiticity of $\mathbf{n}$)

$$\begin{aligned} W_0 &= n_{xx} + n_{yy}, \\ \mathbf{W} = \left\{W_x, W_y, W_z\right\} &= \left\{n_{xx} - n_{yy}, 2\,\mathrm{Re}\,n_{xy}, -2\,\mathrm{Im}\,n_{xy}\right\} \end{aligned} \tag{28}$$

Note that W_0 depends on the isotropic part of the refractive index of the medium, W_x and W_y to its birefringence, and W_z to its optical activity (chirality).

The rate of change in S_z is easily shown to be

$$\begin{aligned} \partial_z S_z &= 2k\langle\psi|\mathbf{W} \times \sigma \cdot \mathbf{e}_z|\psi\rangle \\ &= 2k\langle\psi|\begin{pmatrix} 0 & -\mathrm{i}\left(n_{xx} - n_{yy}\right) - 2\,\mathrm{Re}\,n_{xy} \\ \mathrm{i}\left(n_{xx} - n_{yy}\right) - 2\,\mathrm{Re}\,n_{xy} & 0 \end{pmatrix}|\psi\rangle \end{aligned} \tag{29}$$

Thus the spin angular momentum can be changed by the birefringence of the medium but not by its optical activity.

Similarly, the rate of change of L_z is determined by the commutator in

$$\partial_z L_z = -\mathrm{i}k\langle\psi|\left[W_0(\mathbf{r}) + \mathbf{W}(\mathbf{r}) \cdot \sigma,\ l_z\right]|\psi\rangle \tag{30}$$

Since, from the definition (12),

$$l_z = -\mathrm{i}\partial_\phi \tag{31}$$

where ϕ is the polar angle of $\mathbf{r}$, we recover the obvious result that the orbital angular momentum can be changed only by propagation in a medium that is not azimuthally symmetric, regardless of whether the asymmetry is in the isotropic part W_0 of the medium or its anisotropic (birefringent or chiral) part $\mathbf{W}$.

During any such changes of its angular momentum, the beam will, of course, exert equal and opposite reactive torques on the medium

Acknowledgements

I thank Professor M.S. Soskin for generous hospitality at the conference on Singular Optics, and Professors N.R. Heckenberg, A. Alexeyev and C. Alexeyev for useful discussions.

References

1. Allen, L., Beijersbergen, M. W., Spreew, R. J. C. & Woerdman, J. P.,1992, Orbital angular momentum and the transformation of Gauss-Laguerre modes *Phys. Rev. Lett. A* **45**, 8185-8189.

2. Simpson, N. B., Dholakia, A., Allen, L. & Padgett, M. J.,1997, Mechanical equivalent of spin and orbital angular momentum of light: an optical spanner *Optics Letters* **22**, 52-54.

3. Friese, M. E. J., Rubinsztein-Dunlop, H., Heckenberg, N. R. & Enger, J.,1996, Optical angular momentum transfer to trapped absorbing particles *Phys. Rev. A* **54**, 1593-1596.

4. Friese, M. E. J., Nieminen, T. A., Heckenberg, N. R. & Rubinsztein-Dunlop, H.,1997, Controlled optical torque by elliptical polarization *Optics Letters* submitted.

5. Barnett, S. M. & Allen, L.,1994, Orbital angular momentum and nonparaxial light beams *Opt. Comm.* **110**, 670-678.

6. Jackson, J. D. ,1975, *Classical electrodynamics* (Wiley, New York).

7. Soskin, M. S., Gorshkov, V. N., Vasnetsov, M. V., Malos, J. T. & Heckenberg, N. R.,1997, Topological charge and angular momentum of light beams carrying optical vortices *Phys. Rev. A* **56**, 4064-4075.

8. Nye, J. F. & Berry, M. V.,1974, Dislocations in wave trains *Proc. Roy. Soc. Lond.* **A336**, 165-90.

9. Berry, M. V. ,1981, Singularities in Waves and rays, in *Les Houches Lecture Series Session 35*, eds. Balian, R., Kléman, M. & Poirier, J.-P. (North-Holland: Amsterdam, pp. 453-543.

10. Nye, J. F.,1997, Singularities in wave fields *Phil. Trans. Roy. Soc. Lond.* **355**, 2065-2069.

11. Berry, M. V.,1998, Much ado about nothing: optical dislocation lines (phase singularities, zeros, vortices...), in Proceedings of conference on *Singular Optics* (Frunzenskoe, Crimea), eds. M. S. Soskin & M.V.Vasnetsov (SPIE **3487**, pp1-5).

12. Abramochkin, E. & Volostnikov, V.,1993, Spiral-type beams *Opt. Comm.* **102**, 336-350.

13. Abramochkin, E. & Volostnikov, V.,1996, Spiral-type beams: optical and quantum aspects *Opt. Comm* **125**, 302-323

The Poynting vector in Laguerre–Gaussian beams and the interpretation of their angular momentum density

L. Allen *, M.J. Padgett

Department of Physics and Astronomy, University of Glasgow, Glasgow G12 8QQ, UK

Received 4 July 2000; accepted 16 August 2000

Abstract

We use the local linear momentum density in Laguerre–Gaussian beams to investigate classically the trajectory of the Poynting vector for light of any polarisation. Some of the difficulties of interpretation of the local angular momentum density are discussed. © 2000 Published by Elsevier Science B.V.

Keywords: Poynting vector; Angular momentum density; Polarised light

1. Introduction

The rate of electromagnetic energy flow per unit area, the Poynting vector, is defined and discussed in all books on electromagnetism (see, e.g., Ref. [1]). The vector and its associated linear momentum density is, however, invariably examined for plane waves or at least waves with a plane wavefront. Recent interest in orbital angular momentum of light beams [2] has led to a considerable literature relating to Laguerre–Gaussian beams which have helical wavefronts. The angular momentum density, $\boldsymbol{j}$, is related [1] to the linear momentum density, $\boldsymbol{p}$, through

$$\boldsymbol{j} = \boldsymbol{r} \times \boldsymbol{p}, \tag{1}$$

where

$$\boldsymbol{p} = \varepsilon_0 \boldsymbol{E} \times \mathbf{B}, \tag{2}$$

and the total angular momentum is naturally derived from the linear momentum [3]. The emphasis in the literature is, for obvious reasons, on orbital angular momentum, but as light of arbitrary polarisation is considered, it also accounts for spin angular momentum. In this work, we use the components of the linear momentum density to determine the behaviour of the Poynting vector for Laguerre–Gaussian modes and then examine the meaning of the angular momentum density.

2. Linear momentum density

It is easy to show within the paraxial approximation [2,3] that for any mode with an amplitude given by

$$u(r, \phi, z) = u(r, z) \exp(\mathrm{i}l\phi), \tag{3}$$

the r, φ and z components of linear momentum density, $\varepsilon_0 \boldsymbol{E} \times \boldsymbol{B}$, for a circularly polarised beam propagating in the z-direction are given by

* Corresponding author. Fax: +44-1359-258-314.
E-mail address: lesallen@btinternet.com (L. Allen).

$$p_r = \varepsilon_0 \frac{\omega k r z}{(z_R^2 + z^2)} |u|^2,$$

$$p_\phi = \varepsilon_0 \left[\frac{\omega l}{r} |u|^2 - \frac{1}{2} \omega \sigma \frac{\partial |u|^2}{\partial r} \right],$$

$$p_z = \varepsilon_0 \omega k |u|^2. \tag{4}$$

The component p_r, relates to the spread of the beam as it propagates. The first term of the azimuthal component p_ϕ depends on l, where $l\hbar$ has been identified as the orbital angular momentum per photon [3]. Its second term relates to the spin contribution, where $\sigma = \pm 1$ for left-handed and right-handed circularly polarised light, respectively, and $-1 < \sigma < 1$ describes elliptically polarised light. It is this contribution which leads to the spin angular momentum of the beam. The final component, p_z, is the linear momentum in the direction of propagation.

The relative size of the components determines the trajectory of the Poynting vector (Fig. 1). We have investigated elsewhere this trajectory with respect to the orbital contribution alone [4]. This was for linearly polarised light where $\sigma = 0$; most of the interesting terms investigated here were therefore zero and played no role. When both orbital and spin momenta are included, the rate of change of azimuthal angle of the trajectory of the Poynting vector with z is given by

$$\frac{\partial \theta}{\partial z} = \frac{1}{r} \frac{p_\phi}{p_z} = \frac{l}{kr^2} - \frac{\sigma}{2kr} \frac{1}{|u|^2} \frac{\partial |u|^2}{\partial r}. \tag{5}$$

A Laguerre–Gaussian mode has an amplitude of

$$u_{pl}^{\mathrm{LG}} = \frac{C_{pl}^{\mathrm{LG}}}{w(z)} \left(\frac{r\sqrt{2}}{w(z)} \right)^{|l|} \exp\left[-\frac{r^2}{w^2(z)} \right] L_p^{|l|} \times \exp\left(-\frac{\mathrm{i}kr^2 z}{2(z^2 + z_R^2)} \right) \exp(-\mathrm{i}l\phi) \times \exp\left(\mathrm{i}(2p + |l| + 1) \tan^{-1} \frac{z}{z_R} \right), \tag{6}$$

where C_{pl}^{LG} is the normalisation constant; $L_p^{|l|}(2r^2/w^2(z))$ is a generalised Laguerre polynomial; the radius of the beam at position z is $w(z)$ where

$$w(z)^2 = \frac{2}{k} \frac{z_R^2 + z^2}{z_R}, \quad \text{and} \quad (2p + |l| + 1) \tan^{-1} \frac{z}{z_R}$$

is the Gouy phase, and z_R, the Rayleigh range. Here, p and l are mode indices, where l is the number of intertwined helices, and p, the number of radial nodes. It follows that for such a mode, Eq. (5) becomes

$$\frac{\partial \theta}{\partial z} = \frac{l}{kr^2} - \frac{\sigma |l|}{kr^2} + \frac{2\sigma}{kw^2(z)} + \frac{4\sigma}{kw^2} \frac{L_{p-1}^{|l-1|}(2r^2/w^2(z))}{L_p^{|l|}(2r^2/w^2(z))}, \tag{7}$$

where k may be expressed in terms of $w(z)$, as above.

For modes with $p = 0$, the final term in Eq. (7) is always zero. When $l = 0$ and $p = 0$ and the Laguerre–Gaussian is identical to the Hermite Gaussian (0,0)-mode, the orbital angular momentum is zero and the rotation of the Poynting vector for a polarised beam arises from the effect of spin alone. The first two terms are also zero and $\theta = \sigma \arctan z/z_R$ and $\theta \to \pi/2$ as $z \to \infty$.

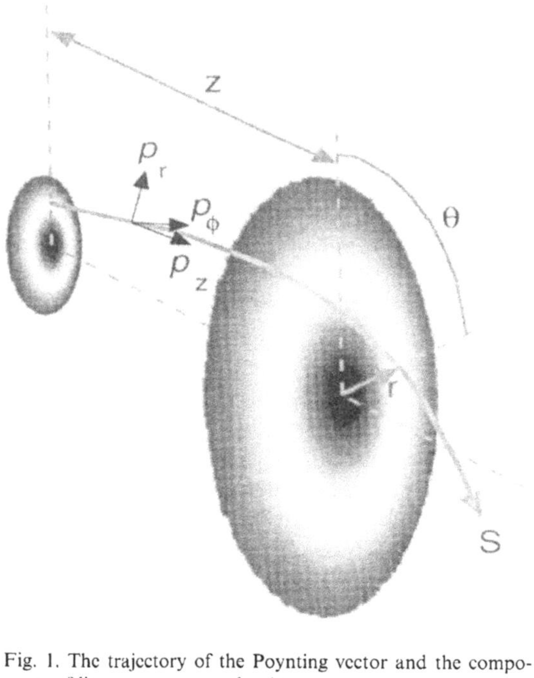

Fig. 1. The trajectory of the Poynting vector and the components of linear momentum density.

For a single ringed Laguerre–Gaussian mode when $p = 0$ but $l \neq 0$,

$$\frac{\partial \theta}{\partial z} = \frac{l}{kr^2}\left(1 - \frac{\sigma|l|}{l}\right) + \frac{2\sigma}{kw^2}. \tag{8}$$

At peak intensity when $r^2 = w^2(z)|l|/2$, we find for all values of l and σ that $\theta = \pm\arctan z/z_R$. The trajectory of the Poynting vector is in general a spiral with a pitch that depends on the radius. Surprisingly, at peak intensity, the pitch is such that it becomes a straight line skewed with respect to the beam axis. Such a trajectory is consistent with a ray optical model where orbital angular momentum may be represented by skew rays in the optical beam.

When $\sigma = \pm 1$, that is for circularly polarised light, we see that if the handedness of spin and orbit are the same, then $\theta = \arctan z/z_R$ for all l and for all positions across the beam amplitude distribution, not just at the peak intensity.

In general, for a mode of arbitrary index p and any value of l and σ,

$$\theta = \left\{\frac{l - \sigma\frac{|l|}{l}}{2}\frac{w^2(0)}{r^2(0)} + \sigma + 2\sigma\frac{L_{p-1}^{|l+1|}(2r^2/w^2(z))}{L_p^{|l|}(2r^2/w^2(z))}\right\} \times \arctan\frac{z}{z_R}. \tag{9}$$

All previous values may be found from this, as well as the solely spin-dependent contribution which arises for higher-order modes. As any arbitrary light beam can be expressed in terms of a set of Laguerre–Gaussian modes, this result is of general applicability.

3. Angular momentum density

We now consider the angular momentum density, which is closely related to the azimuthal component of the Poynting vector. It follows from the linear momentum densities already given in Eq. (4), that the angular momentum density in the direction of propagation z derived from $\boldsymbol{j} = \boldsymbol{r} \times \boldsymbol{p}$ is

$$j_z = \varepsilon_0\left\{\omega l|u|^2 - \frac{1}{2}\omega\sigma r\frac{\partial|u|^2}{\partial r}\right\}. \tag{10}$$

When this quantity is integrated over the beam and divided by the energy, we find [3]

$$\frac{J_z}{W} = \frac{\int j_z r\,\mathrm{d}r\,\mathrm{d}\phi}{\int w_z r\,\mathrm{d}r\,\mathrm{d}\phi} = \frac{l\int|u|^2 r\,\mathrm{d}r\,\mathrm{d}\phi}{\omega\int|u|^2 r\,\mathrm{d}r\,\mathrm{d}\phi} - \frac{\sigma}{2\omega}\frac{\int r^2\frac{\partial|u|^2}{\partial r}\mathrm{d}\phi}{\int|u|^2 r\,\mathrm{d}r\,\mathrm{d}\phi}. \tag{11}$$

When the integration is over the whole cross-section of the beam, the total angular momentum J_z divided by total energy W reduces simply to

$$\frac{J_z}{W} = \frac{l}{\omega} + \frac{\sigma}{\omega}. \tag{12}$$

When the integration extends over only part of the beam, the first term is always constant, l/ω, while the second term depends on the intensity gradient.

If the energy is radiated in photons of energy $\hbar\omega$, then the angular momentum is transferred in units of $\hbar$. As Biedenharn and Louck [5], "It is in this way that Maxwell's equations, which do not contain Planck's constant, are nonetheless compatible with the quantum mechanics of photons. (The agreement fails for order $\hbar^2$.)" It is this simple relationship which has led to the quantity $l\hbar$ being designated as the orbital angular momentum [3]. There are, however, problems. A classical treatment is not straightforwardly in accord with the quantum theory. For example, the square of the angular momentum divided by the square of the energy for a classical electromagnetic multipole (J, J_z) turns out to be J_z^2/ω^2 and not, as might have been expected, $J(J+1)/\omega^2$. In fact, Morette-de Witt and Jensen [6] showed using quantised fields that the correct value is $\{N^2J_z^2 + N[J(J+1) - J_z^2]\}/N^2\omega^2$, where N is the number of quanta in the mode (J, J_z). For large N, the classical value J_z^2/ω^2 occurs, but at the single photon level, we get $J(J+1)/\omega^2$.

A more serious problem is one which has been tackled ever since Beth's famous experimental determination of $\hbar$ [7]. A circularly polarised plane wave with electric and magnetic fields only in the x- and y-directions has a linear momentum density only in the z-direction. When this is crossed with r to give the angular momentum density, there is no contribution in the z-direction. Thus, such a beam has no angular momentum to transfer to a waveplate, yet, Beth was able to make such a transfer – a paradox.

The explanation has a considerable literature (see list of references spanning 1936–54 given in the book by Jauch and Rohrlich [8]). Probably Simmons and Guttmann [9] give the most readily accessible discussion of the problem in their unusual, but fascinating, book. They do not solve the wave equation directly but their rejection of second-order derivatives when examining Maxwell's equations, makes their argument equivalent to the paraxial approximation. The explanation of the paradox is that the finite extent of the waveplate, or the finite extent of the field whichever is smaller, creates an aperturing that prevents the field being a plane wave as its extent is no longer infinite. It then follows that in the direction of propagation, there are electric and magnetic field components, which give azimuthal and radial components to the momentum density. When these are crossed with r, there is an angular momentum density in the z-direction. The essential part of the Simmons and Guttmann's argument is that the aperturing at the edge of the waveplate creates a large intensity gradient, and in this region, the angular momentum components are, therefore, very strong. The term involving the intensity gradient shown in Eq. (10) is to be found in the book in the integrands of the two unnumbered equations prior to Eq. (9.22). Remarkably, when averaged with the zero angular momentum over the area of the waveplate, the result is that the spin angular momentum divided by energy is σ/ω per photon, as expected.

The problem of Gaussian beams is, however, not quite the same. They already have a natural radial intensity gradient because they go to zero intensity at infinity. Such fields, irrespective of the presence of any waveplate, have field components in the direction of propagation. It is because these fields exist that there is a permanent spin angular momentum density, and the fundamental paradox associated with the idealised plane wave is absent. For Laguerre–Gaussian beams, the fields are more complex because of the helical wavefronts and their associated azimuthal phase term. Indeed, it is these more complex axial fields which specifically create the orbital angular momentum with which L–G modes are now associated. The key question is whether the edge effect of a waveplate, or some other aperture or obstacle, still plays as significant a role as that played in the case of a plane wave? The question is important because, as the spin angular momentum density depends on intensity gradient and because these fields do not have uniform intensity, a different amount of angular momentum might be expected to be transferred at different positions in the wavefront.

For the commonly occurring $p = 0$ modes, the expression for total angular momentum divided by total energy Eq. (11) is readily shown to be

$$\frac{J_z}{W} = \frac{l}{\omega} - \frac{\sigma}{\omega} \times \frac{\int_0^b \frac{1}{2} r^2 \left\{ \frac{2|l|}{r} - \frac{4r}{w^2(z)} \right\} r^{2|l|} \exp\left[-\frac{2r^2}{w^2(z)} \right] \mathrm{d}r}{\int_0^b r\, r^{2|l|} \exp\left[-\frac{2r^2}{w^2(z)} \right] \mathrm{d}r}. \tag{13}$$

The local value of the orbital angular momentum given by the first term is of little interest to this discussion, as it is a constant. We see that the second term although associated with spin also depends on the magnitude of l. This does not imply a coupling between spin and orbital angular momentum, it is merely a reflection of the fact that the spin contribution depends on the field intensity gradient and this depends on $|l|$. The integral is easily evaluated to show that when the beam is apertured at any radius b to give a rapid decline in intensity to zero, we find the result given in expression (12) in just the same way as for a plane wave. If the expression equivalent to Eq. (13) is derived for modes with a value of $p > 0$, then it may be readily shown that the parts of the wavefront that lie between any of the regions of zero intensity also integrate neatly to the same result; there is, of course, no plane wave equivalent.

What, however, remains of interest is the possibility that there is some regime where there might be a difference in the detailed behaviour of the spin angular momentum of Laguerre–Gaussian beams compared with plane waves. The intensity-gradient-dependent local value of spin is given by the second term of Eq. (10). If we write the local spin angular momentum per energy, we find

$$\frac{j_{z,\text{spin}}}{w} = -\frac{\sigma r}{2\omega}\frac{\left(\partial|u|^2/\partial r\right)}{|u|^2} \tag{14}$$

which for a $p = 0$ mode, gives a local spin angular momentum per photon of

$$j_{z,\text{spin}} = -\sigma\hbar\left(|l| - \frac{2r^2}{w^2(z)}\right). \tag{15}$$

Even though the beam has been macroscopically polarised with a unique value of σ, the z-component of spin per photon in a $(p = 0, l)$-mode clearly depends in this picture on the position across the beam; in some places, it is negative and in others it is positive, and its absolute value is unlimited. It is easy to see that there is a spread of values of $j_{z,\text{spin}}$ and that it is not limited to a maximum value of unity. The spin angular momentum goes from $-|l|\sigma$ near to the beam axis, to zero at the peak intensity and thereafter becomes positive and arbitrarily high. Nevertheless, on average, the spin per photon remains $\sigma\hbar$.

Is it really possible that matter can respond to the local value of angular momentum, which may not only be different in magnitude but have a different sign from that expected by the state of polarisation of the beam? Clearly any attempt to separate the different regions by means of an aperture is doomed to failure, because the rapid intensity variation of the edge effect will again produce the expected, "correct" result. It would seem that an experiment involving atoms might be more promising. If there were still an equivalent to the edge effect, the dipole approximation would seem to have no meaning, and yet, it is the basic assumption of virtually all theoretical approaches to atom–field interactions.

Another concern is that it appears that a net spin component per photon greater than unity is possible; it is hard to see what this really means. Nevertheless, in the Simmons and Guttmann exposition for a plane wave at the edge of the waveplate, the same thing is true, otherwise the net angular momentum per unit energy cannot average to unity, as in some areas the density is zero. But, as they write, "a classical quantity associated with the electromagnetic field does not necessarily indicate the value of that quantity which will be measured." There is considerable scope for the classical argument to fail; it could well be that there are significantly different observable effects at the single photon level. Nevertheless, the classical approach although simple, has been remarkably successful so far [2] in accounting for so many phenomena. It would be very exciting to find a regime where the experimental results manifest the sign and the varying value of the local angular momentum density in the way the above discussion suggests.

Acknowledgements

L. Allen is pleased to thank the Leverhulme Trust for the award of an Emeritus Fellowship. M.J. Padgett thanks the Royal Society for its support.

References

[1] J.D. Jackson, Classical Electrodynamics, Wiley, New York, 1962.
[2] L. Allen, M.J. Padgett, M. Babiker, in: E. Wolf (Ed.), Progress in Optics XXXIX, Elsevier, Amsterdam, 1999, p. 291.
[3] L. Allen, M.W. Beijersbergen, R.J.C. Spreeuw, J.P. Woerdman, Phys. Rev. A 45 (1992) 8185.
[4] M.J. Padgett, L. Allen, Opt. Commun. 121 (1995) 36.
[5] L.C. Biedenharn, J.D. Louck, Angular Momentum in Quantum Physics, Addison-Wesley, New York, 1980.
[6] C. Morette-de Witt, J.H.D. Jensen, Z. Naturforsch 8a (1953) 267.
[7] R.A. Beth, Phys. Rev. 50 (1936) 115.
[8] J.M. Jauch, F. Rohrlich, The Theory of Photons and Electrons, Addison-Wesley, New York, 1955.
[9] J.W. Simmons, M.J. Guttmann, States, Waves and Photons, Addison-Wesley, Reading, MA, 1970.

PAPER 2.7

Intrinsic and Extrinsic Nature of the Orbital Angular Momentum of a Light Beam

A. T. O'Neil, I. MacVicar, L. Allen, and M. J. Padgett
Department of Physics and Astronomy, University of Glasgow, Glasgow, G12 8QQ, Scotland
(Received 28 June 2001; published 16 January 2002)

We explain that, unlike the spin angular momentum of a light beam which is always intrinsic, the orbital angular momentum may be either extrinsic or intrinsic. Numerical calculations of both spin and orbital angular momentum are confirmed by means of experiments with particles trapped off axis in optical tweezers, where the size of the particle means it interacts with only a fraction of the beam profile. Orbital angular momentum is intrinsic only when the interaction with matter is about an axis where there is no net transverse momentum.

DOI: 10.1103/PhysRevLett.88.053601 PACS numbers: 42.50.Ct

Introduction.—Some 65 years ago Beth [1] demonstrated that circularly polarized light could exert a torque upon a birefringent wave plate suspended in the beam by the transfer of angular momentum. The angular momentum associated with circular polarization arises from the spin of individual photons and is termed spin angular momentum.

More recently, Allen *et al.* [2] showed that for beams with helical phase fronts, characterized by an $\exp(il\phi)$ azimuthal phase dependence, the orbital angular momentum in the propagation direction has the discrete value of $l\hbar$ per photon. Such beams have a phase dislocation on the beam axis that in related literature is sometimes referred to as an optical vortex [3]. In general, any beam with inclined phase fronts carries orbital angular momentum about the beam axis which, when integrated over the beam, can be an integer or noninteger [4,5] multiple of $\hbar$.

In this paper, we experimentally examine the motion of particles trapped off axis in an optical tweezers and are able to associate specific aspects of the motion with the distinct contributions of spin and orbital angular momentum of the light beam. The interpretation of the experiments, when combined with a numerical calculation of the spin and orbital contributions derived from established theory, allows a distinction to be made between the intrinsic and extrinsic aspects of the angular momentum of light.

Angular momentum of a light beam.—The cycle-averaged linear momentum density, $\mathbf{p}$, and the angular momentum density, $\mathbf{j}$, of a light beam may be calculated from the electric, $\mathbf{E}$, and magnetic, $\mathbf{B}$, fields [6]:

$$\mathbf{p} = \varepsilon_0 \langle \mathbf{E} \times \mathbf{B} \rangle, \tag{1}$$

$$\mathbf{j} = \varepsilon_0 (\mathbf{r} \times \langle \mathbf{E} \times \mathbf{B} \rangle) = \mathbf{r} \times \mathbf{p}. \tag{2}$$

Equation (2) encompasses both the spin and orbital angular momentum density of a light beam.

Within the paraxial approximation, the local value of the linear momentum density of a light beam is given by [2]

$$\mathbf{p} = i\omega \frac{\varepsilon_0}{2}(u^* \nabla u - u \nabla u^*) + \omega k \varepsilon_0 |u|^2 \mathbf{z} + \omega\sigma \frac{\varepsilon_0}{2} \frac{\partial |u|^2}{\partial r} \boldsymbol{\Phi}, \tag{3}$$

where $u \equiv u(r, \phi, z)$ is the complex scalar function describing the distribution of the field amplitude. Here σ describes the degree of polarization of the light; $\sigma = \pm 1$ for right- and left-hand circularly polarized light, respectively, and $\sigma = 0$ for linearly polarized.

The cross product of this momentum density with the radius vector $\mathbf{r} = (r, 0, z)$ yields an angular momentum density. The angular momentum density in the z direction depends upon the Φ component of $\mathbf{p}$, such that

$$j_z = r p_\phi. \tag{4}$$

The final term in Eq. (3) depends upon the polarization but is independent of the azimuthal phase and, consequently, this term may be linked directly to the spin angular momentum. The first term in Eq. (3) depends upon the phase gradient and not the polarization, and so gives rise to the orbital angular momentum.

For many mode functions, u, such as for circularly polarized Laguerre-Gaussian modes, Eqs. (3) and (4) can be evaluated analytically such that the local angular momentum density in the direction of propagation is given by [2]

$$j_z = \varepsilon_0 \left\{ \omega l |u|^2 - \frac{1}{2}\omega\sigma r \frac{\partial |u|^2}{\partial r} \right\}. \tag{5}$$

The angular momentum integrated over the beam is readily shown to be equivalent to $\sigma\hbar$ per photon for the spin and $l\hbar$ per photon for the orbital angular momentum [2], that is

$$J_z = (l + \sigma)\hbar. \tag{6}$$

A theoretical discussion of the behavior of local momentum densities has been published elsewhere [7], and it should be noted that the local spin and orbital angular momentum do not have the same functional form.

As is well known, spin angular momentum does not depend upon the choice of axis and so is said to be *intrinsic*. The angular momentum which arises for any light beam from the product of the z component of linear momentum about a radius vector, may be said to be an *extrinsic* because its value depends upon the choice of calculation axis.

Berry showed [8] that the orbital angular momentum of a light beam does not depend upon the lateral position of the axis and can therefore also be said to be intrinsic, provided the direction of the axis is chosen so that the transverse momentum is zero. When integrated over the whole beam the angular momentum in the z direction is

$$\mathbf{J}_z = \varepsilon_0 \iint dx\, dy\, \mathbf{r} \times \langle \mathbf{E} \times \mathbf{B} \rangle . \qquad (7)$$

If the axis is laterally displaced by $\mathbf{r}_0 \equiv (r_{0x}, r_{0y})$ it is easy to show that the change in the z component of angular momentum is given by

$$\begin{aligned} \Delta J_z &= (r_{0x} \times P_y) + (r_{0y} \times P_x) \\ &= r_{0x}\varepsilon_0 \iint dx\, dy\, \langle \mathbf{E} \times \mathbf{B} \rangle_y \qquad (8) \\ &\quad + r_{0y}\varepsilon_0 \iint dx\, dy\, \langle \mathbf{E} \times \mathbf{B} \rangle_x . \end{aligned}$$

The angular momentum is intrinsic only if ΔJ_z equals zero for all values of r_{0x} and r_{0y}. This condition is satisfied only if z is stipulated as the direction for which the transverse momenta $\varepsilon_0 \iint dx\, dy\, \langle \mathbf{E} \times \mathbf{B} \rangle_x$ and $\varepsilon_0 \iint dx\, dy\, \langle \mathbf{E} \times \mathbf{B} \rangle_y$ are exactly zero.

For Laguerre-Gaussian light beams truncated by apertures, Eqs. (3) and (8) can only be evaluated numerically. Nevertheless, for all apertures, of whatever size or position, the spin angular momentum remains $\sigma\hbar$ irrespective of the choice of calculation axis and so is, as expected, intrinsic; see Fig. 1. Any beam with a helical phase front apertured symmetrically about the beam axis has zero transverse momentum and, consequently, an orbital angular momentum of $l\hbar$ per photon, independent of the axis of calculation. The orbital angular momentum of the light beam may therefore be described as intrinsic. However, when the beam is passed through an off-axis aperture, its transverse momentum is nonzero and the orbital angular momentum depends upon the choice of calculation axis and so must be described as extrinsic; see Fig. 1. An interesting result occurs when the orbital angular momentum of the apertured beam is calculated about the original beam axis. Even though the transverse momentum is nonzero, the orbital angular momentum remains $l\hbar$ per photon because r_{0x} and r_{0y} are both zero. However, it does not follow that the angular momentum of the apertured beam is intrinsic as the result does depend upon the choice of calculation axis. When any beam is apertured off axis, it is simpler and more accurate to understand its interaction with particles by considering the components of $\mathbf{p}$ in the x-y plane. For beams with helical phase fronts, these transverse components are in the Φ direction with respect to the beam axis. It is this distinction between spin and orbital angular momentum which gives rise to differences in behavior for the interaction of light with matter.

The transfer of spin and orbital angular momentum to small particles.—The interaction of small particle with the angular momentum of a light beam has been investigated by a number of groups with the use of optical tweezers. Usually implemented by use of a high numerical-aperture microscope, optical tweezers rely on the gradient force to confine a dielectric particle near the point of highest light intensity [9]. For particles trapped on the beam axis, both the spin and orbital angular momentum have been shown to cause rotation of birefringent [10] and absorptive [11] particles, respectively. For absorbing particles, both spin and orbital angular momenta are transferred with the same efficiency so that the applied torque is proportional to the total angular momentum [12], that is $(\sigma + l)\hbar$ per photon.

In this present work we also use optical tweezers, but in this instance the particles are trapped away from the beam axis. This allows us to demonstrate the difference between particle interactions with spin and orbital angular momentum. The experimental configuration is shown in Fig. 2. Our optical tweezers are based on a 1.3 numerical aperture, $\times 100$ objective lens, configured with the trapping beam directed upwards, which allows easier access to the sample plane. This beam is generated from the 100 mW output of a commercial Nd:YLF laser transformed, using a computer generated hologram, to give a Laguerre-Gaussian mode of approximately 30 mW. The beam is circularly polarized, $\sigma = \pm 1$ with a high azimuthal mode index, $l = \pm 8$. The sign of the spin or the orbital angular momentum may be reversed by the insertion of a half-wave plate or a Dove prism, respectively [13]. The radius of maximum intensity, $r_{\max}$ of a Laguerre-Gaussian mode is given by [14],

$$r_{\max} = \sqrt{\frac{z_R l}{k}}, \qquad (9)$$

where z_R is the Rayleigh range of the beam. Even under the tight focusing associated with optical tweezers, the peak intensity ring of a Laguerre-Gaussian mode of high index l may be made several μm in diameter and, consequently, be much larger than the particles it is attempting to trap.

It is not surprising, for such conditions, that we observe the particles to be confined by the gradient force at the radius of maximum light intensity and not on the beam axis. When a birefringent particle such as a calcite fragment is trapped, and circularly polarized light is converted to linear, we observe that the particle spins about its own axis. The sense of rotation is governed by the handedness of the circular polarization.

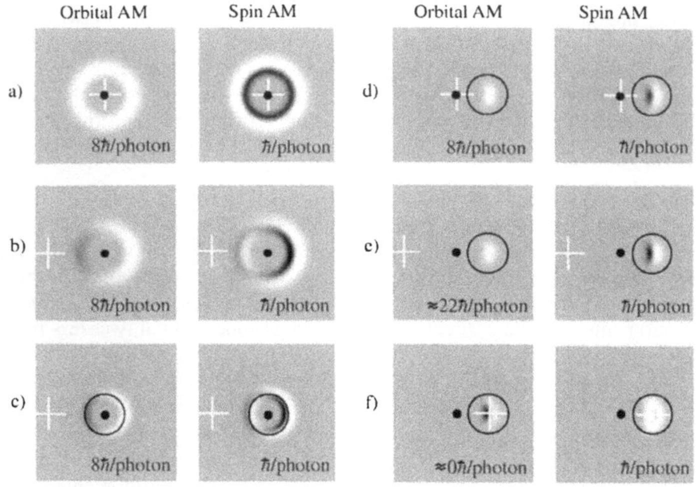

FIG. 1. Numerically calculated local spin and orbital angular momentum densities in the direction of propagation for a $l = 8$ and $\sigma = 1$ Laguerre-Gaussian mode. A positive contribution is shown in white, gray represents zero, and black a negative contribution; the black spot marks the axis of the original beam, the white cross marks the axis about which the angular momenta are calculated and, where appropriate, the black circle marks the position of a soft edged aperture. Note that the spin angular momentum is equivalent to $\sigma\hbar$ per photon irrespective of the choice of aperture or calculation axis, whereas the orbital angular momentum is only $l\hbar$ per photon if the aperture or calculation axes coincide with the axis of the original beam.

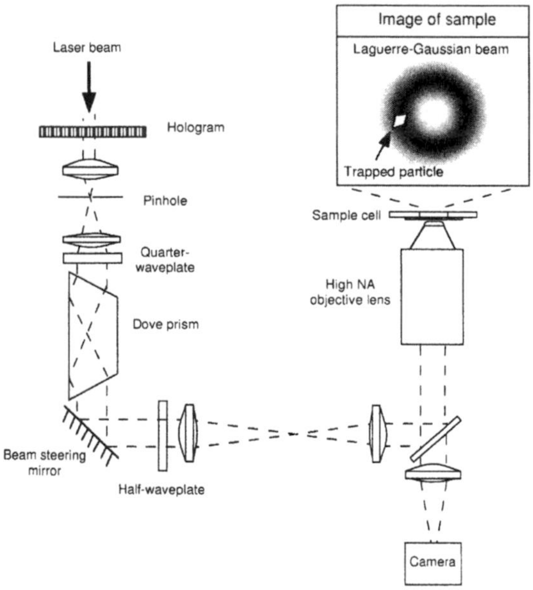

FIG. 2. The optical tweezers use a high-order Laguerre-Gaussian beam to trap the particle in the region of maximum light intensity, away from the beam axis. The circularly polarized Laguerre-Gaussian beam is generated using a quarter-wave plate and a hologram. The sense of the orbital and spin angular momentum can be reversed by the inclusion of a half-wave plate or a Dove prism, respectively.

For small particles the force arising from the light scattering, the momentum recoil force, becomes important. For a tightly focused Laguerre-Gaussian mode, the dominant component of the scattering force lies in the direction of beam propagation. The gradient force again constrains the particle to the annulus of maximum beam intensity. However, as the intensity distribution is cylindrically symmetric, the particle is not constrained azimuthally. Because the particle is trapped off the beam axis, the inclination of the helical phase fronts and the corresponding momentum result in a tangental force on the particle. We observe that a small particle, while still contained within the annular ring of light, orbits the beam axis in a direction determined by the handedness of the helical phase fronts; see Fig. 3. We conclude that the larger calcite and small particles are interacting with intrinsic spin and extrinsic orbital angular momentum, respectively. In principle, it should be possible to observe both the orbital and spin angular momenta acting simultaneously upon the same small birefringent particle. However, our observations have been inconclusive as birefringent particles small enough for the scattering force to induce a rotation about the beam axis are typically too small to see whether they are spinning about their own axis.

This orbital and spin behavior is entirely consistent with the formulation summarized in Eqs. (7) and (8). If one considers the cross section of an off-axis trapped particle to play the rôle of an aperture, then we see that the intrinsic spin of the angular momentum creates a torque about the

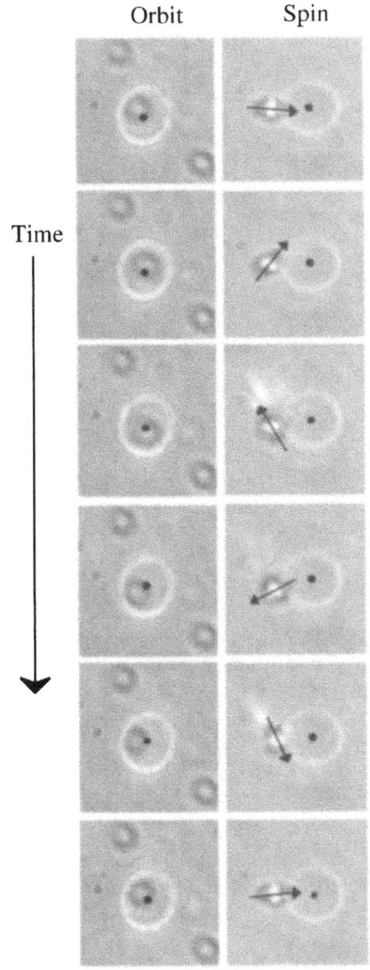

FIG. 3. Successive video frames showing particles trapped near the focus of an $l = 8$ and $\sigma = 1$ Laguerre-Gaussian mode. The left column shows particles of ≈ 1 μm diam. These particles are sufficiently small to be subject to a well-defined scattering force, allowing them to interact with the orbital angular momentum of the beam. They are set in motion, orbiting the beam axis at a frequency of ≈ 1 Hz. The right column shows a calcite fragment with a length of ≈ 3 μm and a width of about ≈ 1.5 μm, which is large enough not to interact detectably with the beam's orbital angular momentum. However, due to its birefringence it interacts with the spin angular momentum of the beam and is set spinning about its own axis at ≈ 0.3 Hz.

particle's own axis, causing it to spin. A calculation of the particle's angular momentum about an arbitrary axis shows a clear distinction between the intrinsic angular momentum associated with its spinning motion and the extrinsic angular momentum associated with its orbital motion. In this situation, orbital angular momentum is better described as the result of a linear momentum component directed at a tangent to the radius vector.

Unlike the spin angular momentum of a light beam which is always intrinsic, the z component of the orbital angular momentum can be described as intrinsic only if the z direction can be stipulated such that the transverse momentum integrated over the whole beam is zero. If an interaction is with only a fraction of the beam cross section, then the orbital angular as measured about the original axis is extrinsic.

L. Allen is pleased to thank the Leverhulme Trust for its support. M. J. Padgett thanks the Royal Society for its support. A. T. O'Neil is supported by the EPSRC.

[1] R. E. Beth, Phys. Rev. **50**, 115 (1936).
[2] L. Allen, M. W. Beijersbergen, R. J. C. Spreeuw, and J. P. Woerdman, Phys. Rev. A **45**, 8185 (1992).
[3] I. V. Basistiy, M. S. Soskin, and M. V. Vasnetsov, Opt. Commun. **119**, 604 (1995).
[4] M. Soljačić and M. Segev, Phys. Rev. Lett. **86**, 420 (2001).
[5] J. Courtial *et al.*, Opt. Commun. **144**, 210 (1997).
[6] J. D. Jackson, *Classical Electrodynamics* (Wiley, New York, 1962).
[7] L. Allen and M. J. Padgett, Opt. Commun. **184**, 67 (2000).
[8] M. V. Berry, in *International Conference on Singular Optics,* edited by M. S. Soskin, SPIE Proceedings Vol. 3487 (SPIE–International Society for Optical Engineering, Bellingham, WA, 1998), p. 6.
[9] A. Ashkin, J. M. Dziedzic, J. E. Bjorkholm, and S. Chu, Opt. Lett. **11**, 288 (1986).
[10] M. E. J. Friese, T. A. Nieminen, N. R. Heckenberg, and H. Rubinsztein-Dunlop, Nature (London) **394**, 348 (1998).
[11] H. He, M. E. J. Friese, N. R. Heckenberg, and H. Rubinsztein-Dunlop, Phys. Rev. Lett. **75**, 826 (1995).
[12] N. B. Simpson, K. Dholakia, L. Allen, and M. J. Padgett, Opt. Lett. **22**, 52 (1997).
[13] M. J. Padgett and J. Courtial, Opt. Lett. **24**, 430 (1999).
[14] M. J. Padgett and L. Allen, Opt. Commun. **121**, 36 (1995).

PAPER 2.8

Spin and Orbital Angular Momentum of Photons.

S. J. VAN ENK(*) and G. NIENHUIS

Huygens Laboratorium, Rijksuniversiteit Leiden
Postbus 9504, 2300 RA Leiden, The Netherlands

(received 1 November 1993; accepted 4 January 1994)

PACS. 32.80 – Photon interactions with atoms.
PACS. 42.50 – Quantum optics.
PACS. 03.50 – Classical field theory.

Abstract. – We consider the separation of the total angular momentum of the electromagnetic field into a «spin» and an «orbital» part. Though this separation is normally considered to be unphysical and not observable, we argue that both members in the separation are separately measurable quantities. However, the commutation relations for the associated quantum operators reveal that neither of them is an angular-momentum operator.

The angular momentum $\boldsymbol{J}$ of the classical electromagnetic field is defined by [1]

$$\boldsymbol{J} = \varepsilon_0 \int \mathrm{d}\boldsymbol{r}\, \boldsymbol{r} \times [\boldsymbol{E} \times \boldsymbol{B}]. \tag{1}$$

It is a conserved quantity for the free field as a result of the invariance of the free Maxwell equations under arbitrary rotations. Two contributions to (1) may be distinguished, originating from the longitudinal and the transverse part of the electric field $\boldsymbol{E}$ [2], respectively. The former gives the angular momentum of the instantaneous Coulomb field of the charges, the latter is the contribution from the radiation field, and is denoted by $\boldsymbol{J}_{\mathrm{rad}}$. It is well known that this $\boldsymbol{J}_{\mathrm{rad}}$ can be separated into two parts, called the «orbital» and the «spin» part [2, 3]. One finds

$$\boldsymbol{J}_{\mathrm{rad}} = \boldsymbol{L}_{\mathrm{rad}} + \boldsymbol{S}_{\mathrm{rad}}, \tag{2}$$

with

$$\boldsymbol{S}_{\mathrm{rad}} = \varepsilon_0 \int \mathrm{d}\boldsymbol{r}\, \boldsymbol{E}_\perp \times \boldsymbol{A}_\perp, \qquad \boldsymbol{L}_{\mathrm{rad}} = \varepsilon_0 \sum_l \int \mathrm{d}\boldsymbol{r}\, E_l^\perp (\boldsymbol{r} \times \nabla) A_l^\perp, \tag{3}$$

(*) Present address: Max-Planck Institut für Quantenoptik, Ludwig-Prandtl-Strasse 10, 85748 Garching, Germany.

where the symbol $\perp$ denotes the transverse part. Since these quantities are defined in terms of the transverse part of the vector potential $\boldsymbol{A}$, they are gauge invariant. The spin part $\boldsymbol{S}_{\text{rad}}$ is also called the intrinsic part, since it is independent of the definition of the origin of the coordinate system. In this sense this separation is similar to the separation for the total angular momentum $\boldsymbol{J}$ of a system of matter particles in external and internal parts.

It may be shown that for a free field $\boldsymbol{L}_{\text{rad}}$ and $\boldsymbol{S}_{\text{rad}}$ are conserved. Only in interaction with matter do they change in time [4]. They can be measured by detecting the angular-momentum effect of the radiation field interacting with matter. Indeed, already in 1936 Beth had observed one component of the spin angular momentum of light [5], by measuring the torque on a birefringent plate exerted by a circularly polarized beam of light. (See also the discussion in [6].) Moreover, it has recently been shown that also the orbital angular momentum of a paraxial laser beam can be measured [7-9]. In both cases the measurement concerns the component along the propagation direction. Both experiments can be understood within a classical framework.

However, it has generally been assumed that the separation (2) is unphysical [2, 3, 10]. In the first place, the spin part should correspond to the total angular momentum of a particle in its rest frame, but this frame does not exist for a photon. This argument does not, of course, prevent one from assigning a definite value to the spin of a photon, but it does preclude the definition of a spin vector. In fact, for a massless spin-s particle the argument implies that one can only define the component of the spin along the propagation direction, which assumes the values $\pm s$. Secondly, the action of the quantum-mechanical operators for spin and orbital angular momentum on a physical state of the radiation field is believed to yield unphysical states (we will elaborate this second argument below), so that $\boldsymbol{S}_{\text{rad}}$ and $\boldsymbol{L}_{\text{rad}}$ would not be observables in the quantum-mechanical sense. We show here that the latter objection is wrong, while the former does not apply to (2). The question remains whether in a quantum-mechanical theory orbital angular momentum and spin are still observables of the radiation field, and what physical meaning they have. In this letter we investigate the quantum-mechanical operators $\boldsymbol{S}_{\text{rad}}$ and $\boldsymbol{L}_{\text{rad}}$.

The transverse electromagnetic field can be quantized by first expanding the fields in any complete set of transverse vector functions. We assume that we have defined a complete orthonormal set of transverse mode functions $\boldsymbol{F}_\lambda$ with λ denoting the quantum numbers of the field, which we take discrete for simplicity. These functions are chosen as solutions of

$$\nabla^2 \boldsymbol{F}_\lambda = -k^2 \boldsymbol{F}_\lambda ; \qquad \nabla \cdot \boldsymbol{F}_\lambda = 0, \tag{4}$$

so that $\hbar\omega_\lambda \equiv \hbar kc$ is the energy of a photon in a mode λ. For our purpose the remaining (three) quantum numbers may be left unspecified. Orthonormality implies that

$$\langle \boldsymbol{F}_\lambda | \boldsymbol{F}_{\lambda'} \rangle \equiv \int \mathrm{d}\boldsymbol{r}\, \boldsymbol{F}_\lambda^* \cdot \boldsymbol{F}_{\lambda'} = \delta_{\lambda\lambda'} . \tag{5}$$

One can expand the transverse part of the vector potential $\boldsymbol{A}_\perp$ in this complete set. After quantization $\boldsymbol{A}_\perp$ and the electric and magnetic fields $\boldsymbol{E}_\perp$ and $\boldsymbol{B}$ become operators,

$$\begin{cases} \boldsymbol{A}_\perp = \sum_\lambda \mathcal{A}_\lambda [a_\lambda \boldsymbol{F}_\lambda + a_\lambda^\dagger \boldsymbol{F}_\lambda^*], \\ \boldsymbol{E}_\perp = \sum_\lambda i\mathcal{E}_\lambda [a_\lambda \boldsymbol{F}_\lambda - a_\lambda^\dagger \boldsymbol{F}_\lambda^*], \\ \boldsymbol{B} = \sum_\lambda \mathcal{A}_\lambda [a_\lambda \nabla \times \boldsymbol{F}_\lambda + a_\lambda^\dagger \nabla \times \boldsymbol{F}_\lambda^*], \end{cases} \tag{6}$$

where a_λ and $a_\lambda^\dagger$ are the annihilation and creation operators for photons in the mode λ, which satisfy the usual commutation rules

$$[a_\lambda, a^\dagger_{\lambda'}] = \delta_{\lambda\lambda'} \,. \tag{7}$$

The normalization factors in (6) are $\mathcal{E}_\lambda = \sqrt{\hbar\omega_\lambda/2\varepsilon_0}$ and $\mathcal{A}_\lambda = \sqrt{\hbar/2\varepsilon_0\omega_\lambda}$. The field operators (6) act, just as a_λ and $a^\dagger_\lambda$, in Fock space [2, 3]. This space is spanned by the Fock state vectors, which describe the quantum state of the radiation field.

The operator for the total angular momentum of the transverse electromagnetic field has the same form (1) as for the classical field. There is no need to symmetrize this expression, since $\boldsymbol{E}_\perp$ and $\boldsymbol{B}$, when evaluated at the same position and time, commute. In the same way as for the clasical field this operator can be decomposed into two terms, corresponding to «spin» and «orbital» angular momentum. By substituting the expansions (6), one finds

$$\begin{cases} \boldsymbol{S}_{\rm rad} = \varepsilon_0 \int {\rm d}\boldsymbol{r}\, \boldsymbol{E}_\perp \times \boldsymbol{A}_\perp = \dfrac{1}{2} \sum\limits_{\lambda\lambda'} [a^\dagger_\lambda a_{\lambda'} + a_{\lambda'} a^\dagger_\lambda] \langle \boldsymbol{F}_\lambda | \widehat{\boldsymbol{S}} | \boldsymbol{F}_{\lambda'} \rangle, \\ \boldsymbol{L}_{\rm rad} = \varepsilon_0 \sum\limits_l \int {\rm d}\boldsymbol{r}\, E^\perp_l (\boldsymbol{r} \times \nabla) A^\perp_l = \dfrac{1}{2} \sum\limits_{\lambda\lambda'} [a^\dagger_\lambda a_{\lambda'} + a_{\lambda'} a^\dagger_\lambda] \langle \boldsymbol{F}_\lambda | \widehat{\boldsymbol{L}} | \boldsymbol{F}_{\lambda'} \rangle. \end{cases} \tag{8}$$

Similar expressions have been derived before in reciprocal (Fourier) space [2, 3]. The operators $\widehat{\boldsymbol{L}}$ and $\widehat{\boldsymbol{S}}$ are defined as

$$\widehat{\boldsymbol{L}} = -i\hbar(\boldsymbol{r} \times \nabla), \qquad (\widehat{\boldsymbol{S}}_k)_{ij} = -i\hbar\varepsilon_{ijk}\,, \tag{9}$$

for $i, j, k = x, y, z$, with ε_{ijk} the Levi-Civitá pseudotensor. Notice that the operator $\widehat{\boldsymbol{L}}$ acts on the position dependence of $\boldsymbol{F}_\lambda$, whereas $\widehat{\boldsymbol{S}}$ operates as a vector of 3×3 matrices on the Cartesian components of $\boldsymbol{F}_\lambda$. These operators, which we distinguish from field operators by a caret, have the same form as the quantum-mechanical operators for orbital angular momentum and spin of a spin-1 particle, respectively. Expressions (8) for $\boldsymbol{S}_{\rm rad}$ and $\boldsymbol{L}_{\rm rad}$ thus closely resemble the expectation values for the quantum-mechanical spin and orbital angular momentum in a given quantum state of a spin-1 particle. Here the state of the radiation field is represented by the transverse mode function $\boldsymbol{F}_\lambda$ of the electromagnetic field. Now the operators $\widehat{\boldsymbol{S}}$ and $\widehat{\boldsymbol{L}}$ do not preserve this transversality. Hence, it appears that $\widehat{\boldsymbol{S}}|\boldsymbol{F}_\lambda\rangle$ and $\widehat{\boldsymbol{L}}|\boldsymbol{F}_\lambda\rangle$ no longer represent physical states of the radiation field. The conclusion would be that $\boldsymbol{S}_{\rm rad}$ and $\boldsymbol{L}_{\rm rad}$ cannot be observables [2, 3]. This is the second objection against the separation (2), as mentioned above. However, the operators $\boldsymbol{S}_{\rm rad}$ and $\widehat{\boldsymbol{S}}$ act in different spaces. The former is an operator that acts in Fock space, whereas the latter operator acts on the classical field modes, and not on quantum states of the radiation field. Since $\boldsymbol{S}_{\rm rad}$ is a gauge-invariant Hermitian operator acting on the quantum states of the radiation field, $\boldsymbol{S}_{\rm rad}$ is an observable. The same conclusion holds for the orbital angular momentum $\boldsymbol{L}_{\rm rad}$. Hence we conclude that the second argument that $\boldsymbol{S}_{\rm rad}$ and $\boldsymbol{L}_{\rm rad}$ are not observable is invalid.

The operators $\widehat{\boldsymbol{S}}$ and $\widehat{\boldsymbol{L}}$ satisfy the standard commutation rules for angular-momentum operators

$$\begin{cases} [\widehat{\boldsymbol{S}}_i, \widehat{\boldsymbol{S}}_j] = \sum\limits_k i\hbar\varepsilon_{ijk} \widehat{\boldsymbol{S}}_k\,, \\ [\widehat{\boldsymbol{L}}_i, \widehat{\boldsymbol{L}}_j] = \sum\limits_k i\hbar\varepsilon_{ijk} \widehat{\boldsymbol{L}}_k\,. \end{cases} \tag{10}$$

Since the total angular-momentum operator $\boldsymbol{J}_{\rm rad}$ generates rotations in space, as was explicitly verified in [11], also $\boldsymbol{J}_{\rm rad}$ obeys this same rule

$$[J_i, J_j] = \sum_k i\hbar\varepsilon_{ijk} J_k \,. \tag{11}$$

Relations (8) may suggest that the operators $\boldsymbol{S}_{\rm rad}$ and $\boldsymbol{L}_{\rm rad}$ obey the same commutation rules (10). This turn out not to be true. This is most easily shown when one chooses circularly polarized plane waves as the field modes, so that

$$\boldsymbol{F}_\lambda = \varepsilon_{\boldsymbol{k},s} \exp[i\boldsymbol{k}\cdot\boldsymbol{r}]/\sqrt{N}\,, \tag{12}$$

with $\boldsymbol{k}$ the wave vector and with $\varepsilon_{\boldsymbol{k},s}$ for $s = \pm 1$ the (orthogonal) right and left circular-polarization vectors. The number N is a normalization factor. In this representation the spin operator $\boldsymbol{S}_{\rm rad}$ takes the simple form [11]

$$\boldsymbol{S}_{\rm rad} = \sum_{\boldsymbol{k}} \frac{\hbar\boldsymbol{k}}{k} (a^\dagger_{\boldsymbol{k},1} a_{\boldsymbol{k},1} - a^\dagger_{\boldsymbol{k},-1} a_{\boldsymbol{k},-1})\,. \tag{13}$$

As this operator contains only number operators, all its components mutually commute, so that

$$[S_i, S_j] = 0\,. \tag{14}$$

This simple, but somewhat surprising result seems to have gone unnoticed before. It implies that $\boldsymbol{S}_{\rm rad}$ cannot generate rotations of the polarization of the field. Therefore, it cannot be considered as the spin angular-momentum operator. Hence the first objection to the separation (2) mentioned in the introduction does not apply. Furthermore, from the representation (13) of $\boldsymbol{S}_{\rm rad}$ it follows that for an arbitrary unit vector $\boldsymbol{u}$, the operator

$$R(\alpha) = \exp[-i\alpha\boldsymbol{u}\cdot\boldsymbol{S}_{\rm rad}/\hbar] \tag{15}$$

rotates the polarization of each $\boldsymbol{k}$-component of the field around its wave vector $\boldsymbol{k}$, over an angle $\alpha \cos(\boldsymbol{u}\cdot\boldsymbol{k}/k)$. Hence the effective rotation vector is the projection of $\boldsymbol{u}$ along the wave vector $\boldsymbol{k}$. This rotation of the polarization preserves the transversality of the field.

In order to find the commutation rules for $\boldsymbol{L}_{\rm rad}$, we first note that both $\boldsymbol{S}_{\rm rad}$ and $\boldsymbol{L}_{\rm rad}$ transform as vectors under rotations, so that

$$[J_i, L_j] = \sum_k i\hbar\varepsilon_{ijk} L_k\,, \qquad [J_i, S_j] = \sum_k i\hbar\varepsilon_{ijk} S_k\,. \tag{16}$$

This can be derived from the general expressions (3). From the commutation relations (14) and (16), it follows immediately that the orbital angular-momentum operator obeys

$$[L_i, L_j] = \sum_k i\hbar\varepsilon_{ijk}(L_k - S_k)\,, \qquad [L_i, S_j] = \sum_k i\hbar\varepsilon_{ijk} S_k\,. \tag{17}$$

This shows explicitly that also $\boldsymbol{L}_{\rm rad}$ is not an angular-momentum operator. It does not generate orbital rotations. Furthermore, $\boldsymbol{L}_{\rm rad}$ and $\boldsymbol{S}_{\rm rad}$ do not commute with each other, unlike $\hat{\boldsymbol{L}}$ and $\hat{\boldsymbol{S}}$, and unlike the corresponding operators for internal and external angular momentum of matter particles.

Finally, the three commuting components of $\boldsymbol{S}_{\rm rad}$ can be simultaneously diagonalized. Indeed, the form (13) of the spin operator makes it clear that in a circularly polarized plane-wave mode the spin vector has a well-defined value $\pm\hbar\boldsymbol{k}/k$. Since this very mode has also a well-defined linear momentum $\hbar\boldsymbol{k}$, $\boldsymbol{S}_{\rm rad}$ commutes with the linear momentum operator $\boldsymbol{P}_{\rm rad}$ of the radiation field, which is in the same representation as (13) given by

$$\boldsymbol{P}_{\text{rad}} = \sum_{\boldsymbol{k}} \hbar \boldsymbol{k} (a^{\dagger}_{\boldsymbol{k},\,1} a_{\boldsymbol{k},\,1} + a^{\dagger}_{\boldsymbol{k},\,-1} a_{\boldsymbol{k},\,-1}) . \tag{18}$$

In conclusion, we have shown that both «spin» $\boldsymbol{S}_{\text{rad}}$ and «orbital» angular momentum $\boldsymbol{L}_{\text{rad}}$ of a photon are well defined and separately measurable. However, these two quantities do not have the meaning of angular momentum. The commutation relations show that neither $\boldsymbol{S}_{\text{rad}}$ nor $\boldsymbol{L}_{\text{rad}}$ generate rotations. Instead, $\boldsymbol{S}_{\text{rad}}$ transforms the polarization of the field such that the transversality condition is preserved. In order to further investigate the physical significance of $\boldsymbol{L}_{\text{rad}}$ and $\boldsymbol{S}_{\text{rad}}$, one should consider the interaction of matter with the radiation field, in particular with a photon in an eigenstate of a given component of $\boldsymbol{L}_{\text{rad}}$ and $\boldsymbol{S}_{\text{rad}}$. This will be discussed elsewhere [4].

* * *

The authors are grateful to M. KRISTENSEN for helpful discussions. This work is part of the research program of the Stichting voor Fundamental Onderzoek der Materie (FOM) which is financially supported by the Nederlandse Organisatie voor Wetenschappelijk Onderzoek (NWO).

REFERENCES

[1] JACKSON J. D., *Classical Electrodynamics* (Wiley, New York, N.Y.) 1975.
[2] COHEN-TANNOUDJI C., DUPONT-ROC J. and GRYNBERG G., *Photons and Atoms* (Wiley, New York, N.Y.) 1989.
[3] SIMMONS J. W. and GUTTMANN M. J., *States, Waves, and Photons* (Addison-Wesley, Reading, Mass.) 1970.
[4] VAN ENK S. J. and NIENHUIS G., to be published in *J. Mod. Opt.*
[5] BETH R. A., *Phys. Rev.*, **50** (1936) 115.
[6] SOKOLOV I. V., *Usp. Fiz. Nauk*, **161** (1991) 175 (*Sov. Phys. Usp.*, **34** (1991) 925).
[7] ALLEN L., BEIJERSBERGEN M. W., SPREEUW R. J. C. and WOERDMAN J. P., *Phys. Rev. A*, **45** (1992) 8185.
[8] VAN ENK S. J. and NIENHUIS G., *Opt. Commun.*, **94** (1992) 147.
[9] BEIJERSBERGEN M. W., ALLEN L., VAN DER VEEN H. E. L. O. and WOERDMAN J. P., *Opt. Commun.*, **96** (1993) 123.
[10] JAUCH J. M. and ROHRLICH F., *The Theory of Photons and Electrons* (Springer-Verlag, Berlin) 1976, and references quoted on p. 34.
[11] LENSTRA D. and MANDEL L., *Phys. Rev. A*, **26** (1982) 3428.

Orbital angular momentum and nonparaxial light beams

Stephen M. Barnett [a], L. Allen [b]

[a] *Department of Physics and Applied Physics, University of Strathclyde, Glasgow G4 0NG, UK*
[b] *Department of Physics, University of Essex, Wivenhoe Park, Colchester CO4 3SQ, UK*

Received 25 January 1994; revised manuscript received 26 April 1994

Abstract

The simple relationship between total angular momentum and energy and the seemingly natural separation of the angular momentum into spin and orbital components in the paraxial approximation, are investigated for a general nonparaxial form of monochromatic beam with near cylindrical symmetry.

1. Introduction

It has been argued [1] that the Laguerre–Gaussian modes [2] possess well-defined angular momentum that can be decomposed into an orbital component and a spin component associated with its polarisation. Such modes are not exact solutions of the full Maxwell theory but rather are solutions of the paraxial wave equation. The well-known analogy between paraxial theory and quantum mechanics [3] has lead to an eigenfunction description of the angular momentum of light [4] and a comparison between paraxial wave optics and quantum harmonic oscillators [5] which exploits the language of ladder operators. The fact that the Laguerre–Gaussian beams may be expressed in terms of, and experimentally produced from, the Hermite–Gaussian beams emitted from real lasers [6] permits experimental examination of the phenomenon.

Allen et al. [1] have noted that paraxial beams possessing cylindrical symmetry possess well defined total angular momentum per unit energy flux and that this angular momentum can be divided into orbital and spin components. It is possible that this is an artefact of the paraxial approximation. This paper seeks to address this question by examining the angular momentum properties of light without invoking the paraxial approximation. We obtain the general form for a beam propagating in the z-direction for which the x- and y-components of the electric field have azimuthal dependence $\exp(il\phi)$. Beyond the paraxial approximation the total angular momentum per unit energy may still be calculated but it is no longer straightforward to identify the orbital and spin components. We also obtain expressions for the energy and angular momentum per unit length and identify a contribution not present in the paraxial approximation. We illustrate our results with an example that reduces to the Laguerre–Gaussian modes in the paraxial limit.

2. The paraxial approximation and angular momentum

The paraxial wave-equation provides an adequate description of nearly all optical resonator and beam propagation problems that arise in the study of real lasers [2]. It may readily be derived from the scalar wave-equation, which for a monochromatic wave

$$\psi(x, y, z, t) = \chi(x, y, z) \exp(-i\omega t) \tag{2.1}$$

reduces to the Helmholtz equation for the space-dependence

$$(\nabla^2 + k^2)\chi(x, y, z) = 0 \,, \tag{2.2}$$

where $k = c/\omega$. We consider an optical beam propagating in the z-direction and introduce the form

$$\chi(x, y, z) = u(x, y, z) \exp(ikz) \,. \tag{2.3}$$

The z-dependence of the wave amplitude $u(x, y, z)$ is essentially due to diffractive effects and is, in general, slow compared with the transverse variation in the width of the beam. The mathematical expression of this slow variation lies in the inequality

$$\left|\frac{\partial^2 u}{\partial z^2}\right| \ll \left|k\frac{\partial u}{\partial z}\right|, \left|\frac{\partial^2 u}{\partial x^2}\right|, \left|\frac{\partial^2 u}{\partial y^2}\right| . \tag{2.4}$$

Dropping the second derivative of u with respect to z leads to the paraxial wave equation

$$i\frac{\partial u}{\partial z} = -\frac{1}{2k}\left(\frac{\partial^2}{\partial x^2} + \frac{\partial^2}{\partial y^2}\right) u \,. \tag{2.5}$$

Lax et al. [7] have pointed out that the paraxial approximation makes the inconsistent assumption that one can have a plane polarised electromagnetic wave whose electric vector depends on transverse distance. They have, however, established that this result is a consistent zero-order solution of Maxwell's equations and identified the higher order correction terms. Davis [8] showed that if the vector potential of the field is assumed to be plane-polarised, the non-vanishing component of the vector potential obeys a scalar wave-equation and produces the same low-order components of the transverse and longitudinal field of a Gaussian beam as derived that by Lax et al. [7]. Pattanayak and Agrawal [9] have also obtained corrections to the paraxial theory and note that: "In the nonparaxial case, polarization is no longer a global property and changes on propagation." The approach of Davis et al., developed by Haus [10], forms the basis of the orbital angular momentum calculation presented by Allen et al. [1].

Poynting [11] inferred by recourse to mechanical analogy that circularly polarised light should exert a torque on a unit area of a quarter wave-plate equal to $\lambda/2\pi$ times the light energy per unit volume, as the plate converts the polarisation of the light from circular to linear polarisation. It follows that the ratio of angular momentum flux to energy flux becomes σ_z/ω where ω is the frequency of the light and σ_z is ± 1 for left- or right-handed light. The torque on a suspended birefringent plate was measured by Beth [12].

The calculation of the resulting torque assumed illumination of the plate by a plane wave and, although the concept of orbital angular momentum of light is well-known, the experiment takes cognisance only of the effect of spin. Rose [13] has considered the role of orbital angular momentum in multipole radiation and connects energy and total angular momentum with the free electromagnetic field. It is common, particularly in nuclear physics, to associate orbital angular momentum with radiative decay but Biedenhahn and Louck [14] warn that: "multipole fields for a given value of the total angular momentum ... do not have ... a definite value of the orbital angular momentum. It is, indeed, not possible to separate the total angular momentum of the photon field into and 'orbital' and a 'spin' part (this would contradict gauge invariance); the best that can be done is to define the helicity operator $\boldsymbol{S}\cdot\hat{\boldsymbol{P}}$ which is an observable (Beth)."

Nevertheless, Allen et al. [1] show for Laguerre–Gaussian beams *within the paraxial approximation* that the z-component of the total angular momentum density per unit power is given by

$$M_z = \frac{l}{\omega}|u|^2 + \frac{\sigma_z \rho}{2\omega}\frac{\partial |u|^2}{\partial \rho} \tag{2.6}$$

and that the ratio of angular momentum flux to energy flux is

$$\frac{J_z}{cP_z}=\frac{(l+\sigma_z)}{\omega}, \tag{2.7}$$

where P_z is the z-component of the momentum and σ_z is ± 1 for circularly polarised light and 0 for linearly polarised light. Here $u(\rho, \phi, z)$ is the complex scalar function, expressed in cylindrical polar coordinates, describing the distribution of the field amplitude of a Laguerre–Gaussian beam and is given by

$$u_{pl}(\rho, \phi, z)=\frac{C}{\sqrt{1+z^2/z_R^2}}\left(\frac{\rho\sqrt{2}}{w(z)}\right)^l L_p^l\left(\frac{2\rho^2}{w^2(z)}\right)\exp\left(\frac{-\rho^2}{w^2(z)}\right)\exp\left(\frac{-\mathrm{i}k\rho^2 z}{2(z^2+z_R^2)}\right)$$
$$\times\exp(\pm \mathrm{i}l\phi)\exp[\mathrm{i}(2p+l+1)\tan^{-1}(z/z_R)]\,. \tag{2.8}$$

where z_R is the Rayleigh range, $w(z)$ is the radius of the beam, L_p^l is the associated Laguerre polynomial, C is a constant and the beam waist is situated at $z=0$. As may be seen, the z-component of both the angular momentum/energy ratio and the angular momentum density naturally devolve into orbital and spin components associated with the values l and σ_z, respectively. The quantity l arises from the azimuthal dependence in the term $\exp(\mathrm{i}l\phi)$ while σ_z, where $\sigma_z=\mathrm{i}(\alpha\beta^*-\beta\alpha^*)$, is readily identifiable with spin when the electric field in the x- and y-directions is proportional to $(\alpha\hat{x}+\beta\hat{y})$. It may be noted that the spin term in the equation for M_z depends on the gradient of the field intensity. A plane wave propagating in the z-direction has no component of the angular momentum density $\epsilon_0\boldsymbol{r}\times(\boldsymbol{E}\times\boldsymbol{B})$ in the z-direction. The correction terms for the field found by Davis [8] would lead to a small modification to expression (2.6) but their effect is not readily summed to all orders.

We conclude this section by examining some general properties of the electromagnetic field that will be useful in our discussion of angular momentum. We write the densities of linear and angular momentum for the field in terms of the electric ($\boldsymbol{E}$) and magnetic ($\boldsymbol{B}$) field as [15,16]

$$\boldsymbol{p}=\epsilon_0\boldsymbol{E}\times\boldsymbol{B},\quad \boldsymbol{j}=\epsilon_0\boldsymbol{r}\times(\boldsymbol{E}\times\boldsymbol{B})\,. \tag{2.9, 2.10}$$

The total linear and angular momenta are found by integrating these densities over all space

$$\boldsymbol{P}=\int \mathrm{d}^3r\,\boldsymbol{p},\quad \boldsymbol{J}=\int \mathrm{d}^3r\,\boldsymbol{j}\,. \tag{2.11, 2.12}$$

Specialising to monochromatic fields we introduce the complex notation

$$\boldsymbol{E}=(\boldsymbol{E}\,\mathrm{e}^{-\mathrm{i}\omega t}+\boldsymbol{E}^*\,\mathrm{e}^{\mathrm{i}\omega t})/2,\quad \boldsymbol{B}=(\boldsymbol{B}\,\mathrm{e}^{-\mathrm{i}\omega t}+\boldsymbol{B}^*\,\mathrm{e}^{\mathrm{i}\omega t})/2\,, \tag{2.13, 2.14}$$

where the asterix denotes complex conjugation. Following van Enk and Nienhuis [4] we eliminate the magnetic field using the Maxwell equation

$$\mathrm{i}\omega\boldsymbol{B}=\nabla\times\boldsymbol{E} \tag{2.15}$$

to write the field linear and angular momenta as

$$\boldsymbol{P}=\frac{\epsilon_0}{2\mathrm{i}\omega}\int \mathrm{d}^3r\,\boldsymbol{E}^*\times(\nabla\times\boldsymbol{E}),\quad \boldsymbol{J}=\frac{\epsilon_0}{2\mathrm{i}\omega}\int \mathrm{d}^3r\,\boldsymbol{r}\times[\boldsymbol{E}^*\times(\nabla\times\boldsymbol{E})]\,. \tag{2.16, 2.17}$$

On partial integration and using the transversality of $\boldsymbol{E}$ we find that for fields that vanish sufficiently quickly as $|\boldsymbol{r}|\to\infty$ we obtain the simpler expressions

$$\boldsymbol{P}=\frac{\epsilon_0}{2\mathrm{i}\omega}\int \mathrm{d}^3r\sum_{j=x,y,z}E_j^*\nabla E_j\,, \tag{2.18}$$

$$\boldsymbol{J}=\frac{\epsilon_0}{2\mathrm{i}\omega}\int \mathrm{d}^3r\sum_{j=x,y,z}E_j^*(\boldsymbol{r}\times\nabla)E_j+\frac{\epsilon_0}{2\mathrm{i}\omega}\int \mathrm{d}^3r\,\boldsymbol{E}^*\times\boldsymbol{E}\,. \tag{2.19}$$

It is tempting to associate the two terms making up the angular momentum with the orbital and spin components. Indeed, within the paraxial approximation these two terms give, for the Laguerre–Gaussian modes, contributions proportional to l and σ_z respectively. Yet as van Enk and Nienhuis note "the interpretation of the first and second terms on the right hand side of Eq. (2.19) as orbital angular momentum and spin is, although seemingly obvious, not without fundamental difficulties."

In addition to the linear and angular momenta we also require the energy associated with the field. The energy density for the free electromagnetic field is [15]

$$U=\tfrac{1}{2}(\epsilon_0 \boldsymbol{E}^2+\mu_0^{-1}\boldsymbol{B}^2)\,, \tag{2.20}$$

giving a total energy

$$U=\frac{1}{2}\int \mathrm{d}^3r\,(\epsilon_0 \boldsymbol{E}^2+\mu_0^{-1}\boldsymbol{B}^2)=\frac{\epsilon_0}{2}\int \mathrm{d}^3r\,\boldsymbol{E}^*\cdot\boldsymbol{E}\,. \tag{2.21}$$

Finally we will examine the linear momentum, angular momentum and energy per unit length along a beam propagating in the z-direction. These quantities are found by integration of the respective densities over the x–y plane perpendicular to the direction of propagation. However, these quantities are not, in general, time-independent and therefore we will perform a cycle average over an optical period $2\pi/\omega$ to give the cycle averaged energy, linear momentum, angular momentum and energy per unit length which we denote $\mathscr{P}$, $\mathscr{J}$ and $\mathscr{E}$, respectively.

$$\mathscr{P}=\iint \mathrm{d}x\,\mathrm{d}y\,\boldsymbol{E}^*\times(\nabla\times\boldsymbol{E})\,, \tag{2.22}$$

$$\mathscr{J}=\frac{\epsilon_0}{2\mathrm{i}\omega}\iint \mathrm{d}x\,\mathrm{d}y\,\boldsymbol{r}\times(\boldsymbol{E}^*\times(\nabla\times\boldsymbol{E}))\,, \tag{2.23}$$

$$\mathscr{E}=\frac{\epsilon_0}{2}\iint \mathrm{d}x\,\mathrm{d}y\,\boldsymbol{E}^*\cdot\boldsymbol{E}\,. \tag{2.24}$$

We emphasise that expressions (2.16) to (2.24) have a validity that does not rely on the paraxial approximation and they form the basis for our analysis of nonparaxial beams.

3. Nonparaxial beams and angular momentum

The principal aim of this paper is to determine the extent to which some of the remarkable properties associated with the angular momentum carried by Laguerre–Gaussian beams are associated with the paraxial approximation. In order to address this question we develop a description of light beams based on the full Maxwell theory. There is, of course, a very wide variety of possible nonparaxial beams, but the considerations of the preceding section lead us to focus on monochromatic beams for which the x- and y-components of the electric field are proportional to $\exp(\mathrm{i}l\phi)$. Accordingly we write the electric field as

$$\boldsymbol{E}=(\alpha\hat{\boldsymbol{x}}+\beta\hat{\boldsymbol{y}})\,E(z,\rho)\,\mathrm{e}^{\mathrm{i}l\phi}+\hat{\boldsymbol{z}}E_z\,, \tag{3.1}$$

where each component satisfies the Helmholtz equation and E_z has to be chosen to ensure the transversality of the electric field. We find that the general solution for $E(z,\rho)$ has the form

$$E(z,\rho)=\int_0^k \mathrm{d}\kappa\,E(\kappa)\,J_l(\kappa\rho)\,\exp(\mathrm{i}\sqrt{k^2-\kappa^2}\,z)\,. \tag{3.2}$$

where $J_l(\kappa\rho)$ is the Bessel function of order l. As we have noted, E_z must also satisfy the Helmholtz equation and must have the correct form to give $\nabla\cdot\boldsymbol{E}=0$. After lengthy but straight forward algebra we find that the electric field for a beam has the form

$$\boldsymbol{E}=\int_0^k \mathrm{d}\kappa\, E(\kappa)\, \mathrm{e}^{\mathrm{i}l\phi} \exp(\mathrm{i}\sqrt{k^2-\kappa^2}\,z)$$
$$\times\left((\alpha\hat{\boldsymbol{x}}+\beta\hat{\boldsymbol{y}})\, J_l(\kappa\rho)+\hat{\boldsymbol{z}}\frac{\kappa}{2\sqrt{k^2-\kappa^2}}\left[(\mathrm{i}\alpha-\beta)\,\mathrm{e}^{-\mathrm{i}\phi} J_{l-1}(\kappa\rho)-(\mathrm{i}\alpha+\beta)\,\mathrm{e}^{\mathrm{i}\phi} J_{l+1}(\kappa\rho)\right]\right). \tag{3.3}$$

A general transverse electric field of the form (3.1) includes an arbitrary superposition of z-polarised plane waves propagating in the x–y plane. The inclusion of such waves is not of interest here as they do not form a z-propagating beam. We note that the z-component of this field has a different dependence on the azimuthal angle to that which has been imposed on the x- and y-components. Our analysis of the properties of nonparaxial beams is based on the general form (3.3). We should note that this beam includes, as a special case, Durnin's non-diffracting Bessel beams [17].

The total energy, linear and angular momenta for the beam formally diverge. However, the corresponding quantities evaluated per unit length are better behaved. In particular, the energy per unit length is

$$\mathscr{E}=\frac{\epsilon_0}{2}\int_0^{2\pi}\mathrm{d}\phi\int_0^{\infty}\rho\,\mathrm{d}\rho\,\boldsymbol{E}^*\cdot\boldsymbol{E}=\pi\epsilon_0\int_0^k\mathrm{d}\kappa\int_0^k\mathrm{d}\kappa'\int_0^{\infty}\rho\,\mathrm{d}\rho\, E^*(\kappa')\,E(\kappa)\exp[\mathrm{i}(\sqrt{k^2-\kappa^2}-\sqrt{k^2-\kappa'^2})z]$$
$$\times\left(J_l(\kappa'\rho)J_l(\kappa\rho)+\frac{\kappa'\kappa}{4\sqrt{k^2-\kappa'^2}\sqrt{k^2-\kappa^2}}\left[|\mathrm{i}\alpha-\beta|^2 J_{l-1}(\kappa'\rho)J_{l-1}(\kappa\rho)+|\mathrm{i}\alpha+\beta|^2 J_{l+1}(\kappa'\rho)J_{l+1}(\kappa\rho)\right]\right). \tag{3.4}$$

Surprisingly, two of these integrals can be performed together using the Fourier–Bessel integral [18]

$$\int_p^q R\,\Phi(R)\,\mathrm{d}R\int_0^{\infty}\lambda\, J_n(\lambda R)\, J_n(\lambda r)\,\mathrm{d}\lambda=\Phi(r)\,, \tag{3.5}$$

for $p<r<q$. The resulting energy per unit length acquires the simple form

$$\mathscr{E}=\frac{\pi\epsilon_0}{2}\int_0^k\mathrm{d}\kappa\frac{|E(\kappa)|^2(2k^2-\kappa^2)}{\kappa(k^2-\kappa^2)}\,. \tag{3.6}$$

The physical requirement that this quantity be finite, places constraints on the possible form of $E(\kappa)$; in particular it must tend to zero sufficiently quickly as $\kappa\to 0$, k. A similar calculation using, in addition, the Bessel function identity

$$\frac{\mathrm{d}}{\mathrm{d}x}J_n(x)\pm\frac{n}{x}J_n(x)=\pm J_{n\mp 1}(x) \tag{3.7}$$

gives equally simple exact forms for the linear and angular momentum per unit length associated with the beam

$$\mathscr{P}_z=\frac{\pi\epsilon_0}{2\omega}\int_0^k\mathrm{d}\kappa\frac{|E(\kappa)|^2\,(2k^2-\kappa^2)}{\kappa\sqrt{k^2-\kappa^2}} \tag{3.8}$$

$$\mathscr{J}_z=[l+\mathrm{i}(\alpha\beta^*-\beta\alpha^*)]\frac{\pi\epsilon_0}{2\omega}\int_0^k\mathrm{d}\kappa\frac{|E(\kappa)|^2\,(2k^2-\kappa^2)}{\kappa(k^2-\kappa^2)}+\mathrm{i}(\alpha\beta^*-\beta\alpha^*)\frac{\pi\epsilon_0}{2\omega}\int_0^k\mathrm{d}\kappa\frac{|E(\kappa)|^2\,\kappa}{k^2-\kappa^2}\,. \tag{3.9}$$

We note that the simple relationships

$$\mathscr{E}=\mathscr{P}_z c, \quad \mathscr{J}_z/\mathscr{E}=(l+\sigma_z)/\omega, \tag{3.10, 3.11}$$

which hold within the paraxial approximation, do not necessarily have general validity. The paraxial limit of the expressions (3.6), (3.8) and (3.9) is recovered, as are the simple relationships between $\mathscr{E}$, $\mathscr{P}_z$ and $\mathscr{J}_z$, if the function $E(\kappa)$ is sharply peaked at small values of κ so that we can neglect terms of order $(\kappa/k)^2$ in the integrands.

The expression for the total angular momentum per unit length, $\mathscr{J}_z$, bears closer examination. The first integral in Eq. (3.9) is simply $(l+\sigma_z)\mathscr{E}/\omega$, while the second term which depends on σ_z, but not on l, is the correction to the simple paraxial result. Thus, for 'linearly polarised' light, with $\sigma_z=0$, the paraxial relation between $\mathscr{J}_z$ and $\mathscr{E}$ is *exact*. It is interesting to examine the origin of the components of $\mathscr{J}_z$. Recall from Eq. (2.23) that $\mathscr{J}_z$ of the angular momentum per unit length is obtained by integrating the z-component of $[\boldsymbol{r}\times\{\boldsymbol{E}^*\times(\nabla\times\boldsymbol{E})\}]$ over the x, y-plane. Application of vector calculus allows us to rewrite this expression in the form

$$[\boldsymbol{r}\times\{\boldsymbol{E}^*\times(\nabla\times\boldsymbol{E})\}]_z=\sum_{j=x,y,z}E_j^*\frac{\partial}{\partial\phi}E_j+(E_x^*E_y-E_y^*E_x)-\sum_{j=x,y,z}E_j^*\frac{\partial}{\partial x_j}(xE_y-yE_x), \tag{3.12}$$

where the last sum produces the final term in Eq. (3.9) on integration across the beam profile. It is tempting, following van Enk and Nienhuis [4] and the discussion in Sec. 2, to ascribe the orbital angular momentum and spin to the first and second terms in Eq. (3.12) respectively as is done in the paraxial limit. However, we find that these two contributions to the angular momentum per unit length are

$$\frac{\epsilon_0}{2i\omega}\int_0^{2\pi}d\phi\int_0^{\infty}\rho\,d\rho\sum_{j=x,y,z}E_j^*\frac{\partial}{\partial\phi}E_j=l\frac{\pi\epsilon_0}{2\omega}\int_0^k d\kappa\frac{|E(\kappa)|^2(2k^2-\kappa^2)}{\kappa(k^2-\kappa^2)}+\sigma_z\frac{\pi\epsilon_0}{2\omega}\int_0^k d\kappa\frac{|E(\kappa)|^2\kappa}{k^2-\kappa^2}. \tag{3.13}$$

$$\frac{\epsilon_0}{2i\omega}\int_0^{2\pi}d\phi\int_0^{\infty}\rho\,d\rho\,(E_x^*E_y-E_y^*E_x)=\sigma_z\frac{\pi\epsilon_0}{2\omega}\int_0^k d\kappa\frac{|E(\kappa)|^2 2}{\kappa}. \tag{3.14}$$

While the second of these terms depends on σ_z but not on l, the first depends on both. In the general nonparaxial case there is no simple separation into an l-dependent orbital and a σ_z-dependent spin component of angular momentum. It is worth noting, however, that the sum of these two contributions is precisely $(l+\sigma_z)\mathscr{E}/\omega$. The contribution to the z-component of the angular momentum per unit length of the final term in Eq. (3.12) is

$$-\frac{\epsilon_0}{2i\omega}\int_0^{2\pi}d\phi\int_0^{\infty}\rho\,d\rho\sum_{j=x,y,z}E_j^*\frac{\partial}{\partial x_j}(xE_y-yE_x)=\sigma_z\frac{\pi\epsilon_0}{2\omega}\int_0^k d\kappa\frac{|E(\kappa)|^2\kappa}{k^2-\kappa^2}. \tag{3.15}$$

To highlight the nonparaxial nature of this term we write it in a slightly different form. Reverting to Cartesian coordinates and performing the integrations containing derivatives with respect to x and y by parts we find

$$\begin{aligned}
&-\frac{\epsilon_0}{2i\omega}\int_{-\infty}^{\infty}dx\int_{-\infty}^{\infty}dy\sum_{j=x,y,z}E_j^*\frac{\partial}{\partial x_j}(xE_y-yE_x)\\
&=-\frac{\epsilon_0}{2i\omega}\int_{-\infty}^{\infty}dy\,[E_x^*(xE_y-yE_x)]_{x=-\infty}^{\infty}-\frac{\epsilon_0}{2i\omega}\int_{-\infty}^{\infty}dx\,[E_y^*(xE_y-yE_x)]_{y=-\infty}^{\infty}\\
&+\frac{\epsilon_0}{2i\omega}\int_{-\infty}^{\infty}dx\int_{-\infty}^{\infty}dy\,(xE_y-yE_x)\left(\frac{\partial}{\partial x}E_x^*+\frac{\partial}{\partial y}E_y^*\right)-\frac{\epsilon_0}{2i\omega}\int_{-\infty}^{\infty}dx\int_{-\infty}^{\infty}dy\,E_z^*\frac{\partial}{\partial z}(xE_y-yE_x).
\end{aligned} \tag{3.16}$$

The first two terms will be zero for physical fields (which fall off sufficiently quickly as $\rho \to \infty$). Writing the final term in Eq. (3.16) as

$$\frac{\epsilon_0}{2i\omega}\left(-\frac{\partial}{\partial z} \int_{-\infty}^{\infty} dx \int_{-\infty}^{\infty} dy\, E_z^* \,(xE_y - yE_x) + \int_{-\infty}^{\infty} dx \int_{-\infty}^{\infty} dy\, (xE_y - yE_x)\frac{\partial}{\partial z} E_z^* \right) \qquad (3.17)$$

and using the fact that $\nabla \cdot \boldsymbol{E}^* = 0$, the third term in Eq. (3.16) combines with Eq. (3.17) to give

$$-\frac{\epsilon_0}{2i\omega} \int_{-\infty}^{\infty} dx \int_{-\infty}^{\infty} dy \sum_{j=x,y,z} E_j^* \frac{\partial}{\partial x_j}(xE_y - yE_x) = -\frac{\epsilon_0}{2i\omega}\frac{\partial}{\partial z} \int_{-\infty}^{\infty} dx \int_{-\infty}^{\infty} dy\, E_z^* \,(xE_y - yE_x)\,. \qquad (3.18)$$

The nonparaxial nature of this contribution is now clear; examining the general field expression (3.3) we see that in the paraxial limit the z-component of the field will be smaller than the x- and y-components by $\sim \kappa/k$. In addition the effect of the derivative with respect to z is to introduce a factor of order κ/k. The net effect is that the contribution (3.18) is of the order of magnitude of terms that are neglected within the paraxial approximation.

In concluding this section we summarise our main results. The general form for a monochromatic field with (near) cylindrical symmetry (3.3) leads to simple expressions for the energy, linear momentum and angular momentum per unit length, given in Eqs. (3.6), (3.8) and (3.9), respectively. These have a validity which does not rely on the paraxial approximation, but the simple relationships between these quantities that holds in the paraxial limit is lost. Moreover, the seemingly natural separation of the angular momentum into σ_z-dependent spin and l-dependent orbital components apparent in the paraxial approximation is no longer possible.

4. A nonparaxial generalisation of the Laguerre–Gaussian mode

The considerations of the preceding section have lead to a general description of the properties of monochromatic beams with (near) cylindrical symmetry. These are the natural, nonparaxial, extension of the familiar Laguerre–Gaussian modes [1,2]. We wish to emphasise that a laboratory laser mode is only accurately described by a solution of the full Maxwell equations. Its representation as a Hermite- or Laguerre–Gaussian mode is only an approximation. It seems likely, therefore, that careful experiments on sufficiently well controlled modes should be capable of detecting departures from the behaviour associated with the approximate form. In order to demonstrate the utility of our general formulae we provide a specific example of a nonparaxial mode and describe some of its properties, paying particular attention to the corrections to the simple paraxial results and to the paraxial limit itself.

The general form of the field given in Eq. (3.3) depends on the function $E(\kappa)$. We have a great deal of freedom in how we choose this function but, as we have noted above, it must tend to zero sufficiently rapidly as $\kappa \to 0, k$ for the energy, momentum and angular momentum per unit length to be finite. As an example (which satisfies these requirements) we examine the field for which

$$E(\kappa) = u(l,p)\exp\left(-\frac{k\kappa^2 z_R}{2(k^2-\kappa^2)}\right)\left(\frac{\kappa^2}{k^2-\kappa^2}\right)^{(2p+l+1)/2}\left(\frac{k^2}{k^2-\kappa^2}\right)^{1/2}, \qquad (4.1)$$

where $u(l,p)$ is a constant for the beam and z_R is a characteristic length scale which, in the paraxial limit, we will associate with the Rayleigh range. This choice has been made to provide simple analytic expressions for the energy and angular momentum per unit length and to give the Laguerre–Gaussian modes directly upon taking the paraxial limit.

Inserting the example (4.1) into our general expressions for the energy and angular momentum per unit length and performing the integrations gives the simple results

$$\mathscr{E}=\frac{\pi\epsilon_0}{2}\,|u(l,p)|^2\,(2p+l)!(kz_{\rm R})^{-(2p+l+1)}\left(1+\frac{2p+l+1}{2kz_{\rm R}}\right) \tag{4.2}$$

$$\mathscr{J}_z=(l+\sigma_z)\,\frac{\mathscr{E}}{\omega}+\frac{\sigma_z\pi\epsilon_0}{2\omega}\,\frac{|u(l,p)|^2}{2}\,(2p+l+1)!(kz_{\rm R})^{-(2p+l+2)}\,. \tag{4.3}$$

Unfortunately, there appears to be no correspondingly simple expression for the linear momentum per unit length. The ratio of the angular momentum per unit length to the energy per unit length is

$$\frac{\mathscr{J}_z}{\mathscr{E}}=\frac{(l+\sigma_z)}{\omega}+\frac{\sigma_z}{\omega}\left(\frac{2kz_{\rm R}}{2p+l+1}+1\right)^{-1}. \tag{4.4}$$

In the paraxial limit, $z_{\rm R}$ is very much greater than the wavelength of the light. This implies that $kz_{\rm R}\gg 1$ and the paraxial result (3.11) is readily regained.

We can obtain an explicit expression for the paraxial limit of the mode with $E(\kappa)$ given by Eq. (4.1) if the exponential is sharply peaked at small values of κ. This corresponds to the real part of z_R being large which, in turn, implies a large beam waist relative to the wavelength. In practice this means dropping terms outside of the exponential that are of order κ^2/k^2 and extending the upper limit of the integral in the general expression for the field (3.3) to infinity. Note, however, that it is necessary to keep terms in the z-dependent exponential of order $\kappa^2 z/k$. The resulting expression for the paraxial approximation to the x-component of the electric field is

$$E_x=\alpha u(l,p)\,{\rm e}^{{\rm i}l\phi}\,{\rm e}^{{\rm i}kz}\int_0^\infty {\rm d}\kappa\,\exp\left(-\frac{\kappa^2}{2k}(z_{\rm R}+{\rm i}z)\right)\left(\frac{\kappa}{k}\right)^{2p+l+1}J_l(\kappa\rho)\,. \tag{4.5}$$

This integral may be evaluated [19] with the result

$$E_x=\frac{\alpha u(l,p)p!}{2}\,{\rm e}^{{\rm i}l\phi}\,{\rm e}^{{\rm i}kz}\,k\left(\frac{k\rho}{2}\right)^l\left(\frac{2}{k(z_{\rm R}+{\rm i}z)}\right)^{p+l+1}\exp\left(\frac{k\rho^2}{2(z_{\rm R}+{\rm i}z)}\right)L_p^l\left(\frac{k\rho^2}{2(z_{\rm R}+{\rm i}z)}\right), \tag{4.6}$$

which we recognise as the 'elegant' form of the Laguerre–Gaussian mode [2]. This expression for the field is indeed a solution of the paraxial wave equation (2.5) as may be verified by direct substitution.

5. Conclusion

We have developed a general nonparaxial form of a monochromatic beam based on the full Maxwell theory. Beams with as near to cylindrical symmetry as allowed by the requirement that the electric field be transverse have a simple form parametrised by a mode function $E(\kappa)$ and a polarisation. We have shown that modes of this large class have simple forms for the energy, linear momentum and total angular momentum per unit length. Appropriate choice of $E(\kappa)$ allows the choice of a form for the modes which gives the Laguerre–Gaussian modes directly upon taking the paraxial limit. This approach allows the systematic investigation of mode effects arising from either paraxial or the nonparaxial form. It has allowed the specific problem of whether the spin and orbit components of the associated angular momentum of the beam may be legitimately separated in an obvious and natural way to be addressed here. We have also shown that corrections to specific paraxial calculations may be readily made within this framework.

The modes treated in this paper do not exhaust the family of near cylindrically symmetric beams. It might be interesting to examine beams for which the form of the field given in Eq. (3.3) is generalised by allowing α and β to be functions of κ. Finally, we note that the formulation of the modes is such as to allow their ready quantisation in order to permit a detailed investigation of atom–mode interactions. Naturally, a quantum description of the full electromagnetic field requires the construction of a complete set of modes orthogonal to that introduced here.

Acknowledgements

SMB thanks the Royal Society of Edinburgh and the Scottish Office Education Department for the award of a Research Fellowship. LA wishes to thank the Leverhulme Trust for the award of a Research Fellowship.

References

[1] L. Allen, M.W. Beijersbergen, R.J.C. Spreeuw and J.P. Woerdman, Phys. Rev. A 45 (1992) 8185.
[2] A.E. Siegman, Lasers (University Science Books, Mill Valley, 1986).
[3] D. Marcuse, Light transmission optics (Van Nostrand, New York, 1972).
[4] S.J. van Enk and G. Nienhuis, Optics Comm. 94 (1992) 147.
[5] G. Nienhuis and L. Allen, Phys. Rev. A 48 (1993) 656.
[6] M.W. Beijersbergen, L. Allen, H.E.L.O. van der Veen and J.P. Woerdman, Optics Comm. 96 (1993) 123.
[7] M. Lax, W.H. Louisell and W.B. McKnight, Phys. Rev. A 11 (1975) 1365.
[8] L.W. Davis, Phys. Rev. A 19 (1979) 1177.
[9] D.N. Pattanayak and G.P. Agrawal, Phys. Rev. A 22 (1980) 1159.
[10] H.A. Haus, Waves and fields in optoelectronics (Prentice-Hall, Englewood Cliffs, NJ, 1984).
[11] J.H. Poynting, Proc. R. Soc. Lond., Ser. A 82 (1909) 115.
[12] R.A. Beth, Phys. Rev. 50 (1936) 115.
[13] M.E. Rose, Multipole fields (Wiley, New York, 1955).
[14] L.C. Biedenharn and J.D. Louck, Angular momentum in quantum physics: volume VIII of the encyclopaedia of mathematics and its applications (Addison-Wesley, New York, 1980).
[15] J.D. Jackson, Classical electrodynamics (Wiley, New York, 1962).
[16] C. Cohen-Tannoudji, J. Dupont-Roc and G. Grynberg, Photons and atoms (Wiley, New York, 1989).
[17] J. Durnin, J. Opt. Soc. Am. A 4 (1987) 651.
[18] A. Gray, G.B. Matthews and T.M. MacRobert, A treatise on Bessel functions (Macmillan, London, 1922).
[19] I.S. Gradshteyn and I.M. Ryzhik, Table of integrals, series and products (Academic, London, 1980).

PAPER 2.10

Optical angular-momentum flux*

Stephen M Barnett

Department of Physics and Applied Physics, University of Strathclyde, Glasgow G4 0NG, UK

Received 26 September 2001
Published 18 December 2001
Online at stacks.iop.org/JOptB/4/S7

Abstract
We introduce the angular-momentum flux as the natural description of the angular momentum carried by light. We present four main results:
(i) angular-momentum flux is the flow of angular momentum across a surface and, in conjunction with the more familiar angular-momentum density, expresses the conservation of angular momentum. (ii) The angular-momentum flux for a light beam about its axis (or propagation direction) can be separated into spin and orbital parts. This separation is gauge invariant and does not rely on the paraxial approximation.
(iii) Angular-momentum flux can describe the propagation of angular momentum in other geometries, but the identification of spin and orbital parts is then more problematic. We calculate the flux for a component of angular momentum that is perpendicular to the axis of a light beam and for the field associated with an electric dipole. (iv) The theory can be extended to quantum electrodynamics.

Keywords: Angular momentum, electromagnetic theory

1. Introduction

Angular momentum has played an important part in electromagnetism since the early writings of Maxwell [1]. Indeed, Maxwell's crowning achievement of identifying light as electromagnetic waves was originally derived in terms of the rotational properties of the luminiferous aether, notably its *torsion* modulus[1]. Maxwell also derived mechanical properties of light in the form of radiation pressure [3]. The mechanical property of momentum was identified by Poynting, who also associated angular momentum with circular polarization [4]. This phenomenon was later demonstrated experimentally [5].

The density of electromagnetic angular momentum is defined as the cross product of the radius vector from the axis of rotation and the Poynting vector [6]. This quantity has been calculated for the light emitted by a rotating dipole [7] and by a multipole [8]. The physical significance and form of the angular-momentum flux were identified by Humblet [9], who applied it to the study of optical angular momentum in the far field. It is also worth noting that the angular momentum of light made an early appearance in quantum theory [10].

Recent interest in optical angular momentum can be traced back to the seminal paper of Allen *et al* [11]. This work showed that the Laguerre–Gaussian modes, familiar from paraxial optics [12], carry a well-defined angular momentum. Moreover, this angular momentum can be decomposed into spin and orbital parts. The research activity stimulated by this paper is reviewed in [13] and a very readable introduction to optical angular momentum is given in [14].

Attempts to extend the ideas of optical angular momentum beyond the paraxial approximation seem to present a more complicated picture [15], in which there does not appear to be a natural or simple separation into spin and orbital parts. This is certainly plausible given the polarization (or spin) dependence of tightly focused, non-paraxial, light beams [16]. We shall show, however, that calculation of the correct quantity (the optical angular-momentum flux) does lead to well-defined fluxes for the spin and orbital angular momenta.

This paper presents the theory of the optical angular-momentum flux. We work with the Maxwell theory for the free electromagnetic field. No paraxial approximation is required and the theory should be considered to be exact within Maxwellian electromagnetism. We work throughout with the SI system of units in which Maxwell's equations for the free electromagnetic field have the form

* This paper is dedicated to Les Allen, physicist, artist and friend, who first introduced me to the beautiful subtleties of optical angular momentum.

[1] A modern account of this can be found in [2].

$$\nabla \cdot \boldsymbol{E} = 0$$
$$\nabla \cdot \boldsymbol{B} = 0$$
$$\nabla \times \boldsymbol{E} = -\frac{\partial \boldsymbol{B}}{\partial t} \quad (1)$$
$$\nabla \times \boldsymbol{B} = \frac{1}{c^2}\frac{\partial \boldsymbol{E}}{\partial t}.$$

Important quantities are reproduced in the Gaussian system of units in appendix A.

The significance of a flux for a physical quantity is most readily appreciated in terms of conservation laws. We illustrate this principle in section 2 by reference to the conservation laws for charge, and for electromagnetic energy and momentum. We derive, in section 3, the form of the electromagnetic angular-momentum flux by appealing to the conservation of angular momentum. The main part of the paper (section 4) is concerned with the angular-momentum flux for light beams and its separation into spin and orbital parts. We show, in sections 5 and 6, that the theory of optical angular momentum can also be applied to the field radiated by an electric dipole and can be extended into the quantum realm.

2. Electromagnetic conservation laws

The physical quantities energy, momentum, angular momentum and, to some extent, charge are important because they are conserved. This conservation can be expressed as a continuity equation relating a density and a flux density, or current, of the conserved quantity. The best known example is the conservation of charge as expressed by the continuity equation [6]

$$\frac{\partial \rho}{\partial t} + \nabla \cdot \boldsymbol{j} = 0, \quad (2)$$

where ρ is the charge density and $\boldsymbol{j}$ is the current density or 'charge flux density'. Integration of this equation over a volume of space and application of Gauss' theorem leads to

$$\frac{\partial}{\partial t}\int_V \rho\, \mathrm{d}V = -\int_S \boldsymbol{j} \cdot \mathrm{d}\boldsymbol{S}, \quad (3)$$

which connects the change in charge in the volume V to the flow of current through its surface S.

It is instructive to express the conservation of other quantities in the same way. The conservation of vector quantities such as momentum then requires a flux in the form of a two-index object. The representation of such quantities in vector notation can be awkward and so we shall work with the index notation commonly used to describe tensors. In this notation, a Roman letter index will take any of the values 1, 2 and 3 corresponding to the x, y and z directions. Repeated indices are summed over all three values so that equation (2) becomes

$$\frac{\partial \rho}{\partial t} + \frac{\partial}{\partial x_i} j_i = 0. \quad (4)$$

A more complete introduction to this notation can be found in [17].

The energy density for the free electromagnetic field is

$$W = \tfrac{1}{2}(\epsilon_0 E^2 + \mu_0^{-1} B^2). \quad (5)$$

It follows from Maxwell's equations (1) that the continuity equation expressing conservation of energy is

$$\frac{\partial W}{\partial t} + \frac{\partial}{\partial x_i}\left[\mu_0^{-1}(\boldsymbol{E} \times \boldsymbol{B})_i\right] = 0, \quad (6)$$

where $\mu_0^{-1}(\boldsymbol{E} \times \boldsymbol{B})_i$ is the energy flux density. We recognize that this quantity is also the momentum density, in the form of Poynting's vector, multiplied by the square of the speed of light. The conservation of momentum can be expressed as three continuity equations, one for each Cartesian component. The conservation of the i-component of momentum, $\epsilon_0(\boldsymbol{E} \times \boldsymbol{B})_i$, is associated with the continuity equation

$$\frac{\partial}{\partial t}\left[\epsilon_0(\boldsymbol{E} \times \boldsymbol{B})_i\right] + \frac{\partial}{\partial x_j} T_{ij} = 0. \quad (7)$$

Here T_{ij} is the momentum flux density

$$T_{ij} = \tfrac{1}{2}\delta_{ij}(\epsilon_0 E^2 + \mu_0^{-1} B^2) - \epsilon_0 E_i E_j - \mu_0^{-1} B_i B_j, \quad (8)$$

where δ_{ij} is the usual Kronecker delta. It is inevitable that the conservation of a vector quantity (like the momentum) should be associated with a two-index flux, just as the conservation of a scalar quantity (like the energy) is associated with a one-index (or vector) flux. The change in the energy in a given volume is associated with the flux of energy across its surface and the integral form of (6) expresses this:

$$\frac{\partial}{\partial t}\int_V W\, \mathrm{d}V = -\int_S \mu_0^{-1}(\boldsymbol{E} \times \boldsymbol{B})_i\, \mathrm{d}S_i. \quad (9)$$

Similarly, the conservation of the i-component of momentum can be expressed in integral form in terms of the flux of momentum across the surface bounding the volume of interest:

$$\frac{\partial}{\partial t}\int_V \epsilon_0(\boldsymbol{E} \times \boldsymbol{B})_i\, \mathrm{d}V = -\int_S T_{ij}\, \mathrm{d}S_j. \quad (10)$$

It is natural to express the conservation of angular momentum in the same way, in terms of an angular-momentum density and a flux density. We shall use the idea of angular-momentum conservation to obtain the correct flux density in the next section.

It will be helpful to be able to express cross-products without explicit use of vector notation. To this end we shall employ the permutation symbol ϵ_{ijk} [17]. This takes the value +1 if $ijk = 123, 312, 231$, the value -1 if $ijk = 321, 132, 213$ and is zero otherwise. With this notation, the i-component of the of the momentum density becomes $\epsilon_0(\boldsymbol{E} \times \boldsymbol{B})_i = \epsilon_0\epsilon_{ijk}E_j B_k$.

3. Electromagnetic angular-momentum flux

The angular-momentum density for the electromagnetic field is obtained, in analogy with mechanics, by forming the cross product of the position vector with the momentum density [6]. We can write the angular-momentum density as a vector in the form[2]

$$\boldsymbol{j} = \epsilon_0 \boldsymbol{r} \times (\boldsymbol{E} \times \boldsymbol{B}). \quad (11)$$

The total electromagnetic angular momentum associated with a volume V is obtained by integration of this quantity over the volume to give

$$\boldsymbol{J} = \int_V \epsilon_0 \boldsymbol{r} \times (\boldsymbol{E} \times \boldsymbol{B})\, \mathrm{d}V. \quad (12)$$

We shall find it convenient to work with the tensor notation in which the i-component of the angular-momentum density is

$$j_i = \epsilon_0 \epsilon_{ijk} x_j \epsilon_{klm} E_l B_m = \epsilon_0(E_i x_j B_j - B_i x_j E_j). \quad (13)$$

[2] This should not be confused with the $\boldsymbol{j}$ that was used to denote current density in section 2. The current makes no further appearances in this paper and henceforth $\boldsymbol{j}$ will always denote the angular-momentum density.

The angular momentum is a conserved quantity and hence we should be able to express its conservation by means of a continuity equation analogous to that for linear momentum (7). Hence, we seek an angular-momentum flux density, which we write as M_{li}, associated with the flux of the i-component of angular momentum through a surface oriented in the direction l. The conservation of the i-component of the electromagnetic angular momentum could then be expressed in the form

$$\frac{\partial}{\partial t} j_i + \frac{\partial}{\partial x_l} M_{li} = 0. \tag{14}$$

A natural quantity to propose for the angular-momentum flux density is the 'cross product' of the position with the momentum-flux density (8). This gives

$$M_{li} = \epsilon_{ijk} x_j \left[\tfrac{1}{2} \delta_{kl} (\epsilon_0 E^2 + \mu_0^{-1} B^2) - \epsilon_0 E_k E_l - \mu_0^{-1} B_k B_l \right]. \tag{15}$$

It is straightforward to show (using Maxwell's equations (1)) that this quantity does satisfy the continuity equation (14) and therefore is the required angular-momentum flux density. It is possible to derive this quantity rather more elegantly and directly by means of the relativistic theory. A brief demonstration of this is presented in appendix B.

It is worth noting that the dimensions of the angular-momentum flux density are angular momentum per unit area per unit time. This suggests that we can interpret it in terms of a *flow* of angular momentum through a surface. This idea is supported by the integral form of the continuity equation for the i-component of angular momentum:

$$\frac{\partial}{\partial t} J_i = \frac{\partial}{\partial t} \int_V j_i \, \mathrm{d}V = - \int_S M_{li} \, \mathrm{d}S_l, \tag{16}$$

where S is the closed surface bounding the volume V. More generally, we shall associate the quantity $\int_S M_{li} \, \mathrm{d}S_l$ with the flux of angular momentum through the surface S even if this surface is not closed.

It is often useful to separate angular momentum into internal and external or spin and orbital parts. The form of this separation for the electromagnetic field has been debated at length and remains unclear [18, 19]. There appear to be fundamental reasons why such a separation is not 'physically observable' [19], although it has been suggested that a physical separation is possible in which neither parts are true angular momenta [20, 21]. Nevertheless, at least for light beams within the paraxial approximation [12], the separation does seem to have a physical and experimentally demonstrable meaning [13, 22, 23]. We shall see in the following section, for the special case of the angular momentum of a light beam about its axis, that it is meaningful to identify spin and orbital parts of the angular momentum. The correct identification of these parts requires the use of the angular-momentum flux.

4. Light beams

The recent surge of interest in optical angular momentum has been driven by studies of suitably prepared laser beams [13]. Indeed the paper that kindled this interest dealt with the orbital angular momentum of Laguerre–Gaussian laser modes [11]. This paper and many subsequent ones considered the angular momentum of a monochromatic, cylindrically symmetric beam about its own axis. For the Laguerre–Gaussian beams, the azimuthal angular dependence of the complex electric field is $\exp(il\phi)$ and the angular momentum of the beam contains a contribution that is proportional to l. Allen *et al*, citing Marcuse [24], exploited the analogy between quantum mechanics and paraxial optics to identify this contribution with the orbital angular momentum of the light beam. This has become the accepted meaning of the expression 'orbital angular momentum' for a light beam. We should note, however, that the angular momentum of the beam has the same value about any axis that is parallel to the axis of the beam [25]. This means that we cannot separate our spin and orbital angular momenta on the basis of dependence on the axis (as is done with mechanical internal and external angular momenta).

Attempts to go beyond the paraxial approximation suggested that the separation of the angular momentum into spin and orbital contributions was an artefact of the paraxial approximation [15]. We shall see that this is not the case and that the problems encountered previously in working outside the paraxial approximation are removed by the correct use of the optical angular-momentum flux.

It is clear from our earlier discussion that the important quantity in discussing the angular momentum for light beams is the optical angular-momentum flux. This is the quantity of angular momentum crossing a surface in unit time. Hence the angular momentum transported by a light beam across a surface will be an optical angular-momentum flux. Earlier studies have concentrated on the angular momentum per unit power or per unit length [11, 15, 18].

4.1. *Angular-momentum flux for a beam about its axis*

The most studied example of optical angular momentum is that of a cylindrically symmetric (or near symmetric) light beam about its axis. Suppose that we have such a beam, the axis of which defines the z-axis. The angular-momentum flux density for this beam through a plane oriented in the z-direction is

$$\begin{aligned} M_{zz} &= M_{33} \\ &= y(\epsilon_0 E_x E_z + \mu_0^{-1} B_x B_z) - x(\epsilon_0 E_y E_z + \mu_0^{-1} B_y B_z). \end{aligned} \tag{17}$$

The total angular-momentum flux through a plane of constant z is $\iint M_{zz} \, \mathrm{d}x \, \mathrm{d}y$. Consider the effect of displacing the axis about which the angular-momentum flux is calculated to $x = -a$, $y = -b$. This changes the total angular-momentum flux density to

$$\begin{aligned} \iint M_{zz} \, \mathrm{d}x \, \mathrm{d}y \to \iint M_{zz} \, \mathrm{d}x \, \mathrm{d}y + a \iint T_{yz} \, \mathrm{d}x \, \mathrm{d}y \\ - b \iint T_{xz} \, \mathrm{d}x \, \mathrm{d}y. \end{aligned} \tag{18}$$

where T_{xz} and T_{yz} are the x- and y-components of the linear-momentum flux through the surface. The total angular-momentum flux is independent of the displacements a and b if $\iint T_{xz} \, \mathrm{d}x \, \mathrm{d}y = 0$ and $\iint T_{yz} \, \mathrm{d}x \, \mathrm{d}y = 0$. This means that all beams, for which the total linear momentum carried through the surface is perpendicular to that surface, will have an angular-momentum flux density that is the same about any axis that is parallel to the axis of the beam. This generalizes

Berry's result for angular momentum to angular-momentum flux [25].

In order to proceed, it is helpful to introduce complex electric and magnetic fields and to limit our attention to monochromatic beams with angular frequency ω. We express the real electric and magnetic field components E_i and B_j in terms of complex field amplitudes $\mathcal{E}_i$ and $\mathcal{B}_j$ as

$$
\begin{aligned}
E_i &= \mathrm{Re}\,[\mathcal{E}_i \exp(-\mathrm{i}\omega t)] \\
B_j &= \mathrm{Re}\,[\mathcal{B}_j \exp(-\mathrm{i}\omega t)].
\end{aligned}
\tag{19}
$$

These complex field amplitudes are related by Maxwell's equations (1) in the form

$$
\begin{aligned}
\mathcal{B}_j &= \frac{1}{\mathrm{i}\omega}\epsilon_{jkl}\frac{\partial}{\partial x_k}\mathcal{E}_l \\
\mathcal{E}_j &= -\frac{c^2}{\mathrm{i}\omega}\epsilon_{jkl}\frac{\partial}{\partial x_k}\mathcal{B}_l.
\end{aligned}
\tag{20}
$$

We are now in a position to derive expressions for the spin and orbital parts of the total angular-momentum flux. The manipulations are a little involved and the reader who is primarily interested in the form and consequences of these contributions may wish to bypass these details by going straight to equations (25) and (26). The cycle-averaged angular-momentum flux density obtained from (17) is

$$
\bar{M}_{zz} = \tfrac{1}{2}\mathrm{Re}\left[y(\epsilon_0\mathcal{E}_x\mathcal{E}_z^* + \mu_0^{-1}\mathcal{B}_x^*\mathcal{B}_z) - x(\epsilon_0\mathcal{E}_y\mathcal{E}_z^* + \mu_0^{-1}\mathcal{B}_y^*\mathcal{B}_z)\right].
\tag{21}
$$

It is instructive to use the Maxwell equations (20) to eliminate the z-components of the electric and magnetic fields to give

$$
\begin{aligned}
\bar{M}_{zz} &= \tfrac{1}{2}\mathrm{Re}\left[\epsilon_0(y\mathcal{E}_x - x\mathcal{E}_y)\frac{c^2}{\mathrm{i}\omega}\left(\frac{\partial}{\partial x}\mathcal{B}_y^* - \frac{\partial}{\partial y}\mathcal{B}_x^*\right)\right. \\
&\quad\left. + \mu_0^{-1}(y\mathcal{B}_x^* - x\mathcal{B}_y^*)\frac{1}{\mathrm{i}\omega}\left(\frac{\partial}{\partial x}\mathcal{E}_y - \frac{\partial}{\partial y}\mathcal{E}_x\right)\right] \\
&= \frac{\epsilon_0 c^2}{2\omega}\mathrm{Re}\left\{-\mathrm{i}\left[-y\frac{\partial}{\partial y}(\mathcal{E}_x\mathcal{B}_x^*) - x\frac{\partial}{\partial x}(\mathcal{E}_y\mathcal{B}_y^*)\right.\right. \\
&\quad\left.\left. + x\mathcal{E}_y\frac{\partial}{\partial y}\mathcal{B}_x^* + y\mathcal{B}_x^*\frac{\partial}{\partial x}\mathcal{E}_y + y\mathcal{E}_x\frac{\partial}{\partial x}\mathcal{B}_y^* + x\mathcal{B}_y^*\frac{\partial}{\partial y}\mathcal{E}_x\right]\right\}.
\end{aligned}
\tag{22}
$$

The total angular-momentum flux is obtained by integrating this quantity over the whole xy-plane:

$$
\begin{aligned}
\mathcal{M}_{zz} &= \iint \bar{M}_{zz}\,\mathrm{d}x\,\mathrm{d}y = \frac{\epsilon_0 c^2}{2\omega}\mathrm{Re}\left\{-\mathrm{i}\iint \mathrm{d}x\,\mathrm{d}y\right. \\
&\quad\left.\times\left[-y\frac{\partial}{\partial y}(\mathcal{E}_x\mathcal{B}_x^*) + x\mathcal{E}_y\frac{\partial}{\partial y}\mathcal{B}_x^* + x\mathcal{B}_y^*\frac{\partial}{\partial y}\mathcal{E}_x + x \leftrightarrow y\right]\right\} \\
&= \frac{\epsilon_0 c^2}{2\omega}\mathrm{Re}\left\{-\mathrm{i}\iint \mathrm{d}x\,\mathrm{d}y\left[\mathcal{E}_x\mathcal{B}_x^* - \mathcal{B}_x^* x\frac{\partial}{\partial y}\mathcal{E}_y\right.\right. \\
&\quad\left.\left. - \mathcal{E}_x x\frac{\partial}{\partial y}\mathcal{B}_y^* + x \leftrightarrow y\right]\right\},
\end{aligned}
\tag{23}
$$

where we have integrated by parts to obtain the second line and $x \leftrightarrow y$ denotes an exchange of all explicit occurrences of x and y. Each of the terms in the second line of (23) arises from the corresponding term in the first line. Hence we can combine the two expressions by taking the first term from the second line together with one half of the second and third terms from both lines and for the same combinations for the terms with x and y interchanged. This procedure gives the expression

$$
\begin{aligned}
\mathcal{M}_{zz} &= \frac{\epsilon_0 c^2}{2\omega}\mathrm{Re}\left\{-\mathrm{i}\iint \mathrm{d}x\,\mathrm{d}y\right. \\
&\quad\times\left[\mathcal{E}_x\mathcal{B}_x^* + \frac{1}{2}\mathcal{B}_x^*\left(y\frac{\partial}{\partial x} - x\frac{\partial}{\partial y}\right)\mathcal{E}_y\right. \\
&\quad\left.\left. + \tfrac{1}{2}\mathcal{E}_x\left(y\frac{\partial}{\partial x} - x\frac{\partial}{\partial y}\right)\mathcal{B}_y^* + x \leftrightarrow y\right]\right\} \\
&= \frac{\epsilon_0 c^2}{2\omega}\mathrm{Re}\left\{-\mathrm{i}\iint \rho\,\mathrm{d}\rho\,\mathrm{d}\phi\left[(\mathcal{E}_x\mathcal{B}_x^* + \mathcal{E}_y\mathcal{B}_y^*)\right.\right. \\
&\quad\left.\left. + \tfrac{1}{2}\left(-\mathcal{B}_x^*\frac{\partial}{\partial\phi}\mathcal{E}_y + \mathcal{E}_y\frac{\partial}{\partial\phi}\mathcal{B}_x^* - \mathcal{E}_x\frac{\partial}{\partial\phi}\mathcal{B}_y^* + \mathcal{B}_y^*\frac{\partial}{\partial\phi}\mathcal{E}_x\right)\right]\right\},
\end{aligned}
\tag{24}
$$

where we have introduced the cylindrical polar coordinates (ρ, ϕ, z).

It is tempting to express the total angular-momentum flux $\mathcal{M}_{zz}$ as the sum of two contributions

$$
\mathcal{M}_{zz}^{\mathrm{spin}} = \frac{\epsilon_0 c^2}{2\omega}\mathrm{Re}\left[-\mathrm{i}\iint \rho\,\mathrm{d}\rho\,\mathrm{d}\phi(\mathcal{E}_x\mathcal{B}_x^* + \mathcal{E}_y\mathcal{B}_y^*)\right]
\tag{25}
$$

$$
\begin{aligned}
\mathcal{M}_{zz}^{\mathrm{orbit}} &= \frac{\epsilon_0 c^2}{4\omega}\mathrm{Re}\left[-\mathrm{i}\iint \rho\,\mathrm{d}\rho\,\mathrm{d}\phi\right. \\
&\quad\left.\times\left(-\mathcal{B}_x^*\frac{\partial}{\partial\phi}\mathcal{E}_y + \mathcal{E}_y\frac{\partial}{\partial\phi}\mathcal{B}_x^* - \mathcal{E}_x\frac{\partial}{\partial\phi}\mathcal{B}_y^* + \mathcal{B}_y^*\frac{\partial}{\partial\phi}\mathcal{E}_x\right)\right]
\end{aligned}
\tag{26}
$$

and to associate these with the spin and orbital angular-momentum fluxes. These quantities contain only the electric and magnetic fields and hence each is manifestly gauge invariant. This separation is also physically meaningful as we now show. We start by considering paraxial light before turning to the more general non-paraxial case. We consider the effects of a birefringent element and of an element that imparts an azimuthal phase dependence on the beam. Such devices have been employed respectively to modify the polarization (by means of waveplates [26]) and the azimuthal phase dependence of light propagating through them (see [13] and references therein). A birefringent element has a fast and a slow axis for which the refractive indices are n_{f} and n_{s}. For simplicity we shall take these axes to be the x and y axes respectively. The effect on the fields of propagation through the birefringent element is to introduce the phase shifts

$$
\begin{aligned}
\mathcal{E}_x &\to \mathcal{E}_x \exp(\mathrm{i}n_{\mathrm{f}}kL) \qquad & \mathcal{B}_y &\to \mathcal{B}_y \exp(\mathrm{i}n_{\mathrm{f}}kL) \\
\mathcal{E}_y &\to \mathcal{E}_y \exp(\mathrm{i}n_{\mathrm{s}}kL) \qquad & \mathcal{B}_x &\to \mathcal{B}_x \exp(\mathrm{i}n_{\mathrm{s}}kL),
\end{aligned}
\tag{27}
$$

where k is the wavenumber for the beam and L is the propagation length. This transformation changes the spin part of the total angular-momentum flux (25) to

$$
\begin{aligned}
\mathcal{M}_{zz}^{\mathrm{spin}} &\to \frac{\epsilon_0 c^2}{2\omega}\mathrm{Re}\left[-\mathrm{i}\iint \rho\,\mathrm{d}\rho\,\mathrm{d}\phi(\mathcal{E}_x\mathcal{B}_x^* \exp[\mathrm{i}kL(n_{\mathrm{f}} - n_{\mathrm{s}})]\right. \\
&\quad\left. + \mathcal{E}_y\mathcal{B}_y^* \exp[-\mathrm{i}kL(n_{\mathrm{f}} - n_{\mathrm{s}})])\right]
\end{aligned}
\tag{28}
$$

but leaves the orbital part (26) *unchanged*. This supports our contention that $\mathcal{M}_{zz}^{\mathrm{spin}}$ should be identified with the spin, or polarization, part of the angular-momentum flux. If, however, we consider an element that imparts an azimuthal phase shift then our fields will all experience the same shift:

$$\begin{aligned} \mathcal{E}_x &\to \mathcal{E}_x \exp(im\phi) \qquad & \mathcal{B}_y &\to \mathcal{B}_y \exp(im\phi) \\ \mathcal{E}_y &\to \mathcal{E}_y \exp(im\phi) \qquad & \mathcal{B}_x &\to \mathcal{B}_x \exp(im\phi). \end{aligned} \tag{29}$$

This transfomation changes the orbital part of the total angular-momentum flux (26) to

$$\mathcal{M}_{zz}^{\text{orbit}} \to \mathcal{M}_{zz}^{\text{orbit}} + \frac{\epsilon_0 c^2}{2\omega}\text{Re}\left[m \iint \rho\,\text{d}\rho\,\text{d}\phi(\mathcal{B}_y^*\mathcal{E}_x + \mathcal{E}_x\mathcal{B}_y^*)\right] \tag{30}$$

but leaves the spin part (25) *unchanged.* This supports our contention that $\mathcal{M}_{zz}^{\text{orbit}}$ should be identified with the orbital rather than the spin part of the angular-momentum flux. This analysis holds only within the paraxial approximation and in order to demonstrate a more general validity we need to consider the behaviour of non-paraxial beams.

We can investigate the spin and orbit parts of the total angular-momentum flux, beyond the paraxial approximation, by means of the rather general beam introduced in [15]. This describes an electomagnetic field satisfying the full Maxwell equations (1) and corresponds to a beam propagating in the $+z$-direction with azimuthal phase dependence $\exp(il\phi)$. The components of this electric field are

$$\begin{aligned} \mathcal{E}_x &= \alpha \int_0^k \text{d}\kappa\, \mathcal{E}(\kappa)\exp(il\phi)\exp\left(\text{i}\sqrt{k^2-\kappa^2}z\right) J_l(\kappa\rho) \\ \mathcal{E}_y &= \beta \int_0^k \text{d}\kappa\, \mathcal{E}(\kappa)\exp(il\phi)\exp\left(\text{i}\sqrt{k^2-\kappa^2}z\right) J_l(\kappa\rho) \\ \mathcal{E}_z &= \int_0^k \text{d}\kappa\, \mathcal{E}(\kappa)\exp(il\phi)\exp\left(\text{i}\sqrt{k^2-\kappa^2}z\right)\frac{\kappa}{2\sqrt{k^2-\kappa^2}} \\ &\quad \times\big[(\text{i}\alpha-\beta)\exp(-\text{i}\phi)J_{l-1}(\kappa\rho) \\ &\quad -(\text{i}\alpha+\beta)\exp(\text{i}\phi)J_{l+1}(\kappa\rho)\big] \end{aligned} \tag{31}$$

where $J_l(\kappa\rho)$ is the Bessel function of order l and $k = \omega/c$. The complex numbers α and β are chosen so as to satisfy the equation $|\alpha|^2 + |\beta|^2 = 1$. We also require the magnetic field and can find this using the first of the Maxwell equations (20):

$$\begin{aligned} \mathcal{B}_x &= \int_0^k \text{d}\kappa \frac{\mathcal{E}(\kappa)\exp(\text{i}\sqrt{k^2-\kappa^2}z)}{\omega\sqrt{k^2-\kappa^2}}\Bigg[-\beta\frac{2k^2-\kappa^2}{2} \\ &\quad \times \exp(il\phi)J_l(\kappa\rho) + \frac{\kappa^2}{4}\big(\exp(\text{i}(l-2)\phi)(\text{i}\alpha-\beta)J_{l-2}(\kappa\rho) \\ &\quad -\exp(\text{i}(l+2)\phi)(\text{i}\alpha+\beta)J_{l+2}(\kappa\rho)\big)\Bigg] \\ \mathcal{B}_y &= \int_0^k \text{d}\kappa \frac{\mathcal{E}(\kappa)\exp(\text{i}\sqrt{k^2-\kappa^2}z)}{\omega\sqrt{k^2-\kappa^2}}\Bigg[\alpha\frac{2k^2-\kappa^2}{2} \\ &\quad \times \exp(il\phi)J_l(\kappa\rho) + \frac{\kappa^2}{4}\big(\exp(\text{i}(l-2)\phi)(-\text{i}\beta-\alpha) \\ &\quad \times J_{l-2}(\kappa\rho) - \exp(\text{i}(l+2)\phi)(-\text{i}\beta+\alpha)J_{l+2}(\kappa\rho)\big)\Bigg] \\ \mathcal{B}_z &= \int_0^k \text{d}\kappa \frac{\kappa\mathcal{E}(\kappa)\exp(\text{i}\sqrt{k^2-\kappa^2}z)}{2\omega}\big[(\text{i}\beta-\alpha) \\ &\quad \times\exp(\text{i}(l+1)\phi)J_{l+1}(\kappa\rho) - (\text{i}\beta+\alpha) \\ &\quad \times\exp(\text{i}(l-1)\phi)J_{l-1}(\kappa\rho)\big]. \end{aligned} \tag{32}$$

We should note that for these fields $\iint T_{xz}\,\text{d}x\,\text{d}y = 0$ and $\iint T_{yz}\,\text{d}x\,\text{d}y = 0$, so that the total linear momentum carried through a plane oriented in the z-direction is parallel to the beam axis. This means that the total angular-momentum flux will be the same about any axis parallel to the axis of the beam.

From the form of the electric field (31) we expect the spin angular-momentum flux to depend on α and β but not on l. The orbital angular-momentum flux, however, should depend on l but not on α and β. It is straightforward to show that this is indeed the case. If we insert our expressions for the electric and magnetic field amplitudes into our formulae for the spin and orbit parts of the angular momentum flux, (25) and (26), we find

$$\mathcal{M}_{zz}^{\text{spin}} = \text{i}(\alpha\beta^* - \alpha^*\beta)\frac{\pi\epsilon_0 c^2}{2\omega^2}\int_0^k \text{d}\kappa\,|\mathcal{E}(\kappa)|^2\frac{2k^2-\kappa^2}{\kappa\sqrt{k^2-\kappa^2}} \tag{33}$$

$$\mathcal{M}_{zz}^{\text{orbit}} = l\frac{\pi\epsilon_0 c^2}{2\omega^2}\int_0^k \text{d}\kappa\,|\mathcal{E}(\kappa)|^2\frac{2k^2-\kappa^2}{\kappa\sqrt{k^2-\kappa^2}}. \tag{34}$$

In obtaining these expressions we have made use of the Fourier–Bessel integral [27], which we can write in the form

$$\int_p^q \text{d}\kappa'\,\kappa'\Phi(\kappa')\int_0^\infty \text{d}\rho\,\rho J_l(\kappa'\rho)J_l(\kappa\rho) = \Phi(\kappa). \tag{35}$$

for $p < \kappa < q$. It is clear that these quantities have the expected dependences and this justifies making the separation of the angular-momentum flux into spin and orbital parts, even if we are operating outside the paraxial approximation. It is convenient to combine the α and β-dependent factor into a single term in the form

$$\sigma_z = \text{i}(\alpha\beta^* - \alpha^*\beta). \tag{36}$$

This term takes the values ± 1 for circularly polarized light and zero for linear polarization.

In their original paper, Allen *et al* [11] established, within the paraxial approximation, a proportionality between the angular momentum and energy fluxes. This led them to identify $\sigma_z\hbar$ and $l\hbar$ respectively as the spin and orbital angular momentum 'per photon'. We are now in a position to show that this relationship also holds beyond the paraxial approximation. In order to do so, we require the energy flux through the surface. This quantity is the integral over the surface of the cycle-averaged energy flux density, $(2\mu_0)^{-1}\text{Re}\,(\mathcal{E}_x\mathcal{B}_y^* - \mathcal{E}_y\mathcal{B}_x^*)$. Hence the energy flux is [15]

$$\begin{aligned} \mathcal{F} &= \frac{1}{2\mu_0\omega}\text{Re}\iint \rho\,\text{d}\rho\,\text{d}\phi\,(\mathcal{E}_x\mathcal{B}_y^* - \mathcal{E}_y\mathcal{B}_x^*) \\ &= \frac{\pi}{2\omega\mu_0}\int_0^k \text{d}\kappa\,|\mathcal{E}(\kappa)|^2\frac{2k^2-\kappa^2}{\kappa\sqrt{k^2-\kappa^2}}. \end{aligned} \tag{37}$$

The integral in this equation is the same as that occurring in (33) and (34). Hence we can write

$$\begin{aligned} \frac{\mathcal{M}_{zz}^{\text{spin}}}{\mathcal{F}} &= \frac{\sigma_z}{\omega} \\ \frac{\mathcal{M}_{zz}^{\text{orbit}}}{\mathcal{F}} &= \frac{l}{\omega} \\ \frac{\mathcal{M}_{zz}}{\mathcal{F}} &= \frac{l+\sigma_z}{\omega}. \end{aligned} \tag{38}$$

which justifies our assertion that $\mathcal{M}_{zz}^{\text{spin}}$ and $\mathcal{M}_{zz}^{\text{orbit}}$ are indeed the spin and orbit parts of the angular-momentum flux for a paraxial or non-paraxial beam about its axis and through a surface that is perpendicular to the direction of propagation. If we multiply both the numerator and denominator, in each of these expressions, by $\hbar$ then we recover the idea of spin and orbital angular momenta $\sigma_z\hbar$ and $l\hbar$ for each quantum of energy $\hbar\omega$. It should be stressed, however, that these expressions have been derived within classical electromagnetism. Inserting $\hbar$ may do no more than indicate a possibly helpful physical picture.

4.2. *Paraxial approximation*

The angular-momentum flux, within the paraxial limit, has been calculated by integrating the angular-momentum density, rather than the angular-momentum *flux* density, over the plane perpendicular to the direction of propagation [11]. An attempt to extend the calculation beyond the paraxial limit, again using the angular-momentum density [15], led to the erroneous conclusion that a separation of the angular-momentum flux into spin and orbit parts was not possible outside the paraxial approximation. It is now clear that the correct quantity is the angular-momentum flux density. It is interesting, however, to ask why an analysis based on the angular-momentum density gives the correct result within the paraxial approximation. To see why this is so, we consider the cycle-averaged angular-momentum flux density given in (21). The paraxial approximation applies to beams that have a small transverse wavevector. This corresponds to neglecting terms that are of second order in κ/k [15]. Inspection of the electric and magnetic field amplitudes (31) and (32) shows that the z-components of both fields are first order in κ/k and hence the paraxial approximation to $\bar{M}_{zz}$ requires only the x- and y-components of the fields to order zero in κ/k. At this order of approximation we have

$$\mathcal{E}_x \simeq c\mathcal{B}_y \qquad \mathcal{E}_y \simeq -c\mathcal{B}_x. \tag{39}$$

If we make these replacements in (21) then we find

$$\bar{M}_{zz} \simeq \tfrac{1}{2}c\epsilon_0 \mathrm{Re}\left[\mathcal{E}_z^*(y\mathcal{B}_y + x\mathcal{B}_x) - \mathcal{B}_z(y\mathcal{E}_y^* + x\mathcal{E}_x^*)\right]. \tag{40}$$

This quantity is simply the product of the cycle-average of the z-component of the angular-momentum density (13) multiplied by the speed of light. Hence, within the paraxial approximation, this quantity gives the same result as that obtained by integration over the (more strictly correct) quantity given in (21).

We can also make the approximations (39) in our expressions (25) and (26) for the spin and orbit angular-momentum fluxes. This gives

$$\mathcal{M}_{zz}^{\text{spin}} \simeq \frac{\epsilon_0 c}{\mathrm{i}2\omega}\iint \rho\,\mathrm{d}\rho\,\mathrm{d}\phi\,(\mathcal{E}_x^*\mathcal{E}_y - \mathcal{E}_y^*\mathcal{E}_x) \tag{41}$$

$$\mathcal{M}_{zz}^{\text{orbit}} \simeq \frac{\epsilon_0 c}{\mathrm{i}2\omega}\iint \rho\,\mathrm{d}\rho\,\mathrm{d}\phi\left(\mathcal{E}_x^*\frac{\partial}{\partial\phi}\mathcal{E}_x + \mathcal{E}_y^*\frac{\partial}{\partial\phi}\mathcal{E}_y\right). \tag{42}$$

These are the expressions derived by van Enk and Nienhuis [18], within the paraxial apprpoximation, for the spin and orbital angular momenta per unit length, multiplied by the speed of light.

4.3. *Flux for other components of angular momentum*

We are not restricted to calculating the angular-momentum flux for the component of angular momentum that is parallel to the beam axis. The fluxes of the other components of angular momentum also have physical significance. We shall illustrate this significance by calculating the flux through the xy-plane of the y-component of angular momentum for the fields (31) and (32). The y-component of the angular-momentum flux density through a surface oriented in the z-direction is (from (15))

$$\begin{aligned} M_{zy} &= -z(\epsilon_0 E_x E_z + \mu_0^{-1}B_x B_z) \\ &\quad -\frac{x}{2}\left[\epsilon_0(E_x^2 + E_y^2 - E_z^2) + \mu_0^{-1}(B_x^2 + B_y^2 - B_z^2)\right]. \end{aligned} \tag{43}$$

If we integrate this quantity over the entire xy-plane then we find the value zero. This is reasonable as the y-axis is a symmetry axis for the beam and we would not expect to find a net flux of angular momentum. Other axes, parallel to the y-axis, will not be symmetry axes for the beam and we can expect a non-zero angular-momentum flux about these.

Consider the flux of the y-component of angular momentum about the axis $x = -a$. For this axis the angular-momentum flux density has the form (43) with $x \to x + a$ and the total flux of the y-component of angular-momentum is

$$\begin{aligned} \mathcal{M}_{zy}(a) &= \iint \rho\,\mathrm{d}\rho\,\mathrm{d}\phi\, M_{zy} = \frac{a}{2}\iint \rho\,\mathrm{d}\rho\,\mathrm{d}\phi \\ &\quad\times\left[\epsilon_0(E_x^2 + E_y^2 - E_z^2) + \mu_0^{-1}(B_x^2 + B_y^2 - B_z^2)\right]. \end{aligned} \tag{44}$$

We can interpret this expression very simply. It is the product of the distance a between the axis of rotation and the centre of the beam with the flux of the z-component of the *linear* momentum through the surface. Hence it expresses the usual angular momentum induced by a linear momentum off-set from the axis of rotation by the distance a. In this sense, this component of angular-momentum flux associated with the beam is an exterior angular momentum. It is of a different character to the spin and orbit angular-momentum fluxes associated with the beam about its axis. A separation of this angular-momentum flux into spin and orbital components, in the sense discussed above in section 4.1, does not appear to be meaningful.

5. Dipole field

At a fundamental level, light is emitted by atoms (or similar microscopic sources). For this reason it is of interest to ask how angular momentum is transferred from excited atoms to the emitted light. This problem has received considerable attention [8, 9, 21, 28, 29]. Here we shall calculate the rate of radiation of angular momentum by the dipole using the angular-momentum flux. Our approach is similar to that adopted by Humblet [9] except that we do not require the surface, through which the angular-momentum flux is calculated, to be far from the source. Our calculation is entirely classical, but an extension to a fully quantum calculation should be possible using the methods developed in [28].

We consider the complex electric and magnetic fields associated with a point dipole positioned at the origin with complex dipole moment $\mu_j\exp(-\mathrm{i}\omega t)$. The complex electric and magnetic fields at position $\boldsymbol{r}$, generated by the dipole, are [6, 28, 30, 31]

$$\mathcal{E}_j(\mathbf{r}) = -\frac{\exp(\mathrm{i}\omega r/c)}{4\pi\epsilon_0 r^3}\Big[[\mu_j - 3(\mu_l\hat{r}_l)\hat{r}_j]\Big(1 - \frac{\mathrm{i}\omega r}{c}\Big)$$
$$-[\mu_j - (\mu_l\hat{r}_l)\hat{r}_j]\Big(\frac{\omega r}{c}\Big)^2\Big] \tag{45}$$
$$\mathcal{B}_j(\mathbf{r}) = \frac{\mathrm{i}\omega\mu_0\exp(\mathrm{i}\omega r/c)}{4\pi r^2}\epsilon_{jkl}\hat{r}_k\mu_l\Big(1 - \frac{\mathrm{i}\omega r}{c}\Big). \tag{46}$$

where $\hat{\mathbf{r}}$ is the unit vector in the radial direction and the exponential factor takes account of retardation due to the finite speed of light. We can use these fields to calculate the cycle-averaged flux of angular momentum through a sphere of radius r centred on the emitting dipole. The cycle-averaged flux of the z-component of angular momentum through a (radially oriented) element of the surface of the sphere is

$$\bar{M}_{rz} = \tfrac{1}{2}\mathrm{Re}\,[\tfrac{1}{2}(x\hat{\mathbf{y}}\cdot\hat{\mathbf{r}} - y\hat{\mathbf{x}}\cdot\hat{\mathbf{r}})(\epsilon_0\mathcal{E}_k^*\mathcal{E}_k + \mu_0^{-1}\mathcal{B}_m^*\mathcal{B}_m)$$
$$-\epsilon_0(x\mathcal{E}_y^* - y\mathcal{E}_x^*)\mathcal{E}_r - \mu_0^{-1}(x\mathcal{B}_y^* - y\mathcal{B}_x^*)\mathcal{B}_r]$$
$$= -\tfrac{1}{2}\mathrm{Re}\,[\epsilon_0(x\mathcal{E}_y^* - y\mathcal{E}_x^*)\mathcal{E}_r], \tag{47}$$

where the hats again denote unit vectors. Integrating this quantity over the surface of the sphere gives the total flux of the z-component of angular momentum through the surface of the sphere, which we find to be

$$\mathcal{M}_{rz} = \int_0^{\pi}\sin\theta\,\mathrm{d}\theta\int_0^{2\pi}\mathrm{d}\phi\,\bar{M}_{rz}$$
$$= \sigma_z\frac{|\mu|^2\omega^3}{12\pi\epsilon_0 c^3}, \tag{48}$$

where we associate $\sigma_z = \mathrm{i}(\mu_x\mu_y^* - \mu_y\mu_x^*)/|\mu|^2$ with the z-component of angular momentum for the dipole. It is interesting to compare this with the flux of energy across the surface of the sphere:

$$\mathcal{F} = \frac{1}{2\mu_0}\int_0^{\pi}\sin\theta\,\mathrm{d}\theta\int_0^{2\pi}\mathrm{d}\phi\,\mathrm{Re}\,(\epsilon_{ijk}\hat{r}_j\mathcal{E}_k\mathcal{B}_l^*)$$
$$= \frac{|\mu|^2\omega^4}{12\pi\epsilon_0 c^3}. \tag{49}$$

The ratio of the total angular-momentum flux to the energy flux is

$$\frac{\mathcal{M}_{rz}}{\mathcal{F}} = \frac{\sigma_z}{\omega}, \tag{50}$$

which has the same form as we found for light beams (38). It has been suggested that half of this can be attributed to spin and half to orbital angular momentum [29]. We have not been able to find, however, any natural or convincing way to separate the angular-momentum flux into spin and orbital components.

6. Quantum theory

The classical theory presented in this paper has a natural extension to quantum electrodynamics. The electric and magnetic fields become operators in the quantum theory, satisfying the equal-time canonical commutation relation [32]

$$[\hat{E}_j(\mathbf{r},t), \hat{B}_k(\mathbf{r}',t)] = \mathrm{i}\frac{\hbar}{\epsilon_0}\epsilon_{jkl}\frac{\partial}{\partial x_l'}\delta(\mathbf{r} - \mathbf{r}'), \tag{51}$$

where the carets denote quantum operators. Maxwell's equations (1) remain valid but with the classical electric and magnetic fields replaced by these operators. It is helpful to separate the field operators into positive- and negative-frequency parts corresponding to time evolutions of the form $\exp(-\mathrm{i}\omega t)$ and $\exp(\mathrm{i}\omega t)$ respectively:

$$\hat{E}_j^{(+)}(\mathbf{r},t) = \int_0^{\infty}\hat{\mathcal{E}}_j(\mathbf{r},\omega)\exp(-\mathrm{i}\omega t)\,\mathrm{d}\omega \tag{52}$$

$$\hat{E}_j^{(-)}(\mathbf{r},t) = \int_0^{\infty}\hat{\mathcal{E}}_j^{\dagger}(\mathbf{r},\omega)\exp(\mathrm{i}\omega t)\,\mathrm{d}\omega, \tag{53}$$

with similar expressions for the magnetic fields. These positive- and negative-frequency parts of the fields are associated with photon annihilation and creation operators respectively. The commutator (51) implies the existence of a divergent zero-point energy associated with the vacuum state of the field [30, 32]. A number of physical quantities discussed in this paper will be ill defined in the quantum theory because of this zero-point energy. These include the energy density (5), the momentum-flux density (8) and the angular-momentum flux density (15).

We can obtain well-behaved quantities by subtracting the vacuum contributions from these quantities. This corresponds to the familiar approach in which field quantities are expressed in normal order [30]. The normal-ordered form of operator products has annihilation operators to the right and creation operators to the left so that the normal-ordered square of the electric field is

$$:\hat{E}^2: \; = \; :(\hat{E}_j^{(-)} + \hat{E}_j^{(+)})(\hat{E}_j^{(-)} + \hat{E}_j^{(+)}):$$
$$= 2\hat{E}_j^{(-)}\hat{E}_j^{(+)} + \hat{E}_j^{(-)}\hat{E}_j^{(-)} + \hat{E}_j^{(+)}\hat{E}_j^{(+)}. \tag{54}$$

Normal-ordered products are denoted by a pair of colons so our normal-ordered angular-momentum flux density operator is

$$:\hat{M}_{li}: \; = \; \epsilon_{ijk}x_j\big[\tfrac{1}{2}\delta_{kl}(\epsilon_0 :\hat{E}^2: + \mu_0^{-1}:\hat{B}^2:)$$
$$-\epsilon_0:\hat{E}_k\hat{E}_l: - \mu_0^{-1}:\hat{B}_k\hat{B}_l:\big]. \tag{55}$$

This normal-ordered operator satisfies the operator analogue of the conservation of angular momentum (14) with the angular-momentum density operator also expressed in normal order in the form

$$:\hat{j}_i: = \epsilon_0 :(\hat{E}_i x_j\hat{B}_j - \hat{B}_i x_j\hat{E}_j):. \tag{56}$$

Hence these normal-ordered quantities do represent a conservation law for the normal-ordered angular momentum and the choice of normal-ordered quantities is physically meaningful.

The quantities introduced in this section allow us to extend our treatments of the angular momentum of light beams and dipole fields into the quantum regime and to discuss problems such as the transfer of angular momentum in elementary absorption and emission by atoms [28]. We shall develop these ideas further elsewhere.

7. Conclusion

In this paper, we have presented the theory of optical angular-momentum flux. This quantity can be derived from consideration of the conservation of angular momentum for the free electromagnetic field. Expression (15) for the angular-momentum flux density is exact in that it does not depend on any paraxial (or related) approximation. We have applied this quantity to calculate the flux of a light beam about its axis (or propagation direction). In this case, we found that it is possible to separate the angular momentum into well-defined spin and orbital contributions. This separation is gauge invariant and the fluxes of spin and orbital angular momentum reduce to the known paraxial forms [18] in the appropriate limit. We have also shown that the optical angular-momentum flux can be applied to the study of other components of optical angular momentum and to the radiation associated with an electric dipole. The theory of optical angular-momentum flux follows from Maxwell's equations and these are valid as operator equations in quantum electrodynamics. It is no surprise, therefore, that the theory presented here can also be applied to the quantum theory of light.

The theory presented in this paper opens new possibilities for the study of the angular momentum of light. We conclude with a brief indication of three such possibilities: (i) the form of the linear electromagnetic momentum inside material media has long been debated (for a review see [33]). It is possible that the study of electromagnetic angular momentum may provide a clearer picture of the momentum and momentum transfer inside such media. (ii) The separation of angular-momentum flux into spin and orbital parts remains a problem for all but the simplest geometries. It is possible, however, that a study of angular-momentum flux for more complicated light sources may provide insight into this problem. In particular we envisage extending the analysis of section 5 to multipole sources [8, 34]. (iii) The quantum theory of optical angular-momentum flux remains to be explored. It is likely that recent ideas relating to the entanglement of orbital angular momentum [35, 36] can be extended to obtain more exotic states in which spin and orbital angular momenta are entangled. We intend to pursue these ideas elsewhere.

Acknowledgments

I am grateful to Les Allen, Colin Baxter, Sonja Franke-Arnold, Claire Gilson and Miles Padgett for encouraging and helpful remarks. This work was supported by the Royal Society of Edinburgh and the Scottish Executive Education and Lifelong Learning Department through the award of a Support Research Fellowship, by the Leverhulme Trust and by the UK Engineering and Physical Sciences Research Council.

Appendix A. Gaussian units

In this paper we have worked with the SI system of units. The Gaussian system is also widely used and so we present here the main formulae in Gaussian units. A more complete discussion of electromagnetic units can be found in [6].

Maxwell's equations for the free electromagnetic field (1):

$$\begin{aligned} \nabla \cdot \boldsymbol{E} &= 0 \\ \nabla \cdot \boldsymbol{B} &= 0 \\ \nabla \times \boldsymbol{E} &= -\frac{1}{c}\frac{\partial \boldsymbol{B}}{\partial t} \\ \nabla \times \boldsymbol{B} &= \frac{1}{c}\frac{\partial \boldsymbol{E}}{\partial t}. \end{aligned} \tag{A.1}$$

Energy density for the free electromagnetic field (5):

$$W = \frac{1}{8\pi}(E^2 + B^2). \tag{A.2}$$

The energy-flux density for the free electromagnetic field is $(c/4\pi)(\boldsymbol{E} \times \boldsymbol{B})_i$ and the density of the i-component of the momentum density is $(4\pi c)^{-1}(\boldsymbol{E} \times \boldsymbol{B})_i$. The momentum-flux density (8) becomes

$$T_{ij} = \frac{1}{4\pi}\left[\frac{1}{2}\delta_{ij}(E^2 + B^2) - E_i E_j - B_i B_j\right]. \tag{A.3}$$

Angular-momentum density (11):

$$\boldsymbol{j} = \frac{1}{4\pi c}\boldsymbol{r} \times (\boldsymbol{E} \times \boldsymbol{B}). \tag{A.4}$$

Angular-momentum flux density (15):

$$M_{li} = \frac{1}{4\pi}\epsilon_{ijk}x_j\left[\frac{1}{2}\delta_{kl}(E^2 + B^2) - E_k E_l - B_k B_l\right]. \tag{A.5}$$

The ratios of the total angular-momentum flux to the energy flux calculated in this paper are unaffected by the change of units.

Finally, the electric and magnetic fields associated with an electric dipole (45) and (46), become

$$\mathcal{E}_j(\boldsymbol{r}) = -\frac{\exp(\mathrm{i}\omega r/c)}{r^3}\left[[\mu_j - 3(\mu_l\hat{r}_l)\hat{r}_j]\left(1 - \frac{\mathrm{i}\omega r}{c}\right) - [\mu_j - (\mu_l\hat{r}_l)\hat{r}_j]\left(\frac{\omega r}{c}\right)^2\right] \tag{A.6}$$

$$\mathcal{B}_j(\boldsymbol{r}) = \frac{\mathrm{i}\omega\exp(\mathrm{i}\omega r/c)}{cr^2}\epsilon_{jkl}\hat{r}_k\mu_l\left(1 - \frac{\mathrm{i}\omega r}{c}\right). \tag{A.7}$$

Appendix B. Relativistic construction

The most elegant and natural theory of the free electromagnetic field is that provided by the special theory of relativity. In this appendix we construct the theory of electromagnetic angular momentum using the tools of relativity. Physical quantities are represented as tensors with indices (written as Greek letters) ranging over four values 0, 1, 2 and 3 corresponding to the t-, x-, y- and z-components. As is usual in relativity, we shall retain the summation convention with repeated Greek indices implying summation over the four dimensions of spacetime. We shall continue to use Roman indices to denote the three spatial dimensions. We shall also work with a rationalized system of units in which $c = \epsilon_0 = \mu_0 = 1$. The analysis presented in this appendix follows from that given by Weinberg [37]. Similar analyses can be found in a number of texts [6, 38, 39].

The symmetric energy–momentum tensor for the electromagnetic field is [37]

$$T^{\alpha\beta} = F^{\alpha}{}_{\gamma}F^{\beta\gamma} - \tfrac{1}{4}\eta^{\alpha\beta}F_{\gamma\delta}F^{\gamma\delta}. \tag{B.1}$$

This quantity is denoted $\Theta^{\alpha\beta}$ in Jackson [6]. Here $\eta^{\alpha\beta}$ is the Minkowski metric, which takes the value +1 for $\alpha = \beta = 1, 2, 3$, the value −1 for $\alpha = \beta = 0$ and is zero otherwise. The antisymmetric field tensor $F^{\alpha\beta}(= -F^{\beta\alpha})$ has the non-zero components

$$F^{12} = B_3 = B_z \qquad F^{23} = B_1 = B_x \qquad F^{31} = B_2 = B_y$$
$$F^{01} = E_1 = E_x \qquad F^{02} = E_2 = E_y \qquad F^{03} = E_3 = E_z. \tag{B.2}$$

The energy and momentum densities are

$$T^{00} = \tfrac{1}{2}(E^2 + B^2) \tag{B.3}$$

$$T^{i0} = T^{0i} = (\boldsymbol{E} \times \boldsymbol{B})_i = \epsilon_{ijk} E_j B_k \tag{B.4}$$

and the remaining components are

$$T^{ij} = T^{ji} = [\tfrac{1}{2}\delta_{ij}(E^2 + B^2) - E_i E_j - B_i B_j]. \tag{B.5}$$

The conservation of energy and momentum becomes, in the relativistic theory, the conservation of energy–momentum and the mathematical expression of this conservation is as the simple equation

$$\frac{\partial}{\partial x^\alpha} T^{\alpha\beta} = 0. \tag{B.6}$$

This leads us to associate T^{i0} with the energy-flux density (as well as with the momentum density) and T^{ij} with the flux of the j-component of the momentum through a surface oriented in the direction i.

The angular momentum is introduced in terms of the rank three tensor

$$M^{\gamma\alpha\beta} = x^\alpha T^{\beta\gamma} - x^\beta T^{\alpha\gamma}. \tag{B.7}$$

This tensor, like the energy–momentum tensor, is a conserved quantity in that it obeys the continuity equation

$$\frac{\partial}{\partial x^\gamma} M^{\gamma\alpha\beta} = 0. \tag{B.8}$$

The angular momentum tensor may be defined as

$$J^{\alpha\beta} = \int \mathrm{d}^3x\, M^{0\alpha\beta}, \tag{B.9}$$

the spatial components of which are the familiar angular momenta. The i-component of the angular momentum is $J_i = \frac{1}{2}\epsilon_{ijk} J^{jk}$. In particular the z-component is

$$J_z = J^{12} = \int \mathrm{d}^3x\, (x^1 T^{20} - x^2 T^{10})$$
$$= \int \mathrm{d}^3x\, [x(\boldsymbol{E} \times \boldsymbol{B})_y - y(\boldsymbol{E} \times \boldsymbol{B})_x]. \tag{B.10}$$

The x- and y-components are J^{23} and J^{31} respectively. The quantity

$$j_i = \tfrac{1}{2}\epsilon_{ijk} M^{0jk} = \epsilon_{ijk} x_j (\boldsymbol{E} \times \boldsymbol{B})_k = [\boldsymbol{r} \times (\boldsymbol{E} \times \boldsymbol{B})]_i \tag{B.11}$$

is the i-component of the angular-momentum density with M^{023}, M^{031} and M^{012} being its x-, y- and z-components respectively. If j_i is the i-component of the angular momentum density, then it follows from the conservation law (B.8) that $M^l{}_i = \frac{1}{2}\epsilon_{ijk} M^{ljk}$ is the angular-momentum flux density for a surface pointing in the direction l. Hence the total flux for the i-component of angular momentum through a surface S is $\int_S M^l{}_i\, \mathrm{d}S^l = \frac{1}{2}\epsilon_{ijk} \int_S M^{ijk}\, \mathrm{d}S_i$. The i-component of the angular-momentum flux density is

$$M^l{}_i = \epsilon_{ijk} x^j T^{kl} = \epsilon_{ijk} x_j [\tfrac{1}{2}\delta_{kl}(E^2 + B^2) - E_k E_l - B_k B_l]. \tag{B.12}$$

If we rewrite this using SI units then we recover the form (15) given earlier.

We see that the form of the angular-momentum flux density appears naturally in the relativistic formulation as a consequence of angular-momentum conservation as expressed in (B.8).

Note added in proof. The significance of optical angular-momentum flux has also been noted by Alexeyev *et al* [40], who applied the idea to study angular momentum in optical fibres. These authors also suggested a possible form for the flux of the spin part of the angular-momentum flux [41]. I am grateful to Michael Berry for bringing this work to my attention.

References

[1] Maxwell J C 1861 *Phil. Mag.* **21** 161
Maxwell J C 1861 *Phil. Mag.* **21** 281
Maxwell J C 1861 *Phil. Mag.* **21** 338
[2] Siegel D M 1991 *Innovation in Maxwell's Electromagnetic Theory* (Cambridge: Cambridge University Press)
[3] Maxwell J C 1998 *A Treatise on Electricity and Magnetism* (Oxford: Oxford University Press)
[4] Poynting J H 1909 *Proc. R. Soc.* A **82** 560
Poynting J H 1920 *Collected Scientific Papers* (Cambridge: Cambridge University Press) (reprint)
[5] Beth R A 1936 *Phys. Rev.* **50** 115
[6] Jackson J D 1999 *Classical Electrodynamics* 3rd edn (New York: Wiley)
[7] Abraham M 1914 *Phys. Zeits.* **15** 914 (in German)
[8] Heitler W 1936 *Proc. Camb. Phil. Soc.* **37** 112
[9] Humblet J 1943 *Physica* **10** 585 (in French)
[10] Darwin C G 1932 *Proc. R. Soc.* A **136** 36
[11] Allen L, Beijersbergen M W, Spreeuw R J C and Woerdman J P 1992 *Phys. Rev.* A **45** 8185
[12] Siegman A E 1986 *Lasers* (Mill Valley, CA: University Science)
[13] Allen L, Padgett M J and Babiker M 1999 *Prog. Opt.* **39** 291
[14] Padgett M and Allen L 2000 *Contemp. Phys.* **41** 275
[15] Barnett S M and Allen L 1994 *Opt. Commun.* **110** 670
[16] Quabis S, Dorn R, Eberler M, Glökl O and Leuchs G 2000 *Opt. Commun.* **179** 1
Quabis S, Dorn R, Eberler M, Glökl O and Leuchs G 2001 *Appl. Phys.* B **72** 109
[17] Stephenson G and Radmore P M 1990 *Advanced Mathematical Methods for Engineering and Science Students* (Cambridge: Cambridge University Press)
[18] van Enk S J and Nienhuis G 1992 *Opt. Commun.* **94** 147
[19] Cohen-Tannoudji C T, Dupont-Roc J and Grynberg G 1992 *Photons and Atoms: Introduction to Quantum Electrodynamics* (New York: Wiley)
[20] van Enk S J and Nienhuis G 1994 *Europhys. Lett.* **25** 497
[21] van Enk S J and Nienhuis G 1994 *J. Mod. Opt.* **41** 963
[22] He H, Friese M E J, Heckenberg N R and Rubinsztein-Dunlop H 1995 *Phys. Rev. Lett.* **75** 836
[23] Simpson N B, Dholakia K, Allen L and Padgett M J 1997 *Opt. Lett.* **22** 52
[24] Marcuse D 1972 *Light Transmission Optics* (New York: Van Nostrand-Reinhold)
[25] Berry M V 1998 *Singular Optics (SPIE vol 3487)* (Bellingham, WA: SPIE) p 6
[26] Hecht E and Zajac A 1980 *Optics* (Reading, MA: Addison-Wesley)
[27] Gray A, Mattews G B and MacRobert T M 1922 *A Treatise on Bessel Functions* (London: Macmillan)
[28] Franke S and Barnett S M 1996 *J. Phys. B: At. Mol. Opt. Phys.* **29** 2141
[29] Crichton J H and Marston P L 2000 *Electron. J. Diff. Eqns* Conf. **04** 37 webpage http://edje.math.swt.edu
[30] Milonni P W 1994 *The Quantum Vacuum* (San Diego, CA: Academic)

[31] Stratton J A 1941 *Electromagnetic Theory* (New York: McGraw-Hill)
[32] Loudon R 2000 *The Quantum Theory of Light* 3rd edn (Oxford: Oxford University Press)
[33] Brevik I 1979 *Phys. Rep.* **52** 133
[34] Rose M E 1955 *Multipole Fields* (New York: Wiley)
[35] Mair A, Vaziri A, Weihs G and Zeilinger A 2001 *Nature* **412** 313
[36] Franke-Arnold S, Barnett S M, Padgett M J and Allen L *Phys. Rev.* A in press
[37] Weinberg S 1972 *Gravitation and Cosmology: Principles and Applications of the General Theory of Relativity* (New York: Wiley)
[38] Barut A O 1964 *Electrodynamics and Classical Theory of Particles and Fields* (New York: Macmillan)
[39] Yilmaz H 1965 *Introduction to the Theory of Relativity and the Principles of Modern Physics* (New York: Blaisdell)
[40] Alexeyev C N, Alexeyev A N, Fridman Y A and Volyar A V 1998 *SPIE Proc.* **3487** 94
[41] Alexeyev C N, Fridman Y A and Alexeyev A N 2001 *SPIE Proc.* **4403** 71

Section 3

LABORATORY BEAMS CARRYING ORBITAL ANGULAR MOMENTUM

Allen *et al.* (3.1, **Paper 2.1**) in 1992 were the first to recognise that a beam with helical wavefronts has an orbital angular momentum in the direction of propagation. They showed that all beams with helical phasefronts characterised by an $\exp(il\phi)$ azimuthal phase have an orbital angular momentum equivalent to $l\hbar$ per photon.

Laser beams are usually described by use of a Hermite–Gaussian basis set, but a beam with helical wavefronts is best described in terms of the Laguerre–Gaussian basis set which has an explicit $\exp(il\phi)$ phase term. Hermite–Gaussian ($HG_{m,n}$) modes are characterised in terms of the mode indices m and n which denote the order of the Hermite polynomials in the x and y directions. Laguerre–Gaussian (LG_p^l) modes are described by the indices l and p where the l-index relates to the azimuthal phase and the p-index to the number of additional concentric rings around the central zone. The form for the LG mode field amplitude is given by

$$u_p^l(r,\phi,z) = C_{pl}e^{-ikr^2/2R}e^{-r^2/w^2}e^{-i(2p+|l|+1)\tan^{-1}(z/z_r)}e^{-il\phi}(-1)^p(r\sqrt{2}/w)^{|l|}L_p^{|l|}(2r^2/w^2)$$

When $l \neq 0$, the helical phasefronts result in a phase discontinuity on the beam axis and a corresponding zero in the beam intensity. Consequently, the $l \neq 0$ modes have the appearance of an annular ring for $p = 0$, or rings when $p > 0$. Note also the Gouy phase term $\exp(-i(2p + |l| + 1)\tan^{-1}(z/z_r))$ which describes the phase change as a beam moves through the beam waist. Padgett *et al.* (3.2, **Paper 3.1**) illustrate a number of examples of HG and LG modes and show interferograms formed between the LG modes and a plane wave where the helical phasefronts of the modes are revealed in the form of spiral fringes.

As both Hermite–Gaussian and Laguerre–Gaussian are complete orthonormal sets, any arbitrary beam may be described by a linear superpositon of either set. As long ago as 1963 it was recognised that an LG_0^1 mode could be expressed as the phase-quadrature superposition of $HG_{1,0}$ and $HG_{0,1}$ modes (3.3). However for most actual laser cavities, astigmatism breaks the frequency degeneracy between the $HG_{1,0}$ and $HG_{0,1}$ modes and, in practice, an LG output is never observed. Deliberate removal of the cavity astigmatism allows the two modes to become degenerate and frequency lock together. If an on-axis intensity block is introduced, the modes become locked in phase-quadrature to give an LG output. Just as the handedness of circular polarisation depends upon the relative phase of the two composite linear polarisations, so the handedness of the LG mode depends upon the relative phase of the composite HG modes.

Vaughan and Willets (3.4, **Paper 3.2**) produced a laser mode with an annular intensity profile in 1983, but made no observation concerning the potential of the beam to carry orbital angular momentum. In 1990 Tamm and Weiss (3.5, **Paper 3.3**) designed an astigmatism free laser cavity to produce an LG_0^1 mode and used selective feedback to define its handedness. However, again the orbital angular momentum properties of the resulting beam were not recognised. Slightly more recently, Harris *et al.* (3.6, **Paper 3.4**) succeeded in interfering an LG output from a laser with its own mirror image to produce spiral interference fringes, which showed the helical phase structure of the mode.

Laguerre–Gaussian modes were first produced in order to demonstrate their orbital angular momentum content, by Beijersbergen *et al.* in 1993 (3.7, **Paper 3.5**). This work used the results of Abramochkin and Volostnikov (3.8) in their investigation of the near field for astigmatic optical elements, to establish a general algebraic relationship with which any LG mode can be expressed as a linear superpositon of HG modes and vice versa. Beijersbergen *et al.* demonstrated a lossless mode converter, consisting of a pair of cylindrical lenses, which transformed any HG mode into an LG mode. It relies on the fact that when a beam is focused between two cylindrical lenses a Gouy phase shift is introduced that depends upon the indices of the mode. This generalised the approach of Tamm and Weiss (3.5) but now, just as appropriate focusing of the incident beam and the canonical separation of the lenses introduces a $\pi/2$ phase shift between $HG_{1,0}$ and $HG_{0,1}$ modes to produce an LG_0^1 mode, so the same lens configuration converts every HG mode into an LG mode with $l = m - n$ and $p = \min(m, n)$. The action of the cylindrical lens $\pi/2$-mode converter for orbital angular momentum is, therefore, analogous to that of a quarter wave plate for spin. Beijersbergen *et al.* also explained how the angular momentum transfer between the lens and the light could be explained in terms of the asymmetry of the incident HG mode. This can be shown to produce a gradient force which exerts a torque on the cylindrical lens and a subsequent transfer of angular momentum.

An increase in the separation between the lenses increases the Gouy phase shift between the modes. When the spacing is twice the focal length, the phase shift becomes very close to π and any incident LG mode is converted to the same mode with the opposite sign of l. This transformation is effectively a mirror inversion of the beam and has associated with it a reverse of the orbital angular momentum content. It should be noted here that a Dove prism achieves the same inversion without any assumption of diffraction free propagation (3.9). Recently, Molina-Terriza *et al.* (3.10, **Paper 3.6**) pointed out that on either side of a cylindrically focused LG mode, the sign of the vortex is reversed. As no angular momentum transfer can occur in free space, this emphasises that the angular momentum content is associated with the inclined phasefronts and not with the vortex itself. The phase distribution and transfer mechanism has been further analysed by Padgett and Allen (3.11, **Paper 3.7**).

Cylindrical lenses have also been used to impart orbital angular momentum to elliptical Gaussian beams. This produces an intensity profile which rotates on propagation (3.12). In such beams, the orbital angular momentum density is non-uniform and increases with increasing distance from the beam axis. Despite their very high total angular momentum, any resulting forces or torques are difficult to observe because of the presence of the gradient force, see section 4, which results from the beam profile.

The idea of mode transformation arising from the introduction of a phase delay between orthogonal components has also been exploited in a stressed fibre, where the resulting change in the guide index can introduce a phase delay between $HG_{1,0}$ and $HG_{0,1}$ modes, to again produce an LG mode (3.13).

At around the same time as the early work on the orbital angular momentum of LG modes, a number of groups were using computer-generated holograms for the production of beams containing optical vortices. A hologram can be considered as a modified diffraction grating, where the first order diffracted beam has a specific phase and amplitude structure. The required form of the diffraction grating can be produced experimentally by interfering the light reflected from the desired object with a reference beam. Once photographically developed, the film is again placed in the reference beam and the resulting diffraction pattern gives a beam identical to that reflected from the original object. When the desired beam has a simple form, the first step in this process can be avoided as it is straightforward to calculate the interference pattern directly. If a calculated interference pattern is transferred to holographic film, or another recordable phase medium, a computer generated hologram or diffractive optic is produced. As with all holograms, and indeed diffraction gratings, the efficiency with which light can be diffracted into the chosen order is enhanced if the hologram is blazed. This modifies the phase of the transmitted light rather than just its amplitude. A perfect phase hologram is, in principle, 100% efficient and efficiencies in excess of 50% have been experimentally demonstrated.

The use of computer generated holograms for the generation of optical beams containing vortices was reported in 1990 by Bazhenov *et al.* (3.14, **Paper 3.8**) and in 1992 by Heckenberg *et al.* (3.15, **Paper 3.9**). As is now common practice, their designs were calculated for the two interfering beams meeting at an angle. This gives a diffraction grating appearance to the interference pattern, and results in the various diffraction orders appearing at differing angles, allowing easy selection of the desired beam. An alternative approach adopted by Heckenberg *et al.* (3.16) calculated the hologram for the case when the desired beam is focused. The result is that when the hologram is illuminated with a plane wave, the various diffraction orders focus to different axial positions, again allowing the desired diffraction order to be selected.

Holograms are now widely used to produce beams containing any manner of phase singularities for the studies of orbital angular momentum and related effects. Programmable spatial light modulators are now commercially available and can be configured to act as holograms that can be changed under software control. This allows virtually every aspect of the beam, including the orbital angular momentum, to be modified in real-time. In 1996 Soskin *et al.* (3.17, **Paper 3.10**) used computer generated holograms to create beams containing optical vortices and studied both vortex interaction and the effects of interference on the beam profile.

The spiral phase plate gives results similar to the computer generated hologram. It is an optical element with an optical thickness that increases with azimuthal angle such that, upon transmission, an incident quasi-plane wave emerges with a helical phasefront. It does in fact perform an identical operation to the ideal phase hologram although it requires sub-wavelength mechanical tolerance to ensure that the step height corresponds exactly to a whole number of optical wavelength. The first use of a spiral phase plate was by Beijersbergen *et al.* (3.18, **Paper 3.11**), but the extreme tolerances required in the optical regime meant that the quality of the resulting mode was below that which could be obtained using the cylindrical lens mode converter. In the mm-wave region of the spectrum, however, the required accuracy is obtainable and Turnbull *et al.* (3.19, **Paper 3.12**) successfully used this approach to generate LG modes using a waveplate fabricated from PTFE with a step height of several mm. This paper also includes a simple derivation based on refraction at the inclined surface of the waveplate to show that even a ray optic model predicts that the orbital angular momentum is indeed $l\hbar$ per photon. The same technique was later used by Courtial *et al.* (3.20, **Paper 6.6**) in their experiments to demonstrate and measure precisely the rotational Doppler effect (see Section 6). A spiral phaseplate has also been used recently as one of the mirrors in a laser cavity, resulting in the formation of a helical mode within the laser cavity itself (3.21).

Although both the holograms and spiral phaseplates produce LG modes with a single l value, it should be appreciated that even when using an HG ($m = 0, n = 0$) mode as the input light, the resulting radial intensity distribution does not correspond to that of a single LG mode. The resulting beam is a superposition of LG modes with the same l-index but with a range of p-indices. Under optimum conditions approximately 90% of the energy can be contained in the $p = 0$ mode (3.22).

High-order Bessel beams also have helical wavefronts characterised by an $\exp(il\phi)$ phase term and so carry an orbital angular momentum of $l\hbar$ per photon. Such beams can be produced using computer generated holograms, or by passing an LG mode through a glass cone known as an axicon (3.23). Recently, in an experiment similar to those performed with LG modes (3.24, Paper 2.7), high-order Bessel beams have been used to set particles into orbital motion around the beam axis (3.25).

REFERENCES

3.1 L Allen, MW Beijersbergen, RJC Spreeuw and JP Woerdman, 1992, *Phys. Rev. A* **45** 8185.
3.2 M Padgett, J Arlt, N Simpson and L Allen, 1996, *Am. J. Phys.* **64** 77.
3.3 WW Rigrod, 1963, *Appl. Phys. Lett.* **2** 51.
3.4 JM Vaughan and DV Willets, 1983, *J. Opt. Soc. A* **73** 1018.
3.5 C Tamm and CO Weiss, 1990, *J. Opt. Soc Am. B* **7** 1034.
3.6 M Harris, CA Hill and JM Vaughan, 1994, *Opt. Commun.* **106** 161.
3.7 MW Beijersbergen, L Allen, HELO van der Veen and JP Woerdman, 1993, *Opt. Commun.* **96** 123.
3.8 M Abramochkin and V Volostnikow, 1991, *Opt. Commun.* **83** 123.
3.9 M Padgett and L Allen, 2000, *Contemp. Phys.* **41** 275.
3.10 G Molina-Terriza, J Recolons, JP Torres and L Torner, 2001, *Phys. Rev. Lett.* **87** 023902.
3.11 MJ Padgett and L Allen, 2002, *J. Opt. B.* **4** S17.
3.12 J Courtial, K Dholakia, L Allen and MJ Padgett, 1997, *Opt. Commun.* **144** 210.
3.13 D McGloin, NB Simpson and MJ Padgett, 1998, *Appl. Opt.* **37** 469.
3.14 V Yu Bazhenov, MV Vasnetsov and MS Soskin, 1990, *JETP Letts.* **52** 429.
3.15 NR Heckenberg, R McDuff, CP Smith, H Rubinsztein-Dunlop and MJ Wegener, 1992, *Opt. Quantum Electron.* **24** S951.
3.16 NR Heckenberg, R McDuff, CP Smith, and AG White, 1992, *Opt. Lett.* **17** 221.
3.17 MS Soskin, VN Gorshkov, MV Vasnetsov, JT Malos and NR Heckenberg, 1997, *Phys. Rev. A* **56** 4046.
3.18 MW Beijersbergen, RPC Coerwinkel, M Kristensen and JP Woerdman, 1994, *Opt. Commun.* **112** 321.
3.19 GA Turnbull, DA Robertson, GM Smith, L Allen and MJ Padgett, 1996, *Opt. Commun.* **127** 183.
3.20 J Courtial, DA Robertson, K Dholakia, L Allen and MJ Padgett, 1998, *Phys. Rev. Lett.* **81** 4828.
3.21 R Oron, N Davidson, AA Friesem and E Hasman, 2000, *Opt. Commun.* **182** 205.
3.22 MA Clifford, J Arlt, J Courtial and K Dholakia, 1998, *Opt. Commun.* **156** 300.

3.23 J Arlt and K Dholakia, 2000, *Opt. Commun.* **177** 297.
3.24 AT O'Neil, I MacVicar, L Allen and MJ Padgett, 2001, *Phys. Rev. Lett.* **88** 053601.
3.25 K Volke-Sepulveda, V Garcés-Chávez, S Chávez-Cerda, J Arlt and K Dholakia 2002, *J. Opt. B* **4** S82

PAPER 3.1

An experiment to observe the intensity and phase structure of Laguerre–Gaussian laser modes

M. Padgett, J. Arlt, and N. Simpson
J. F. Allen Research Laboratories, Department of Physics and Astronomy, The University of St. Andrews, North Haugh, St. Andrews, Fife, KY16 9SS, United Kingdom

L. Allen
Department of Physics, University of Essex, Colchester, Essex CO4 3SQ, United Kingdom

(Received 14 June 1994; accepted 3 April 1995)

We outline an easily reproduced experiment that allows the student to investigate the intensity and phase structure of transverse laser modes. In addition to discussing the usual Hermite–Gaussian laser modes we detail how Laguerre–Gaussian laser modes can be obtained by the direct conversion of the Hermite–Gaussian output. A Mach–Zehnder interferometer allows the phase structure of the Laguerre–Gaussian modes to be compared with the phase structure of a plane wave with the same frequency. The resulting interference patterns clearly illustrate the azimuthal phase dependence of the Laguerre–Gaussian modes, which is the origin of the orbital angular momentum associated with each of them. © *1996 American Association of Physics Teachers.*

I. INTRODUCTION

Lasers form a key topic within an increasing number of undergraduate physics and optoelectronic degrees. This paper presents an easily reproduced experiment that allows the student to investigate the amplitude and phase structure of various transverse laser modes. In general, transverse laser modes are best described by a product of a Hermite polynomial and a Gaussian and are known as Hermite–Gaussian (HG) modes. However, in this paper we show how other modes can be generated and their phase distribution analyzed. Specifically, we present results relating to the Laguerre–Gaussian (LG) transverse laser modes. Laguerre polynomials are more frequently encountered within quantum mechanics as the radial term in the solution to Schrödinger's time-independent wave equation for a harmonic oscillator potential (e.g., the hydrogen atom problem). As in the quantum mechanical example the presence of an azimuthal phase term may be interpreted as indicating the presence of orbital angular momentum, hence the current interest in these unusual laser modes[1–4]

II. LONGITUDINAL AND TRANSVERSE MODES WITHIN LASER OSCILLATORS

All lasers essentially consist of a gain material within which stimulated emission amplifies the light and an optical resonator to provide the feedback needed to form an oscillator. The electromagnetic field within the laser resonator must satisfy a number of boundary conditions. In the steady state, there is the requirement that it reproduce itself in phase after one round trip of the laser cavity. This gives rise to the longitudinal mode structure of the laser. For a two-mirror linear cavity of optical length L, the operating wavelength of the laser, λ, must satisfy the equation

$$m(\lambda/2)=L, \tag{1}$$

where m is an integer.

The second boundary condition constrains the transverse nature of the electromagnetic field. In free space, there is a requirement that the electromagnetic field falls to zero away from the axis of the laser cavity. This gives rise to the transverse mode structure within the laser cavity and the observed intensity profile of the laser output. The analytical form of the allowed laser modes must satisfy these conditions and be a scalar solution to the paraxial limit of Maxwell's wave equation. (The paraxial limit assumes that the light diverges at only a small angle with respect to the optical axis of the system.[5]) This requires both the field amplitude and its derivative to fall to zero at a large distance from the axis of the cavity.

The rectangularly symmetric HG modes are described in part by the product of two-independent Hermite polynomi-

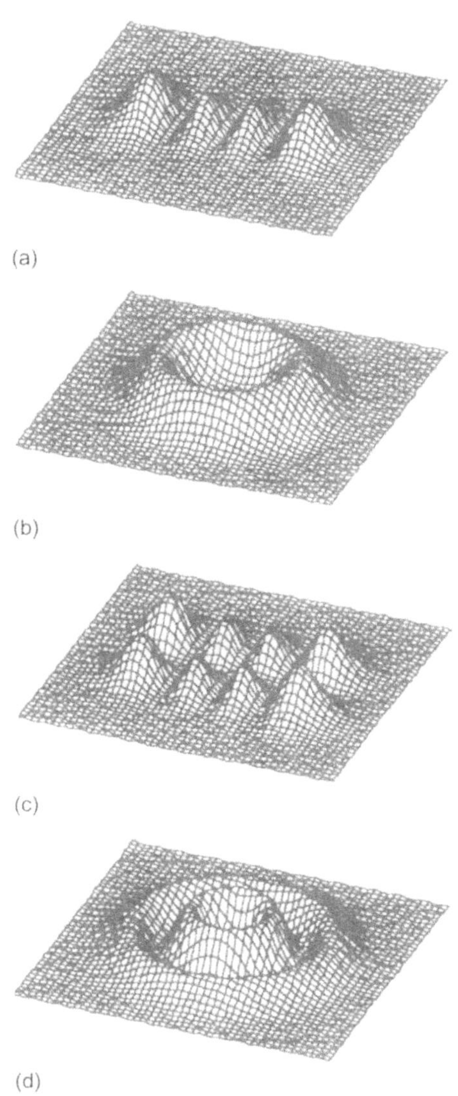

Fig. 1. Intensity ($|E|^2$) distribution of (a) a Hermite–Gaussian mode $m=3$, $n=0$, (b) a Laguerre–Gaussian model $l=3$, $p=0$, (c) a Hermite-Gaussian mode $m=3$, $n=1$, and (d) a Laguerre–Gaussian mode $l=2$, $p=1$.

als, for the field distribution in the x and y directions, respectively. They are characterized by the integer subscripts m and n which give the order of the two polynomials. The values of m and n correspond to the number of nodes in the electromagnetic field. For example, the $HG_{3,1}$ mode has three field nodes in the x direction and one in the y direction [See Figs. 1(a) and 1(c)].

The circularly symmetric LG modes are similarly described by a single Laguerre polynomial with the superscript l and the subscript p. The index l gives the number of 2π cycles of phase in the azimuthal direction around the circumference of the mode, while $p+1$ gives the number of nodes across the radial field distribution. The LG_1^2 mode has a 4π phase variation around the circumference of the mode and two radial nodes one of which is on axis. [See Figs. 1(b) and 1(d)].

As frequently encountered within quantum mechanics, it is the symmetry of the potential well and associated boundary conditions that determines the set of functions that most conveniently describes the eigenfunctions of the system. In the case of the simple harmonic oscillator, rectangular symmetry gives rise to HG eigenfunctions and circular symmetry results in LG eigenfunctions. In the case of a laser system, it would seem probable that given circular mirrors, the LG functions would provide the most accurate description of the transverse modes in a real laser, but this is not the case. Slight asymmetries in the laser cavity such as the inclusion of a Brewster window give rise to rectangular symmetry which results in a product of Hermite polynomials providing a more appropriate description. A full description with many examples of transverse laser modes can be found in Siegman's book.[6] As an example of what the eigenfunctions look like, the field amplitudes, E, transverse to the direction of propagation for the $HG_{3,1}$ and LG_1^2 modes, by way of example, are given by

$$E(HG_{31}) \propto \exp\left(-ik\frac{(x^2+y^2)}{2R}\right) \times \exp\left(\frac{-(x^2+y^2)}{\omega^2}\right) H_3\left(\frac{\sqrt{2}x}{\omega}\right) H_1\left(\frac{\sqrt{2}y}{\omega}\right)\psi(z),$$

$$E(LG_1^2) \propto \exp\left(-ik\frac{r^2}{2R}\right)\exp\left(\frac{-r^2}{\omega^2}\right)\left(\sqrt{2}\frac{r}{\omega}\right)^2 \times L_1^2\left(\frac{2r^2}{\omega^2}\right)\exp(-2i\phi)\psi(z), \quad (2)$$

where R is the radius of curvature of the near spherical wave front, k is the wave number of the electromagnetic wave, while x, y, and r are the transverse positions and radii within the beam, ϕ is the azimuthal angle within the beam, and ω is the beam radius at which the Gaussian term falls to $1/e$ of its on axis value. In both of the above expressions, the first factor relates to the phase change that results from the curvature of the wave front, and the factor $\psi(z)$ is the Gouy phase which we shall consider in detail later. The polynomials in these cases are

$$H_3\left(\frac{\sqrt{2}x}{\omega}\right) = 8\left(\frac{\sqrt{2}x}{\omega}\right)^3 - 12\left(\frac{\sqrt{2}x}{\omega}\right),$$

$$H_1\left(\frac{\sqrt{2}y}{\omega}\right) = 2\,\frac{\sqrt{2}y}{\omega}$$

and

$$L_1^2\left(\frac{2r^2}{\omega^2}\right) = 3 - \frac{2r^2}{\omega^2}. \quad (3)$$

III. THE GENERATION OF HIGH-ORDER TRANSVERSE LASER MODES

In most uses of a laser, the multilobed or multiringed nature of a high-order transverse mode is undesirable. Steps are taken to force the laser to oscillate in the fundamental mode $m=0$, $n=0$, the amplitude of which has the form of a simple Gaussian centered on axis. As a laser will tend to oscillate in the transverse mode for which the losses are lowest, the fundamental mode is often preferentially selected by the inclusion of an aperture into the cavity. Similarly, if one of the higher-order HG modes is in fact required, then this is most simply achieved by inserting a cross wire into the laser cavity with the wires aligned with the nodes of the desired mode. However, obtaining a pure LG is not so easy because the azimuthal phase term can be either clockwise or anticlockwise. The two modes have identical intensity distributions and a simple mask or aperture is therefore not sufficient to select one mode in preference to the other. Although a LG laser output has been reported, see for example the recent work by Harris *et al.*,[7] the experimental difficulties are beyond what can be expected of a teaching laboratory and we show an alternative way in which such modes may be investigated.

Recent work by Beijersbergen *et al.*[8] has demonstrated, following the work of Tamm and Weiss[9] on low-order modes, that there is an extra-cavity way to obtain pure LG modes of any order. A mode converter consisting of two cylindrical lenses can transform the HG output of a conventional laser into the corresponding LG mode. In the experiment described here, we convert the HG output of a He–Ne laser into the desired LG mode and investigate the relative phase structure by superposing the LG mode on the HG mode derived from the same laser source. The resulting interference pattern allows the direct observation of the azimuthal phase dependence of the LG modes.

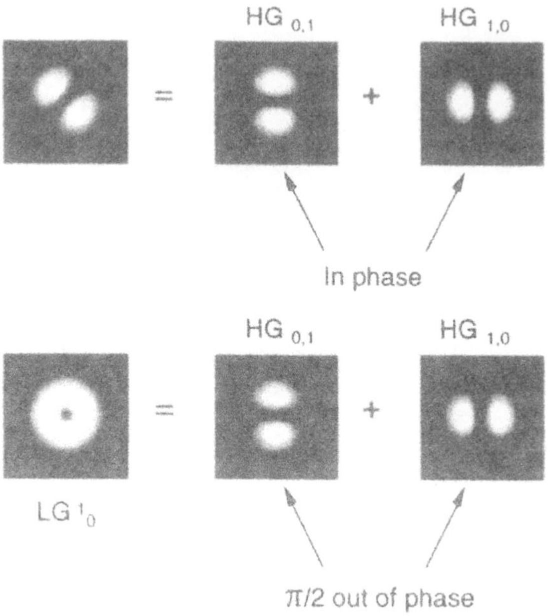

Fig. 2. The LG_0^1 mode can be formed by the superposition of the $HG_{1,0}$ and the $HG_{0,1}$ modes with a phase delay of $\pi/2$.

IV. THE OPERATING PRINCIPLE OF THE MODE CONVERTER

The HG and LG modes both form complete sets of solutions to the paraxial wave equation. Any arbitrary paraxial distribution can be described as a superposition of HG or LG terms with the appropriate weighting and phase factors. It follows that an LG mode can be described as a superposition of various HG modes weighted and phased accordingly, and vice versa. It has been shown[1,10] surprisingly recently, that the relationship between the HG and LG modes is comparatively simple, and examples detailing how various LG modes can be synthesized from combinations of HG modes are presented in Ref. 8. Figure 2 shows how a $HG_{1,0}$ mode rotated at 45° to the $x-y$ axis is equivalent to the sum of a $HG_{1,0}$ and $HG_{0,1}$ modes and how these two modes are related to the LG_0^1 mode. Specifically, the LG_0^1 can be formed by the superposition of the $HG_{1,0}$ and the $HG_{0,1}$ modes with a phase difference of $\pi/2$.

The propagation of laser beams with a HG or LG structure can be described in the usual language of Gaussian beams (see Ref. 5). The radius of the beam, ω, at which the Gaussian factor in the expression for the electric field falls to $1/e$ of its on axis value and the radius of curvature, R, of the near spherical wave fronts are given by,

$$\omega^2(z)=\frac{2(z_r^2+z^2)}{kz_r},$$

$$R(z)=\frac{(z_r^2+z^2)}{z}, \tag{4}$$

where the quantity z_r is the Rayleigh range and k is the wave number of the electromagnetic wave. The position $z=0$ corresponds to the smallest beam radius, and this position is referred to as the beam waist. The radius of the beam waist, ω_0, is related to the Rayleigh range by

$$\omega_0^2=2z_r/k. \tag{5}$$

In the vicinity of the beam waist, a Gaussian beam experiences a phase shift compared to that of a plane wave of the same frequency. This phase shift ψ is termed the Gouy phase.[11] For the simple case of a HG mode, described by mode indices m and n in the x and y directions respectively, and with identical Rayleigh ranges in the $x-z$ and $y-z$ planes, $z_{r(x-z)}=z_{r(y-z)}$, the Gouy phase is given by

$$\psi(z)=(n+m+1)\arctan(z/z_r), \tag{6}$$

while for a LG mode the phase is written

$$\psi(z)=(2p+l+1)\arctan\left(\frac{z}{z_r}\right), \tag{7}$$

where z is the distance along the axis from the beam waist in each case. If the Gaussian beam is focused by a cylindrical lens then the situation becomes more complex as the Rayleigh ranges in the $x-z$ and $y-z$ planes are not equal, $z_{r(x-z)}\neq z_{r(y-z)}$. Such a beam is termed an elliptical Gaussian beam and the corresponding Gouy phase shift for the $HG_{m,n}$ mode is given by

$$\psi(z)=\left(m+\frac{1}{2}\right)\arctan\left(\frac{z}{z_{r(x-z)}}\right)+\left(n+\frac{1}{2}\right)\arctan\left(\frac{z}{z_{r(y-z)}}\right). \tag{8}$$

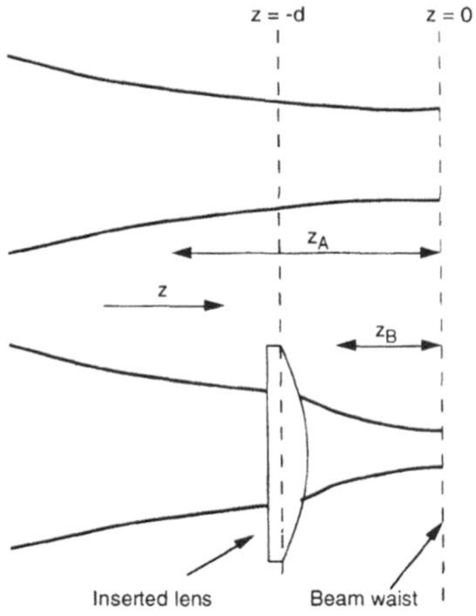

Fig. 3. A lens inserted into a Gaussian beam with a Rayleigh range z_A can be selected to change the Rayleigh range to z_B while leaving the position of the beam waist unaltered.

It is the Gouy phase shift in the presence of a cylindrical lens that forms the basis of the mode converter.

Consider a Gaussian beam traveling in the z direction with a Rayleigh range z_A and a beam waist at $z=0$. A lens inserted at $-d$ would change the radius of curvature of the beam and it proves possible to select a focal length which changes the Rayleigh range from z_A to z_B without affecting the position of the beam waist (see Fig. 3). A second, identical, lens positioned at $+d$ would change the Rayleigh range back to z_A. If the lenses are cylindrical, aligned with their axes parallel to the x direction, then only the Rayleigh range in the $y-z$ plane is affected and consequently in the region between the lenses $z_{r(x-z)}$ is no longer equal to $z_{r(y-z)}$ and the beam waist is elliptical (see Fig. 4). The differing Rayleigh ranges in the $x-z$ and $y-z$ planes means that in transmission through the lens pair, the $HG_{m,n}$ and $HG_{n,m}$ $(m \neq n)$ modes will undergo different Gouy phase shifts. For the generation of the LG_0^1 mode, we require that this phase difference be

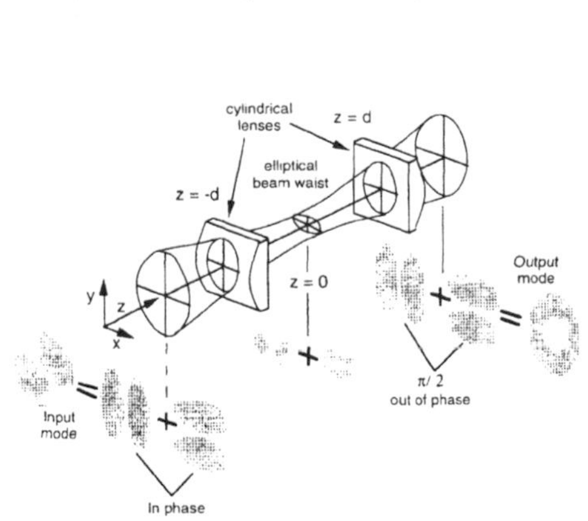

Fig. 4. The cylindrical lens mode converter. If the input $HG_{1,0}$ mode is oriented at 45° with respect to the cylinder axis of lens the mode is converted into the LG_0^1 mode.

$\pi/2$. Beijersbergen *et al.* show that this occurs when the separation, $2d$, of the cylindrical lenses of focal length f is given by

$$2d=f\sqrt{2}, \tag{9}$$

and the Rayleigh range, z_r, of the incident beam is given by

$$z_r=(1+1/\sqrt{2})f. \tag{10}$$

The correct Rayleigh range of the beam entering the mode converter is established using an additional spherical lens of an appropriate focal length, prior to the mode converter, positioned to form a beam waist at $z=0$. For a diverging input beam, with a wave-front radius of R_i and of radius ω, the required focal length, f_{in}, of the input lens is given by

$$\frac{1}{f_{\text{in}}}=\frac{\sqrt{2k\omega^2 z_r-4z_r^2}}{k\omega^2 z_r}+\frac{1}{R_i}, \tag{11}$$

where k is the wave number of the light.

It can be shown that in addition to the transformation of the $HG_{1,0}$ mode, a mode-converter of the above design will transform any $HG_{m,n}$ mode, rotated at 45° to the axis of the lens, into the corresponding LG_p^l mode with $l=|m-n|$ and $p=\min(m,n)$.

V. EXPERIMENTAL APPARATUS

The design of the mode converter used in this work is based on the simple equations given in the previous section and has the following specifications:

25.4 mm focal length cylindrical lenses (e.g., Newport CKX025[12]),

35.9 mm lens separation (between the principal planes) and an input beam with a Rayleigh range of 43.3 nm.

The lenses are mounted with their plane faces cemented to a 19 mm long tube to set the correct separation of their principal planes.

The optimum focal length for the input lens can be calculated from the diameter of the input beam and its wave-front curvature. However, in practice it is simpler to experiment with a number of different focal lengths and select the one that gives the best results. In our case, a single lens of 250 mm focal length is used to focus the output of the HeNe laser into the mode converter.

For the laser source, we use an unmounted He–Ne tube with a Brewster window and an external output coupler.[13] By positioning the output coupler 100 mm from the Brewster window, an $x-y$ translation stage carrying a crosswire (made from 10 μm diam tungsten wire) can be inserted into the laser cavity. With careful adjustment of the cross wire and alignment of the output coupler, the student can force the laser to oscillate in a variety of higher-order HG modes.

In order to observe the phase structure of the LG modes it was necessary to extend the experimental arrangement beyond that of the mode converter, as shown in Fig. 5. The experimental arrangement is based on a Mach–Zehnder interferometer, with the mode converter and input lens ($f=250$ mm) in one arm and a beam expansion telescope in the other. The purpose of the lens ($f=160$ mm) after the mode converter is to form a second beam waist outside the interferometer, so that the mode structure can be examined at various positions either side of a beam waist. The beam expansion telescope consists of two lenses ($f=40$ mm and $f=300$ mm) to give an expanded, collimated output beam. This beam is

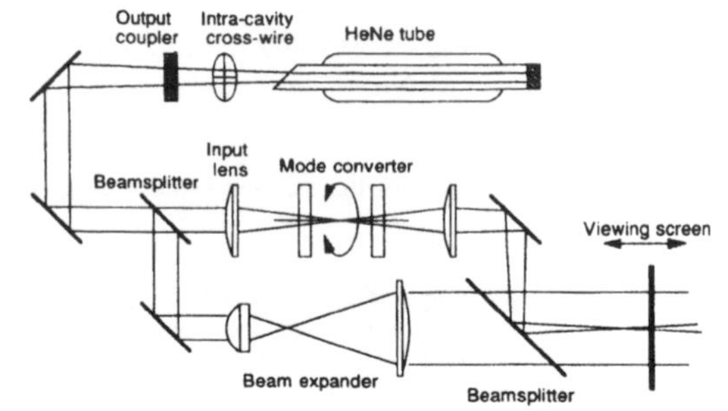

Fig. 5. The experimental arrangement for the observation of the intensity and phase structure of Laguerre–Gaussian transverse laser modes.

an expanded form of the input HG mode. Although having a Hermite Gaussian intensity profile, the large Rayleigh range now achieved for the mode results in a near-planar wave front that acts as a phase reference for the examination of the LG modes.

VI. EXPERIMENTAL OBSERVATION OF THE INTENSITY AND PHASE STRUCTURE OF LG MODES

Initial experiments can be completed with the beam expander arm blocked off. The results presented in Ref. 8 can be repeated; namely after selecting a high-order HG mode from the laser the student can examine the intensity distribution of the beam after it is transmitted by the mode converter. If the cylindrical lenses are rotated about the optical axis so that the axis of the cylinders are at 0° or 90° to the input HG mode then the transmitted light is also an HG mode. If, however, they are rotated to lie at 45°, the input $HG_{n,m}$ mode is transformed to the corresponding LG_p^l mode ($l=|m-n|$ and $p=\min(m,n)$).

After the various intensity patterns have been observed, the student can superimpose the expanded HG mode, aligned so that the LG mode falls within one of the greatly expanded lobes, and the resulting interference pattern can be observed. The expanded HG mode acts as a plane-wave reference and the observed fringes reveal the relative phase variations of the HG or LG mode in question.

Figure 6 shows the intensity distributions and interference patterns of a $HG_{3,0}$ mode and the corresponding LG_0^3 mode at various positions about the beam waist. These intensity distributions can be observed simply by placing a screen in the beam path. For the purpose of publication and student reports, permanent record of the intensity distributions can be obtained by directly exposing a CCD array interfaced to a framegrabbing card running on a desktop computer. Subsequently, the interferogram image can be printed or transferred into a variety of applications running on the same computer.

As can be seen from the figure, at the beam waist the 6π phase change around the circumference of the LG_0^3 mode gives rise to three dark radial fringes. Away from the beam waist, spiral fringes are observed. These arise from the combination of the radial phase variation due to wave-front curvature, and the tangential phase variation due to the azimuthal phase dependence. The sense of the spiral is reversed on either side of the waist in accordance with the curvature

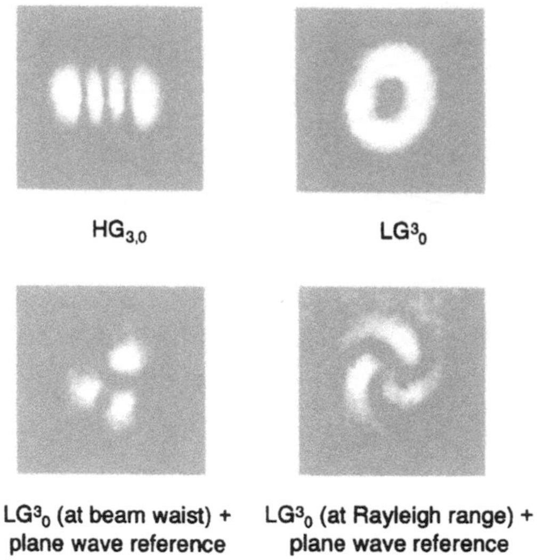

Fig. 6. Photographs of the Hermite–Gaussian $HG_{3,0}$ mode, the transformed Laguerre–Gaussian LG_0^3 mode and interferograms between the LG_0^3 mode and a plane wave.

of the wave fronts. In addition, the sense of the spirals can be reversed by changing the sense of the azimuthal phase variation. This is most readily achieved by rotating the mode converter about the optical axis by 90° (i.e., from +45 to −45° with respect to the input mode).

Figure 7 shows the intensity and interference patterns ob-

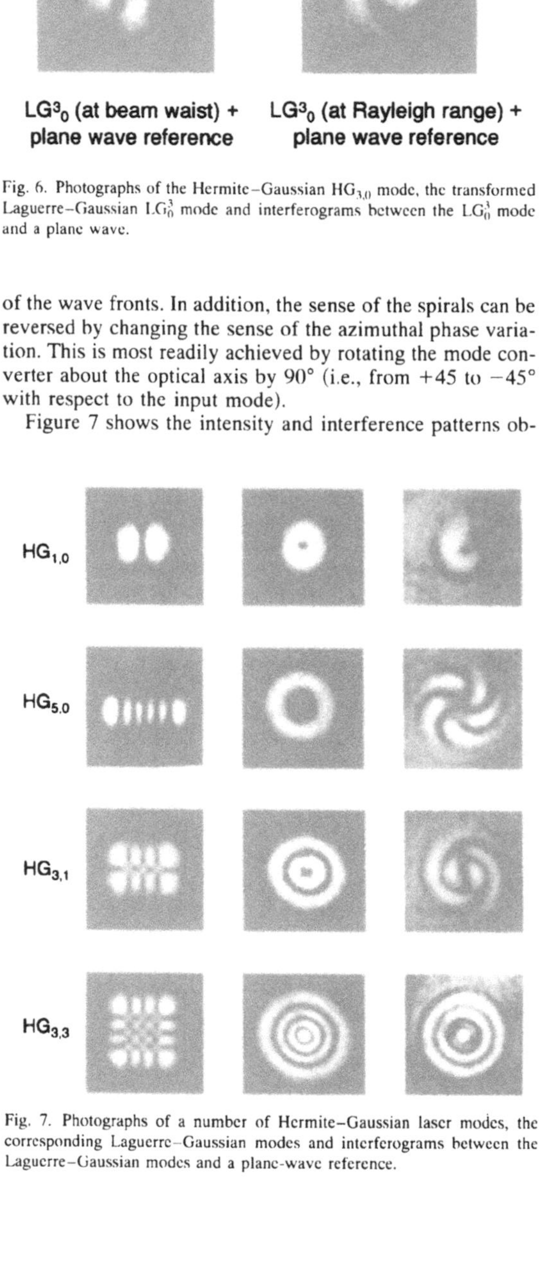

Fig. 7. Photographs of a number of Hermite–Gaussian laser modes, the corresponding Laguerre–Gaussian modes and interferograms between the Laguerre–Gaussian modes and a plane-wave reference.

tained for several higher-order HG and LG modes. In all cases the number of radial fringes is equal to l (i.e., $|m-n|$) and the number of radial nodes is equal to $p+1$ [i.e., $\min(m,n)+1$]. For the LG modes with multiple ring amplitudes (e.g., the LG_1^2, which is obtained from the $HG_{3,1}$), the azimuthal position of the fringe maxima and minima are reversed between successive rings, corresponding to the change of phase of the Laguerre–Gaussian distribution. Note also that in the special case of $m=n$ there is an on axis intensity for the LG mode but no azimuthal phase dependence and hence the fringes are circular and not spiral. The radii of the fringes can be calculated from the wave-front curvature of the beam and are analogous to Newton's rings.

VII. CONCLUSIONS

This paper presents an easily reproduced experiment with which the student can investigate the intensity and phase structure of various transverse laser modes. In addition to the standard Hermite–Gaussian laser modes we detail how Laguerre–Gaussian laser modes can be obtained by the direct conversion of the Hermite–Gaussian laser output using a mode converter utilizing the Gouy phase shift in the region of an elliptical beam waist. Using a Mach-Zehnder interferometer, the phase structure of the Laguerre–Gaussian modes can be compared with that of a plane wave of the same frequency.

One of the current sources of interest in the LG laser modes is that they are predicted to carry orbital angular momentum in addition to the spin momentum associated with the polarization state of the photons.[1] By superimposing the LG mode with a plane-wave reference, the resulting interference patterns clearly illustrate the azimuthal phase dependence of the Laguerre–Gaussian modes which is the origin of the orbital angular momentum associated with these modes.

ACKNOWLEDGMENT

Miles Padgett is a Royal Society of Edinburgh Research Fellow.

[1]L. Allen, M. W. Beijersbergen, R. J. C. Spreeuw, and J. P. Woerdman, "Orbital angular momentum of light and the transformation of Laguerre-Gaussian laser modes," Phys. Rev. A **45**, 8185–8189 (1992).

[2]M. W. Beijersbergen, M. Kristensen, and J. P. Woerdman, "Spiral phase-plate used to produce helical wavefront laser beams," Technical Digest, CLEO. Europe, CFA5 (1994).

[3]M. J. Padgett and L. Allen, "The Poynting vector in Laguerre-Gaussian modes," submitted to Optic Commun.

[4]M. Babiker, W. L. Power,and L. Allen, "Light-induced torque on moving atoms," Phys Rev. Lett. **73**, 1239-1242 (1994).

[5]For example see R. Guenther, *Modern Optics* (Wiley, New York, 1990), pp. 336–343; A. E. Siegman, *Lasers* (University Science Books, Mill Valley, 1986), Sec. 7.3, pp. 276–279.

[6]A. E. Siegman, *Lasers* (University Science Books, Mill Valley, 1986), Sec. 17.5, pp. 685–695.

[7]M. Harris, C. A. Hill, and J. M. Vaughan, "Optical helices and spiral interference fringes," Opt. Comm. **106**, 161–166 (1994).

[8]M. W. Beijersbergen, L. Allen, H. E. L. O. van der Veen, and J. P. Woerdman, "Astigmatic laser mode converters and transfer of orbital angular momentum," Opt. Comm. **96**, 123–132 (1993).

[9]Chr. Tamm and C. O. Weiss, "Bistability and optical switching of spatial patterns in a laser," J. Opt. Soc. Am. B **7**, 1034–1038 (1990).

[10]S. Danakas and P. K. Aravind, "Analogies between two optical systems (photon beam splitters and laser beams) and two quantum systems (the two-dimensional oscillator and the two-dimensional hydrogen atom)," Phys. Rev. A **45**, 1973–1977 (1992).

[11]A. E. Siegman, *Laser* (University Science Books, Mill Valley, 1986), Sec. 17.4, pp. 682–685.

[12]Newport Optics Catalogue.

[13]Melles Griot, He–Ne plasma tube 05LHB570, Output coupler L01002-1.

PAPER 3.2

Temporal and interference fringe analysis of TEM_{01}* laser modes

J. M. Vaughan and D. V. Willetts

Royal Signals and Radar Establishment, St. Andrews Road, Great Malvern, Worcestershire WR14 3PS, UK

Received February 11, 1983

The properties of TEM_{01}* doughnut modes have been examined by frequency analysis and by two-beam interference. Interpretation in terms of an evolving helix, made up to two orthogonal TEM_{01} modes of different frequency, is supported by computer simulation of fringe patterns. These patterns are shown to correspond closely with photographic recordings; the implications of the phenomena for the crossed-beam technique in laser velocimetry are outlined. Finally, the possibility of developing a beam of pure helical cophasal surface is discussed and its interference patterns analyzed.

INTRODUCTION

In previous publications[1,2] we have investigated two-beam interference patterns that are due to TEM_{00} and TEM_{01} laser modes of different relative power. It was suggested that the TEM_{01}* doughnut mode, composed of two TEM_{01} modes in phase and space quadrature, may possess a cophasal surface of helical form. However, in order to explain the observed interference patterns derived from several high-gain Ar^+- and Kr^+-laser lines, it was necessary to suppose that the observed doughnut mode was an evolving helix, which switched rapidly back and forth between a left-hand and a right-hand character. In our earlier papers the interference theory of these modes was performed analytically in terms of phase contours; we now present new results on the temporal behavior of these modes that explain the previous assumptions of helical hand-switching. These results are incorporated into computer calculations and graphical constructions of the fringe profiles, which are found to be in good agreement with experimental photographic recordings. The calculations have been carried out both for evolving helices of the TEM_{01}* mode alone and for various admixtures of the TEM_{00} mode. In the final section we propose a simple means of producing a stationary helical wave of either hand and draw attention to unusual interference properties that it should possess.

It may be remarked that interest in these modes is not confined solely to the unusual character of the wave surfaces and their interference patterns. These modes also have considerable practical importance in photon-correlation spectroscopy and in crossed-beam fringe velocimetry. In the first case considerable reduction in the strength of fluctuations may occur, leading to a reduced signal-to-noise ratio; this important problem is discussed in detail, both experimentally and theoretically, in Ref. 3. In velocimetry the reduced fringe visibility and changing phase may introduce considerable distortion of experimental data; this is outlined in the present paper. These effects are of course particularly troublesome if such modes are additionally present in a nominally TEM_{00} laser. In the past such effects were widely found, particularly when high-gain ion lasers were used, but they have not hitherto been explained or evaluated in detail.

TEMPORAL MEASUREMENTS

The source employed in this work was a Kr^+-ion laser oscillating at 476 nm, with an intracavity étalon for longitudinal-mode selection. It must be emphasized, however, that the phenomena are quite general for high-gain laser lines. Similar two-beam interference patterns have been seen in many lines from Ar^+- and Kr^+-ion lasers and have also been observed in the interference patterns that are due to a doughnut mode from a CO_2 laser operating at 10.6 μm, viewed with a thermal imager. For the present work the laser output frequency was analyzed in the 10–500-MHz range with a 20-cm scanning, confocal, Fabry–Perot interferometer. If the whole mode were allowed to enter the interferometer, it was found to consist of a single frequency when the laser was emitting a TEM_{00} mode; by adjustment of étalon tilt and discharge current a pure TEM_{01}* mode, with characteristic dark center, could be obtained and was found to consist of two frequencies ω_2 and ω_3 separated by 128 MHz. The spatial distribution of these frequencies was investigated by isolating a small fraction of the mode in the (ρ, θ) plane perpendicular to the propagation direction. This was accomplished by scanning a small aperture in this plane and subjecting the light that passed through it to spectral analysis. Frequency ω_2 was found to be concentrated near the horizontal and ω_3 near the vertical axis, qualitatively consistent with spatial distributions of the kind $\rho^2 \exp(-\rho^2/\sigma^2)\cos^2\theta$ and $\rho^2 \exp(-\rho^2/\sigma^2)\sin^2\theta$ to be expected of TEM_{01} modes.[4]

These observations were confirmed by beat-frequency analysis in the region between dc and 1.8 GHz by using a small silicon avalanche photodiode coupled to a Tektronix 7L12 spectrum analyzer. The magnitude of the beat-frequency signal at $|\omega_3 - \omega_2|$ (equal to 128 MHz) maximized along the $\theta = \pm 45°$ directions and was zero along the vertical and horizontal axes, distributed qualitatively as $\rho^2 \exp(-\rho^2/\sigma^2) \sin 2\theta$. We conclude from the spectral data that the TEM_{01}* mode does indeed result from the superposition of two TEM_{01} modes in space quadrature but that in our case they are nondegenerate in frequency, being adjacent longitudinal modes of the laser resonator whose length L is 117 cm ($c/2L$ = 128 MHz). There appears to be no fundamental reason why both modes should not be degenerate; in the present case the degeneracy is almost certainly lifted by cavity asymmetry introduced by the Brewster windows, dispersive prism, etc. The fact that the polarization direction, defined by the Brewster windows and the prism, is vertical and coincident with the axis of one of the TEM_{01} modes lends support to this notion.

A consequence of the different frequencies of the two TEM_{01} modes is that their resultant evolves from a left- to right-handed helix and back again via intermediate TEM_{01} modes orientated along the bisectors of the ω_2 and ω_3 modes; the evolution repeats at 128 MHz. This fact explains the earlier assumption of an evolving helical mode of changing hand that was required to give the observed interference patterns. It is straightforward to show that a simple helix is equivalent to two TEM_{01} modes of the same frequency in phase and space quadrature. More generally, two orthogonal TEM_{01} modes of the same frequency form an equivalent basis set to two helical modes of opposite hand, also at the same frequency.

FRINGE-PATTERN FORMULATION

Calculations of fringe intensity are carried out with the coordinate basis shown in Fig. 1. The centers of the two interfering beams are at $(\pm R, 0)$, and the two beams are propagating into the plane of the paper to intersect at an angle 2α. We consider the general case in which the beams are composed of both a Gaussian (TEM_{00}) mode at frequency ω_0 and a doughnut ($TEM_{01}{}^*$) mode oscillating initially at a single frequency ω_1, i.e., operating as a pure helix. Then the expression for the electric field at any point in the plane is given by

$$\begin{aligned} E(\rho, \theta, t) = {} & C[\exp(-\rho_1{}^2/2\sigma^2)\sin(\omega_0 t - \phi_1) \\ & + \exp(-\rho_2{}^2/2\sigma^2)\sin(\omega_0 t + \phi_2)] \\ & + (\rho_1/\sigma)\exp(-\rho_1{}^2/2\sigma^2)\sin(\omega_1 t \pm \theta_1 - \phi_1) \\ & + (\rho_2/\sigma)\exp(-\rho_2{}^2/2\sigma^2)\sin(\omega_1 t + \theta_2 + \phi_2), \end{aligned} \quad (1)$$

where the term within brackets represents the Gaussian contribution and the remainder the helical wave; we have used the fact that the phase of a helical wave is equal to the azimuthal angle θ. C is a constant representing the relative TEM_{00} field strength, and σ describes the mode radius. In the third term, the plus applies when the helices are uninverted with respect to each other, the minus when they are inverted (Figs. 1b and 1c). The phase angles ϕ_i arise from the

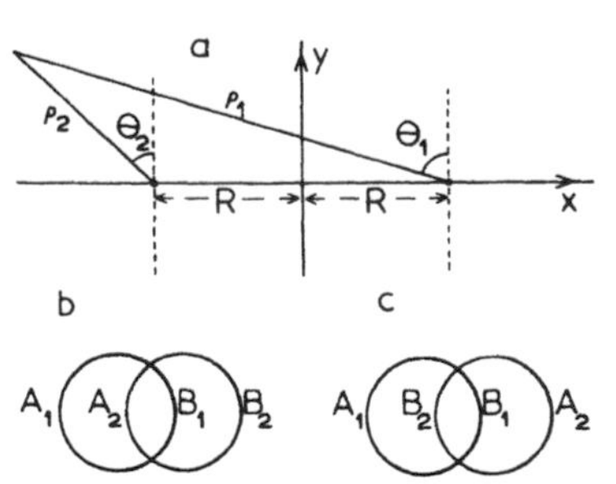

Fig. 1. (a) Coordinate system used for fringe intensity calculations. The angles θ_1 and θ_2 are constrained within the limits $-(\pi/2) < \theta_1 \leqslant (3\pi/2)$ and $-(3\pi/2) < \theta_2 \leqslant (\pi/2)$ chosen to simplify the analysis. (b) Uninverted beams, (c) inverted beams, both shown displaced sideways.

inclination α of the beams to the (ρ, θ) plane and are given by $(2\pi/\lambda)\sin\alpha \times \rho_i \sin\theta_i$ $(i = 1, 2)$. Since the beams are inclined in opposite directions, the ϕ_i appear in Eq. (1) with opposite signs. The intensity arising from the field of Eq. (1) is obtained by taking the square modulus and time averaging in the usual way, with the result that

$$\begin{aligned} I \propto {} & \tfrac{1}{2} \sum_{i=1}^{2} (a_i{}^2 + b_i{}^2) + a_1 a_2 \cos(\phi_1 + \phi_2) \\ & + b_1 b_2 \cos(\phi_1 + \phi_2 \mp \theta_1 - \theta_2), \end{aligned} \quad (2)$$

where

$$a_i = C\exp(-\rho_i{}^2/2\sigma^2) \qquad i = 1, 2,$$
$$b_i = (\rho_i/\sigma)\exp(-\rho_i{}^2/2\sigma^2) \qquad i = 1, 2.$$

This is the result for an unchanging helix. Our previous observations show that the helix switches hand at the intermode frequency $|\omega_2 - \omega_3|$ of 128 MHz. Writing the complete expression for this case shows that the third term in expression (2) should be replaced with

$$\tfrac{1}{2} b_1 b_2[\cos(\phi_1 + \phi_2 \mp \theta_2 - \theta_1) + \cos(\phi_1 + \phi_2 \pm \theta_1 + \theta_2)].$$

Under these circumstances the intensity becomes

$$\begin{aligned} I \propto {} & \tfrac{1}{2} \sum_{i=1}^{2} (a_i{}^2 + b_i{}^2) \\ & + [a_1 a_2 + b_1 b_2 \cos(\theta_2 \pm \theta_1)]\cos(\phi_1 + \phi_2). \end{aligned} \quad (3)$$

COMPUTATIONAL PROCEDURE

A Cartesian basis set was chosen that is related to the polar coordinates of Fig. 1 by

$$x = (-1)^{i+1}R - \rho_i \sin\theta_i, \qquad i = 1, 2,$$
$$y = \rho_i \cos\theta_i, \qquad i = 1, 2.$$

A program to display $I(x, y)$ was written in BASIC by using the Hewlett-Packard 9800 Series calculator and plotter. For fixed $y = y_0$, $I(x, y)$ was computed at a series of points in x differing by Δx, and these points were plotted to give a vertical section through the surface $I(x, y)$ at $y = y_0$; y_0 was then incremented to $y_0 + \Delta y$, and the computation was repeated. The results were plotted with a vertical shift on the chart of $K \times \Delta y$ $(K < 1)$ and a horizontal stagger s. The process was repeated until the whole pattern was built up from a series of vertical sections to give the illusion of an isometric projection of the fringe pattern, as shown in Figs. 2a–2d. For small s, the apparent incidence angle of view is $\cos^{-1}K$. The input data to the program were the limits in x and y and the respective increments Δx and Δy, K and s, and the fringe data C, σ, R, and $(2\pi/\lambda)\sin\alpha$. In the examples shown, the centers of the beams are marked if they are separated; a scale of σ is also provided. $(2\pi/\lambda)\sin\alpha$ was chosen to give 20 fringes across the diagram, and K was 0.8, so that the apparent incidence angle was 37°.

The computations for various configurations may be compared with the experimentally obtained interferograms of Figs. 3a–3f. These photographs were obtained by two-beam interference, which provides a useful examination of spatial coherence.[5,6] The inclusion of an intracavity étalon in the laser permitted longitudinal-mode selection and some degree of tuning of the relative intensity of the two TEM_{01} modes.

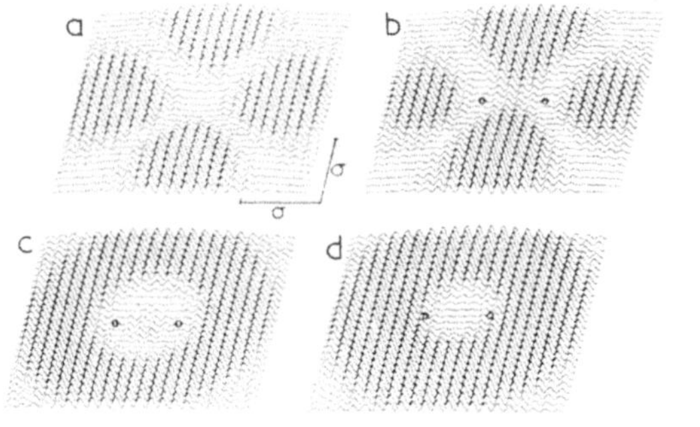

Fig. 2. Computer reconstruction of two-beam interferograms with a TEM_{01}* laser mode: (a) beams inverted (i.e., Fig. 1c) but not displaced, (b) beams inverted and displaced horizontally, (c) beams uninverted and displaced horizontally (i.e., Fig. 1b); (d) like (c) but with the addition of 14% TEM_{00} mode. Note the half-cycle displacements in the fringes in different regions; see text for details of the perspective. The centers of the separated beams are marked by circles.

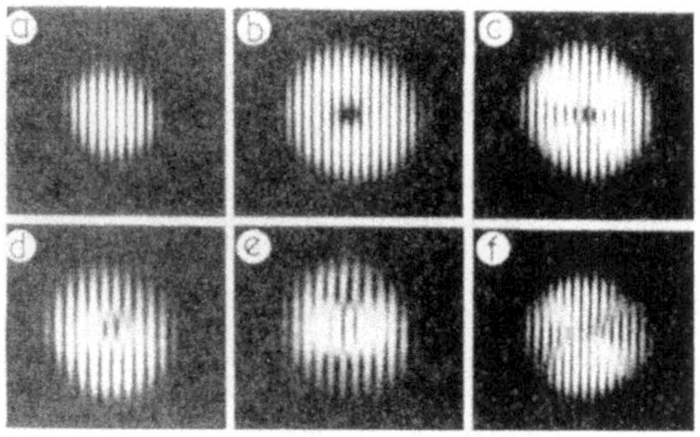

Fig. 3. Two-beam interference fringes. The top row shows the beams overlapped exactly without displacement: (a) Gaussian TEM_{00} mode, (b) doughnut TEM_{01}* mode uninverted (i.e., Fig. 1b), (c) doughnut TEM_{01}* mode inverted (i.e., Fig. 1c). The lower row shows the effect of lateral displacement of the beams: (d) doughnut TEM_{01}* mode uninverted and displaced horizontally, (e) doughnut TEM_{01}* mode uninverted and displaced vertically, (f) doughnut TEM_{01}* mode inverted and displaced horizontally. Note in particular the half-cycle changes on the vertical fringes and compare them with Fig. 2.

Details of the procedure for photographing the fringe patterns have been described in Refs. 1 and 2. In essence, the original laser beam was divided at a prism beam splitter, and the two beams were enlarged and recombined on a translucent screen for ease of examination. The angle between the beams and the degree of overlap could be varied in both the inverted and the uninverted arrangements. The clearest fringe patterns, as might be expected, were obtained with the two TEM_{01} modes of nearly equal intensity. In these two figures the corresponding theoretical and experimental recordings are those of Figs. 2a and 3c, 2c and 3d, and 2b and 3f, respectively. It is clear that there is good correspondence between the appropriate pairs of diagrams; the fringe-smearing region where the phase difference between helices approaches $\pm\pi/2$ is clearly reproduced in Fig. 2 as an area of constant nonzero intensity separating regions in which the fringes are of opposite phase. Figure 2d illustrates that addition of TEM_{00} mode contracts the phase-changed circle of Fig. 2c; in contrast, for the inverted case it pushes the limbs of the phase-change cross in Fig. 2a apart vertically. If the helices are separated to the extent shown in Fig. 2b, the addition of the same amount of TEM_{00} power ($C = 0.4$) used in these calculations gives rise to a pattern similar to that of Fig. 2a.

The precautions needed in using the crossed-beam technique in laser velocimetry have been outlined in Refs. 1 and 2 and are graphically illustrated in Figs. 2 and 3. The problem is readily appreciated by considering a small particle traversing the fringes shown in these figures. If the beams are recombined in an uninverted manner, then at the region of exact overlap (e.g., Fig. 3b), there is at least no change of phase, although the beam profile varies with the particular trajectory and is far from Gaussian. However, even in this uninverted case, there are changes of phase away from exact overlap (Fig. 3d); on the other hand, if the beams are inverted and recombined, phase changes occur throughout (Figs. 3c and 3f). In practice it is unlikely that such a pure doughnut mode would pass unnoticed. Much more insidious is the possibility of a residual TEM_{00} mode's filling in the hole of the TEM_{01}* mode to give an apparently smooth Gaussian-like profile. This may be illustrated by the example with as low as 14% of the power in the TEM_{00} mode. At first sight it might be thought that, with 86% of the power in the TEM_{01}* mode, a beam would appear clearly to be a doughnut; in fact, the central intensity is as much as 37% of the maximum, and, given the inherent nonlinearity of the eye (and photographic emulsion) particularly at high intensity, this central minimum might well pass unnoticed, with the beam being classed as Gaussian. Indeed, several workers have commented on the occasional sudden enlargement of a beam as it snaps into a larger, but apparently Gaussian, mode. It can now be recognized that this is almost certainly the growth, as the laser gain conditions become favorable, of a TEM_{01}* mode around the lowest-order mode. Techniques of laser velocimetry employing the Doppler difference or fringes technique usually rely on accurate measurement of the time intervals between fringe peaks (for strong scattering) or on accumulating an intensity-correlation function (most particularly for weak scattering). Both methods rely on accurate knowledge of the fringe geometry; in the latter case, for work of the highest accuracy, reliable information is needed on the beam profile, usually considered to be Gaussian (indeed, this assumption, that all trajectories of particles through the beam give rise to Gaussian scattering profiles of equivalent form, is usually fundamental). Obviously the distortion of the fringe patterns introduced by higher-order laser modes must introduce considerable loss of precision and speed. The conclusion must be that stringent precautions should be taken when two-beam fringe velocimetry is used to ensure that the laser beam is in the lowest-order TEM_{00} mode. The precautions should include, at least for high-gain lasers, adjustment of an internal cavity aperture and, most importantly, detailed optical examination of the fringes under the actual conditions of use.

GENERATION OF HELICAL WAVE OF CONSTANT HAND

In an earlier section of this paper it was shown that the TEM_{01}* mode described in our experiments evolved back and forth between left- and right-hand helices since the two constituent TEM_{01} modes were oscillating at different frequencies. In Fig. 4 we show an arrangement designed to produce

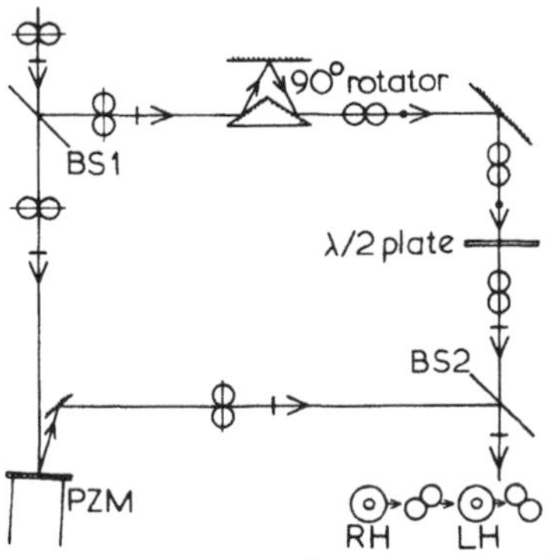

Fig. 4. Optical arrangement to produce a beam of pure helical cophasal form from a beam initially in a two-spot TEM_{01} mode of single frequency. The two paths between the beam splitters BS1 and BS2 should be arranged to be approximately equal. The axis of the two-spot mode lies in the plane of the paper when the modes are shown normal to the beam and is normal to the plane of the paper when the mode axis lies along the beam. The optical path and phase between the two beams may be varied by the piezoadjustable mirror at PZM. As the phase is changed the beam varies between a left-hand and a right-hand helix, as shown.

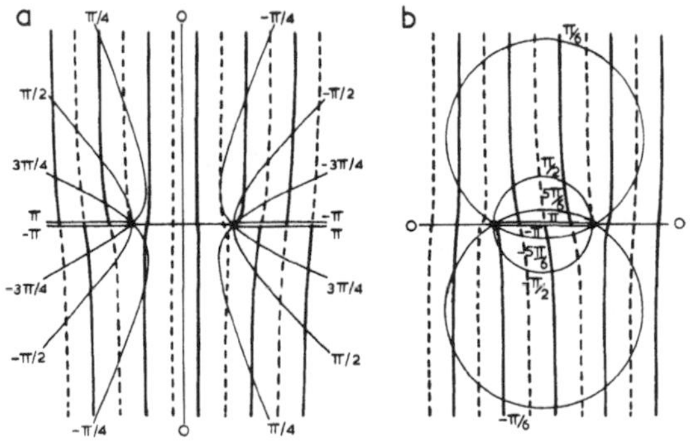

Fig. 5. Phase-difference contour maps (thin lines) and fringe patterns (thick lines, continuous and dashed) for two-beam interference of a true helical TEM_{01}^* laser mode: (a) inverted and separated beams, (b) uninverted and separated beams. The contours are labeled in phase difference.

a helical TEM_{01}^* mode of either hand. A single-mode linearly polarized laser is forced to oscillate on a TEM_{01} mode by inclusion of a pin within its resonator. Its output is fed into a Mach–Zehnder interferometer having in one arm an optical beam rotator, such as a Dove prism, so that the two beams are in space quadrature. Collinearity of output polarization is ensured by an appropriately orientated half-wave plate, and fine length tuning of one of the interferometer arms (e.g., by piezoelectric means) is provided so that the desired phase relation between the two TEM_{01} beams may be produced. As this phase relationship is varied, it is readily seen that the resultant mode moves through a series of patterns. Starting in phase to produce a pure right-hand helix, a $\pi/2$ change produces a two-lobe distribution (with its axis lying between the axes of the two primary beams). A further $\pi/2$ change gives a left-hand helix, and another $\pi/2$ change produces a two-lobe distribution with its axis at right angles to the previous one. A final $\pi/2$ displacement, making 2π, brings the mode back to the original right-hand helix.

The fringe patterns to be expected from such a single-hand helical mode may be analyzed by recourse to the phase-difference diagram method developed in Ref. 1. In previous work calculations were performed modulo π since hand-switching phenomena were in evidence; for a single-hand helix the effects of the multivalued nature of the phase-difference surface must be included. Singularities called branch points[7] arise at the centers of the helical modes, joined by a discontinuity in the surface, which is a branch line. Figure 5 shows the phase-difference contour maps for both uninverted and inverted displaced TEM_{01}^* modes with predicted fringe patterns overlaid. Branch lines are shown as double (where there is a cliff of altitude 2π), arranged to contract when a TEM_{00} mode is mixed in. The fringes are drawn continuous and dashed to distinguish between adjacent order numbers; the order number changes by one as a branch line is crossed, corresponding to the 2π phase change. A fringe terminates on each of the branch points, but, since these points are at zero intensity, the fringe will not terminate abruptly. Investigations of single-hand interference patterns will be the subject of a future publication.

ACKNOWLEDGMENTS

We are indebted for useful discussions to several colleagues at the Royal Signals and Radar Establishment and in particular to W. A. Cambridge, formerly of Trinity College, Cambridge, for pointing out the existence of branch lines and for the computer programming.

REFERENCES

1. J. M. Vaughan and D. V. Willetts, "Interference properties of a light beam having a helical wave surface," Opt. Commun. **30**, 263–267 (1979).
2. D. V. Willetts and J. M. Vaughan, "Properties of a laser mode with a helical cophasal surface," in *Laser Advances and Applications*, B. S. Wherrett, ed. (Wiley, New York, 1980), pp. 51–56.
3. P. N. Pusey, J. M. Vaughan, and D. V. Willetts, "Effect of spatial incoherence of the laser in photon-correlation spectroscopy," J. Opt. Soc. Am. **73**, 1012–1017 (1983).
4. H. Kogelnik and T. Li, "Laser beams and resonators," Appl. Opt. **5**, 1550–1566 (1966).
5. M. Bertolotti, B. Daino, F. Gori, and D. Sette, "Coherence properties of a laser beam," Nuovo Cimento **38**, 1505–1514 (1965).
6. M. Carnevale and B. Daino, "Spatial coherence analysis by interferometric methods," Opt. Acta **24**, 1099–1104 (1977).
7. P. M. Morse and H. Feshbach, *Methods of Theoretical Physics* (McGraw-Hill, New York, 1953), Part I, Sec. 4.4.

PAPER 3.3

Bistability and optical switching of spatial patterns in a laser

Chr. Tamm

Physikalisch-Technische Bundesanstalt, Braunschweig, Federal Republic of Germany

C. O. Weiss

Physics Department, University of Queensland, St. Lucia/Brisbane, Australia

Received September 29, 1989; accepted December 27, 1989

Bistability of the transverse configuration of the optical field is found for a helium–neon laser oscillating in the $TEM_{01}{}^*$ hybrid optical-resonator mode. Optically induced switching between the mirror-symmetric field patterns occurs when a fraction of the spatially rotated output field is reinjected into the laser resonator. These phenomena are described theoretically with laser equations modeling the mutual interaction of the two first-order transverse modes of the optical resonator. The experimentally observed bistability and switching of the field patterns is in full qualitative agreement with the theoretical results.

1. INTRODUCTION

The formation and competition of radiation patterns in nonlinear-optical systems is one manifestation of self-organization in nonequilibrium systems with spatial degrees of freedom.[1] In the case of transverse field patterns in nonlinear-optical resonators, the transverse variation of amplitude and phase of an outcoupled light field can yield complete information on the nature of the radiation patterns formed inside the cavity. Conversely, the conditions of pattern formation in nonlinear-optical resonators may be manipulated by injecting field patterns into the resonator.

For the interaction of nearly degenerate transverse modes of the optical field in nonlinear resonators, the occurrence of symmetry breaking and Turing instabilities and the formation of stationary transverse radiation patterns have been analyzed theoretically.[2,3] It can be concluded from these studies, and also from one early publication,[4] that lasers operating in a state of cooperative frequency locking[3] of transverse resonator modes may show bistability or multistability, in the sense that it is possible to have stable laser output of any pattern from among a discrete set of different transverse field patterns for the same operating conditions.

This paper, for the first time to our knowledge, gives experimental evidence for the bistability of a transverse field pattern of a laser. Moreover, optically induced switching between the two stable field configurations is demonstrated. The experiment uses a laser that emits in the $TEM_{01}{}^*$ hybrid mode, i.e., in a superposition of TEM_{10} and TEM_{01} Gauss–Hermite optical-resonator modes[5] of equal optical frequency that maintain a constant optical phase difference of either $+\pi/2$ or $-\pi/2$.[6] It has been noted that these two values of optical phase difference lead to two physically different transverse field patterns that are mirror images of each other.[7] It is shown here that laser oscillation in the $TEM_{01}{}^*$ hybrid mode is in fact bistable in these two field patterns. Hysteresis and switching between the stable states is observed when the transversally rotated output field pattern of the laser is reinjected into the laser resonator with a changing optical phase.

The experimental scheme is shown in Fig. 1. A single-longitudinal-mode helium–neon laser operates in a stationary $TEM_{01}{}^*$ hybrid-mode configuration.[7] The gain medium of this laser is inhomogeneously (Doppler) broadened; the optical resonator is of the Fabry–Perot type. The laser output is split into three beams. One beam is used for viewing the $TEM_{01}{}^*$ emission with its annular, rotationally symmetric intensity pattern. Another beam passes through two cylindrical lenses; this arrangement allows discrimination between the two possible spatial field configurations of the $TEM_{01}{}^*$ laser that have identical transverse intensity distributions (see below). A third beam passes through a combination of reflectors that has the property of rotating a given field or intensity pattern by 90° in the transverse plane. This beam is matched to the laser resonator mode with respect to polarization and beam parameters and is attenuated and fed back into the laser with a phase difference determined by the optical path length.

2. THEORETICAL MODEL

In order to describe the main features of the system, a simplified theoretical model is used that assumes a traveling-wave laser with a homogeneously broadened gain medium and, corresponding to the experimental situation, with a resonator geometry that supports only equally polarized TEM_{10} and TEM_{01} Gauss–Hermite transverse modes. When beam divergence and the curvature of wave fronts are neglected, the resulting optical field $\mathbf{E}(\mathbf{r}, t)$ inside the cavity is written as

$$\mathbf{E}(\mathbf{r}, t) = \hat{e}_p N \exp[-(x^2 + y^2)/w^2](A + A^*)$$
$$A = [F_{10}(x/w) + F_{01}(y/w)]\exp[i(\omega t - kz + \phi_0)]. \quad (1)$$

The transverse distribution of intensity $I(x, y)$ is then given by

$$I(x, y) = K \exp[-2(x^2 + y^2)/w^2]$$
$$\times [|F_{10}|^2(x^2/w^2) + |F_{01}|^2(y^2/w^2) + (F_{10}F_{01}{}^* + F_{10}{}^*F_{01})(xy/w^2)]. \quad (2)$$

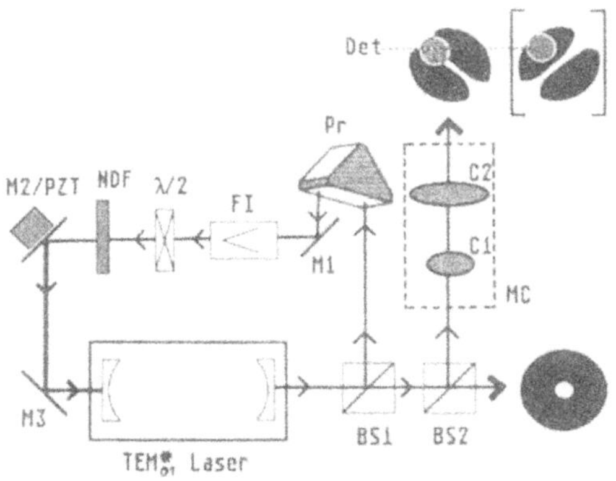

Fig. 1. Experimental setup: BS1, BS2, beam splitters; C1, C2, cylindrical lenses, forming the transverse-mode converter MC; Det, detector monitoring the local intensity in the bistable transverse intensity pattern—the detector position coincides with the nodal line of one pattern; Pr, rectangular reflecting prism—the plane of reflection of Pr is rotated out of the drawing plane by 45°; FI, Faraday isolator (optical diode); λ/2, λ/2 retardation plate; NDF, neutral density filter; PZT, piezoceramic transducer; M1, M2, M3, mirrors. For simplicity, mode matching and beam expanding lenses are not shown.

Here the unit vector $\hat{e}_p$ denotes the polarization direction, N and K are real constants, ϕ_0 is an arbitrary phase, z is the longitudinal coordinate, ω is the optical frequency, k denotes the modulus of the wave vector, w scales the transverse dimensions of the beam, and F_{10} and F_{01} are normalized complex field amplitudes of the TEM_{10} and TEM_{01} modes, respectively.

With the modal decomposition performed in Eqs. (1), one may now derive from the Maxwell–Bloch laser equations a system of coupled-mode equations[3] for the field amplitudes F_{10} and F_{01} and for the atomic variables of a homogeneously broadened two-level medium. For the case when the feedback of a fraction of the rotated output field pattern is taken into account in the evolution equations for F_{10} and F_{01}, and assuming a cylindrically symmetric pumping profile, resonant tuning of the two first-order transverse modes, and negligible excitation of other transverse modes of field and polarization, the laser equations are of the form

$$\frac{\partial F_{10}}{\partial t} = -\kappa[F_{10} + \beta \exp(i\delta')F_{01} - 2CP_{10}], \tag{3a}$$

$$\frac{\partial F_{01}}{\partial t} = -\kappa[F_{01} - \beta \exp(i\delta')F_{10} - 2CP_{01}], \tag{3b}$$

$$\frac{\partial P_{10}}{\partial t} = -\gamma_\perp(P_{10} - \tfrac{1}{2}F_{10}D_0 - \tfrac{1}{4}F_{10}D_1 - \tfrac{1}{4}F_{01}D_2), \tag{3c}$$

$$\frac{\partial P_{01}}{\partial t} = -\gamma_\perp(P_{01} - \tfrac{1}{2}F_{01}D_0 - \tfrac{1}{4}F_{01}D_1 - \tfrac{1}{4}F_{10}D_2), \tag{3d}$$

$$\frac{\partial D_0}{\partial t} = -\gamma_\parallel[D_0 - 2\sqrt{\pi} + \tfrac{1}{4}(F_{10}P_{10}^* + F_{10}^*P_{10} + F_{01}P_{01}^* + F_{01}^*P_{01})], \tag{3e}$$

$$\frac{\partial D_1}{\partial t} = -\gamma_\parallel[D_1 + \tfrac{1}{4}(F_{10}P_{10}^* + F_{10}^*P_{10} - F_{01}P_{01}^* - F_{01}^*P_{01})], \tag{3f}$$

$$\frac{\partial D_2}{\partial t} = -\gamma_\parallel[D_2 + \tfrac{1}{4}(F_{10}P_{01}^* + F_{10}^*P_{01} + F_{01}P_{10}^* - F_{01}^*P_{10})]. \tag{3g}$$

Here $1/\kappa$ denotes the time constant of the laser resonator, β^2 is the ratio of feedback intensity to output intensity, δ' is the phase difference between the feedback field and cavity field, and C is the pumping parameter. P_{10} and P_{01} and $\gamma_\perp$ are the two normalized mode amplitudes and the decay rate of the atomic polarization, respectively. D_0, D_1, and D_2 denote spatial modes of the normalized atomic inversion $D(x, y)$ (decay rate $\gamma_\parallel$) and its possible azimuthal variation that is due to the coupling to the field and polarization modes:

$$\begin{bmatrix} D_0 \\ D_1 \\ D_2 \end{bmatrix} = \frac{2}{\pi w^2}\int dx dy \times D(x, y)\exp[-2(x^2 + y^2)/w^2]\begin{bmatrix} \sqrt{\pi}(x^2 + y^2) \\ \sqrt{2}(x^2 - y^2) \\ 2\sqrt{2}xy \end{bmatrix}. \tag{3h}$$

An analytical examination of Eqs. (3a)–(3g) in the limit of adiabatic elimination of the atomic variables shows that above laser threshold and for $\beta = 0$ (i.e., without feedback) the solution of the complete system of laser equations (3a)–(3g) is always stationary and bistable, with relations between the field amplitudes given by $F_{01} = +iF_{10}$ or $F_{01} = -iF_{10}$. This implies that for stable laser oscillation in frequency-degenerate TEM_{10} and TEM_{01} modes, both mode fields are locked in phase quadrature, a configuration that is known as the $TEM_{01}{}^*$ hybrid mode.[6] Numerical solutions of the laser equations (3a)–(3g) outside the adiabatic limit show that these characteristics of the bistable solutions persist for a broad range of typical operation parameters and that they do not change substantially for small perturbations of the cylindrical symmetry of the laser resonator.[7]

The two field configurations that are found as stationary solutions of Eqs. (3a)–(3g) are identical in their intensity patterns, which are rotationally symmetric in traveling and standing waves. With the use of Eqs. (1) it can be shown, however, that the equiphase surfaces of the traveling-wave fields are helices that are mirror images of each other. This corresponds to a breaking of space-inversion symmetry that makes the two field patterns physically distinguishable (see below). It may be noted that in an earlier work the presence of helical components in the output field of a laser was inferred from an analysis of interference patterns.[8] Helical equiphase surfaces were also found in a recent study of optical vortices.[9]

The effect of the perturbation of the laser caused by feedback of the rotated output field pattern (see Fig. 1) is best shown in a new mode system that is adapted to the two steady-state field patterns of the $TEM_{01}{}^*$ laser. With the mode amplitudes $F_+ = 2^{-(1/2)}(F_{10} + iF_{01})$ and $F_- = 2^{-(1/2)}(F_{10} - iF_{01})$ and a corresponding definition of P_+ and P_-, and with $d = D_1 + iD_2$, one obtains from Eqs. (3a)–(3g) ($\delta = \delta' + \pi/2$):

$$\frac{\partial F_+}{\partial t} = -\kappa[F_+ + \beta \exp(i\delta)F_+ - 2CP_+], \tag{4a}$$

$$\frac{\partial F_-}{\partial t} = -\kappa[F_- - \beta \exp(i\delta)F_- - 2CP_-], \tag{4b}$$

$$\frac{\partial P_+}{\partial t} = -\gamma_\perp(P_+ - \tfrac{1}{2}F_+D_0 - \tfrac{1}{4}F_-d), \tag{4c}$$

$$\frac{\partial P_-}{\partial t} = -\gamma_\perp(P_- - \tfrac{1}{2}F_-D_0 - \tfrac{1}{4}F_+d^*), \tag{4d}$$

$$\frac{\partial D_0}{\partial t} = -\gamma_\parallel[D_0 - 2\sqrt{\pi} + \tfrac{1}{4}(F_+P_+^* + F_+^*P_+ + F_-P_-^* + F_-^*P_-)], \tag{4e}$$

$$\frac{\partial d}{\partial t} = -\gamma_\parallel[d + (1/2)(F_+P_-^* + F_-^*P_+)]. \tag{4f}$$

Equations (4a) and (4b) show that for the two steady-state modes F_+ and F_- the optical feedback leads to decoupled interference terms of opposite sign in the evolution equations. It is interesting to note that without these interference terms the system of Eqs. (4a)–(4f) shows a strong similarity to the theoretical model of a bidirectional ring laser.[10]

The optical switching of transverse patterns in our experiment relies on selecting the transverse field configuration given by F_+ (F_-) by creating conditions of destructive interference for the competing mode F_- (F_+) through the choice of δ. To illustrate this point, Fig. 2 shows the result of a numerical experiment based on Eqs. (4a)–(4f) that demonstrates the hysteretic switching of the mode intensities $|F_+|^2$ and $|F_-|^2$ that is due to the variation of δ; here the atomic relaxation parameters are chosen to be near the limiting case, where the atomic variables may be adiabatically eliminated. In the adiabatic limit, an (analytical) linear-stability analysis of Eqs. (4a)–(4f) yields the result that switching occurs under the conditions

$$|\beta \cos\delta| \geq \frac{1}{2}\left(\frac{C}{C_{\text{th}}} - 1\right),$$

$$\cos\delta > 0 \qquad \text{for } F_+ \neq 0, F_- = 0 \rightarrow F_+ = 0, F_- \neq 0,$$

$$\cos\delta < 0 \qquad \text{for } F_+ = 0, F_- \neq 0 \rightarrow F_+ \neq 0, F_- = 0, \tag{5}$$

where C_{th} denotes the pump parameter at laser threshold.

The above suggests that we should also expect an analogous selective excitation of spatial field patterns for the case when the self-injection scheme considered here is replaced by the injection of field patterns from a separate coherent source.

In order to discriminate between the two stable spatial field configurations of the TEM_{01}^* laser, a mode converter relying on the astigmatic imaging of higher-order transverse modes[11] is used in the experimental setup shown in Fig. 1: A cylindrical lens, oriented with its cylinder axis parallel to the x axis, creates a beam waist for the TEM_{01}^* field in the y–z plane, leaving propagation in the x–z plane unaffected. As a result the TEM_{10} and TEM_{01} components of the TEM_{01}^* field propagate here with different Guoy phase shifts[12] $\phi_{10}(z)$ and $\phi_{01}(z)$. A second cylindrical lens, oriented with its axis parallel to and placed at a distance l from the first lens, is used to limit the region of astigmatic propagation. It is easy to show[13] that the case $\phi_{10}(d) - \phi_{01}(l) = \pi/2$ corresponds to the conversion of input fields described by the steady-state mode amplitudes F_+ and F_- into output fields given by $F_+' = 2^{-(1/2)}(F_{10} + F_{01})$ and $F_-' = 2^{-(1/2)}(F_{10} + F_{01})$, respectively. (Common phase factors related to the optical path length l have been omitted.) With the use of Eqs. (1) and (2), it can be shown that the field amplitudes F_+' and F_-' correspond to TEM_{10} and TEM_{01} fields, respectively, with nodal planes that are rotated by 45° with respect to the orientation of the cylindrical lenses. Thus, the two field patterns of a TEM_{01}^* laser appear as mirror-symmetric intensity patterns at the output of the mode converter.

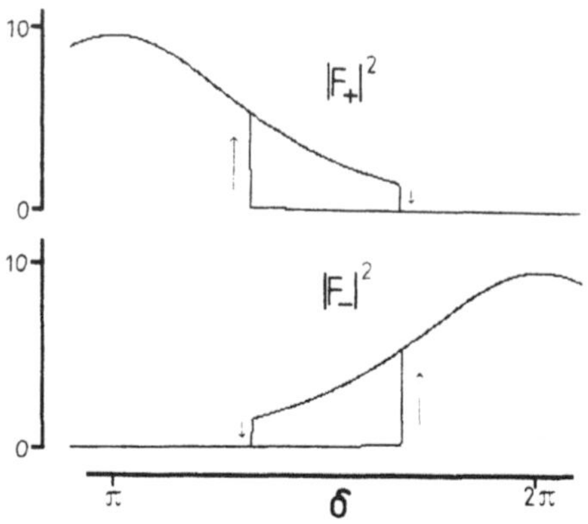

Fig. 2. Calculated normalized intensities $|F_+|^2$ and $|F_-|^2$ as a function of slowly varying feedback phase angle δ. Parameters: $\beta = 0.5$, $\gamma = 10\kappa$, $\gamma_\perp = 20\kappa$, $C = 2C_{\text{th}}$, sweep rate $d\delta/dt \approx \kappa \times 10^{-5}$ rad; for definition of symbols see text. Small random fluctuations were introduced into the numerical calculation to stimulate the decay of unstable states. The arrows indicate the path followed in the hysteresis cycle.

3. EXPERIMENT

The laser used in the experimental setup shown in Fig. 1 consists of a 0.22-m-long, $\lambda = 633$-nm, helium–neon laser tube (Melles–Griot type 05-LHB-290) terminated by Brewster windows. The Fabry–Perot laser resonator has a length of 0.25 m and uses mirrors with a reflectivity of 99.2% and a radius of curvature of 0.6 m. The eigenfrequency difference between the TEM_{10} and TEM_{01} modes of the optical resonator resulting from the astigmatism of the Brewster windows of the laser tube is compensated for by diffraction at the laser tube bore aperture, which is placed slightly noncentrally with respect to the optical axis of the resonator; TEM_{01}^* oscillation is obtained by placing a small absorbing dot (approximate size, 50 μm) close to the optical axis of the resonator.[7] Under these conditions the typical output power from each of the outcoupling mirrors is in the range of 30–100 μW, with the actual value depending on alignment, tuning, and the discharge current of the laser tube.

The effect of a mode converter of the type described above and the bistability of a TEM_{01}^* laser is illustrated by the photographs in Fig. 3; the practical realization of the mode converter is described in detail in Ref. 13. The change of the transverse field configuration of the laser that shows up in Figs. 3(b) and 3(c) is induced without external feedback by moving a beam stop through the laser resonator perpendicu-

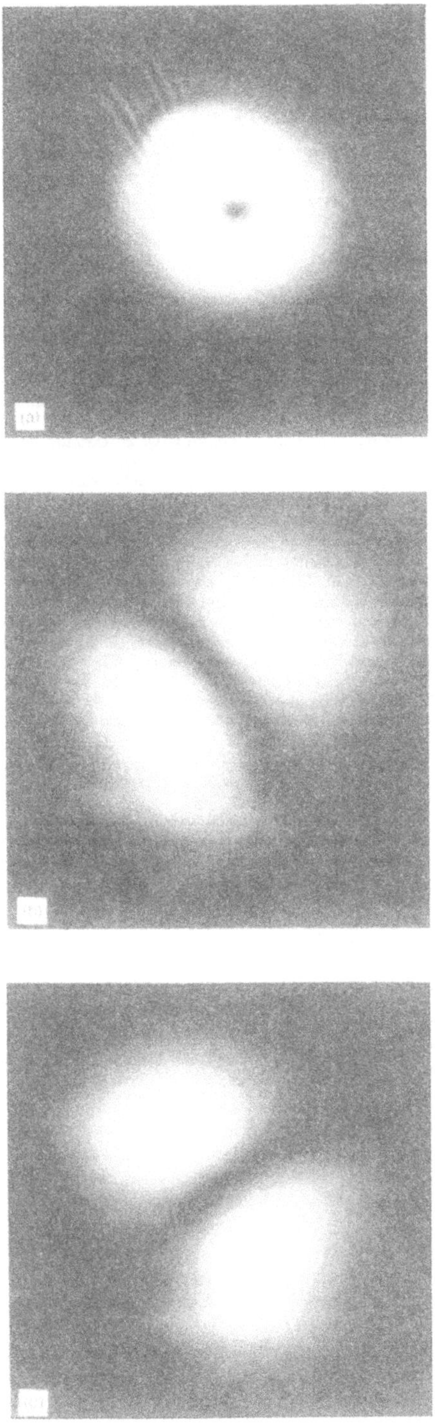

Fig. 3. Photographs of (a) transverse intensity distribution of the laser output and (b), (c) bistable intensity patterns at output of mode converter. The cylindrical lenses of the mode converter (see text) are oriented horizontally with respect to patterns (b) and (c).

lar to the optical axis: It is observed that the field pattern emitted by the laser can be selected by the direction of movement of the edge of the obstacle when the beam path becomes unblocked. This effect probably is caused by diffraction of the intracavity field at the edge of the beam stop: The corresponding perturbation of the cylindrical symmetry of the laser resonator may favor the onset of lasing in one of the two field patterns.

Hysteresis and switching between the two different field patterns is observed by monitoring the local (see Fig. 1) optical power density at the output of the mode converter as the optical feedback phase δ is swept back and forth by translation of a mirror in the feedback path. One experimental result that corresponds to a recording of a single hysteresis cycle in $\approx$0.1 sec is shown in Fig. 4(a). For unchanged experimental conditions, Figs. 4(b) and 4(c) show the switching transients with higher temporal resolution. Here the feedback coupling strength corresponds to $\beta \approx 0.12$, and the laser tube discharge current is approximately two times the laser threshold value.

It is found that the width of the observed hysteresis loop strongly depends on the discharge current; the width appears to be approximately proportional to the laser output power. No switching is observed for $\beta \lesssim 0.04$; above the switching threshold, the variation of the feedback coupling strength leads only to small changes of the shape of the hysteresis loop. The switching transients are always found, with typical rise and fall times of 2–5 μsec. This value is

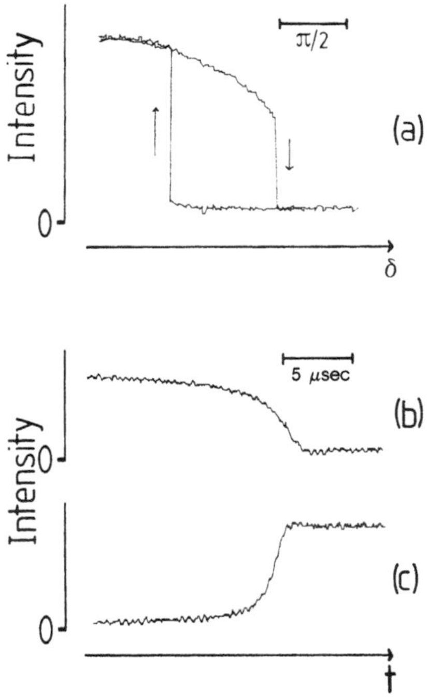

Fig. 4. (a) Experimentally determined local intensity at output of the transverse-mode converter (see Fig. 1) as a function of feedback phase δ. (b), (c) The same signal as a function of time t in the transient regimes. In all cases the sweep rate is $d\delta/dt \approx 0.1$ rad/msec. For other experimental parameters, see text.

probably associated with the buildup time of the optical field in the laser resonator: with $1/\kappa \approx 0.1$ μsec as the time constant of the laser resonator, derived from the resonator data given above, estimates of the field buildup time[14] in our case yield values in the range of 1–10 μsec, which covers the range of experimentally observed switching times.

4. CONCLUSIONS AND OUTLOOK

The experimental findings are in full qualitative agreement with the results of the theoretical model described in Section 2. A significant quantitative difference is that in the experiment the fraction of the laser output field that is necessary to switch the system is much smaller than predicted. This seems to indicate a decreased stability margin of the experimental steady-state field patterns with respect to the perturbations induced by the injected field. It may be noted here that in the theoretical model atomic diffusion in directions perpendicular to the optical axis has not been taken into account. This process may play a crucial role for the stability of transverse optical patterns in gas lasers, since it washes out the differences in the transverse variation of inversion associated with different transverse intensity distributions.

The experimental techniques described in this paper may also be useful for more general studies on radiation patterns in nonlinear-optical systems. Complex stationary transverse patterns and multistability are to be expected when frequency-locked transverse modes of higher order are excited in a laser; until now, multiple-transverse-mode lasers have been studied mainly with respect to their dynamical behavior, i.e., under nonstationary conditions.[15] Optical feedback techniques similar to the one described here may serve as a tool to probe the stability and to manipulate the generated transverse optical patterns.

Finally, one might consider a nonlinear-optical resonator whose transverse field structure is determined by an externally injected field pattern. Systems of this type could be used in optical data processing and might even be capable of performing pattern recognition and associative memory operations.

ACKNOWLEDGMENTS

This work was supported by the Deutsche Forschungsgemeinschaft and by the ESPRIT grant of the Commission of the European Communities.

REFERENCES

1. H. Haken, *Advanced Synergetics* (Springer-Verlag, Berlin, 1987), p. 1.
2. L. A. Lugiato and R. Lefever, Phys. Rev. Lett. **58,** 2209 (1987).
3. L. A. Lugiato, C. Oldano, and L. M. Narducci, J. Opt. Soc. Am. B **5,** 879 (1988); L. A. Lugiato, G. L. Oppo, M. A. Pernigo, J. R. Tredicce, L. M. Narducci, and D. K. Bandy, Opt. Commun. **68,** 63 (1988).
4. I. M. Belousova, G. N. Vinokurov, O. B. Danilov, and N. N. Rozanov, Sov. Phys. JETP **25,** 761 (1967).
5. H. Kogelnik and T. Li, Appl. Opt. **5,** 1550 (1966).
6. W. W. Rigrod, Appl. Phys. Lett. **2,** 51 (1963).
7. C. Tamm, Phys. Rev. A **38,** 5960 (1988).
8. J. M. Vaughan and D. V. Willets, Opt. Commun. **30,** 263 (1979).
9. P. Coullet, L. Gil, and F. Rocca, Opt. Commun. **73,** 403 (1989).
10. H. Zeghlache, P. Mandel, N. B. Abraham, L. M. Hoffer, G. L. Lippi, and T. Mello, Phys. Rev. A **37,** 470 (1988).
11. D. C. Hanna, IEEE J. Quantum Electron. **QE-5,** 483 (1969).
12. The Guoy phase effect and its relation to the propagation properties of laser beams are discussed in, e.g., A. E. Siegman, *Lasers* (University Science Books, Mill Valley, Calif., 1986), p. 645–685.
13. C. Tamm, "The field patterns of a hybrid mode laser: detecting the 'hidden' bistability of the optical phase pattern," in *Proceedings of the Workshop on Quantitative Characterization of Dynamical Complexity in Nonlinear Systems*, N. B. Abraham and A. Albano, eds. (Plenum, New York, 1989).
14. See, e.g., Reference 11, p. 491–495.
15. R. Hauck, F. Hollinger, and H. Weber, Opt. Commun. **47,** 141 (1983); D. J. Biswas, V. Dev, and U. K. Chatterjee, Phys. Rev. A **38,** 555 (1988); W. Klische, C. O. Weiss, and B. Wellegehausen, Phys. Rev. A **39,** 919 (1989); J. R. Tredicce, E. J. Quel, A. M. Ghazzawi, C. Green, M. A. Pernigo, L. M. Narducci, and L. A. Lugiato, Phys. Rev. Lett. **62,** 1274 (1989).

PAPER 3.4

Optical helices and spiral interference fringes

M. Harris, C.A. Hill and J.M. Vaughan
DRA Malvern, St Andrews Road, Gt Malvern, Worcs, WR14 3PS, UK

Received 11 October 1993

Very pure optical helices have been generated in an argon ion laser of low Fresnel number. The beam character, with continuous cophasal surface of helical form, is clearly demonstrated by spiral interference fringes produced in a novel interferometric arrangement. In addition to single-start helices the multistart fringe patterns establish both two-start and three-start helices (of pitch two and three wavelengths, respectively), and also the state of helicity (i.e. rotational hand) of the beams.

In a collimated monochromatic beam of light the surfaces of constant phase (modulo 2π) are usually considered to be a series of discs, normal to the beam axis, spaced one wavelength λ apart. However it has been known for some time [1–4], although not widely appreciated even among optical scientists, that monochromatic beams can possess a continuous cophasal surface of helical form extending through space. Such helical waves must of course satisfy Maxwell's equations; not surprisingly the phase discontinuity on axis necessitates a null in the intensity distribution, as readily seen in so-called laser "doughnut" modes. In the present work, techniques have been applied for generating very pure helices in an argon ion laser of low Fresnel number. Single-start helices (of pitch λ) and both two-start and three-start helices (pitch 2λ and 3λ, respectively) have been created. Novel interference methods demonstrate this helical character in a remarkable range of multistart spiral interference fringes.

The concept of screw and edge dislocations in wave fronts was first introduced by Nye and Berry [1] in 1974 originating in attempts to understand radio echoes from the bottom of Antarctic ice sheets; analogous experiments were carried out in the laboratory using ultrasound. In the first demonstration at optical frequencies [2,3], comparable helices in the lowest order doughnut mode of an argon ion laser were shown to be evolving between left- and right-hand helicity, at the frequency difference of the two constituent TEM_{01} modes. The impact of such spatial coherence properties on various laser applications, including photon correlation spectroscopy, fringe velocimetry, speckle and coherent processing was also studied [4]. Subsequently frequency locking of two transverse modes in a He-Ne laser was achieved [5], and complex spatial and temporal instabilities in a CO_2 laser were analysed [6] in terms of nonlinear interaction of transverse cavity modes with the active medium. At the same time, in a theoretical study [7] based on the Maxwell–Bloch model, laser cavities of large Fresnel number were shown to exhibit optical vortices. In the most recent developments, studies of these optical helix/dislocation/phase-singularity/bifurcation/vortex-phenomena include bistability and optical switching of spatial patterns [8], waves of considerable spatiotemporal complexity in a CO_2 laser [9], the analysis of regular patterns of singularity arrays [10], the dynamics of optical vortices [11–14] and the creation of topologically stable patterns [15], and the theory of propagation in nonlinear media [16] and of arrays in free space [17]. Experimental studies have been made on a single longitudinal mode Na_2 ring laser [10,18–20] that generates families of transverse modes. In addition the design of optical and holographic phase converters for the generation of phase singularities has been developed [20–23]. Much of this work has concentrated on the properties of large arrays; the recent experimental evidence [14,17,19,20,21] for embedded vortices of first order consists of forked fringes employing plane-wave and displaced-beam interference techniques (of intrinsically poor visibility). In addition single spiral fringes have been shown in an optical fibre [17] and by mixing with a spherical reference wave [24]. One of the first suggestions of multistart helices, and the only experimental evidence of a two-start, or charge 2, helix (derived from plane-wave fringes in a holographically generated doughnut), is contained in ref. [20]. The present work establishes that pure, higher order, helical modes may be generated in an active laser cavity.

The experimental arrangement is shown in fig. 1. The argon ion laser was a modified Lexel Model 95 operating double-ended at 488 nm wavelength, with maximum Fresnel number ~10, and a longitudinal mode spacing Δ of 156.02 MHz. Tuning to the required mode follows a technique related to that of Tamm [5]. Suppression of the on-axis modes is achieved by insertion of a small circular absorber into the cavity; a set of these with diameters ranging from 50 to 500 μm is deposited on a glass plate mounted on an X–Y translation stage. Similarly a set of clear apertures is inserted into the optical cavity. A range of transverse laser modes may be excited simultaneously; these are evident both from strong intermode beats and from the direct spectral display of the Fabry–Pérot interferometer. With suitable alignment of absorber and aperture these may be reduced to just two modes. For example these might be the Laguerre-gaussian 01 (two-lobe) modes in phase and

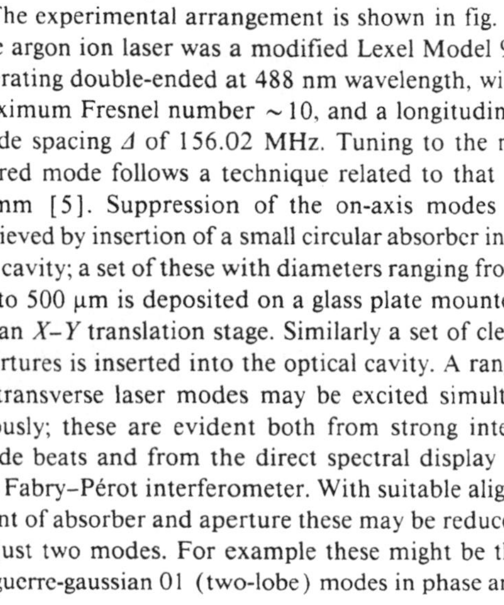

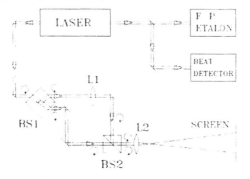

Fig. 1. Experimental arrangement. A circular absorber and aperture are inserted into the laser cavity and the laser output is monitored on the Fabry–Pérot interferometer and beat frequency detector. In the lower interferometric arrangement lens L_1 sets up a difference of curvature in the two beams which are recombined at beam splitter BS2. The helicity of each beam at different points is indicated.

space quadrature which form the circular-symmetric first-order 01* doughnut mode. Characteristically these give a strong intermode beat signal within a few MHz of the longitudinal mode spacing Δ. The frequency of this beat signal may be tuned quite close to Δ by small lateral adjustments of the intracavity absorber and/or aperture, thus modifying the astigmatism of the optical cavity. The two modes thus approach degeneracy and at a certain point, usually within ~0.5 MHz of Δ, the beat disappears and the modes are locked to a single frequency which is typically stable over a wide range of laser power (up to 80 mW at full current of the laser discharge). A similar procedure, with suitably selected absorbers and apertures, permits selection of the higher-order, circular-symmetric, 02* and 03* doughnut modes.

The interferometric arrangement that unequivocally manifests the pure helical character of these beams by producing spiral interference patterns is shown in the lower half of fig. 1. The aim of this arrangement is three fold: after splitting at beam splitter BS1 the two equal beams are recombined at BS2 with (a) different curvature, (b) reversed handedness (i.e. helicity) and (c) equal size at the screen. Variations of curvature are achieved with a selection of lenses L_1 of different focal length whose precise position is adjusted to achieve condition (c) above. The overall change of handedness arising from the different reflections at beamsplitters and mirrors is indicated in fig. 1. It must be emphasised that this refers to the helicity of the beams; they remain linearly polarised throughout. As shown the two beams arrive with opposite helicity at the screen; common helicity is easily achieved with an additional mirror in the lower leg of the interferometer. The fringes are displayed on the screen, where they are recorded photographically. With beams of equal size and intensity the fringe visibility is 100%; for optimum fringe stability the optical paths between the beamsplitters are made as equal as possible and shielded from air movements.

Sets of photographs are shown in figs. 2 and 3. Of these fig. 2a shows the simple circular fringes expected for a lowest-order 00 gaussian mode. Arising as it does from beams of different curvature, the form of the fringes is essentially that of Newton's rings. Figure 2b similarly shows circular fringes for a 01* doughnut mode in which the two beams have been

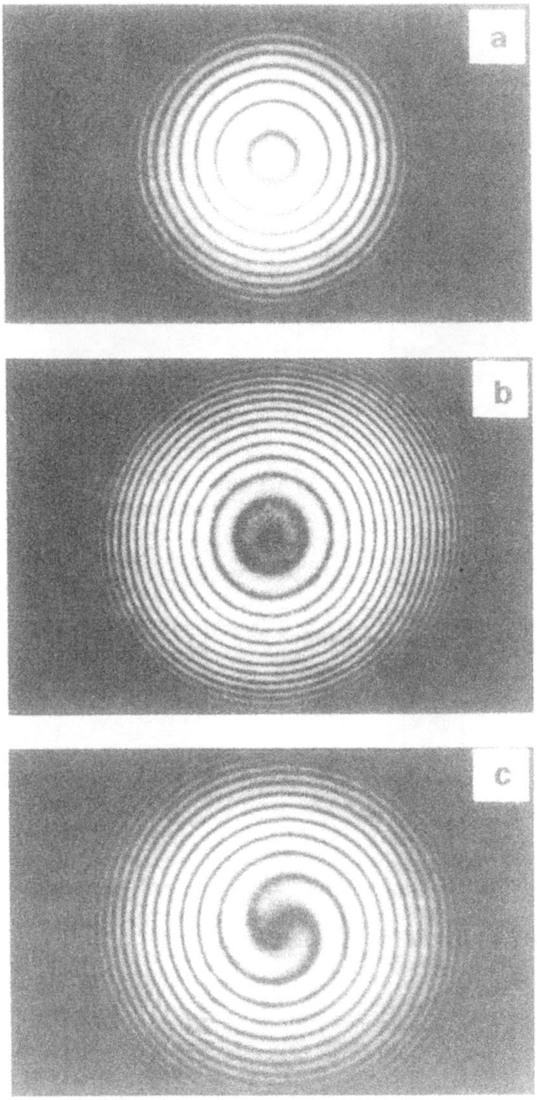

Fig. 2. Photographs of interference fringes: (a) lowest order gaussian mode, with circular fringes similar to Newton's rings; (b) circular fringes with a 01* doughnut mode in which the beams have been recombined with the same helicity; (c) the same as (b) but with the beams brought together with opposite helicity (as in fig. 1). Note the two-start spiral fringes.

brought together with the same handedness (i.e. an extra mirror has been introduced into the interferometer). In contrast fig. 2c shows the remarkable effect of bringing the beams together with opposite helicity; the two-start spiral interference fringes directly manifest the helical character of the component beams. The sets of fringes in fig. 3 have been selected to demonstrate the effect of varying curvature difference for 01* and the higher order 02* and 03* modes. In the upper row the curvatures of the two beams are well matched. In consequence the fringes closely approximate the intensity patterns of individual $TEM_{01,02,03}$ spot modes; their orientation is determined by the relative phase difference of the interfering beams. By contrast, in the middle row the small curvature difference amounts to less than one wavelength across the beams; only a fraction of a spiral extends across the beam front. In the lower row the curvature difference is a few wavelengths, leading to several complete spirals. The two-, four- and six-start spirals for the higher order modes are strikingly evident in each set of photographs.

The electric field at the screen that describes the interference of two helical beams of opposite hand is given for the 01* mode by

$$E=E_0 r \exp(-r^2/2a^2)\{\exp\{i[\theta+\phi(r)]\} + \exp\{-i[\theta+\phi(r)]\}\} \quad (1)$$

$$=2E_0 r \exp(-r^2/2a^2)\cos[\theta+\phi(r)]\,, \quad (2)$$

where $2\phi(r)$ is the phase term shared equally between the curved wavefronts and r is the radial distance from the common beam axis. If the radii of wavefront curvature are R_1 and R_2 the phase difference $2\phi(r)$ is given by

$$2\phi(r)=r^2|R_1-R_2|/2\lambda R_1 R_2=2k^2r^2\,, \quad (3)$$

where k is a constant of the optical geometry. On replacing $\phi(r)$ by k^2r^2, and writing the intensity distribution for the 01* mode as $I_0(1)$, the fringe intensity derived from eq. (2) is

$$I=I_0(1)\cos^2(\theta+k^2r^2)\,. \quad (4)$$

Starting from the dark centre, dark fringes are given by $\cos(\theta+k^2r^2)$ equal to zero, from which they are defined by

$$kr=[(n+1/2)\pi-\theta]^{1/2}\,,$$

$$n=\ldots-2,\ -1, 0, 1, 2\ldots \quad (5)$$

Examination shows two distinct spiral branches, one for n odd and one for even. Thus two-start spiral

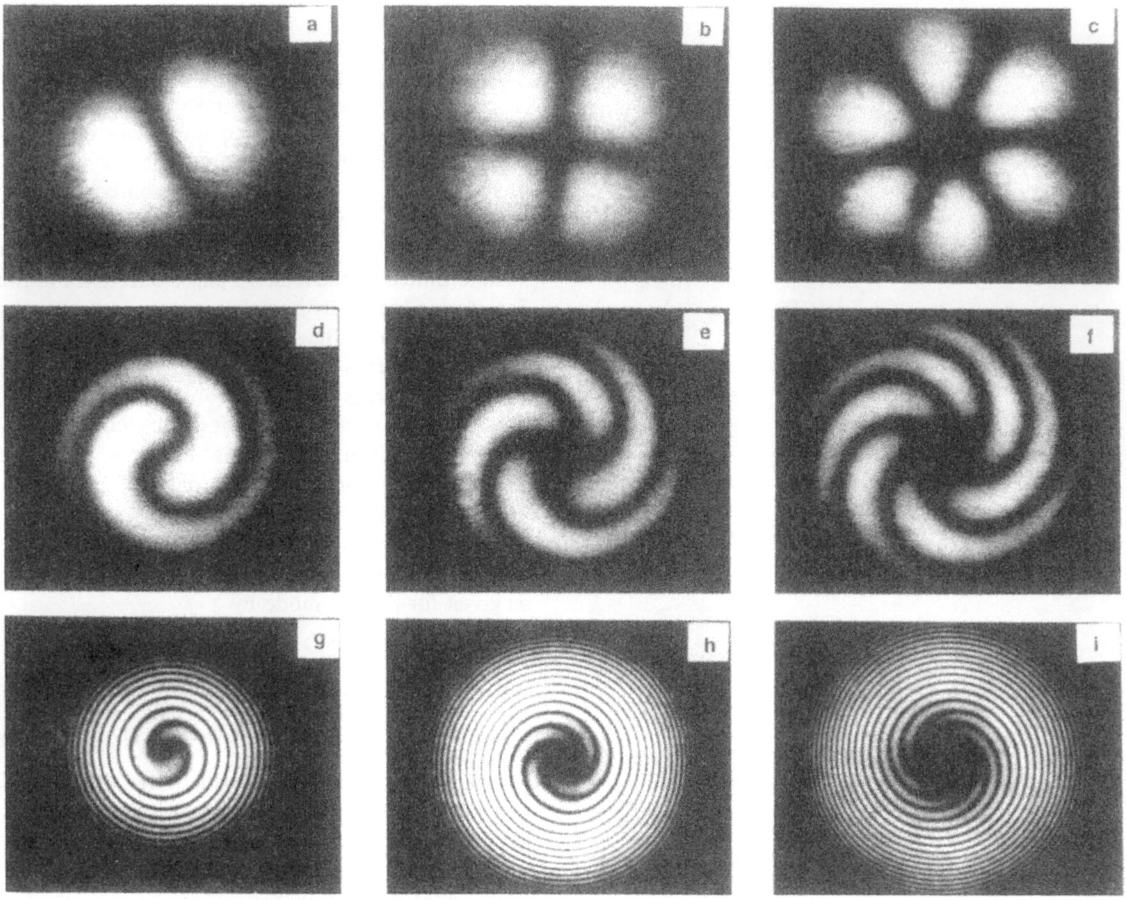

Fig. 3. Spiral fringe patterns with the laser operating on doughnut modes of different order: 01* ((a), (d), (g)); 02* ((b), (e), (h)) and 03* ((c), (f), (i)). The beam curvatures are well matched in the upper row resulting in fringes that closely approximate the intensity patterns of individual $TEM_{01,02,03}$ spot modes. In contrast the beam curvature difference for the middle row is small (less than one wavelength) and for the lower row is several wavelengths. Note the multi-start spirals (two, four and six, respectively) and also the differing handedness of (g) and (i) (clockwise) from which the absolute helical sense can be established.

fringes are superimposed on the intensity distribution $I_0(1)$ of the single-start helical mode. Corresponding expressions for the higher-order modes are similarly derived; for the mth order (m-start) helix the fringe distribution is $I_0(m)\cos^2(m\theta+\phi)$ thus giving a $2m$-start spiral. Computer-generated contour plots of these expressions show patterns in excellent agreement with figs. 2 and 3. Such multistart cophasal helices obviously require that the pitch of each individual helix must be $m\lambda$. Detailed examination of the fringes in fig. 3 shows also that for d, e, f and h the spirals are anti-clockwise and for g and i clockwise. This reflects the fact that such modes have two possible states (of roughly equal probability) of left- and right-handed helicity.

In a further investigation the first few resonator modes have been calculated for a spherical-mirror open resonator, closely similar to the argon laser cavity, with the introduction of circular pinhole apertures and circular absorbers. The model is based on resonator matrix codes for CO_2 lasers [25] and has been adapted for Laguerre–gaussian TEM_{lp} basis modes at 488 nm. For the calculations closest to the experimental arrangement a fixed pinhole aperture of 2.0 mm diameter was used and all basis modes up to $l=7$ and $p=7$ were included; the resultant round trip losses versus absorber diameter are shown in fig. 4. In practice it may be expected that the laser will operate on the lowest-loss resonator mode. Thus over a certain range of absorber/pinhole diameter one particular circular doughnut will be preferred; in effect modes of lower and higher order are significantly blocked by the absorber and pinhole, respectively. Not surprisingly the exact shapes of the predicted loss curves (and their crossing points) are rather sensitive to the choice of pinhole diameter; the calculation is also simplified by assuming that the absorber and pinhole lie at a mirror plane. The curves in fig. 4 should therefore not be interpreted too strictly. Nevertheless they agree qualitatively with the observations and show clearly the transitions (marked by the arrows) from TEM_{00} (very small absorber) through the first three doughnut modes.

In summary, single, pure 1- 2- and 3-start optical helices have been generated in an argon ion laser with a low Fresnel number cavity in reasonable accord with resonator mode calculations. The modes are stable over a wide range of power and it is likely that yet higher order modes and considerably higher power could be generated by suitable discharge and cavity design. A novel interferometric arrangement has been developed that immediately manifests the

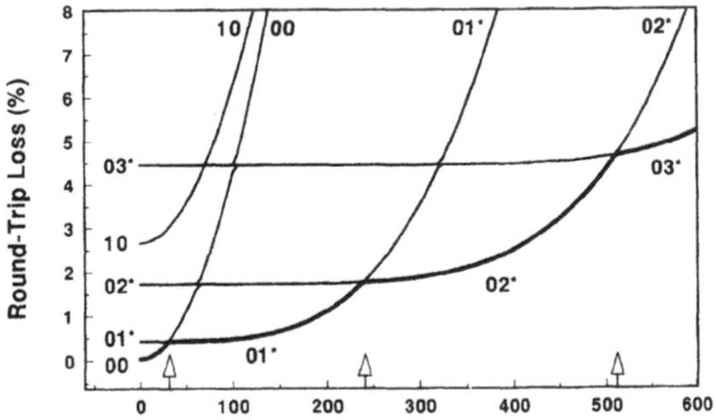

Fig. 4. Calculated round-trip loss versus absorber diameter for various resonator modes and with a 2.0 mm diameter pinhole aperture. The transitions (marked by arrows) from the TEM_{00} mode through the first three doughnut modes agree reasonably well with the observations.

helical character of the beams, and characterises both the state of helicity and the multi-start order number. Possible applications of helical modes have recently been listed by Heckenberg et al. [20], and include small particle levitation and focusing of atomic beams. Transfer of angular momentum [23] and rotational sensing are other possibilities. The stable, well-defined, multi-start helices presently demonstrated in an argon ion laser, together with the potential high power, should be valuable to such applications.

We thank D.V. Willetts, P. Tapster, T.J. Shepherd and E. Jakeman of DRA (Malvern) and R. Loudon and L. Allen of Essex University for valuable discussions.

References

[1] J.F. Nye and M.V. Berry, Proc. R. Soc. Lond. A 336 (1974) 165.
[2] J.M. Vaughan and D.V. Willetts, Optics Comm. 30 (1979) 263.
[3] J.M. Vaughan and D.V. Willetts, J. Opt. Soc. Am. 73 (1983) 1018.
[4] P.N. Pusey, J.M. Vaughan and D.V. Willetts, J. Opt. Soc. Am. 73 (1983) 1012.
[5] C. Tamm, Phys. Rev. A 38 (1988) 5960.
[6] J.R. Tredicce, E.J. Quel, A.M. Ghazzawi, C. Green, M.A. Pernigo, L.M. Narducci and L.A. Lugiato, Phys. Rev. Lett. 62 (1989) 1274.
[7] P. Coullet, L. Gil and F. Rocca, Optics Comm. 73 (1989) 403.
[8] C. Tamm and C.O. Weiss, J. Opt. Soc. Am. B 7 (1990) 1034.
[9] C. Green, G.B. Mindlin, E.J. D'Angelo, H.G. Solari and J.R. Tredicce, Phys. Rev. Lett. 65 (1990) 3124.
[10] M. Brambilla, F. Battipede, L.A. Lugiato, V. Penna, F. Prati, C. Tamm and C.O. Weiss, Phys. Rev. A 43 (1991) 5090.
[11] D. Hennequin, C. Lepers, E. Louvergneaux, D. Dangoisse and P. Glorieux, Optics Comm. 93 (1992) 318.
[12] K. Staliunas, Optics Comm. 90 (1992) 123.
[13] C.O. Weiss, H.R. Telle, K. Staliunas and M. Brambilla, Phys. Rev. A 47 (1993) R1616.
[14] G. Indebetouw and S.R. Liu, Optics Comm. 91 (1992) 321; Optics Comm. 101 (1993) 442.
[15] L. Gil, K. Emilsson and G.L. Oppo, Phys. Rev. A 45 (1992) R567.
[16] V.I. Kruglov, Yu.A. Logvin and V.M. Volkov, J. Mod. Optics 39 (1992) 2277.
[17] G. Indebetouw, J. Mod. Optics 40 (1993) 73.
[18] C. Tamm and C.O. Weiss, Optics Comm. 78 (1990) 253.

[19] A.G. White, C.P. Smith, N.R. Heckenburg, H. Rubinsztein-Dunlop, R. McDuff, C.O. Weiss and C. Tamm, J. Mod. Optics 38 (1991) 2531.

[20] N.R. Heckenburg, R. McDuff, C.P. Smith, H. Rubinsztein-Dunlop and M.J. Wegener, Opt. and Quant. Elec. 24 (1992) S951.

[21] N.R. Heckenberg, R. McDuff, C.P. Smith and A.G. White, Optics Lett. 17 (1992) 221.

[22] L. Allen, M.W. Beijersbergen, R.J.C. Spreeuw and J.P. Woerdman, Phys. Rev. A 45 (1992) 8185.

[23] M.W. Beijersbergen, L. Allen, H.E.L.O. van der Veen and J.P. Woerdman, Optics Comm. 96 (1993) 123.

[24] L.E. Grin, P.V. Korolenko and N.N. Fedotov, Opt. Spectrosc. 73 (1992) 604.

[25] C.A. Hill in: The physics of laser resonators, eds. D.R. Hall and P.E. Jackson (Adam Hilger, 1989) pp. 40–61.

Astigmatic laser mode converters and transfer of orbital angular momentum

M.W. Beijersbergen, L. Allen, H.E.L.O. van der Veen and J.P. Woerdman
Huygens Laboratory, University of Leiden, P.O. Box 9504, 2300 RA Leiden, Netherlands

Received 17 August 1992

We present the design of a mode converter which transforms a Hermite–gaussian mode of arbitrarily high order to a Laguerre–gaussian mode and vice versa. The converter consists of two cylindrical lenses and is based on appropriate use of the Gouy phase. We demonstrate mode conversion experimentally and consider where the concomitant transfer of orbital angular momentum is localized.

1. Introduction

A paraxial light beam of circular polarization is known to carry *spin* angular momentum of $+\hbar$ or $-\hbar$ per photon, for σ^+ or σ^- polarized beams, respectively. Interaction of such a light beam with a birefringent plate may lead to a mechanical torque, as was first demonstrated by Beth [1]. Recently we have shown theoretically that *orbital* angular momentum of light is also a useful concept for a paraxial light beam, particularly for a Laguerre–gaussian beam, which has a $(\exp il\phi)$ azimuthal dependence [2]. Explicit calculation based on Maxwell's equations shows that such a beam, when linearly polarized, carries $l\hbar$ orbital angular momentum per photon. When the beam is circularly polarized, it carries $(l\pm 1)\hbar$ as total angular momentum per photon. We have speculated that conversion of a paraxial beam with specific orbital angular momentum into another beam, with a different orbital angular momentum, will give rise to a torque on the converter. In the previous paper [2] the properties of the converter were only briefly alluded to. In this paper we discuss the design of the mode converter in detail, and report experimental demonstration of its ability to transform modes. We also analyze theoretically how the converter takes up the change in orbital angular momentum of the light beam.

The first part of our paper is, in fact, a generalization of previous work by Tamm on gaussian mode conversion [3,4]; we extend the results obtained by Tamm for low-order modes to modes of arbitrary order. There is also a strong connection with recent work by Abramochkin and Volostnikov [5]; they also deal with the influence of astigmatism on gaussian modes. They consider cases in which the astigmatism does not conserve the gaussian mode character of the incoming beam, in that the transverse intensity pattern of the outcoming beam changes upon propagation. We deal with suitable astigmatic elements which conserve the mode character and thus operate as mode *converters*; this restriction greatly simplifies the theoretical discussion.

2. Mode decomposition

In this section we introduce expansion formulas for a Hermite–gaussian (HG) and a Laguerre–gaussian (LG) mode which will turn out to be essential for an understanding of the mode converter. We use the following definitions for the amplitude of the Hermite–gaussian (HG) and Laguerre–gaussian (LG) laser modes which propagate along the z axis

$$u_{nm}^{\mathrm{HG}}(x,y,z)=C_{nm}^{\mathrm{HG}}(1/w)\exp[-ik(x^2+y^2)/2R] \times\exp[-(x^2+y^2)/w^2]\exp[-i(n+m+1)\psi] \times H_n(x\sqrt{2}/w)\,H_m(y\sqrt{2}/w)\,, \quad (1)$$

$$u^{\mathrm{LG}}_{nm}(r,\phi,z) = C^{\mathrm{LG}}_{nm}(1/w)\exp(-ikr^2/2R)\exp(-r^2/w^2) \times \exp[-i(n+m+1)\psi]\exp[-i(n-m)\phi] \times (-1)^{\min(n,m)}(r\sqrt{2}/w)^{|n-m|} \times L^{|n-m|}_{\min(n,m)}(2r^2/w^2)\,, \tag{2}$$

with

$$R(z) = (z_R^2 + z^2)/z\,, \tag{3}$$

$$\tfrac{1}{2}kw^2(z) = (z_R^2 + z^2)/z_R\,, \tag{4}$$

$$\psi(z) = \arctan(z/z_R)\,. \tag{5}$$

$H_n(x)$ is the Hermite polynomial of order n, $L_p^l(x)$ is the generalized Laguerre polynomial [6], k is the wave number, and z_R is the Rayleigh range (half the confocal parameter) of the mode. We introduce $N=n+m$ as the order of the mode. Normalization of the amplitude such that $\int\int dx\, dy\, |u|^2 = 1$ yields

$$C^{\mathrm{HG}}_{nm} = \left(\frac{2}{\pi n! m!}\right)^{1/2} 2^{-N/2}\,, \tag{6}$$

$$C^{\mathrm{LG}}_{nm} = \left(\frac{2}{\pi n! m!}\right)^{1/2} \min(n,m)!\,. \tag{7}$$

Note that the indices we use for the LG mode differ from those normally used. The radial index p normally used is $\min(n,m)$, the minimum of n and m; the azimuthal index l is $n-m$. Our notation brings advantages in the context of the present paper: we shall show that a mode converter can transform a HG_{nm} mode into a LG_{nm} mode or vice versa.

By using relations between Hermite and Laguerre polynomials (see e.g. refs. [2,5]) one can show that a LG mode can be decomposed into a set of HG modes of the same order:

$$u^{\mathrm{LG}}_{nm}(x,y,z) = \sum_{k=0}^{N} i^k\, b(n,m,k)\, u^{\mathrm{HG}}_{N-k,k}(x,y,z)\,, \tag{8}$$

with real coefficients

$$b(n,m,k) = \left(\frac{(N-k)!k!}{2^N n! m!}\right)^{1/2} \times \frac{1}{k!}\frac{d^k}{dt^k}[(1-t)^n(1+t)^m]|_{t=0}\,. \tag{9}$$

The factor i^k in eq. (8) corresponds to a $\pi/2$ relative phase difference between successive components. Perhaps surprisingly, a HG mode whose principal axes make an angle of 45° with the (x, y) axes (a 'diagonal' mode) can be decomposed, using relations between products of Hermite polynomials [2,5], into exactly the same constituent set:

$$u^{\mathrm{HG}}_{nm}\left(\frac{x+y}{\sqrt{2}}, \frac{x-y}{\sqrt{2}}, z\right) = \sum_{k=0}^{N} b(n,m,k)\, u^{\mathrm{HG}}_{N-k,k}(x,y,z)\,, \tag{10}$$

with the same real coefficients $b(n, m, k)$ as above. In this expansion, however, the successive components are in phase. In fig. 1 some examples of a mode decomposition of order 2 are given in diagrammatic form. In table 1 the coefficients $b(n, m, k)$ are given for modes up to order 3. For completeness we note that the relationship between the LG and HG modes can also be established via operator algebra [7] and by direct comparison with a 2D quantum harmonic oscillator [8].

3. Astigmatic Gouy phase

From eqs. (8),(10) it is clear that in order to perform the conversion from a HG mode to a LG mode one has to rephase the terms in the decomposition. This can be done by exploiting the Gouy phase $\psi(z)$

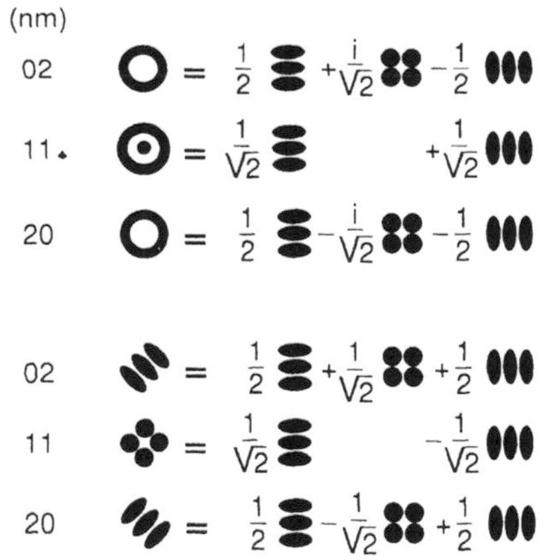

Fig. 1. Examples of the decomposition of HG_{nm} and LG_{nm} modes of order 2.

Table 1
The coefficients $b(n, m, k)$ which occur in eqs. (8), (10).

n	m	$k=0$	1	2	3
0	0	1			
0	1	$1/\sqrt{2}$	$1/\sqrt{2}$		
1	0	$1/\sqrt{2}$	$-1/\sqrt{2}$		
0	2	$1/2$	$1/\sqrt{2}$	$1/2$	
1	1	$1/\sqrt{2}$	0	$-1/\sqrt{2}$	
2	0	$1/2$	$-1/\sqrt{2}$	$1/2$	
0	3	$1/\sqrt{8}$	$\sqrt{3/8}$	$\sqrt{3/8}$	$1/\sqrt{8}$
1	2	$\sqrt{3/8}$	$1/\sqrt{8}$	$-1/\sqrt{8}$	$-\sqrt{3/8}$
2	1	$\sqrt{3/8}$	$-1/\sqrt{8}$	$-1/\sqrt{8}$	$\sqrt{3/8}$
3	0	$1/\sqrt{8}$	$-\sqrt{3/8}$	$\sqrt{3/8}$	$-1/\sqrt{8}$

of a gaussian mode which appears in eqs. (1), (2), i.e. the phase shift that the beam undergoes when going through a waist as compared to that of a plane wave. Tamm has recognized that one can use the difference in Gouy phase between an astigmatic HG_{01} and HG_{10} mode to convert a first-order LG mode to a HG mode [3]. Here we show that an arrangement similar to that used by Tamm can transform modes of arbitrary order.

For an isotropic (i.e., non-astigmatic) gaussian beam the Gouy phase appears in eqs. (1), (2) as

$$(n+m+1)\,\psi(z)\,, \tag{11}$$

with $\psi(z)=\arctan(z/z_R)$ for a waist at position $z=0$. For an astigmatic beam the situation is different. Consider first an astigmatic HG beam which has its nodal lines parallel to the axes of the astigmatism. Such a beam can be produced by passing a HG beam through a cylindrical lens aligned along the axes of the mode pattern. The amplitude of this mode can be considered separately in the two transverse planes (x, z) and (y, z); in each plane the beam is characterized by the z-coordinate and the Rayleigh range of the waist. The resulting Gouy phase has two contributions, one from each transverse direction [9,10], and may be written as

$$(n+1/2)\,\psi_x(z)+(m+1/2)\,\psi_y(z)\,, \tag{12}$$

with

$$\psi_z=\arctan[(z-z_{0x})/z_{Rx}]\,, \tag{13}$$

$$\psi_y=\arctan[(z-z_{0y})/z_{Ry}]\,, \tag{14}$$

where z_{0x} and z_{0y} are the positions of the waists and z_{Rx} and z_{Ry} the corresponding Rayleigh ranges in the (x, z) and (y, z) planes, respectively. For an isotropic HG beam the two waists coincide and have equal diameters, so that eq. (12) reduces to eq. (11). It follows that isotropic HG beams of the same order $n+m$ have the same Gouy phase. But for the astigmatic HG beam ψ_x and ψ_y are different functions of z so that the relative phase of HG modes of the same order, but with different n and m, is a function of z.

It should be realized that neither a LG beam, nor a HG beam when passed through a cylindrical lens at an *arbitrary* angle relative to the mode pattern, can be described in the same way, since the amplitude of these beams is not separable in x and y. For such beams (and also for more general beams) one should decompose the transverse pattern into HG modes oriented along the axes of the lens. Since in an astigmatic beam the relative phase of these HG components changes upon propagation, the transverse pattern of these astigmatic beams will change accordingly. This applies to the cases considered in ref. [5].

4. Mode converter

In order to exploit the Gouy phase to construct a mode converter, the beam should be made astigmatic in a confined region only, while it is isotropic outside this region. When the beam is passed through this region, a definite phase difference will be introduced between the HG components which are oriented along the axes of astigmatism. Consider for example an astigmatic beam for which the waists coincide but have different Rayleigh ranges, z_{Rx} and z_{Ry}, respectively (fig. 2a). At the position where the two transverse radii of the astigmatic beam are equal, a cylindrical lens may be placed to match the radii of curvature of the beam such that the beam outside the lens is no longer astigmatic (fig. 2b). When the same cylindrical lens is also placed on the other side of the waist and if the input beam is properly mode-matched, the beam is astigmatic only between the two lenses (fig. 2c). The condition that the transverse radii of the beam (given by eq. (4)) are equal at the position of the lens $z=\pm d$ leads to

$$\frac{z_{Rx}^2+d^2}{z_{Rx}}=\frac{z_{Ry}^2+d^2}{z_{Ry}}, \tag{15}$$

and the condition that the input beam is mode-matched is fulfilled if the focal distance f of the cylindrical lenses satisfies (see eq. (3)),

$$\frac{1}{f}=\frac{1}{R_x(d)}-\frac{1}{R_y(d)}=\frac{d}{z_{Rx}^2+d^2}-\frac{d}{z_{Ry}^2+d^2}. \tag{16}$$

It is useful to introduce a parameter p such that

$$p=\sqrt{\frac{1-d/f}{1+d/f}}. \tag{17}$$

Equations (15), (16) then lead to

$$z_{Rx}=dp\,, \tag{18}$$

$$z_{Ry}=d/p\,. \tag{19}$$

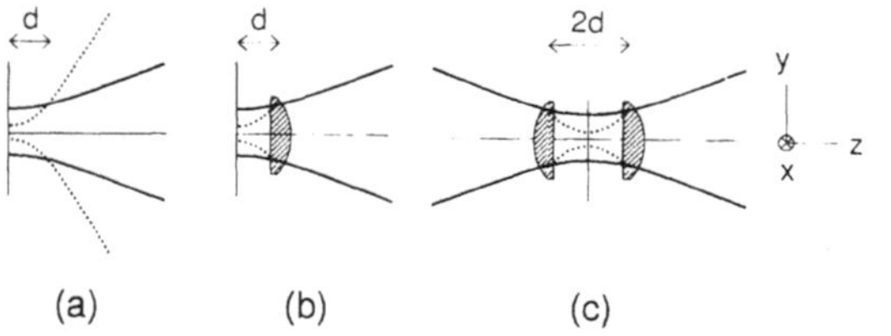

Fig. 2. Sketch of a symmetric mode converter. The dashed curve denotes the gaussian beam envelope in the (x, z) plane, and the solid curve that in the (y, z) plane. (a) An astigmatic waist at $z=0$. (b) A cylindrical lens matches the radii of curvature at $z=d$. (c) Two cylindrical lenses act as a converter on a mode-matched beam.

The change in Gouy phase $\psi(z)$ of a HG mode oriented along the axes of the lenses, when passed through this region, is, using eq. (12),

$$\Delta\psi=(n+m+1)(\Delta\psi_x+\Delta\psi_y)/2$$
$$+(n-m)(\Delta\psi_x-\Delta\psi_y)/2\,, \tag{20}$$

with

$$\Delta\psi_x=\psi_x(d)-\psi_x(-d)=2\arctan(d/z_{Rx}) \tag{21}$$

and analogously for $\Delta\psi_y$.

We now consider a 'diagonal' HG mode which is passed through the converter of fig. 2c and expand the input mode into HG modes of the same order $n+m$ oriented along the lens axes (eq. (10)). The successive terms in this expansion differ by two in the value of $(n-m)$. Therefore the total phase difference which is introduced *between* successive terms is

$$\theta=2[\arctan(d/z_{Rx})-\arctan(d/z_{Ry})]$$
$$=2[\arctan(1/p)-\arctan p]\,. \tag{22}$$

The phase difference is thus determined by the parameter p only and ranges from 0 to π.

If this phase difference is set equal to $\pi/2$ the system introduces a factor i^k in front of each term in the expansion of eq. (10), so that the HG mode is converted into a LG mode with the same indices n, m (eq. (8)). The condition $\theta=\pi/2$ is fulfilled if $p=-1+\sqrt{2}$, which leads to

$$d=f/\sqrt{2}\,. \tag{23}$$

Mode-matching (eq. (19)) requires that the input beam has a Rayleigh range (cf. fig. 2c)

$$z_{Ry}=f+d=(1+1/\sqrt{2})f. \tag{24}$$

We will call this converter, which converts a diagonal HG mode to a LG mode, a '$\pi/2$ converter', referring to the value of θ. Of course the argument can also be reversed: the $\pi/2$ relative phase different between the components of the LG mode can be removed with a $\pi/2$ converter, so that a HG mode is produced.

The converter with $\theta=\pi$, the 'π converter', implies $p=0$, and therefore

$$d=f,\quad z_{Rx}/f=0,\quad z_{Ry}/f=\infty\,, \tag{25}$$

which corresponds to a confocal configuration of the

cylindrical lenses with a collimated incident beam. Note that the ideal π converter exists only in the geometrical optical limit. In practice, i.e. in a wave-optical description, z_{Rx} and z_{Ry} are always finite, so that $\theta=\pi-\epsilon$ instead of π. In a diffraction-limited system, ϵ can be made arbitrarily small by making the system (and incident-beam diameter) sufficiently large. In the geometrical optical limit, the lens system exchanges the left and the right side of the beam, so that a diagonal HG mode u_{nm}^{HG} is converted to u_{mn}^{HG}, and a LG mode u_{nm}^{LG} is converted to u_{mn}^{LG}, which has an azimuthal dependence of the opposite sign.

The $\pi/2$ and π converters are compared in fig. 3; here we have varied the distance between the cylindrical lenses, keeping their focal length constant. It illustrates that a $\pi/2$ converter generally requires a tightly focussed input beam, whereas a π converter operates on a collimated beam.

Obviously, such converters can also be constructed by using two cylindrical *mirrors*. Another possibility would be to exploit the astigmatism furnished by an off-axis configuration of spherical lenses and/or mirrors.

5. Experiments

For testing the mode converters described in sect. 4, we used a HeNe laser consisting of a gain tube of 35 cm length with Brewster windows (Spectra Physics 120S) in a two-mirror cavity (fig. 4). As we wanted this laser to operate in a higher-order transverse mode, we had to make the Fresnel number of the cavity as large as possible. Since the bore of the gain tube (∅ 1.8 mm) is the effective aperture in the cavity, this implies a small beam diameter at the Brewster windows. From eq. (4) it follows that the beam radius at a distance d from the waist has a minimum for $z_R=d$. Therefore we chose the position of the waist in the middle of the tube, with a Rayleigh range of about half the length of the tube ($z_R\approx 18$ cm). This was obtained with two commercial HeNe laser mirrors, one with 600 mm radius of curvature, placed at 525 mm from the middle of the gain tube, and the other with 437 mm radius of curvature, placed at 306 mm from the middle of the tube. The first mirror was a high-reflector for 633 nm, the second mirror was an output coupler with 1.2% transmission. In order to force the laser to operate in a higher-order Hermite–gaussian mode, two metal wires of 20 μm diameter were carefully positioned inside the cavity in front of the high-reflector, one vertical and the other horizontal, both perpendicular to the beam. The wires force the laser to operate in the higher-order HG mode which has nodal lines at the position of the wires. Any mode of order 0 up to 3 could be made this way, with output powers between 0.5 and 5 mW.

We first tested the operation of the $\pi/2$ converter by mode-matching a ‘diagonal’ HG laser beam to such a converter by means of two spherical lenses. The cylindrical lenses of the converter had a focal distance of 19 mm, while the Rayleigh range of the input beam and the distance between the lenses were chosen according to eqs. (23), (24). A lens with a short focal length projected the output mode on to a screen, and this output pattern was photographed. The results shown in fig. 5 illustrate the conversion of several input HG modes to corresponding LG modes. When the $\pi/2$ converter was rotated around the propagation axis of the beam, an incoming HG beam was transformed to a HG mode when the converter was at 0 or 90 degrees and to a LG mode when the converter was at ±45 degrees (fig. 6a). In another experiment the HG laser mode was converted into a LG mode with a fixed $\pi/2$ mode converter and then passed through a rotatable $\pi/2$ converter, which transformed the mode back to a HG mode. The output mode was photographed with the second converter at various angles (fig. 6b). The output is a HG

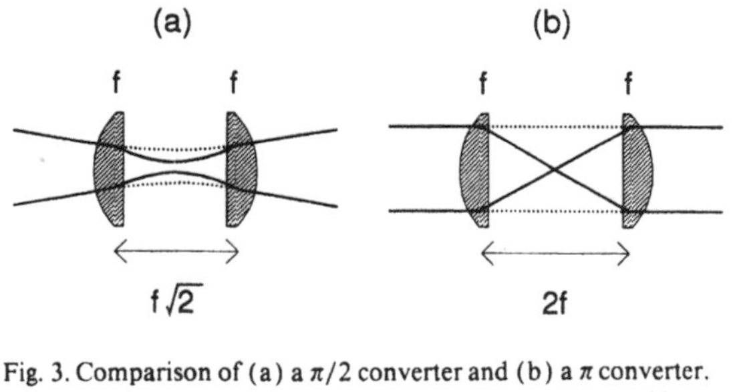

Fig. 3. Comparison of (a) a $\pi/2$ converter and (b) a π converter. Both converters consist of two identical cylindrical lenses of focal length f; they focus in the plane of the paper. The distance between the lenses is $f\sqrt{2}$ for the $\pi/2$ converter and $2f$ for the π converter. Dashed lines indicate the propagation of the beam in the other transverse direction.

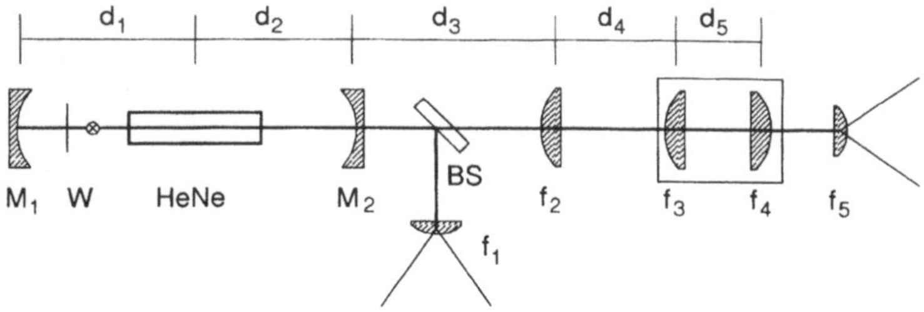

Fig. 4. Experimental arrangement used to demonstrate the operation of the $\pi/2$ mode converter. Two mutually perpendicular wires (W) are placed perpendicular to the mode axis inside a HeNe laser consisting of two mirrors (M_1, $R=600$ mm and M_2, $R=437$ mm) and a HeNe gain tube (HeNe). With a beam splitter (BS) part of the output is split off and projected onto a screen with a lens of short focal length (f_1). The beam is mode-matched with lens f_2 into a $\pi/2$ converter built with two cylindrical lenses (f_3 and f_4). The output is projected onto a screen with a lens of short focal length (f_5). Dimensions are: $d_1=525$ mm, $d_2=306$ mm, $d_3=225$ mm, $d_4=176$ mm, $d_5=27$ mm; $f_1=f_5=20$ mm, $f_2=160$ mm, $f_3=f_4=19$ mm (cylindrical).

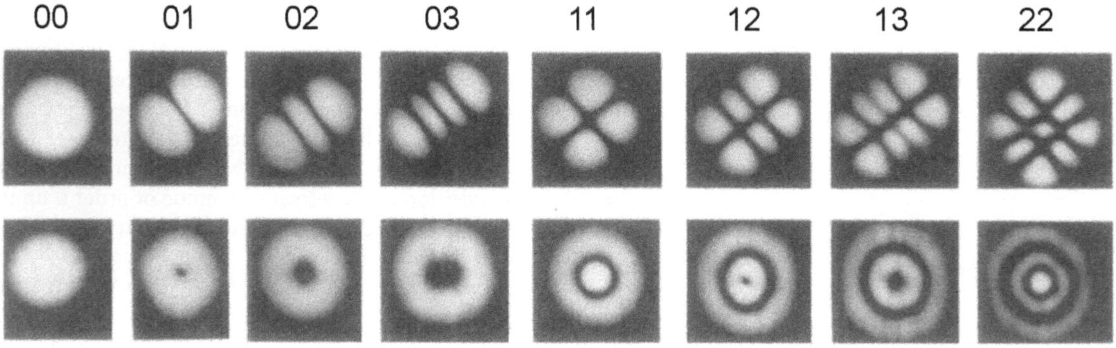

Fig. 5. Experimental results obtained with the $\pi/2$ converter. The top row shows the input HG_{nm} mode; the bottom the output LG_{nm} mode, where n, m is indicated above the modes.

mode which is always at 45 degrees with respect to the axes of the lenses of the converter.

To test the π converter we mounted such a converter in the rotatable mount and passed a collimated HG beam through it. The output is again a HG mode which rotates twice as fast as the converter, as shown in fig. 7.

To interpret these results it is helpful to keep in mind the resemblance between decomposition of a mode pattern on the one hand and decomposition of polarization on the other hand, as pointed out in ref. [2]. A quarter-wave plate converts linearly to circularly polarized light by introducing a $\pi/2$ phase difference between the linearly polarized components and is therefore analogous to the $\pi/2$ mode converter, which converts a HG mode into a LG mode by introducing a $\pi/2$ phase difference between the HG components. A half-wave plate converts left-handed to right-handed circularly polarized light by introducing a π phase difference and is therefore similar to the π converter.

Using this analogy, the experiment of fig. 6a is similar to passing a linearly polarized beam through a rotating quarter-wave plate. Depending on the orientation of the plate, this would result in a circularly or linearly polarized beam. The experiment of fig. 6b is similar to passing a circularly polarized beam through a rotating quarter-wave plate. In this case

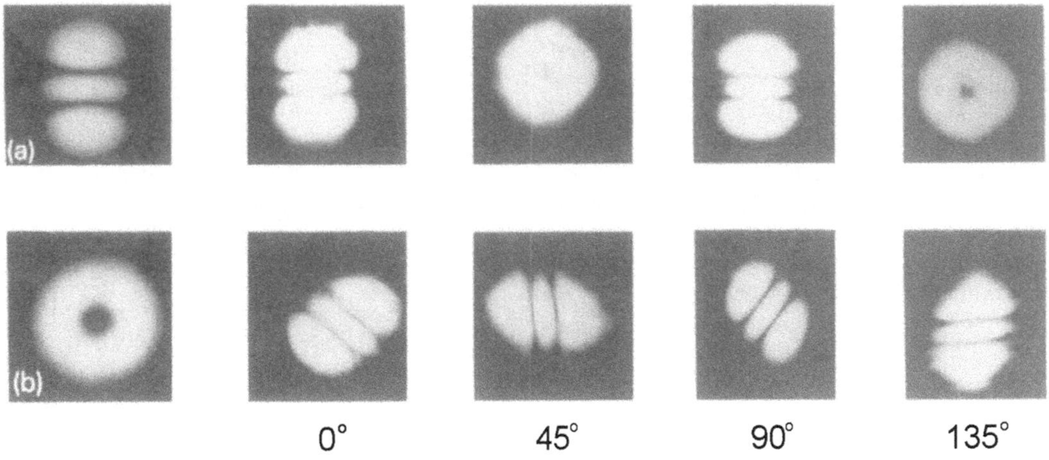

Fig. 6. Experimental results obtained with the $\pi/2$ converter at different angles, (a) with an input HG_{02} mode and (b) with an input LG_{02} mode. The left-most pattern in each row is the input, on the right are the output patterns with the angle of the converter indicated below.

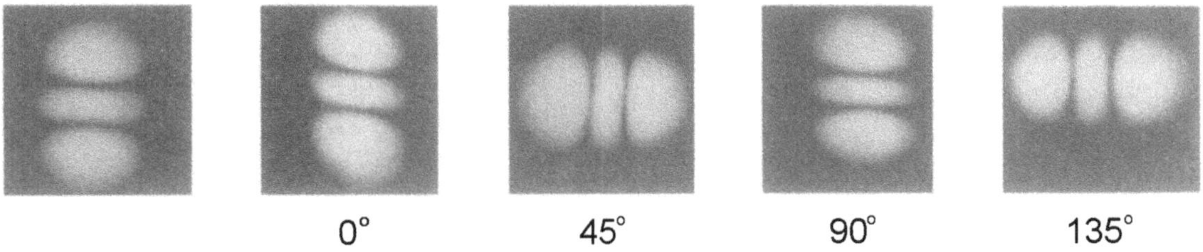

Fig. 7. Experimental results obtained with the π converter at different angles. On the left is the input HG_{02} mode, on the right the output patterns.

the output would be a linearly polarized beam whose polarization axis is at 45 degrees with respect to the axes of the plate. The last experiment (fig. 7) is similar to passing a linearly polarized beam through a half-wave plate. This would result in a linearly polarized beam whose polarization axis rotates twice as fast as the plate.

Note that in the paraxial approximation the polarization decomposition is two-dimensional, spanned by two polarization vectors. The mode decomposition has dimension $N+1$ for a mode of order N since such a mode can be expressed as a superposition of $N+1$ modes of the same order.

6. Angular momentum transfer

Now that we have studied in detail how a HG mode may be converted to a LG mode we remind the reader that it has been established previously that a LG beam carriers orbital angular momentum whereas a HG beam does not [2]. It may thus be expected that the transformation leads to a mechanical torque on the converter. We are preparing an experiment to investigate this issue. Here we wish to address theoretically the question where precisely, that is at which cylindrical lens, the transfer of angular momentum will take place. We consider a linearly polarized mode with vector potential $\boldsymbol{A}=\boldsymbol{e}\, u(x, y, z)\exp(-ikz)$ in the Lorentz gauge (with $\boldsymbol{e}$ the polarization vector), where $u(x, y, z)$ is a solution of the paraxial wave equation. The time-averaged angular momentum density of this mode has a component in the direction of propagation given by [2]

$$M_z = \mathrm{i}\omega\epsilon_0/2\left[\left(xu^*\frac{\partial u}{\partial y} - xu\frac{\partial u^*}{\partial y}\right) - \left(yu^*\frac{\partial u}{\partial x} - yu\frac{\partial u^*}{\partial x}\right)\right]$$
$$= -\omega\epsilon_0(\partial\alpha/\partial\phi)\,|u|^2\,, \tag{26}$$

with $\alpha=\arg u$. The angular momentum density per unit length in the direction of propagation is

$$L_z = \int_{-\infty}^{\infty}\int_{-\infty}^{\infty} \mathrm{d}x\,\mathrm{d}y\, M_z\,. \tag{27}$$

For a HG mode we deduce from eq. (1) the phase of u

$$\alpha = -kr^2/2R(z) - (n+m+1)\psi(z) \tag{28}$$

and substitution in eq. (26) yields that the angular momentum of such a mode is zero. For a LG mode we deduce from eq. (2)

$$\alpha = -kr^2/2R(z) - (n+m+1)\psi(z) - (n-m)\phi\,, \tag{29}$$

so that the ϕ derivative leads to an angular momentum proportional to $n-m$. It has been shown previously that it is in fact equal to $(n-m)\hbar$ per photon [2]. As the mode is linearly polarized, the intrinsic spin angular momentum is zero and the angular momentum found here is orbital angular momentum.

The $\pi/2$ converter transforms a HG mode without orbital angular momentum to a LG mode with orbital angular momentum. It might be anticipated that the transfer of angular momentum takes place at both lenses, but it turns out that this is not the case. For any mode $u(x, y, z)$ which is a solution of the paraxial wave equation, the angular momentum is given by eqs. (26), (27). When this mode is passed through a lens it acquires an extra phase χ which is a function of the transverse coordinates [#1], so that the mode function directly after the lens can be written as

$$u' = u\exp[\mathrm{i}\chi(x, y)]\,. \tag{30}$$

In ref. [11] it is shown that

$$L_z(u') = L_z(u) + \delta L_z\,, \tag{31}$$

with

$$\delta L_z = -\omega\epsilon_0\int_{-\infty}^{\infty}\int_{-\infty}^{\infty}\mathrm{d}x\,\mathrm{d}y\left(x\frac{\partial\chi}{\partial y} - y\frac{\partial\chi}{\partial x}\right) \times |u(x, y, z)|^2\,. \tag{32}$$

For a cylindrical lens we have $\chi = -\frac{1}{2}(k/f)x^2$, so that

$$\delta L_z = -\omega\epsilon_0\int_{-\infty}^{\infty}\int_{-\infty}^{\infty}\mathrm{d}x\,\mathrm{d}y\,(k/f)xy\,|u(x, y, z)|^2\,. \tag{33}$$

This may be normalized against the energy in the beam to yield the angular momentum transferred to the lens per photon. For a LG mode, as $|u_{nm}^{\mathrm{LG}}|^2$ is symmetric around the z axis, δL_z is zero. A diagonal HG mode with $m\neq n$, however, has unequal intensities in the four xy quadrants, which leads to $\delta L_z\neq 0$. The conclusion is that in the $\pi/2$ converter, with a HG mode $(m\neq n)$ incident on one side and a LG mode exciting from the other, the angular momentum conversion is expected to take place *only* at the lens which sees the input HG mode. It follows that the astigmatic HG mode just after the first lens already contains the orbital angular momentum of the output LG mode. The function of the second lens is to change the astigmatic beam into a pure LG mode. For the same reason, if a LG mode is incident on a $\pi/2$ converter and is converted into a HG mode, the transfer of orbital angular momentum is expected to take place at the second lens.

In fact, for a macroscopic lossless dielectric object such as a lens the radiative force is a conservative force ("gradient force") which is derived from the

[#1] Writing the operation of the lens as a phase factor is only correct for a lens whose principal planes coincide, i.e. a thin lens.

dielectric polarization energy as the potential [12]. Dielectric matter is drawn to positions where the electric field strength is largest (as an example, when the plates of a capacitor are dipped into a dielectric fluid, the fluid is drawn into the volume between the plates). This explains why the transfer of angular momentum, that is the mechanical torque, occurs at the 'HG side' of the $\pi/2$ converter only. For a cylindrical lens in a HG beam the dielectric polarization energy clearly depends on the angle between the cylinder axis and the HG axes x and y. The cylindrical lens will tend to align with the maximum optical intensity inside the dielectric. A radiative torque on the lens results unless the polarization energy is minimum or maximum, that is unless the cylinder axis is aligned with x or y. For a cylinder lens in a LG beam the radiative torque is zero as the polarization energy is independent of its orientation.

It follows from eq. (33), and also from the energy arguments above, that the change in orbital angular momentum may in principle be made arbitrarily high by choosing the lens arbitrarily strong ($f \rightarrow 0$). It follows similarly that the orbital angular momentum may be changed by astigmatic elements other than a cylinder lens (e.g. a prism). Also, for similar reasons, a LG mode with an angular sector removed with a suitable mask *will* transfer orbital angular momentum to a cylindrical lens. However, all such operations will not lead to a gaussian mode at the output and thus do not correspond to a well-defined mode conversion.

The π converter seems an anomaly in this context. If the perfect π converter has a LG mode incident at the first lens, a LG mode of opposite handedness will leave the second lens. According to the arguments given above it appears that there is no orbital angular momentum exchange on either of the lenses, although the orbital angular momentum of the beam changes sign. As noted in sect. 4, however, a perfect π converter does not exist. If a perfect LG mode is incident on a practical converter, the output mode will always, however slightly, differ from a perfect LG mode, which invalidates the analysis give above. An appropriate analysis of this converter has been given by Van Enk and Nienhuis [11].

7. Conclusions

We have shown that it is possible to build a mode converter with two cylindrical lenses which converts a Hermite–gaussian mode or arbitrary order into a Laguerre–gaussian mode of the same order and vice versa. The conversion is described using a mode analysis, based on the decomposition of a LG mode and a diagonally oriented HG mode into Hermite–gaussians. Two converters are introduced: the $\pi/2$ converter which converts a HG to a LG mode or vice versa, and the π converter which exchanges the indices of the incoming mode and thereby converts a LG mode into one with opposite azimuthal dependence. We have demonstrated these conversions experimentally and have shown the analogy between the mode converters and half- and quarter-wave plates. The transfer of orbital angular momentum, occurring when a HG mode without orbital angular momentum is converted into a LG mode with orbital angular momentum, is shown to take place at the lens onto which the HG mode is incident.

Acknowledgements

We thank R.J.C. Spreeuw, G. Nienhuis and S.J. van Enk for helpful discussions. This work is part of the research program of the Foundation for Fundamental Research on Matter (FOM) and was made possible by financial support from the Netherlands Organization for Scientific Research (NWO).

References

[1] R.A. Beth, Phys. Rev. 50 (1936) 115.
[2] L. Allen, M.W. Beijerbergen, R.J.C. Spreeuw and J.P. Woerdman, Phys. Rev. A 45 (1992) 8185.
[3] Chr. Tamm, in: Proc. the workshop on quantitative characterization of dynamical complexity in non-linear systems, eds. N.B. Abraham and A. Albano (Plenum, New York, 1989) p. 465.
[4] Chr. Tamm and C.O. Weiss, J. Opt. Soc. Am. B 7 (1990) 1034.
[5] E. Abramochkin and V. Volostnikov, Optics Comm. 83 (1991) 123.
[6] M. Abramowitz and I.A. Stegun, Handbook of mathematical functions (Dover, New York, 1965).
[7] G. Nienhuis and L. Allen, to be published.

[8] A. Danakis and P.K. Aravind, Phys. Rev. A 45 (1992) 1972.
[9] D.C. Hanna, IEEE J. Quantum Electron. 5 (1969) 483.
[10] A.E. Siegman, Lasers (University Science Books, Mill Valley, 1986).
[11] S.J. van Enk and G. Nienhuis, Optics Comm. 94 (1992) 147.
[12] S. Stenholm, Contemp. Phys. 29 (1988) 105.

Observation of the Dynamical Inversion of the Topological Charge of an Optical Vortex

Gabriel Molina-Terriza, Jaume Recolons, Juan P. Torres, and Lluis Torner
Laboratory of Photonics, Department of Signal Theory and Communications, Universitat Politecnica de Catalunya, Gran Capitan UPC-D3, Barcelona, ES 08034, Spain

Ewan M. Wright
Optical Sciences Center, University of Arizona, Tucson, Arizona 85721
(Received 10 January 2001; published 22 June 2001)

We report what is believed to be the first detailed experimental observation of the dynamic inversion of the topological charge of an optical vortex under free-space propagation. The vortex self-transformation occurs through continuous deformation of the noncanonical strength of the corresponding screw wave front dislocation, and is mediated by the occurrence of an extremely sharp turn in a Berry vortex trajectory, which observed at a Freund critical foliation appears as an edge-line dislocation orthogonal to the propagation direction, at a crucial point of the light evolution.

DOI: 10.1103/PhysRevLett.87.023902 PACS numbers: 42.25.–p, 42.65.–k

Topological deformation of fields is at the heart of the description of a vast variety of natural phenomena ranging from cosmology, to oceanography, chemistry, and biosciences. Its most extreme, yet ubiquitous manifestation occurs when the field is so strongly folded that it bends back upon itself to form a topological dislocation, or defect [1–5]. Screw dislocations, or vortices, are a common defect type. The multiplicity of the folding around the defect determines its topological charge. The net charge within a region of space must be conserved under continuous evolution provided that no charges enter or leave the region [6]. Therefore, dynamic vortex creation typically occurs through the nucleation of twin pairs of opposite charge. However, here we uncover a striking mechanism that inverts the topological charge of a single vortex under propagation. This vortex self-transformation is mediated by the occurrence of an extremely sharp turn in a Berry vortex trajectory [7,8], which in the experimental plane of observation appears as an edge-line dislocation orthogonal to the light propagation direction, at a crucial point of the light evolution. Observations were performed with vortices nested in light beams [5–14]. The phenomenon can have direct implications to quantum photon states with angular momentum [15,16], to superfluidity [17,18], and to weakly interacting Bose-Einstein condensates (BEC's) [19,20], and is believed to be generic to diverse vortex entities.

The key ingredient of the new phenomenon is the existence of vortices with intrinsic spatial structure [6], hereafter referred to as noncanonical vortices. In the case of optical beams, vortices are spiral phase ramps around a dark spot where the phase of the wave is undefined and thus the light intensity vanishes. The slowly varying field envelope E of a single-charge vortex nested in a paraxial light beam propagating along the Z axis has the form

$$E = \{x - x_1(z) + iA(z)[y - y_1(z)]\}F(x,y,z), \quad (1)$$

where (x_1, y_1) stands for the vortex location in the X-Y plane and F is the host beam. The value of the complex quantity A determines the noncanonical strength of the vortex: Only when $A = \pm 1$ does the vortex have a traditional, or canonical, structure. Vortices are located at the complex zeros of E. In the case of canonical vortices, the lines in the transverse plane where the real and imaginary parts of E vanish cross at right angles. When $A \neq \pm 1$, the intensity pattern of the light beam E has no radial symmetry around the vortex core, and at the location of the phase dislocation the real and imaginary parts of E cross each other at general angles. The light intensity and phase structure of a single-charge noncanonical vortex with strength $A = 1 + i$ are shown in Fig. 1. The sign of the topological charge of the vortex, given by the circulation of the gradient of the phase of E around the dislocation, is given by the sign of Re(A). Therefore, a family of different noncanonical vortices that carry the same topological charge can be constructed by varying the complex value of A while keeping the sign of Re(A) fixed.

However, the dynamical evolution followed by the vortices, including the value of its topological charge, can depend critically on the noncanonical strength of the dislocations. In particular, self-inversion of the topological charge of a noncanonical vortex can occur. Furthermore, in contrast to traditional views, such phenomenon can take place in a variety of physical settings, including properly tailored light propagation in vacuum, counterintuitive though it might seem. The principle of the experimental realization of the dynamic inversion of the topological charge of a vortex is illustrated by the paraxial evolution of a single-charge vortex nested in the elliptical Gaussian beam $F = C(z)\exp[ix^2/q_x(z) + iy^2/q_y(z)]$, where $C(z)$ is the complex beam amplitude and $q_x(z)$ and $q_y(z)$ are the complex beam widths. In free space, the beam widths evolve as $q_x = 2(z - z_x) - iw_{0x}^2$, $q_y = 2(z - z_y) - iw_{0y}^2$, where z_x and z_y stand for the locations

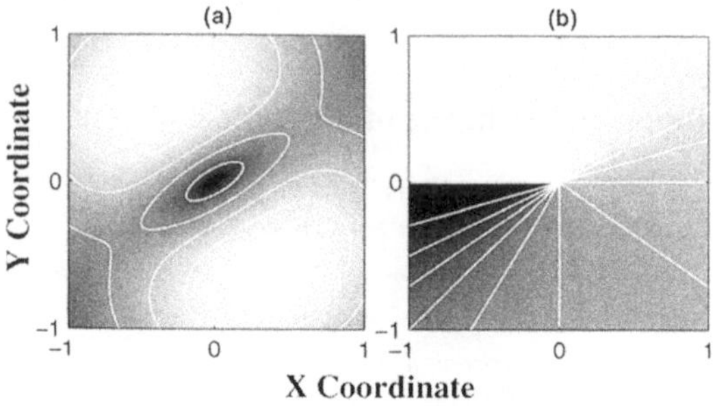

FIG. 1. Core of a noncanonical vortex with strength $A = 1 + i$. (a) Amplitude. (b) Phase. Contour lines are spaced 0.11 dimensionless units in (a), and $\pi/8$ units in (b).

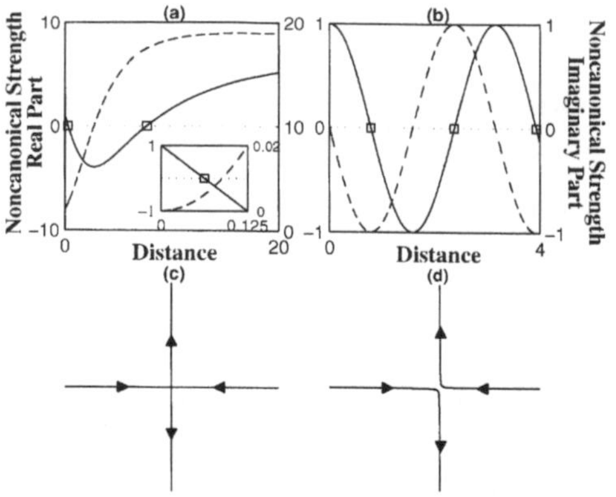

FIG. 2. Predicted evolution of the noncanonical strength of a vortex under different propagation conditions. Continuous and dashed lines show the evolution of the real and imaginary parts of A, respectively. Axes are in dimensionless units. The squares highlight the critical propagation distances where the inversion of the topological charge of the vortex occurs. (a) Evolution of a noncanonical vortex with $A(0) = 1 + 2i$, nested in an elliptical Gaussian beam with $w_{0x} = 1w_{0y} = 3$. Inset: Predicted evolution of the noncanonical strength of the vortex under the conditions of our experimental setup. (b) Evolution of an input canonical vortex nested in an elliptical beam propagating in a GRIN medium with refractive index $n(x, y) = 2x^2 + 8y^2$. The widths of the Gaussian beam are the stationary modes of the medium: $w_x = 1$, $w_y = 1/\sqrt{2}$. (c),(d) Sketch of the Berry trajectories followed by on-axis and slightly off-axis vortices, respectively, when they undergo a topological charge inversion.

of the beam waists along the propagation axis Z and w_x, w_y are the corresponding waists. Under such conditions, one finds that the noncanonical strength of the vortex varies during propagation according to the expression [21]

$$A(z) = A_0 \frac{2(z - z_x) - iw_{0x}^2}{2(z - z_y) - iw_{0y}^2}, \tag{2}$$

where A_0 is a constant. The vortex moves during propagation and it is located at the transverse coordinates that satisfy

$$\frac{x_1(z)}{q_x} + iA_0 \frac{w_{0y}^2}{w_{0x}^2} \frac{y_1(z)}{q_y} = \frac{x_{1,0}}{iw_{0x}^2}. \tag{3}$$

When $z_x = z_y$ and $w_{0x} = w_{0y}$, $A(z)$ is constant during propagation and hence so is the topological charge of the vortex. Nevertheless, as shown in Fig. 2a, this is not the case when the noncanonical character of the vortex is taken into account, and the charge of the vortex inverts sign. When the vortex is located exactly on axis (i.e., $x_{1,0} = 0$), the inversion of the vortex charge occurs through a vortex trajectory [7] orthogonal to the light propagation direction (Fig. 2c). In the experimental plane of observation, which corresponds to Freund's *critical foliation* for this case [8], such trajectory appears as an edge-line dislocation. When either $x_{1,0} \neq 0$ or the beam is not exactly elliptical, as is the case in any experiment, the vortex trajectory is predicted to feature two extremely sharp turns towards and from the transverse infinity (Fig. 2d). An analogous behavior as above is obtained in more complex linear and nonlinear media, where, in general, the vortex evolution can be traced by solving numerically the paraxial wave equation, and accurate approximations of the evolution of $A(z)$ can be derived using multiple-scale expansion techniques around the vortex core [21,22]. As an illustrative example with relevance to evolution of vortices in trapped weakly interacting Bose-Einstein condensates [19,20], shown in Fig. 2b, is the evolution of an initially canonical vortex in an asymmetric graded-index (GRIN) linear medium (the GRIN medium plays the role of a magnetic trap for the BEC). In this case, the charge of the noncanonical vortex is predicted to change periodically along propagation.

The experimental realization of an evolving noncanonical vortex can be performed in different settings. In an effort to highlight the universality of the phenomenon and to rule out complexity that might make the origin of the observations controversial, we performed our experiment to realize the evolution shown in Fig. 2a using an astigmatic optical scheme in free space. A linearly polarized, collimated light beam produced by a He:Ne laser at a wavelength 632.8 nm was split into two beams. Single-charge screw wave front dislocations were nested in one of the beams by using computer-generated holograms [11]. We used off-axis holograms, which were designed to generate traditional canonical vortices, to optimize the separation of the resulting Fresnel diffraction orders. The other beam was used as a reference wave. The beam after the holographic mask was spatially filtered to extract the diffraction order containing the single-charge vortex and then passed through a lens of 10 cm focal length. The wave front of the resulting beam was analyzed by superposing the beam with the vortex nested and the reference wave tilted slightly relative to the propagation axis. The interference pattern arising exhibits a characteristic fork at the position of the dislocation and its topological charge is visually evaluated. The pattern of output light was analyzed by a charge-coupled device (CCD) camera. Noncanonical vortex dynamics were generated by operating the system with a cylindrical lens. The left and central rows of Fig. 3 show the observed outcome at different planes after the lens. As the ellipticity of the beam increases while it focuses, the vortex core becomes increasingly eccentric. At a crucial instant of the evolution the vortex converts itself into what, according to Figs. 2c and 2d, is believed to be a transverse vortex trajectory which inside the finite experimental accuracy always appears as an edge-line dislocation. Subsequently the quasiedge is observed to transform itself into a vortex of opposite topological charge relative to the input value. The orientation of the edge line in the plane is dictated by the value of the noncanonical strength. Notice that the inversion of the helicity of the spiral phase front of the vortex field is fully consistent with the imaging properties of the lens. We verified that the charge of a traditional canonical vortex does not vary by operating the system with a spherical lens (Fig. 3, right row).

We note the one-to-one map existing between the astigmatic scheme employed in the experiment and the evolution displayed in Fig. 2a. We stress that the observed inversion of the topological charge of the vortex does not occur at the surface of the lens, i.e., it does not involve a wave front dislocation induced by external discontinuities. We also stress that it occurs in free space, thus it is not related to any instabilities mediated with nonlinear effects or vortex solitons [10]. Rather, on the light of the relationship between general screw and edge dislocations studied by Nye [9], and of the generic vortex trajectories reported by Freund [8], it is believed to be a genuine and fundamental dynamic self-effect of noncanonical vortices.

Because of the associated spiral wave front, paraxial light beams with nested vortices carry orbital angular momentum [12] that can be transferred to trapped suitable material particles causing them to rotate [5]. The total angular momentum can also contain a spin contribution, associated with polarization [13]. In the case of canonical vortices nested in symmetric beams, the orbital angular momentum, in units of $\hbar\omega$, is given by $L = mN$, where m is the topological charge of the vortex and N is the number of photons. In our experiment $m = 1$, thus for the beam incident upon the lens $L/N = 1$. We verified that in spite of the inversion of the topological charge of the vortex, the angular momentum of the beam is conserved during its evolution. Such angular momentum could be modified by passing the noncanonical vortex through additional optical elements [14], but otherwise in free space L is constant so that $L/N = 1$ everywhere. Therefore, contrary to traditional views, beyond the point where the charge of the vortex is inverted, the orbital angular momentum carried by the beam and the topological charge of the vortex nested in it have opposite signs. This fact has a simple interpretation. The beam incident upon the cylindrical lens is an almost pure Laguerre-Gaussian mode. The astigmatic lens redistributes the incoming orbital angular momentum into a superposition of Laguerre-Gaussian modes carrying different charges (Fig. 4), while keeping the total value constant. Therefore, the density of angular momentum is dynamically redistributed across the beam during its evolution.

We now consider the implications of the above to quantum photon states. The Laguerre-Gaussian modes form a complete Hilbert set and can thus be used to represent the quantum photon states within the paraxial theory of light propagation. The quantum angular momentum

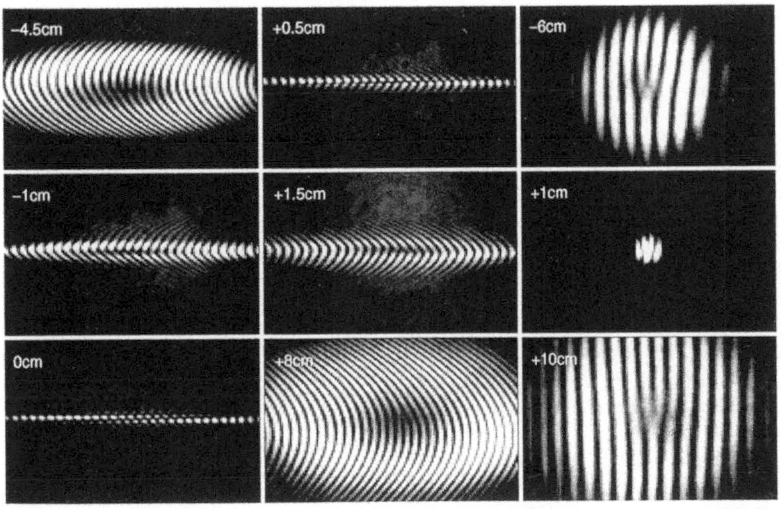

FIG. 3. Observed evolution of the topological charge of a vortex during a noncanonical evolution. The pictures show the evolution on propagation of the interference between the vortex beam and a reference beam tilted relative to the propagation axis. The images were acquired by a CCD camera placed at different positions beyond the lens. Labels inside the pictures stand for the location of the CCD camera relative to the lens focal plane. Left and center rows: Experimental realization of the noncanonical evolution generated by an input canonical vortex nested in a radially symmetric Gaussian beam with a 5 mm waist, passed through a cylindrical lens with a 10 cm focal length. Right row: Observed light pattern under identical conditions but with a spherical lens.

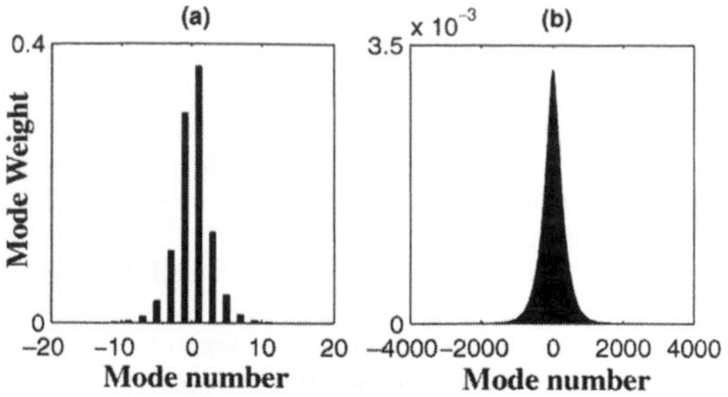

FIG. 4. Decomposition into normal modes with a well-defined angular momentum. Modes are numbered according to their topological charge, or azimuthal index. (a) Mode decomposition for the beam parameters of Fig. 2a. (b) Mode decomposition for the light beam with the noncanonical vortex nested used in the experimental observations, revealing that it contains 10^3 modes.

number carried by the photon is then represented by the charge, or azimuthal index m, of the corresponding mode. The photons incident upon the lens are in an almost pure angular momentum state $|m = 1\rangle$, but after the lens the photon state is a superposition of orbital angular momenta $|m\rangle$ (Fig. 4). The outcome is that, even though a full quantum average over many realizations yields the classical expectation $\langle L\rangle/n = 1$, with n the mean photon number, the quantum angular momentum number obtained in most of the realizations is not $m = 1$. We thus notice the direct implications to the generation and teleportation of quantum entangled photons with angular momentum [15,16] by parametric down-conversion of light in quadratic nonlinear crystals pumped with noncanonical vortex beams [23].

We finally point out that our observations here showing vorticity transformations of complex fields through noncanonical vortex trajectories and edge dislocations may provide new insight on the issue of how vorticity enters a different physical system governed by similar evolution equations. In particular, they may illuminate an old issue in superfluidity [17,18], namely, how do quantized vortices enter into weakly interacting Bose-Einstein condensates trapped in a container that is stirred or rotated [19,20]. Our observations on the self-transformation of noncanonical vortices open up a new broad scenario where general dislocations, including edge dislocations, might evolve dynamically from the edge of the container towards the interior where the vortices are stabilized by repulsive many-body nonlinearities. They also suggest the possibility of dynamic charge inversion of a weakly interacting Bose-Einstein condensate vortex under the action of an asymmetric trapping potential.

In conclusion, we emphasize that the observed self-inversion of the topological charge of the vortex occurs through its dynamic propagation, and it does not involve discontinuities or nonlinearities or the nucleation of twin vortex pairs. Our observations reported here were performed with vortices nested in light beams, but the phenomenon is believed to be a generic effect of noncanonical vortices, thus the results have implications to all systems employing noncanonical vortex entities.

This work was supported by the Generalitat de Catalunya and by TIC2000-1010.

[1] R. Thom, *Structural Stability and Morphogenesis* (Benjamin, Reading, MA, 1975).
[2] J. F. Nye and M. V. Berry, Proc. R. Soc. London A **336**, 165 (1974).
[3] *Physics of Defects*, edited by R. Balian, M. Kleman, and J.-P. Poirier (North-Holland, Amsterdam, 1981).
[4] A. T. Winfree, *The Geometry of Biological Time* (Springer, Berlin, 1990).
[5] A. Ashkin, Opt. Photonics News **10**, 41 (1999); M. E. J. Friese *et al.*, Nature (London) **394**, 348 (1998); M. J. Padgett and L. Allen, Contemp. Phys. **41**, 275 (2000).
[6] I. Freund, Opt. Commun. **159**, 99 (1999).
[7] M. V. Berry, in *Singular Optics*, edited by M. S. Soskin, SPIE Proceedings Vol. 3487 (SPIE–International Society for Optical Engineering, Bellingham, WA, 1998), pp. 1–5.
[8] I. Freund, Opt. Commun. **181**, 19 (2000); Opt. Lett. **26**, 545 (2001); I. Freund and D. A. Kessler, Opt. Commun. **187**, 71 (2001).
[9] J. F. Nye, Proc. R. Soc. London **378**, 219 (1981).
[10] G. A. Swartzlander and C. T. Law, Phys. Rev. Lett. **69**, 2503 (1992); A. W. Snyder, L. Poladian, and D. J. Mitchell, Opt. Lett. **17**, 789 (1992); G. S. McDonald, K. S. Syed, and W. J. Firth, Opt. Commun. **94**, 469 (1992).
[11] V. Yu. Bazhenov, M. V. Vasnetsov, and M. S. Soskin, JETP Lett. **52**, 429 (1991); N. R. Heckenberg *et al.*, Opt. Lett. **17**, 221 (1992).
[12] L. Allen, M. J. Padget, and B. Babiker, in *Progress in Optics*, edited by E. Wolf (Elsevier, Amsterdam, 1999), Vol. XXXIX, pp. 291–372.
[13] R. A. Beth, Phys. Rev. **50**, 115 (1936).
[14] E. Abramochkin and V. Volostnikov, Opt. Commun. **83**, 123 (1991); S. J. van Enk and G. Nienhuis, Opt. Commun. **94**, 147 (1992); M. W. Beijersbergen *et al.*, Opt. Commun. **96**, 123 (1993).
[15] A. Mair and A. Zeilinger, "Entangled States of Orbital Angular Momentum," http://info.uibk.ac.at/c/c7/c704/qo/photon/_vortices
[16] A. Furusawa *et al.*, Science **282**, 706 (1998).
[17] S. J. Putterman, M. Kac, and G. E. Uhlenbeck, Phys. Rev. Lett. **29**, 546 (1972).
[18] R. J. Donnelly, *Quantized Vortices in Hellium II* (Cambridge University Press, Cambridge, 1991).
[19] D. A. Butts and D. S. Rokhsar, Nature (London) **397**, 327 (1999); J. E. Williams and M. J. Holland, Nature (London) **401**, 568 (1999).
[20] M. R. Matthews *et al.*, Phys. Rev. Lett. **83**, 2498 (1999); K. W. Madison *et al.*, Phys. Rev. Lett. **84**, 806 (2000).
[21] G. Molina-Terriza, E. M. Wright, and L. Torner, Opt. Lett. **26**, 163 (2001).
[22] Y. S. Kivshar *et al.*, Opt. Commun. **152**, 198 (1998).
[23] P. Di Trapani *et al.*, Phys. Rev. Lett. **81**, 5133 (1998); J. Arlt *et al.*, Phys. Rev. A **59**, 3950 (1999).

PAPER 3.7

Orbital angular momentum exchange in cylindrical-lens mode converters

M J Padgett and L Allen

Department of Physics and Astronomy, Kelvin Building, University of Glasgow, Glasgow G12 8QQ, UK

Received 2 November 2001
Published 29 January 2002
Online at stacks.iop.org/JOptB/4/S17

Abstract
Cylindrical-lens mode converters (Beijersbergen M W, Allen L, van der Veen H E L O and Woerdman J P 1993 *Opt. Commun.* **96** 123–32) are used to transform between Hermite–Gaussian and Laguerre–Gaussian modes with a resulting transfer of angular momentum to the light beam and a corresponding torque on the lenses. By numerically analysing both the total and local angular momentum of the light beam, we explain the origin of this torque and confirm that is not evenly distributed between the lenses. We also confirm that any vortex contained within the beam may change sign even when the orbital angular momentum of the beam remains constant.

Keywords: Orbital angular momentum density, cylindrical-lens mode converter, vortex inversion

It is now well understood that light beams can carry an orbital angular momentum about the beam axis associated with helical wavefronts. Allen *et al* [1] showed that any beam, of which the Laguerre–Gaussian mode is perhaps the best known example, which has an $\exp(il\phi)$ helical phase structure carries an orbital angular momentum of $l\hbar$ per photon.

One method for producing, or transforming, pure modes is to use a pair of cylindrical lenses to convert from high-order Hermite–Gaussian to a Laguerre–Gaussian mode of the same mode order. This form of mode converter has been fully described by Beijersbergen *et al* [2]. Two forms of the mode converter are particularly useful. For a beam with Rayleigh range $(1 + 1/\sqrt{2})f$, and for lenses separated by $\sqrt{2}f$, a $\pi/2$ converter transforms any Hermite–Gaussian mode with indices m, n to a Laguerre–Gaussian mode with indices $l = m - n$ and $p = \min(m, n)$. The other, arguably notional, is a π converter, where two identical cylindrical lenses, of focal length f, are separated by $2f$. For a collimated beam, a π converter forms a mirror image of the beam such that, for a Laguerre–Gaussian mode, the azimuthal mode index l is transformed to $-l$. As these mode converters introduce phase shifts of $\pi/2$ or π between orthogonal modes, they are analogous to waveplates which introduce phase shifts between orthogonal polarization. The analogy extends to geometrical construction of a Poincaré sphere [3] and mathematical representations in the form of the Jones matrices [4].

In general, the passage of a light beam through either of these mode converters results in a change in the azimuthal mode index of the constituent Laguerre–Gaussian modes and hence in a change of the orbital angular momentum of the light beam. Clearly, such a change in angular momentum results in a torque on one, or both, of the lenses. It is the purpose of this paper to examine the physical origin of this torque and to identify how it is divided between the lens pair.

To examine the nature of the light beam at each point in the optical system we employ a plane wave decomposition of the type used by Sziklas and Siegman [5] and used recently to look at imperfections within mode converter design [6]. This allows the calculation of the field in any subsequent plane for an arbitrary field in any plane. Once the amplitude of the wave is known, the momentum density can be calculated from [7]

$$\begin{aligned} p &= \frac{\varepsilon_0}{2}(\boldsymbol{E}^* \times \boldsymbol{B} + \boldsymbol{E} \times \boldsymbol{B}^*) \\ &= \mathrm{i}\omega\frac{\varepsilon_0}{2}(u^*\nabla u - u\nabla u^*) + \omega k\varepsilon_0|u|^2\boldsymbol{z} + \omega\sigma\frac{\varepsilon_0}{2}\frac{\partial|u|^2}{\partial r}\boldsymbol{\Phi} \end{aligned} \quad (1)$$

where $u \equiv u(r, \phi, z)$ is the complex scalar function describing the distribution of the field amplitude and σ describes the degree of polarization of the light; $\sigma = \pm 1$ for right- and left-hand circularly polarized light, respectively, and $\sigma = 0$ for linearly polarized. For a Laguerre–Gaussian mode, $u(r, \phi, z)$ is given by

$$\begin{aligned} u(r, \phi, z) = {} & \frac{C_{pl}^{LG}}{w(z)}\frac{z_R}{(z_R^2 + z^2)}\left[\frac{r\sqrt{2}}{w(z)}\right]^l L_p^l\left[\frac{2r^2}{w^2(z)}\right] \\ & \times \exp\left(\frac{-r^2}{w^2(z)}\right)\exp\left(\frac{-\mathrm{i}kr^2z}{2(z^2 + z_R^2)}\right) \\ & \times \exp(-\mathrm{i}l\phi)\exp\left(\mathrm{i}(2p + l + 1)\tan^{-1}\frac{z}{z_R}\right) \end{aligned} \quad (2)$$

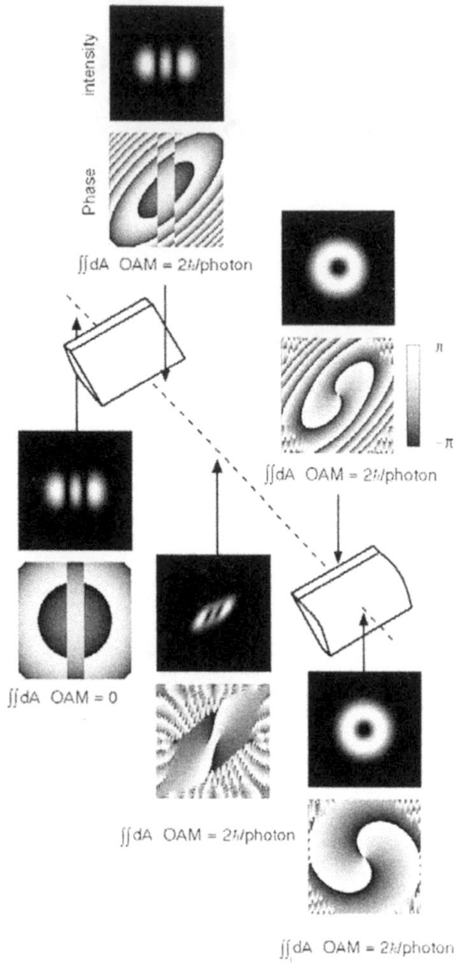

Figure 1. A $\pi/2$ converter comprising two cylindrical lenses of focal length f separated by $\sqrt{2}f$. An input Hermite–Gaussian mode of Rayleigh range $(1 + 1/\sqrt{2})f$ is converted to a Laguerre–Gaussian. Numerical methods are applied to calculate the intensity and phase structure at any point in the system. The angular momentum transfer takes place exclusively at the left-hand lens. (Note that, far from the beam axis in regions of low intensity, the numerical calculation of the phase is subject to noise.)

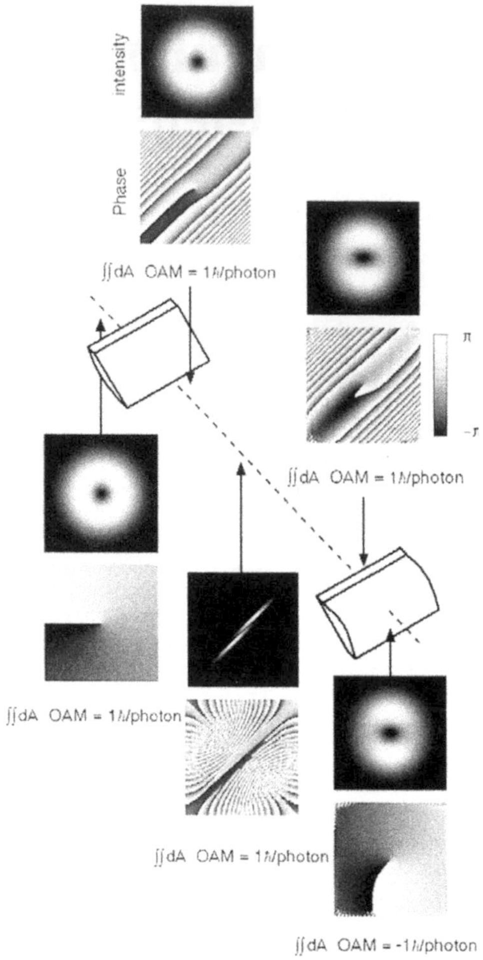

Figure 2. A π converter comprising two cylindrical lenses of focal length f separated by $2f$. A collimated Laguerre–Gaussian mode with azimuthal index l is converted to a Laguerre–Gaussian with index $-l$. Numerical methods are applied to calculate the intensity and phase structure at any point in the system. Despite the reverse in vortex sign which occurs about the centre position, the angular momentum transfer takes place at the right-hand lens. (Note that, far from the beam axis in regions of low intensity, the numerical calculation of the phase is subject to noise.)

where C_{pl}^{LG} is the normalization constant, L_p^l is a generalized Laguerre polynomial, $w(z)$ is the radius of the beam at position z and z_R is the Rayleigh range. In addition to the azimuthal mode index, l, there is a second mode index, p, which describes the number of radial nodes in the field. Note that all modes of the same mode order, $N = (2p + l)$, have the same Gouy phase.

The cross product of the momentum density (equation (1)) with the radius vector $\boldsymbol{r} \equiv (r, 0, z)$ yields an angular momentum density. The angular momentum density in the z direction depends upon the Φ component of $\boldsymbol{p}$, such that

$$j_z = rp_\phi. \qquad (3)$$

The expression for j_z can be numerically evaluated to give both a local value of the angular momentum density and an integrated value over the whole beam. Figure 1 shows a schematic of a $\pi/2$ converter and the numerically calculated intensity, phase and total orbital angular momentum in different planes.

As argued by Beijersbergen *et al* [2], we see that the angular momentum transfer occurs only at the lens where the Hermite–Gaussian mode enters or leaves the mode converter. The reasoning behind their observation was that the mechanism for such a momentum transfer lies in the gradient force that acts on any dielectric object, producing a force directed towards the high intensity region of the beam. A beam with a cylindrically

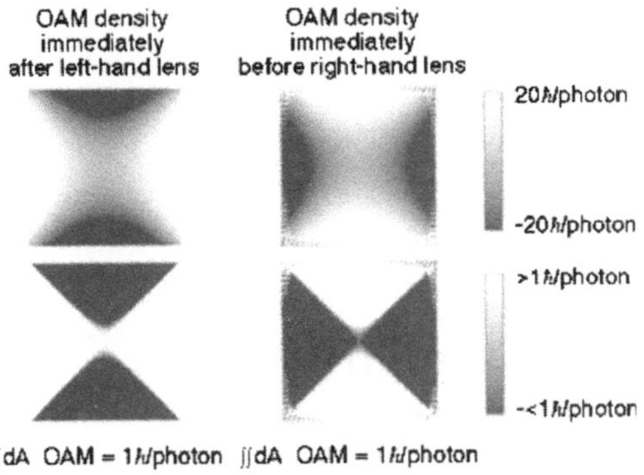

Figure 3. The local orbital angular momentum per photon, calculated in the planes immediately after and before the left and right lenses, respectively. For these conditions, the density reaches $+/-20\hbar$ (upper plots). When integrated over the whole beam cross section the total orbital angular momentum in both planes corresponds to an average of $\hbar$ per photon. However, careful inspection (lower plots) in the region of the vortex shows that its sign has changed from +1 to −1.

symmetric intensity cannot therefore create a torque upon a transparent object. In contrast, the rectangularly symmetric Hermite–Gaussian modes, when rotated by 45° with respect to the axis of the cylindrical lens, do create a torque on the lens which acts to align the optically thick axis of the lens with the long axis of the mode.

A physical understanding of the π converter is somewhat more difficult. In the case of a collimated Laguerre–Gaussian input mode we expect the converter to transform the azimuthal mode index from l to $-l$, thereby introducing an angular momentum exchange of $2l\hbar$ per photon. However, as both the input and output Laguerre–Gaussian beams have a circularly symmetric intensity, how can either exert a torque on the lenses? The resolution of this paradox was suggested by both Beijersbergen *et al* [2] and van Enk and Nienhuis [8]. Due to the diffractive spreading of the beam between the lenses, even a perfectly aligned mode converter still results in a residual ellipticity of the output Laguerre–Gaussian mode. The slightly elliptic 'Laguerre–Gaussian' mode has its major axis at 45° to the lens axis and hence does exert a torque. Figure 2 shows a schematic of a π converter and the numerically calculated intensity, phase and total orbital angular momentum at different positions. The slight ellipticity of the transformed beam is evident in the calculated intensity distributions. In general, the degree of ellipticity is reduced by using a beam of larger diameter, where the diffractive spreading between the lenses is reduced. Indeed the larger diameter means that the degree of ellipticity needed to produce the required torque is smaller.

As can be seen from the calculated phase distributions, figure 2, as the beam propagates between the lenses the sign of the optical vortex is reversed. The inversion of the topological charge by cylindrical focusing has recently been observed by Molina-Terriza *et al* [9] and might be thought counter-intuitive, as the vortex within the optical field is often described as carrying the orbital angular momentum. However, orbital angular momentum is not linked directly to vortices [10] but rather to an azimuthal phase term. Figure 3 shows the calculated orbital angular momentum density immediately after and immediately before the first and second lenses, respectively. If integrated over the whole of the beam, the total orbital angular momentum between the lenses remains constant. We do, however, note that in the vicinity of the beam axis the optical angular momentum density reverses sign, as does the sign of the optical vortex.

The numerical calculations we have performed within this work are compatible with the reasoning of Beijersbergen *et al* [2] and van Enk and Nienhuis [8]. It confirms that the inversion of the topological charge upon cylindrical focusing observed by Molina-Terriza *et al* [9] still occurs, even when the overall orbital angular momentum of the beam is conserved.

References

[1] Allen L, Beijersbergen M W, Spreeuw R J C and Woerdman J P 1992 Orbital angular-momentum of light and the transformation of Laguerre–Gaussian laser modes *Phys. Rev.* A **45** 8185–9

[2] Beijersbergen M W, Allen L, van der Veen H E L O and Woerdman J P 1993 Astigmatic laser mode converters and transfer of orbital angular momentum *Opt. Commun.* **96** 123–32

[3] Padgett M J and Courtial J 1999 A Poincaré-sphere equivalent for light beams containing orbital angular momentum *Opt. Lett.* **24** 430–2

[4] Allen L, Courtial J and Padgett M J 1999 A matrix formulation for the propagation of light beams with orbital and spin angular momenta *Phys. Rev.* E **60** 7497–503

[5] Sziklas E A and Siegman A E 1975 Mode calculations in unstable resonators with flowing saturable gain: fast Fourier transform method *Appl. Opt.* **14** 1874–89

[6] Courtial J and Padgett M J 1999 Performance of a cylindrical lens mode converter for producing Laguerre–Gaussian laser modes *Opt. Commun.* **159** 13–18

[7] Allen L and Padgett M J 2000 The Poynting vector in Laguerre–Gaussian beams and the interpretation of their angular momentum density *Opt. Commun.* **184** 67–71

[8] van Enk S J and Nienhuis G 1992 Eigenfunction description of laser beams and orbital angular momentum of light *Opt. Commun.* **94** 147–58

[9] Molina-Terriza G, Recolons J, Torres J P, Torner L and Wright E M 2001 Observation of the dynamical inversion of the topological charge of an optical vortex *Phys. Rev. Lett.* **87** 023902

[10] Berry M V 1998 Much ado about nothing: optical dislocation lines (phase singularities, zeros, vortices ...) *Singular Optics* vol 3487, ed M S Soskin (Frunzenskoe: SPIE) pp 1–5

Laser beams with screw dislocations in their wavefronts

V.Yu. Bazhenov, M.V. Vasnetsov, and M.S. Soskin
Institute of Physics, Academy of Sciences of the Ukrainian SSR

(Submitted 28 August 1990)

Pis'ma Zh. Eksp. Teor. Fiz. **52**, No. 8, 1037–1039 (25 October 1990)

Coherent optical fields with wavefront dislocations of various orders have been produced experimentally and studied during the passage of a laser beam through a multimode waveguide and during diffraction by some holograms which have been synthesized.

Wavefront dislocations[1] or optical vortices have attracted interest as some of the new entities which exist in optical fields with a complex spatial structure or in laser cavities with a large Fresnel number.[2] When there is a screw dislocation, the wavefront is a common helical surface (right-handed or left-handed) with a singularity. When this singularity is circumvented, there is a phase shift of some multiple of 2π. The wavefront surface may have a singularity only where the modulus of the complex field amplitude E vanishes, i.e., where the real and imaginary parts are simultaneously zero. First-order dislocations (those for which the change in phase is 2π) have been observed in the speckle field of scattered coherent light.[1]

Our purpose in the present study was to produce, and learn about the properties of, regular optical fields with screw dislocations, including dislocations of higher orders.

To study the structure of the field of an isolated dislocation, we used an experimental apparatus (Fig. 1) consisting of a Mach-Zehnder interferometer with a length of braided multimode optical fibers in one arm. The beam from a helium–neon laser with a Gaussian mode is focused into one of the fibers of the braid. At the exit from this fiber we observe a pattern consisting of two or three spots of irregular shape. A diverging lens is placed in the reference arm of the interferometer in order to equalize the divergences of the interfering beams. By rotating the semitransparent exit mirror of the interferometer, we are able to observe an interference pattern in various regions of the beam emerging from the waveguide. The customary annular interference pattern is observed at the light intensity maxima. If there is instead a dislocation at the center of the interference region, the pattern becomes a helix (Fig. 1b), clearly demonstrating the existence of a helical wave surface. At the center of the helix there is always an intensity zero of the beam emerging from the waveguide.

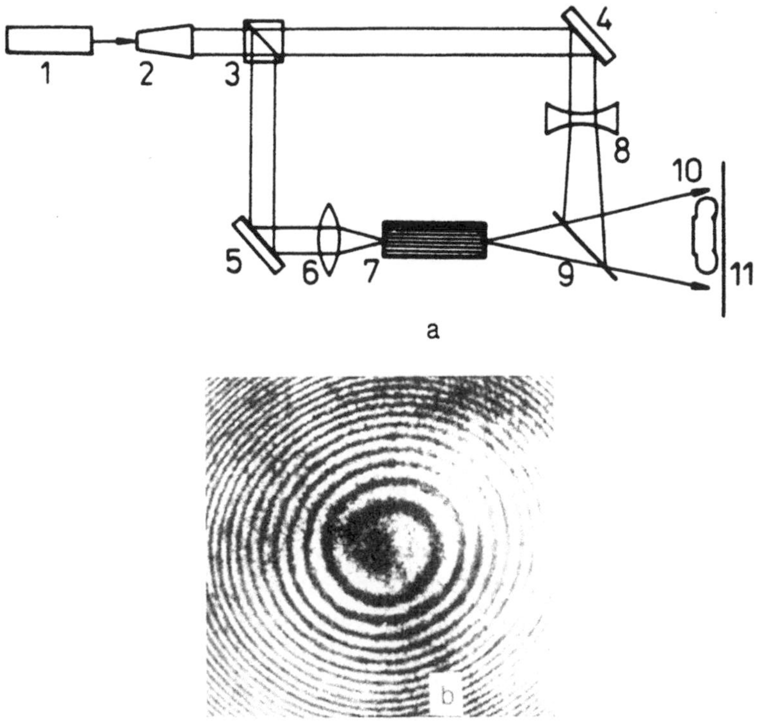

Figure 1. a: Experimental setup. 1 – He–Ne laser; 2 – beam expander; 3 – beam splitter; 4, 5 – mirrors; 6 – objective; 7 – braided optical fibers; 8 – diverging lens; 9 – half-silvered mirror; 10 – camera; 11 – screen for observing the interference of the beams. b: Interference pattern formed by the spherical reference wave and a wave with a screw wavefront dislocation.

When the interfering beams are incident on a screen at a small relative angle, we observe a splitting or disappearance of an interference fringe, similar to the event described in Ref. 1 for dislocations in speckle fields.

The dislocations observed experimentally in the setup in Fig. 1a had only the plus first or minus first order (the change in phase upon a circumvention was $+2\pi$ or -2π, respectively). The numbers of right-handed and left-handed screw dislocations were the same. It is extremely unlikely that dislocations of higher orders would be observed in this setup, since such an observation would require the intersection at one point of more than two pairs of lines on which the conditions $\mathrm{Re}\,E = 0$ and $\mathrm{Im}\,E = 0$ hold. In the interference pattern of the beams this situation would correspond to a splitting of the interference fringe into four or more new fringes. In order to produce beams with dislocations of higher orders, we accordingly synthesized some amplitude holograms to simulate the interference field of a plane wave and a wave carrying a dislocation of the desired order. After a numerical calculation was carried out, and the results displayed on a monitor, reduced copies were made on photographic film with a period of 0.1 mm and a size of 2×2.5 mm. Figure 2 shows the gratings which were synthesized.

The properties of the beams diffracted by these holograms were studied in a setup similar to that shown in Fig. 1a, in which the braided fiber and the focusing objective were replaced by this grating. A beam expander was not used in this case. An annular structure of the beams was clearly observed in the diffraction orders. With increasing

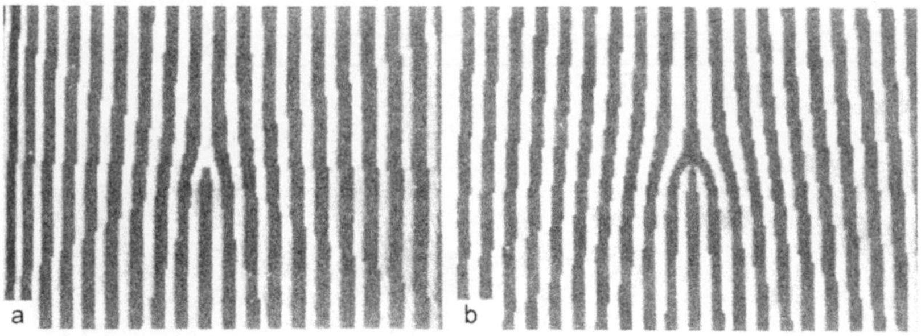

Figure 2. a – The hologram which was synthesized in order to produce beams with a first-order dislocation; b – the same, for a second-order dislocation.

order of the dislocation, the intensity dip at the center of the beam acquired progressively sharper edges. For the dislocations of higher orders, the pattern of the interference with the reference wave takes the form of several nested helices, in a number equal to the order of the dislocation. The helices corresponding to positive and negative diffraction orders wind in opposite directions. The transverse structure of the beams is not substantially altered as they propagate. This result means that the conclusion that there is an instability of optical fields with higher-order dislocations[1] requires refinement.

We have also observed an increase in the order of the dislocation in diffracted beams of higher orders (in precise correspondence with the order of the diffraction), but the quality of the beams produced in this manner turned out to be extremely low.

In summary, this study has demonstrated that it is possible to produce coherent light beams with a regular transverse structure which have screw wavefront dislocations. Beams with high-order dislocations have been produced for the first time with the help of some holograms which have been synthesized.

We wish to thank SG Odulov and VV Shkunov for useful discussions.

REFERENCES

1 B.Ya. Zel'dovich, N.F. Pilipetskiĭ, and V.V. Shkunov, *Phase Conjugation*, Nauka, Moscow, 1986.
2 P. Coullet, L. Gil, and F. Rocoa, *Opt. Commun.* **73** 403 (1989).

Translated by D. Parsons

PAPER 3.9

Laser beams with phase singularities

N. R. HECKENBERG, R. McDUFF, C. P. SMITH,
H. RUBINSZTEIN-DUNLOP, M. J. WEGENER
Department of Physics, The University of Queensland, Brisbane, Austrialia

Received 20 January 1991; accepted 23 April 1992

Phase singularities in an optical field appear as isolated dark spots and can be generated in active laser cavities or by computer generated holograms. Detection and categorization of these singularities can easily be achieved either by interferometry or Fourier transform pattern recognition using a computer generated hologram.

1. Introduction

Most laser beams have essentially spherical wavefronts and any deviation from sphericity constitutes a degradation of beam quality. However, the TEM_{01}^* 'doughnut' mode, often observed in high-power lasers, can exhibit a *helical* wavefront structure, associated with a phase singularity on the beam axis [1]. This occurs when the frequency-degenerate TEM_{01} and TEM_{10} modes oscillate simultaneously in phase quadrature.

The electric field will have the form

$$\begin{aligned} E_{01}^* &= E_0[(x \pm iy)/\omega]\, e^{-(x^2+y^2)/\omega^2}\, e^{-ikr^2/2R}\, e^{i(kz+\Phi)} \\ &= E_0(r/\omega)\, e^{\pm i\theta}\, e^{-(x^2+y^2)/\omega^2}\, e^{-ikr^2/2R}\, e^{i(kz+\Phi)} \end{aligned} \quad (1)$$

where Φ is the Guoy phase shift, R is the wavefront radius of curvature, ω is the spot size and r and θ are polar coordinates in the X–Y plane. Notice that during any complete circuit around the axis the phase changes by $\pm 2\pi$, expressing the helical form of the wavefronts, and that the field goes to zero on the axis, as it must where the phase is undefined.

The doughnut mode is a stable cavity mode because it is a linear combination of Hermite–Gaussian TEM_{01} and TEM_{10} modes and propagates in a self-similar way in free space and through optical systems for the same reason. In fact other more complex combinations containing phase singularities can be produced from higher-order modes. According to resonator theory [2] Hermite–Gaussian modes TEM_{mn} with the same total $m + n$ (or Laguerre–Gaussian modes TEM_{pl} with the same total $2p + l$) are frequency degenerate and can form such combinations. The singularities show up as multiple isolated irradiance zeros in the modal spot pattern.

One example is the TEM_{02}^* hybrid or 'optical leopard' [3] which can be constructed from the Gaussian–Hermite modes TEM_{02} and TEM_{20}. The pattern has a central irradiance peak surrounded by four smaller peaks and four zeros where there are two positive and two negative singularities diagonally opposed. Figure 1a shows the irradiance distribution and Fig. 1b the form of the wavefronts with four interconnected helices.

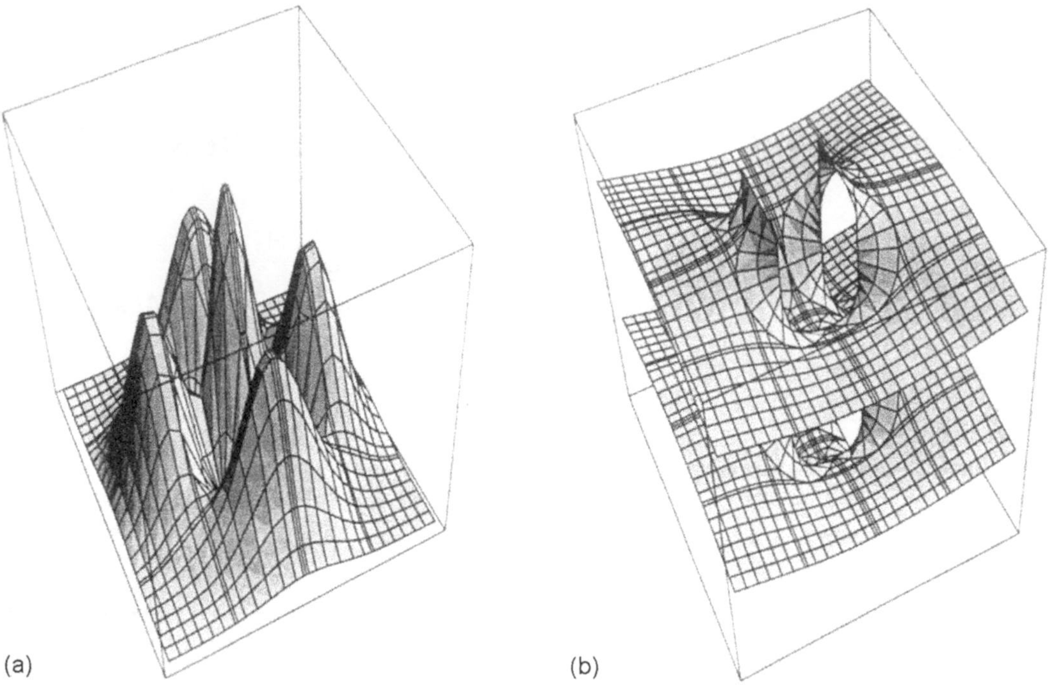

(a) (b)

Figure 1 TEM_{02}^* or 'optical leopard'. (a) Irradiance distribution (vertical) with spatial position (horizontal). (b) Form of wavefronts. In the diagram, phase increases in the vertical direction. Two surfaces of constant phase differing by 2π, corresponding to a plane wave at large distances from the beam axis, are connected around the singularities.

Other examples contain higher-order singularities. The simplest cases are the higher-order doughnuts (which are just Gaussian–Laguerre TEM_{0n} modes)

$$E_{0n}^* = E_0(r/\omega)^n \, e^{-(x^2+y^2)/\omega^2} \, e^{\pm in\theta} \, e^{-ikr^2/2R} \, e^{i(kz+\Phi)} \tag{2}$$

The parameter n is often referred to as the 'charge' of the singularity, with zero charge corresponding to a Gaussian TEM_{00} beam. The irradiance profiles of several higher-order doughnuts are shown in Fig. 2. The wavefronts form multistart helices.

Of course, in a real laser, astigmatism in the cavity often removes the frequency degeneracy between such modes, but Brambilla *et al.* [3] have shown theoretically and experimentally that, so long as the astigmatism is not too severe, a cooperative frequency-locking process can occur, leading to a range of stable patterns.

However, it is important to remember that not all dark spots in patterns are necessarily phase singularities. If the frequency degeneracy of the contributing modes is broken, a rapidly time-varying pattern will result, the time average of which may still contain dark spots [1]. In some applications, involving only average irradiances, that will not matter, but in others it will be important. We show below how such cases can be distinguished experimentally.

2. Production of beams with singularities

There are several ways in which beams with singularities might be produced. As mentioned above, thanks to the process of cooperative frequency locking [3] such modes can arise

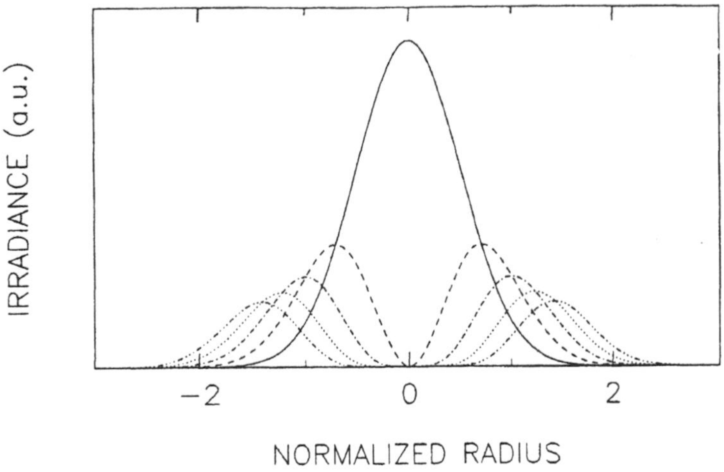

Figure 2 Irradiance profiles for Gaussian and doughnut modes up to charge 4. Each has the same total power.

spontaneously within a laser and in some cases a considerable degree of control can be exercised by the experimenter. Alternatively, a hologram can be used to convert part of the output of an existing laser into the desired beam. We consider both approaches below.

The 525 nm Na_2 vapour laser pumped by the Ar-ion laser [4] is ideal for studies of combination-mode formation because its velocity-selective optical pumping mechanism leads to very narrow gain linewidths so that only one 'family' of transverse modes with a certain $m + n$ or $2p + l$ sum can oscillate at a time, and the otherwise dominant TEM_{00} Gaussian can be suppressed. We have used a system very similar to that used by Brambilla *et al.* [3] to generate a range of patterns and, as described below, to study methods to detect and classify modes with singularities.

A less efficient but a more flexible way to produce modes with singularities is through the use of computer-generated holograms. In our first experiments we used on-axis holograms which have the form of spiral Fresnel zone plates [5], but these suffer the same problem as Gabor's original holograms, i.e. lack of separation between reconstructed beam and incident beam. It turns out that even crude off-axis binary holograms are quite effective, as will now be explained.

A hologram is really just a recording of the interference pattern between a field of interest and some simple reference field. For the relatively simple fields involved in modes with singularities it is possible to calculate the form of such patterns and plot them out. Let us take as an example a charge-one doughnut (Equation 1), at a beam-waist ($R \to \infty$) for simplicity. Consider the interference pattern on a screen in the X–Y plane when a plane reference beam

$$R = R_0\, e^{ik_x x + ik_z z} \tag{3}$$

is incident at an angle $\phi = \sin^{-1}(k_x/k)$. The irradiance on a screen at $z = 0$ will be

$$\begin{aligned} I &= |(R_0\, e^{ik_x x} + E_0(r/\omega)\, e^{i\theta}\, e^{-r^2/\omega^2}|^2 \\ &= R_0^2 + E_0^2(r/\omega)^2\, e^{-2r^2/\omega^2} + 2R_0E_0(r/\omega)\, e^{-r^2/\omega^2} \cos(k_x x - \theta) \end{aligned} \tag{4}$$

It is the last term which expresses the interference pattern. A photographic recording of this

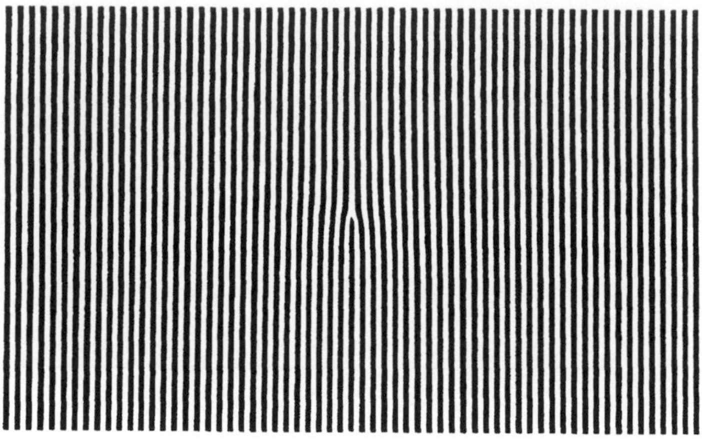

Figure 3 Pattern for off-axis binary hologram for charge-one singularity. The origin is at the tip of the fringe defect.

pattern can now act as a hologram capable of 'reconstructing' the original doughnut (and its complex conjugate) when illuminated by a wave given by Equation 3.

In what follows it will be convenient to work with a simplified pattern which ignores the amplitude variation of the doughnut beam and retains only the important phase information in the form of a spatially varying transmissivity

$$T = \tfrac{1}{2}(1 - \cos(k_x x - \theta)) \tag{5}$$

Now consider the effect of illuminating this pattern with a Gaussian beam propagating along the axis. Just after the hologram the field will be

$$E_T = TA_0\, e^{-r^2/\Omega_0^2} \tag{6}$$

where A_0 is the central amplitude and Ω_0 is the spot size of the beam, assumed plane at this point. Substituting for T, we find

$$E_T = (A_0/2)\, e^{-r^2/\Omega_0^2} - (A_0/4)\, e^{-r^2/\Omega_0^2}\, e^{i(k_x x - \theta)} - (A_0/4)\, e^{-r^2/\Omega_0^2}\, e^{i(-k_x x + \theta)} \tag{7}$$

This field can be recognized as consisting of a zero-order beam propagating along the axis, and two (conjugate) first-order diffracted beams, each of them containing a singularity of opposite charge.

In fact it is much easier to print binary holograms than ones incorporating the sinusoidal variations in optical density implicit in Equation 5. Thus we actually use a 'square wave' transmissivity function which can be expressed in the form

$$T = \tfrac{1}{2} - \sum_{n=1}^{\infty} \operatorname{sinc}(n\pi/2) \cos[n(k_x x - \theta)] \tag{8}$$

Its appearance is shown in Fig. 3. It has the appearance of a grating with a defect where a stripe branches.

When this is illuminated, the output field will contain terms of the form

$$E_n = (A_0/4) \operatorname{sinc}(n\pi/2)\, e^{-r^2/\Omega_0^2}\, e^{i(nk_x x - n\theta)} \tag{9}$$

Each is of course a diffraction order from the 'grating' and can be recognized as being

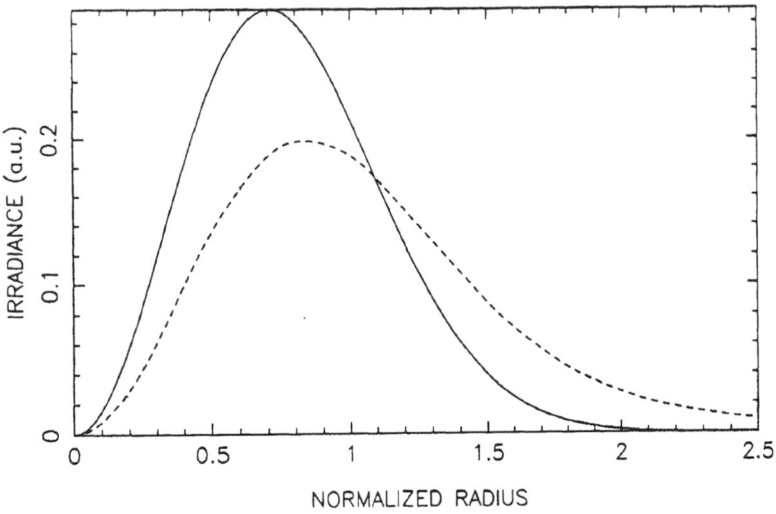

Figure 4 Comparison of the far-field spatial profile of the TEM_{01}^* mode (full line) with a first-order beam from a hologram illuminated by a Gaussian beam with the same parameters (dashed line). The TEM_{01}^* mode has the same power as the zeroth term of the series in Equation 13.

closely related to an nth-order doughnut propagating at an angle

$$\phi_n = \sin^{-1}(nk_x/k) \simeq n \sin^{-1}(k_x/k) \tag{10}$$

to the axis. Actually, Equation 9 has the form of a charge-n singularity embedded in a Gaussian beam and as such will not propagate in a self-similar way. To investigate the far-field spatial profile of this beam, we can decompose Equation 9 in terms of the orthogonal set of Gaussian–Laguerre modes:

$$\psi_{pl} = \sqrt{\left(\frac{2p!}{\pi(p+l)!}\right)} \left(\frac{\Omega_0}{\Omega_1}\right) \left(\frac{\sqrt{2}r}{\Omega_1}\right)^{|l|} L_p^{|l|}\left(\frac{2r^2}{\Omega_1^2}\right) \times e^{-r^2/\Omega_1^2} e^{il\theta} e^{-ikr^2/2R} e^{i(2p+l+1)\Phi_1} e^{i(nk_x x + k_z z)} \tag{11}$$

where Ω_1 is the spot size, R_1 is the radius of curvature of the beam and Φ_1 is the Guoy phase shift. As the beam propagates, the parameters Ω_1, R_1 and Φ_1 are related to the waist spot size Ω_0 via the standard Gaussian–Laguerre propagation laws [6]. The amplitudes of the terms of the series expressing the field at the hologram are given by

$$E_{pl} = 0 \quad \text{if} \quad l \neq n$$
$$E_{pn} = \frac{A_0}{4} \operatorname{sinc}(n\pi/2)\Omega_0^2 \sqrt{\left(\frac{\pi p!}{2(p+n)!}\right)} \frac{n}{2} \frac{\Gamma(p+n/2)}{p!} \tag{12}$$

In the far-field the spatial amplitude distribution is

$$E_{far} = \sum_{p=0}^{\infty} E_{pn}\psi_{pn} \tag{13}$$

where $\Phi_1 = \pi/2$. Figure 4 shows that the far-field amplitude distribution for the first-order beam closely resembles a TEM_{01}^* doughnut of slightly increased spot size. The helical structure of the wavefronts is identical. Similar results are obtained for higher orders. Thus,

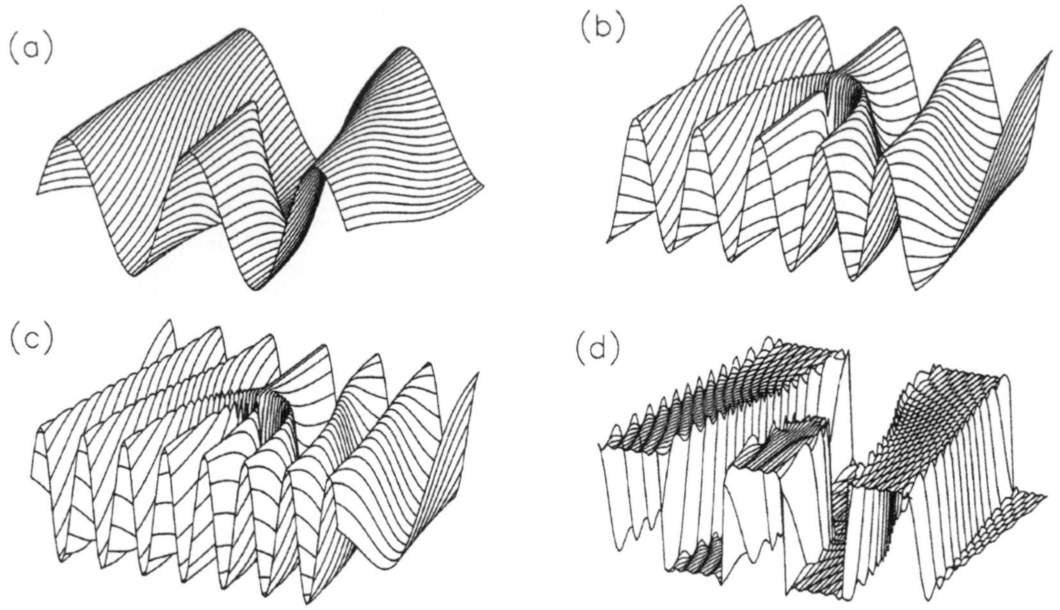

Figure 5 Decomposition of binary hologram pattern into harmonic components: (a–c) first, second and third components; (d) synthesis of first 30 components.

although the beams produced by our holograms are not strictly TEM^*_{0n} doughnuts, in the far-field they are, for all practical purposes, equivalent and for simplicity we refer to them as such.

The way in which the binary pattern produces multiply charged singularities can be understood by reference to Fig. 5, where the first three harmonic components of the very centre of the pattern in Fig. 3 are plotted. Each of Figs 5a to 5c can be recognized as the hologram for a successively higher-order singularity. Figure 5d shows the square-wave pattern synthesized from 30 components.

Figure 3 showed an example of a computed binary off-axis hologram pattern. This was printed by a laser printer on A4 paper and reduced by photographing onto half of a 35-mm slide. Figures 6a and 6b show charge-one and charge-two doughnuts produced as first and second orders, using a HeNe laser for illumination. The patterns, like all subsequent experimental ones, were recorded by an Electrim EDC-1000 CCD camera. About 5% of the incident power was coupled into the first order. This could be improved considerably by converting the present hologram into a phase hologram if facilities were available. Although the charge-two pattern is somewhat distorted, it shows clearly the larger diameter and narrower bright annulus expected. In fact, a symmetric grating like ours would be expected to produce a dim second-order term at best as it should effectively be a 'missing order'. We have also used holograms with four fringe defects to generate an 'optical leopard' pattern and still more complex beam shapes could be produced. It would even be possible to generate combinations involving transverse modes of different $m + n$ (or $2p + l$) sum (e.g. $TEM_{00} + TEM^*_{01}$) which could not be generated at a single frequency in a laser oscillator. These would suffer some change of shape during propagation owing to differential Guoy phase shift, but perhaps that could be put to some use.

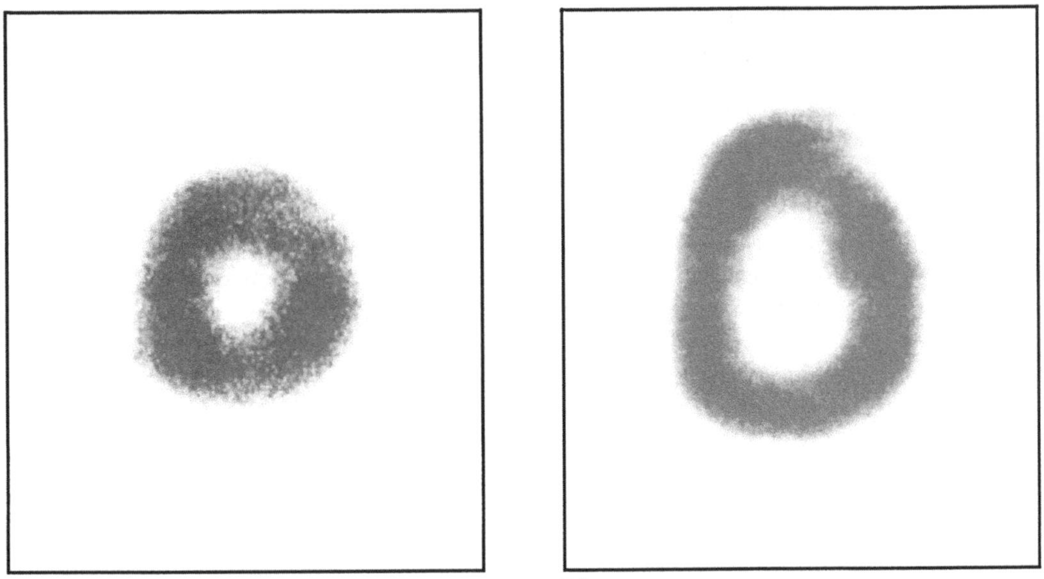

(a) (b)

Figure 6 Doughnut beams produced by hologram using HeNe laser light: (a) charge one; (b) charge two.

3. Detection of phase singularity modes

In their studies of doughnut modes, Tamm and Weiss [7] used an astigmatic imaging technique to determine the charges of the singularities, but a more versatile approach is to measure the phase structure of the beam by interference, as had been done by Vaughan and Willetts [1]. Ideally, one should have available a coherent plane wave to act as reference, and in the vicinity of each singularity a fringe pattern with a defect, like that in Fig. 3, would be observed. Exactly this behaviour is demonstrated in Fig. 7, where interference patterns

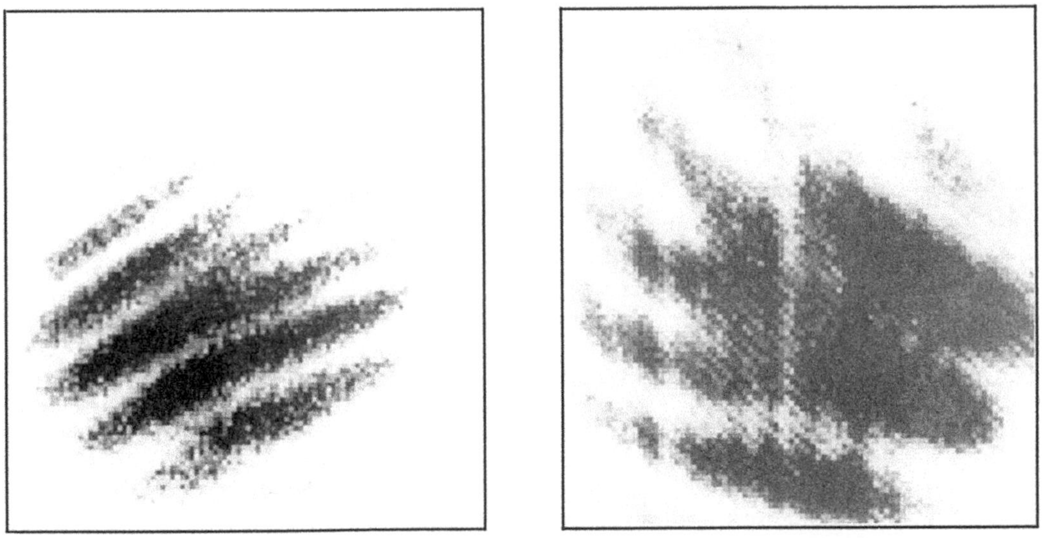

(a) (b)

Figure 7 Interference of doughnuts of Fig. 6 with plane wave: (a) charge one; (b) charge two.

for the two holographically produced doughnuts shown in Fig. 6 are displayed. In the charge-one case, a bright fringe (dark in this print) splits into two; in the charge-two case a fringe splits into three.

When the singularity mode pattern is produced in a laser, it is not so easy to obtain a coherent plane wave for use as reference, and a simple split-beam technique is useful [1]. The beam of interest is split in a Mach–Zehnder interferometer and recombined with sufficient misalignment to produce straight fringes and to displace the two patterns so that the singularities in one fall in relatively uniform areas of the other where the phase varies only slowly. The result is a defect in the combined fringe pattern at each singularity position. Opposite charges in the same pattern fork in opposite directions, and corresponding singularities behave oppositely at the two points where they appear. Although the interference pattern becomes complicated, this works well when the pattern does not contain too many singularities [8].

In interpreting these patterns it is important to keep in mind that the visibility of the fringes will be small near the singularity as the irradiance there is small, and that the exact form of the fringe splitting depends on the position of the fringes.

Figure 8a shows a slightly asymmetric doughnut produced by the Na_2 laser and Fig. 8b shows the corresponding split-beam interference pattern, indicating the presence of a charge-one phase singularity in the beam. For comparison Fig. 8c shows a split-beam interference pattern for an 'unlocked' doughnut, i.e. one where cavity astigmatism has broken the frequency degeneracy of the constituent modes. This shows the characteristic 'sideways shift' of the fringes inside a circular region as explained by Vaughan and Willetts [1]. The loss of frequency degeneracy leads to mode beating at a frequency of a few megahertz which can be detected with a photodiode.

An alternative means of detecting and classifying phase singularities in a beam is by optical Fourier transform recognition techniques where the holograms discussed above can be used as matched filters [9].

This has been demonstrated using the arrangement shown in Fig. 9. The Fourier transform of an input pattern is formed at the focal plane of a lens where the hologram for a pattern of interest, say a charge-one singularity, is placed. The transmitted field is Fourier transformed again by a second lens. At the output plane three beams can be distinguished – a central magnified image of the field at the input plane, and two 'first-order' fields. (Owing to the binary nature of our holograms, higher-order fields also appear but are generally too weak to be useful.) One first-order field gives the cross-correlation between the input field and the field used to make the hologram, and the other the convolution of the input field with the hologram field [9]. In this case it is more helpful to realize that this is also the cross-correlation between the input field and the conjugate of the hologram field. Thus, if a hologram for a charge-one singularity is used as a filter, a bright spot will appear in one field at each point where a positive charge-one singularity appears in the input field, and in the other field a bright spot will appear at each point where a negative charge-one singularity appears in the input.

This is shown in Fig. 10a where the singularity in a charge-one doughnut produced by the Na_2 laser is recognized. At the centre of the picture is the image of the input doughnut, flanked by the recognition fields (which are slightly magnified as a result of lens aberrations). The intense spot on the right indicates a charge-one singularity. Figure 10b shows the result when the input field is a 'leopard'. Here the diagonal placement of the four charge-one singularities, two of each sign, shows up clearly. An unlocked doughnut gives

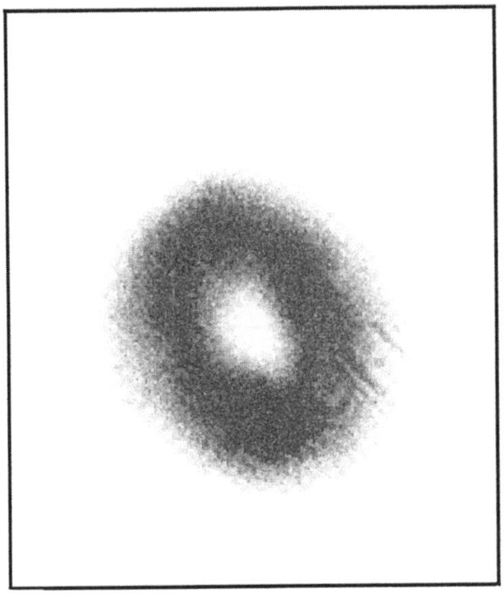

(a)

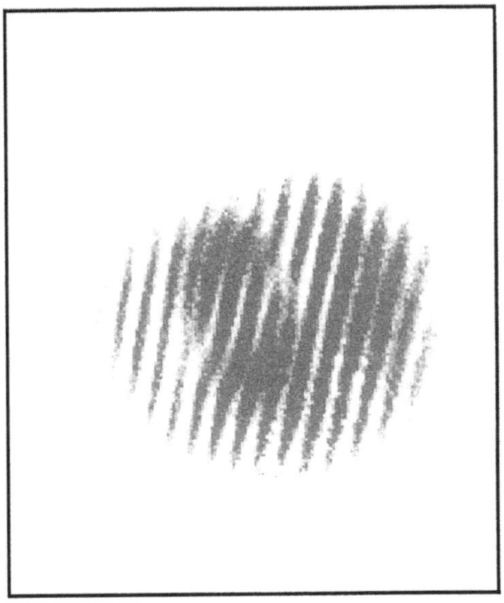

(b)

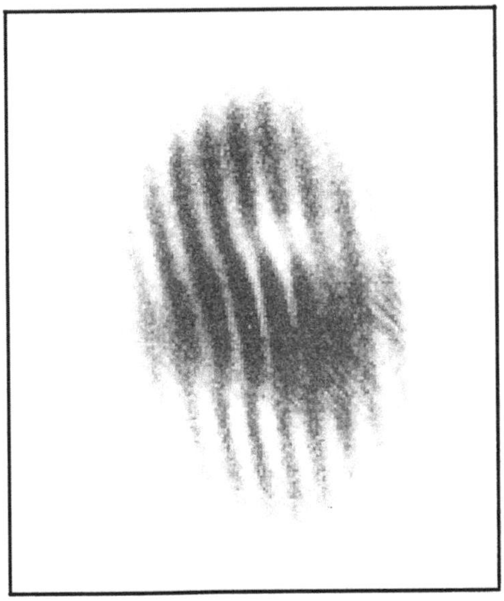

(c)

Figure 8 Doughnuts produced by Na_2 laser. (a) Phase-locked doughnut. (b) Split-beam interference pattern for (a). Note two forks in the fringe pattern, indicating the presence of a singularity. (c) Split-beam interference pattern for unlocked doughnut.

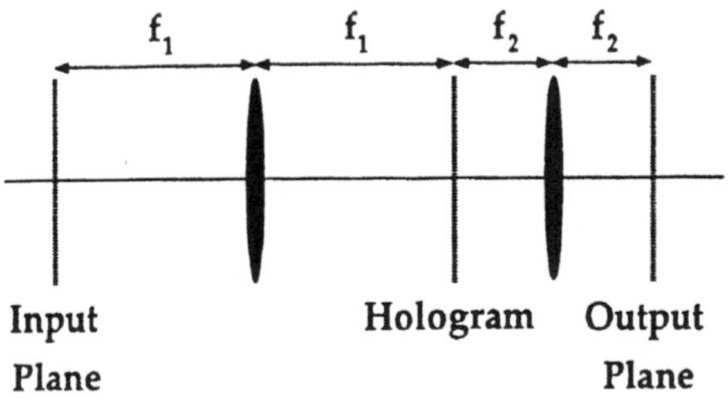

Figure 9 Arrangement for Fourier transform recognition of singularities in beams.

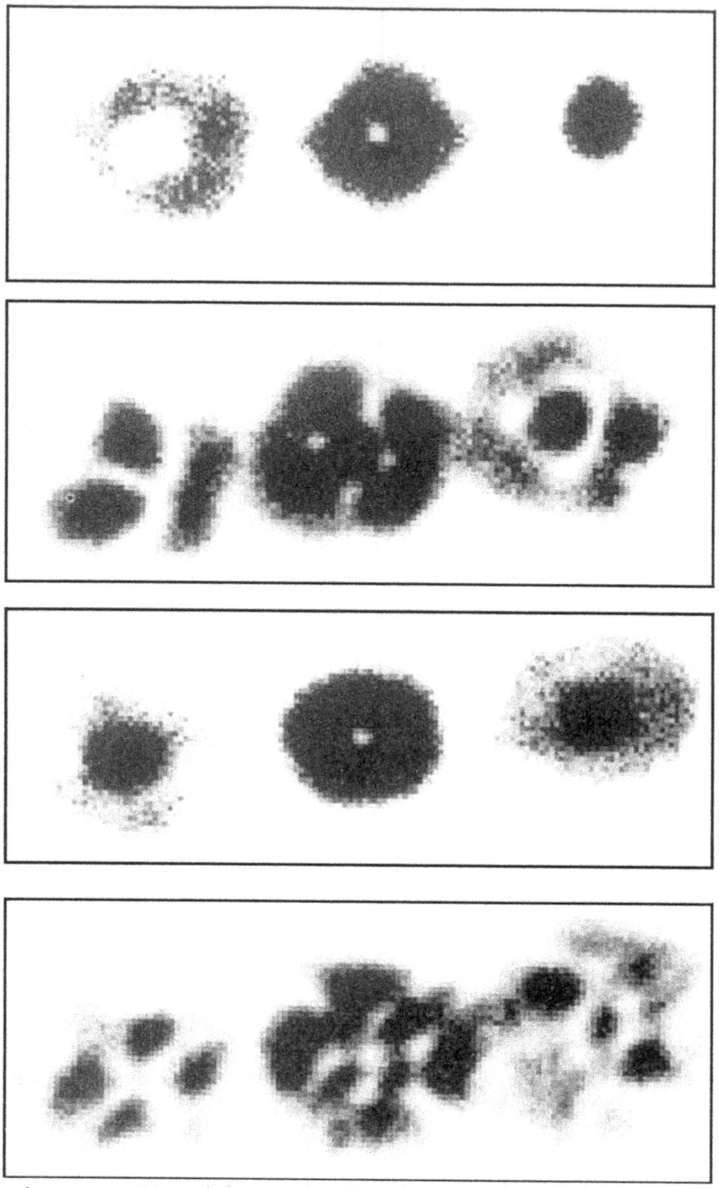

(a)

(b)

(c)

(d)

Figure 10 Fourier transform recognition of singularities. In each case an image of the input field appears at the centre, flanked by recognition fields for ±1 charged singularities: (a) Charge-one doughnut from the Na_2 laser. (b) 'Leopard' from the Na_2 laser. (c) Unlocked $m + n = 1$ doughnut from Na_2 laser. (d) Seven-spot beam from Na_2 laser.

the result shown in Fig. 10c, as the actual pattern is rapidly changing, spending part of each beat period as a doughnut of each sign [1], so that bright spots appear in the recording on both sides. A more complex pattern from the Na_2 laser, with seven dark spots, is analysed in Fig. 10d, which shows that four charge-one singularities of one sign form a square with the three others of opposite sign lying along a line. Such a pattern, called a 'seven-hole' by Brambilla *et al.* [3], would be difficult to analyse using split-beam interference.

Although there is clearly room for improvement, the optical Fourier recognition technique has the advantage of needing only a single beam and providing output in the form of readily recognized spots rather than 'forks' in an interference pattern. Indeed, it is interesting to note that, presumably owing to the local rotational symmetry and self similarity of the regions surrounding singularities, the technique is less affected by variations in scale-size or orientation than is, say, recognition of alphanumeric characters.

4. Applications

This work arose originally out of an interest in pattern formation and optical turbulence but it is clear that a 'technology' now exists for the production, detection, and classification of beams with phase singularities associated with isolated irradiance zeros. Many lasers can be run on the TEM_{01}^{*} mode, and well-designed phase holograms could convert the light from almost any laser into such a form, or a more complex pattern [10].

In some cases the helical phase structure is of importance, as in switching helicities as a means of information processing [7], but in many other cases any beam with a central minimum in irradiance (thus including unlocked doughnuts) could be of use. Note that the structure we have been discussing is entirely independent of polarization. Such a shape could be advantageous when it is desired to launch a beam through a reflecting telescope with a central obstruction. A higher-order doughnut would be appropriate for this.

Another area of potential application is in light–particle interactions [11]. The use of doughnut beams has been suggested for small-particle levitation [12] and more recently for the focusing of atomic beams [13].

Finally, it should be pointed out that speckle patterns formed by laser beams passing through inhomogeneous media have been shown to contain many phase singularities [14] presumably produced by random structures approximating the holograms discussed above.

5. Conclusions

We have shown that laser beams with phase singularities can be generated in lasers or can be produced from normal Gaussian beams using computer-generated holograms. The presence of the singularities can be detected and they can be classified either by reference to defects in interference patterns or by optical Fourier transform pattern-recognition techniques using computer-generated holograms as matched filters. Possible applications of the simplest of such beams, the TEM_{01}^{*} doughnut, include efficient launching of single-mode beams through telescopes and atom and particle trapping.

Acknowledgement

This work was supported by the Australian Research Council.

References

1. J. H. VAUGHAN and D. V. WILLETTS, *J. Opt. Soc. Am.* **73** (1983) 1018.
2. H. KOGELNIK and T. LI, *Proc. IEEE* **54** (1966) 1312.
3. M. BRAMBILLA, F. BATTIPEDE, L. A. LUGIATO, V. PENNA, F. FRATI, C. TAMM and C. O. WEISS, *Phys. Rev.* **A43** (1991) 5090.
4. B. WELLEGEHAUSEN, *IEEE J. Quantum Electron.* **QE-15** (1979) 1108.
5. N. R. HECKENBERG, R. McDUFF, C. P. SMITH and A. G. WHITE, *Opt. Lett.* **17** (1992).
6. J. P. TACHÉ, *Appl. Opt.* **26** (1987) 2698.
7. CHR. TAMM and C. O. WEISS, *J. Opt. Soc. Am.* **B7** (1990) 1034.
8. A. G. WHITE, C. P. SMITH, N. R. HECKENBERG, H. RUBINSZTEIN-DUNLOP, R. McDUFF, C. O. WEISS and CHR. TAMM, *J. Mod. Opt.* **38** (1991).

9. J. W. GOODMAN, *Introduction to Fourier Optics* (McGraw-Hill, San Francisco, 1968).
10. Phase holograms for charge-one singularities have already been produced by P. Vanetti, Alenia Sistemi Difesi.
11. S. CHU, *Science* **253** (1991) 253.
12. A. ASHKIN and J. M. DZIEDZIC, *Appl. Phys. Lett.* **24** (1974) 586.
13. J. J. McCELLAND and M. R. SCHEINFEIN, *J. Opt. Soc. Am.* **B8** (1991) 1974.
14. N. B. BARANOVA, B. YA. ZEL'DOVICH, A. V. MAMAEV, N. F. PILIPETSKIL and V. V. SHKUKOV, *JETP Lett.* **33** (1981) 195.

PAPER 3.10

Topological charge and angular momentum of light beams carrying optical vortices

M. S. Soskin, V. N. Gorshkov, and M. V. Vasnetsov
Institute of Physics, National Academy of Sciences of the Ukraine, Kiev 252650, Ukraine

J. T. Malos and N. R. Heckenberg
Department of Physics, University of Queensland, Brisbane 4072, Australia
(Received 17 March 1997)

We analyze the properties of light beams carrying phase singularities, or optical vortices. The transformations of topological charge during free-space propagation of a light wave, which is a combination of a Gaussian beam and a multiple charged optical vortex within a Gaussian envelope, are studied both in theory and experiment. We revise the existing knowledge about topological charge conservation, and demonstrate possible scenarios where additional vortices appear or annihilate during free propagation of such a combined beam. Coaxial interference of optical vortices is also analyzed, and the general rule for angular-momentum density distribution in a combined beam is established. We show that, in spite of any variation in the number of vortices in a combined beam, the total angular momentum is constant during the propagation.
[S1050-2947(97)09910-1]

PACS number(s): 42.65.Sf, 42.50.Vk

INTRODUCTION

Light beams possessing phase singularities, or wave-front dislocations [1,2] have been studied intensively in linear and nonlinear optics [3–21], to reveal their basic properties, and for the sake of possible applications. At the singularity the phase becomes undetermined and the wave amplitude vanishes, resulting in a "dark beam" within a light wave. Phase singularities appear on wave fronts in diffuse light scattering, when a light wave has a form of speckle field, and each speckle has one screw wave-front dislocation in the vicinity [3], also known as an optical vortex [4]. At present, several different techniques are used to generate "singular beams:" synthesized holograms [5,6], phase masks [7–9], active laser systems [10–13], and low-mode optical fibers [14]. Recently the generation of phase singularities was observed as a result of light wave-front deformations caused by self-action in nonlinear media [15,16].

The general result of these investigations is the origin of a new chapter of modern optics and laser physics—singular optics, which operates with terminology and laws quite new to traditional optics. Phase singularities are topological objects on wave-front surfaces, and possess topological charges which can be attributed to the helicoidal spatial structure of the wave front around a phase singularity. This structure is similar to a crystal lattice defect, and therefore was at first known as a wave-front screw dislocation [1,2]. The interference of a wave possessing such wave-front screw dislocation with an ordinary reference wave produces a spiral fringe pattern [12,17,18], or, in the case of equal wave-front curvatures, radial fringes [19]. The number of fringes radiating from the center of the interference pattern equals the modulus of the topological charge, and the direction of the spiraling is determined by the sign of the charge and relative curvature of the wave fronts.

Optical vortices embedded in a host light beam behave in some degree as charged particles. They may rotate around the beam axis, repel and attract each other, and annihilate in collision [20,21]. Other types of wave-front defects also may occur, such as edge or mixed screw-edge dislocations [22]. The possible transformations between edge and screw dislocations (in laser mode terms, 01 and doughnut modes) may be performed by modal converters [11,23,24].

The light field of a singular beam carries angular momentum [23] which may be transferred to a captured microparticle causing its rotation in a direction determined by the sign of the topological charge [25]. In nonlinear optics, so-called vortex solitons, which are singular beams in nonlinear medium, are a subject of growing interest [26,27].

However, singular beams demonstrate unusual properties even in linear optics, in free-space propagation. Optical phase singularities as morphological objects (tears of wave front) are robust with respect to perturbations. For instance, addition of a small coherent background does not destroy a vortex, but only shifts its position to another place where the field amplitude has a zero value. For vortices with multiple charge this operation will split an initially m-charged vortex into $|m|$ single-charge vortices [21]. Intuition suggests that the total topological charge would be conserved in a beam propagating in free space [1,2]. Our goal is to analyze the main properties of beams containing phase singularities in a general way, and demonstrate the limitations on the topological charge conservation principle for real beams, both theoretically and in experiment. On the basis of the present study, we establish the rules of angular-momentum transformations in light beams with phase singularities.

WAVE EQUATION SOLUTIONS POSSESSING PHASE SINGULARITIES

In this section we demonstrate some particular solutions possessing phase singularities of the scalar wave equation for a uniform isotropic medium

$$\nabla^2 E = \frac{1}{c^2}\frac{\partial^2 E}{\partial t^2}, \tag{1}$$

where E is the wave amplitude, c is the speed of light, and t is time. The existence of these solutions for a monochromatic light wave was first emphasized by Nye and Berry [1]. The derived complex amplitude of a light wave with frequency ω, wavelength λ, and wave vector $k=2\pi/\lambda$ oriented along z axis has the form

$$E(\rho,\varphi,z,t)\propto\begin{cases}\rho^{|m|}\exp(im\varphi+ikz-i\omega t), & \text{(2a)}\\ \rho^{-|m|}\exp(im\varphi+ikz-i\omega t), & \text{(2b)}\end{cases}$$

where ρ, φ, and z are cylindrical coordinates. Both solutions have a phase depending on the azimuth angle φ multiplied by an integer m (positive or negative) called the topological charge of the phase singularity (optical vortex). The wavefront (equal phase surface) forms in space part of a helicoidal surface given by the equality $m\varphi+kz=\text{const}$. After one round trip of the wave front around the z axis there is a continuous transition onto the next (or preceding) wave-front sheet separated by $m\lambda$, which results in a continuous helicoidal wave-front surface. The topological charge attributed to this wave-front structure is positive for a right-screw helicoid ($m>0$), and vice versa. In the case $|m|>1$, the wavefront structure is composed from $|m|$ identical helicoids nested on the z axis and separated by the wavelength λ.

The solutions in forms (2a) and (2b) cannot describe any real wave because of the radial amplitude dependence which grows proportionally to ρ for Eq. (2a), and tends to infinity when $\rho\to 0$ for Eq. (2b). To avoid unwanted amplitude growth, we may combine the solution in Eq. (2a) with a Gaussian beam. Using the paraxial approximation of the scalar wave equation in the form

$$\frac{1}{\rho}\frac{\partial}{\partial\rho}\left(\rho\frac{\partial E}{\partial\rho}\right)+\frac{1}{\rho^2}\frac{\partial^2 E}{\partial\varphi^2}-2ik\frac{\partial E}{\partial z}=0, \tag{3}$$

we obtain the corresponding solution for a "singular" wave in a Gaussian envelope carrying an optical vortex with charge m [17]:

$$E(\rho,\varphi,z)=E_s\frac{\rho_s}{w_s}\left(\frac{\rho}{w_s}\right)^{|m|}\exp\left(-\frac{\rho^2}{w_s^2}\right)\exp[i\Phi_s(\rho,\varphi,z)], \tag{4}$$

where E_s is the amplitude parameter, and ρ_s is the beam waist parameter. The phase Φ_s is

$$\Phi_s(\rho,\varphi,z)=-(|m|+1)\arctan\frac{2z}{k\rho_s^2}+\frac{k\rho^2}{2R_s(z)}+m\varphi+kz, \tag{5}$$

the transversal beam dimension is

$$w_s=\sqrt{\rho_s^2+(2z/k\rho_s)^2}, \tag{6}$$

and the radius of the wave front curvature is

$$R_s(z)=z+k^2\rho_s^4/4z. \tag{7}$$

The amplitude distribution in a transverse cross section of the beam has a form of an annulus, and the waist parameter ρ_s is connected with the radius of maximum amplitude at $z=0$ by a relation

$$\rho_{\max}=\rho_s\left(\frac{|m|}{2}\right)^{1/2}, \tag{8}$$

The maximum amplitude value of a singular wave at $z=0$, $\rho=\rho_{\max}$, therefore amounts to

$$E_{sm}=E_s\left(\frac{|m|}{2e}\right)^{|m|/2}. \tag{9}$$

The phase singularity disappears when $m=0$, and solution (4) becomes an ordinary Gaussian beam:

$$E(\rho,z)=E_g\frac{\rho_g}{w_g}\exp\left(-\frac{\rho^2}{w_g^2}\right)\exp[i\Phi_g(\rho,z)], \tag{10}$$

where the propagating parameters for Gaussian beam correspond to Eqs. (5)–(7) with $m=0$.

As an example we show that a solution in form (2b) having an amplitude singularity at $\rho\to 0$ may be used to create a solution with only a phase singularity. For this reason we take a similar solution with an amplitude singularity within Gaussian envelope,

$$E(\rho,\varphi,z)=E_s\frac{\rho_s}{\rho}\exp\left(-\frac{\rho^2}{w_s^2}\right)\exp[i\Phi_s(\rho,\varphi,z)], \tag{11}$$

where the phase Φ_s is

$$\Phi_s(\rho,\varphi,z)=\frac{k\rho^2}{2R_s(z)}+\varphi+kz. \tag{12}$$

The combination of solutions (11) and (2b) for $m=1$ removes the amplitude singularity and gives a wave which has a phase singularity at $\rho=0$:

$$E(\rho,\varphi,z)=E_s\frac{\rho_s}{\rho}\{1-\exp[-\rho^2/w_s^2+ik\rho^2/2R_s(z)]\}\times\exp(i\varphi+ikz). \tag{13}$$

The amplitude of a wave created this way is zero at the center ($\rho=0$) and decreases on periphery $\propto 1/\rho$. Other possible solutions of the scalar wave equation are Bessel beams [28] and Bessel-Gauss beams [29] carrying optical vortices.

OPTICAL VORTICES IN COMBINED BEAMS

In any practical realization of singular beams by use of synthesized holograms or special optical elements, a small coherent background is always present in a singular beam. The origin of this background may be a scattering in the direction of the singular beam propagation or readout beam diffraction by the fundamental spatial frequency of an imperfect hologram. This background causes splitting of an optical vortex with charge $|m|>1$ into $|m|$ single-charged vortices [21]. Another case is the interference between a singular wave and copropagating reference wave which is used in analyzing the value and sign of the topological charge of the phase singularity [18,21]. We shall now generalize the problem of coherent coaxial addition of singular beams carrying optical vortices with different charges (including the vortex-free wave, $m=0$). Our goal is to establish principles of topological charge addition and subtraction, and revise the existing knowledge about topological charge conservation in a light beam propagating in free space.

Without loss of generality, we shall take a coaxial Gaussian wave as a coherent background for a singular wave. By varying the amplitude E_g and waist parameter ρ_g of the Gaussian beam, we may examine its influence in near and far zones, as plane or spherical waves in limiting cases. The sum of the singular beam with the coaxial Gaussian beam (or another singular beam) we shall call a combined beam.

The presence of a coherent background changes the position of a vortex which was initially localized at the center of singular beam Eq. (4). To find the positions of the vortices in the combined beam, we need to write two equations, one giving the radius of the zero-amplitude point (amplitudes of the singular and Gaussian waves are equated), and another giving the angular coordinate φ, which corresponds to the destructive interference between the singular and Gaussian waves:

$$E_g \frac{\rho_g}{w_g} \exp\left(-\frac{\rho^2}{w_g^2}\right) = E_s \frac{\rho_s}{w_s}\left(\frac{\rho}{w_s}\right)^{|m|} \exp\left(-\frac{\rho^2}{w_s^2}\right),$$

$$\Phi_g(\rho,z) = \Phi_s(\rho,\varphi,z) \pm \pi. \qquad (14)$$

Equations (14) are the basis for analysis of vortex behavior in a combined beam. The first equation is easy to analyze to show the number of possible amplitude zeros in a combined beam.

To simplify the calculations, we suppose both beams have a waist at $z=0$, and use a normalized transverse coordinate $r=\rho/\rho_s$ and distance $\xi=z/L_R$, where L_R is the Rayleigh length of the singular beam, $L_R=k\rho_s^2/2$. The first equation of system (14) may be rewritten as

$$r^{|m|} = \frac{E_g}{E_s} C(\xi)\exp(\alpha r^2), \qquad (15)$$

where

$$C(\xi) = (1+\xi^2)^{|m|/2}\left(\frac{1+\xi^2}{1+\xi^2/\kappa^4}\right)^{1/2}, \qquad (16)$$

$$\alpha = \frac{1}{1+\xi^2} - \frac{1}{\kappa^2+\xi^2/\kappa^2}, \qquad (17)$$

and κ is the ratio of waist parameters $\kappa=\rho_g/\rho_s$.

If $\alpha\leqslant 0$, Eq. (15) has only one root ($|m|$ times degenerate), as the left side is a function of r growing from zero to infinity, and the right side is a function decreasing from E_g/E_s to zero. The condition $\alpha\leqslant 0$ corresponds to the following relations for κ and ξ:

$$\kappa \geqslant 1,$$

$$\xi \geqslant \kappa, \qquad (18a)$$

$$\kappa \leqslant 1,$$

$$\xi \leqslant \kappa, \qquad (18b)$$

and

$$\kappa = 1,$$

$$\alpha = 0, \quad 0<\xi<\infty. \qquad (18c)$$

All zeros of amplitude in the combined beam are located at the same distance from the center. The number of zeros, n, is equal to $|m|$ and each amplitude zero is a center of a single vortex. The total topological charge is conserved in the combined beam when $\alpha\leqslant 0$.

Another situation occurs when $\alpha>0$. The function of r in right side of Eq. (15) grows from E_g/E_s to infinity. Analysis shows three possibilities: no real roots of Eq. (15), two roots (each $|m|$ times degenerate), and one root ($2|m|$ times degenerate). The case of one root corresponds to the touching of lines representing the left and right sides of Eq. (15). This condition is determined by taking derivatives on r, which in combination with Eq. (15) gives a solution

$$\alpha r^2 = \frac{|m|}{2}, \qquad (19)$$

which gives the critical ratio between the amplitude of the Gaussian beam E_g and maximum amplitude of the singular beam E_{sm}, Eq. (9):

$$\left(\frac{E_g}{E_{sm}}\right)_{cr}^2 = \eta_{cr}^2 = \frac{1+\xi^2/\kappa^4}{1+\xi^2}\left(1-\frac{1+\xi^2}{\kappa^2+\xi^2/\kappa^2}\right)^{-|m|}. \qquad (20)$$

When the amplitude ratio $\eta=E_g/E_{sm}$ is higher than η_{cr}, no vortices exist in the combined beam, as Eq. (15) has no real roots. The resulting topological charge is zero. If the ratio η is smaller than η_{cr}, additional $|m|$ single-charged vortices appear with charge opposite to the original m-charged vortex, and the resulting total charge is zero again.

As the parameter α changes its sign during propagation of the combined beam both for cases $\kappa>1$ and $\kappa<1$, we may expect a variation in the number of vortices n in a combined beam propagating in free space, which means a change of the topological charge of the beam. Only the case $\kappa=1$ will conserve the initial topological charge unchanged from $\xi=0$ to infinity, independent of the beam amplitude ratio.

Figure 1 demonstrates the variation of the number of vortices in a combined beam during propagation along the ξ axis. The solid curve dividing the diagram is the dependence of η_{cr} vs ξ for a particular value of κ, $\kappa=0.5$ [Fig. 1(a)] and $\kappa=2$ [Fig. 1(b)]. The part of the diagram above the curve corresponds to zero number of vortices in a combined beam. In the case $\kappa<1$ [Fig. 1(a)], the combined beam conserves topological charge until $\xi=\kappa$ ($\alpha\leqslant 0$). This area on the diagram is separated by a vertical line. Outside this region the number of vortices may vary between zero and $2|m|$, depending on the ratio η, with total topological charge equal to zero. A similar situation occurs for $\kappa>1$, but now conservation of initial topological charge of singular beam will apply at $\xi>\kappa$ ($\alpha\leqslant 0$).

Amplitude profiles of Gaussian and singular beams are plotted in Fig. 2 for different distances ξ. The choice of parameters $m=1$, $\kappa=2$, and $\eta=1.05$ corresponds to the diagram shown in Fig. 1(b). Two points of intersection exist in a region $0<\xi<0.655$ (two oppositely charged vortices in the combined beam), no intersections in the region $0.655<\xi$

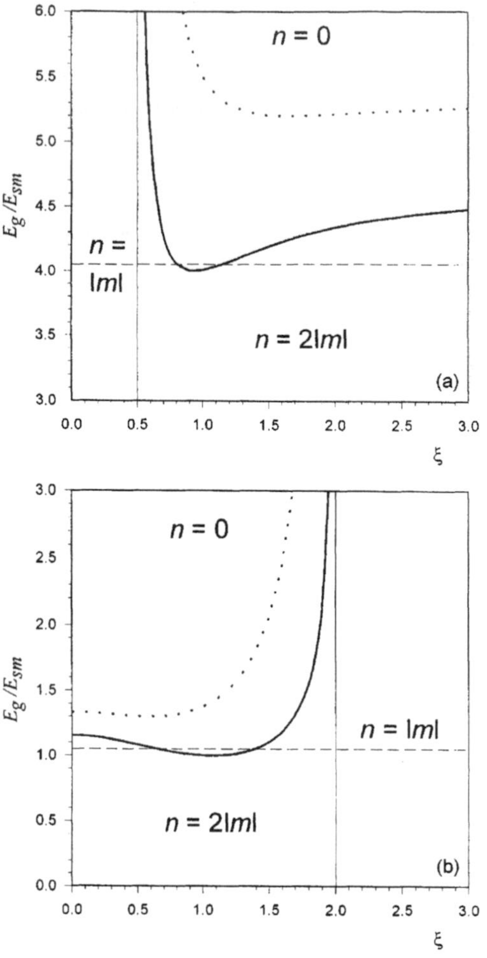

FIG. 1. Diagrams (a) and (b) show the number of vortices in a combined beam. (a) The solid curve is the dependence of the critical amplitude ratio $(E_g/E_{sm})_{cr}$ on the normalized distance ξ plotted for $\kappa=0.5$ and $|m|=1$. The area above the curve contains no vortices. The region $\xi<\kappa$ separated by the vertical line contains $|m|$ vortices, and the part below the curve corresponds to $2|m|$ vortices in a combined beam. The horizontal dashed line corresponds to a particular value of the amplitude ratio, $\eta=4.05$. This line crosses in turn regions with $|m|$, $2|m|$, 0, $2|m|$ vortices. The total topological charge is m, while $\xi\leq\kappa$, and zero outside. (b) The same as (a), but for the case $\kappa>1$. The solid curve is a plot of η_{cr} for $\kappa=2$. The horizontal dashed line corresponds to the amplitude ratio $\eta=1.05$. For higher m values, diagrams (a) and (b) are very similar: dotted lines are plots of η_{cr} for $|m|=2$.

<1.39 (no vortices in the beam), two intersections and two oppositely charged vortices for $1.39<\xi<2$, and finally one intersection and thus a single vortex for $\xi>2$.

In a practical situation we may not have information about the initial beam amplitude ratio and waist parameters ρ_g and ρ_s, or the optical paths of beams from waist to observation plane may be different. The only experimental measured values at the plane of observation are intensity

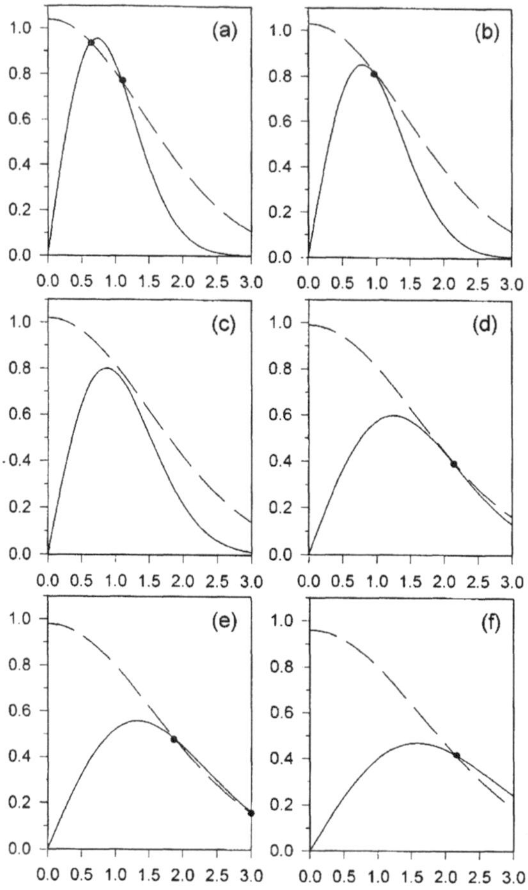

FIG. 2. Amplitude profiles of Gaussian (dashed line) and singular (solid line, $|m|=1$) waves with amplitude ratio $\eta=1.05$ and $\kappa=2$, for different normalized distances ξ. (a) $\xi=0.3$, curves intersect in two points shown by circles. (b) $\xi=0.655$, touching of the curves. (c) $\xi=1$; no intersections. (d) $\xi=1.39$, once again touching. (e) $\xi=1.5$, two intersections. (f) $\xi=2$, only the one intersection. Horizontal axis: normalized transverse coordinate r; vertical axis: wave amplitudes $E(r), E_{sm}=1$.

distributions of the Gaussian and singular beams. The amplitudes $A_g(\rho)$ and $A_s(\rho)$ and transversal parameters w_g and w_s may be calculated from energy distributions, and m is also easy to determine. The amplitudes necessary for existence of vortices in a combined beam may be written in this case as

$$A_g \exp\left(-\frac{\rho^2}{w_g^2}\right)=A_s\left(\frac{\rho}{w_s}\right)^{|m|}\exp\left(-\frac{\rho^2}{w_s^2}\right) \tag{21a}$$

or

$$\left(\frac{\rho}{w_s}\right)^{|m|}=\frac{A_g}{A_s}\exp\left(\frac{\rho^2}{w_s^2}-\frac{\rho^2}{w_g^2}\right), \tag{21b}$$

where A_g is the amplitude of the Gaussian beam at maximum, and A_s is connected with the maximum amplitude of

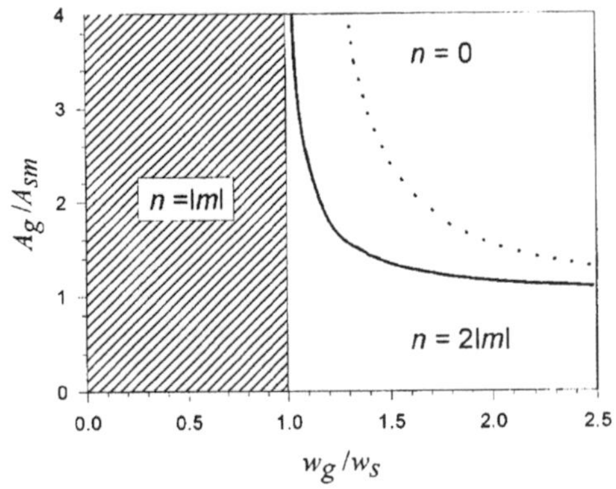

FIG. 3. Diagram showing the number of vortices in a beam combined from arbitrary Gaussian and coaxial singular beams. The horizontal coordinate is the transverse size ratio, the vertical coordinate is the amplitude ratio. The dashed area corresponds to the case of conservation of the total topological charge in the combined beam (the singular beam is wider than the Gaussian). The solid curve at the right side is the critical amplitude ratio for $|m|=1$, and the dotted curve is for $|m|=3$. The area under the curves corresponds to $2|m|$ vortices in the combined beam. No vortices are in the combined beam with parameters in the area above the curves.

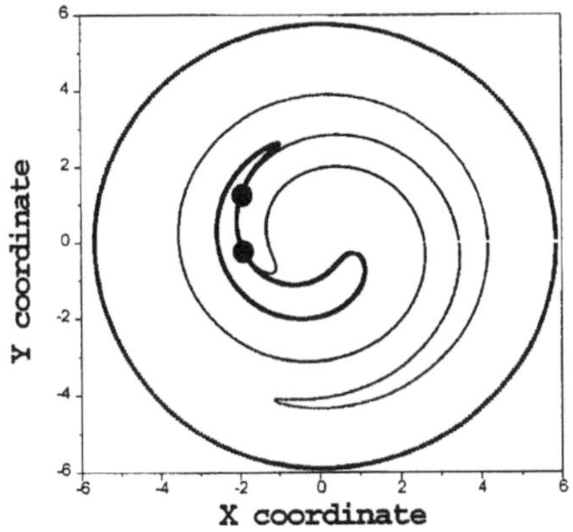

FIG. 4. Phase map shows the lines Re(E)=0 (thick line) and Im(E)=0 (thin line) in (x,y) coordinates normalized on ρ_s at a cross section $\xi=1.405$ of a combined beam with parameters $m=1$, $\kappa=2$, and $\eta=1.05$. Positions of vortices are shown by dots (intersections of zero lines). The original vortex is shifted from the center, and an additional vortex with opposite charge is located at the second intersection point.

the singular wave A_{sm} by the same relation as Eq. (9). Taking into account the previous analysis, we may formulate the following consequences important for experimental observation of vortex transformations in a combined beam: If the singular beam is broader than the Gaussian beam at the observation plane, $w_s > w_g$, Eq. (21b) has one $|m|$-times degenerate root, the total topological charge is conserved, and the original m-charged vortex breaks into $|m|$ vortices. If the Gaussian beam is broader than the singular, $w_g > w_s$, the total topological charge of the combined beam is zero. The number of vortices in a combined beam is determined by the relation between amplitudes. If $A_g/A_{sm} > (A_g/A_{sm})_{cr}$, the combined beams will contain no vortices. If $A_g/A_{sm} < (A_g/A_{sm})_{cr}$, an additional $|m|$ vortices will appear in the combined beam, as discussed above. The critical ratio $(A_g/A_{sm})_{cr}$ is

$$\left(\frac{A_g}{A_{sm}}\right)_{cr} = \left(\frac{w_g^2}{w_g^2 - w_s^2}\right)^{|m|/2}. \qquad (22)$$

The number of vortices in the combined beam for this practical situation may be determined from the diagram shown as Fig. 3. The hatched area corresponds to the case $w_s > w_g$ when the topological charge remains the same as in the singular beam. In the right part of the diagram a solid curve is plotted according to Eq. (22) for $m=1$ and a dotted curve for $m=3$. The region under the curves corresponds to the appearance of $|m|$ additional vortices and zero total topological charge. No vortices exist in the region above the corresponding curves.

To determine the position of vortices on a plane r,φ we need to use the second equation of system (14). Alternatively, we may calculate directly the position of the vortices as points of intersection of lines representing the zeros of the real and imaginary parts of the complex amplitude $E(r,\varphi,\xi)$ of the combined beam: Re(E)=0 and Im(E)=0 [3,15,30,31]. Figure 4 shows these lines in x,y coordinates (normalized on ρ_s) at cross section $\xi=1.405$ of a combined beam with parameters $m=1$, $\kappa=2$, and $E_g/E_{sm}=1.05$. The positions of vortices are shown by dots (intersections of zero lines). The original vortex is shifted from the center and an additional vortex with opposite charge is located at the second intersection point. Both lines Re(E)=0 and Im(E)=0 are closed.

Figure 5 exhibits transformations of zero-amplitude lines and corresponding vortices map for a combined beam with $m=3$, $\kappa=2$, and $\eta=1.75$. At near field ($\xi=0.5$) all vortices are suppressed by Gaussian wave with larger amplitude and waist parameter. The faster transverse spread of singular beam leads to the equalizing of amplitudes at combined beam periphery and appearance of three pairs of opposite charged vortices. Three vortices move away from the beam and disappear when $\xi=2$. Finally, when $\xi>2$, combined beam contains three single vortices located symmetrically with the conservation of the initial topological charge of the singular beam.

The trajectories of vortices within the cross section of the combined beam are shown in Fig. 6 for two main situations, $\kappa>1$ and $\kappa<1$. The combined beam ($m=1$, $\kappa=2$, and $\eta=0.5$) starts with two vortices [Fig. 6(a)] which move on their trajectories as shown in Fig. 6(a) in opposite directions, and the negative vortex leaves the beam at $\xi=2$. For another initial amplitude ratio $\eta=1.05$, the behavior of the vortices is somewhat different. Two vortices annihilate in collision at $\xi\approx 0.655$ [Fig. 6(b)], and then the beam does not contain any singularity until $\xi\approx 1.39$. Born as a pair, two new vortices with opposite charges repel each other and finally one (negatively charged) disappears at infinity ($r\rightarrow\infty$) at $\xi=2$, and the remaining one carries the initial charge of the singular beam. In the case $\kappa<1$ the combined beam ($m=1$, $\kappa=0.5$, and $\eta=4.05$) starts with only the primary vortex (shifted from the center), and an additional negative vortex enters the beam at

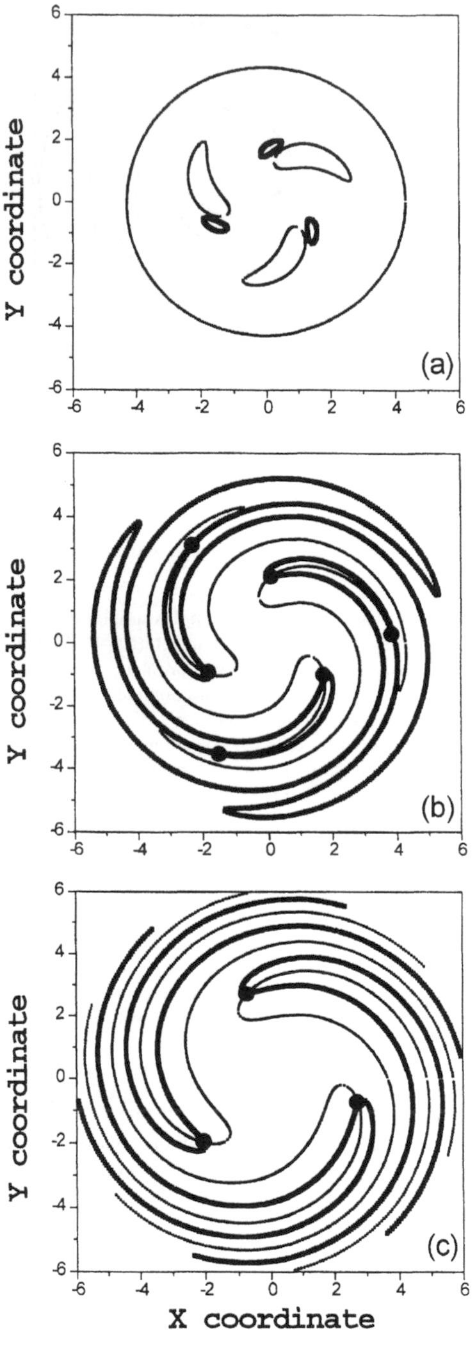

FIG. 5. Phase maps for different distances ξ of a combined beam propagation with parameters $m=3$, $\kappa=2$, and $\eta=1.75$. (a) $\xi=0.5$, lines $\mathrm{Re}(E)=0$ and $\mathrm{Im}(E)=0$ do not intersect, and no vortices exist in the combined beam. (b) $\xi=1.25$, lines $\mathrm{Re}(E)=0$ and $\mathrm{Im}(E)=0$ intersect in points shown by dots, and six single-charged vortices (three pairs) exist in the beam. (c) $\xi=2.5$, three intersection points are shown by dots, and three vortices are located symmetrically around the beam center.

$\xi=0.5$ coming from infinity [Fig. 6(c)]. After several round trips it meets the prime vortex, and both annihilate in collision at $\xi\approx0.81$. A new vortices pair originates at $\xi\approx1.16$, and exists within the beam up to infinity.

The analysis of vortex behavior in a beam combined from two coaxial singular beams may be performed in a similar manner. However, the combined beam may possess some new features in this case. First, both beams have zero amplitude at the center. If $|m_1|\neq|m_2|$, the combined beam will have at the center a vortex with the smaller absolute value of the charge. On the beam periphery, depending on the ratios of amplitudes and sizes, the number of vortices may vary between zero, $|m_1-m_2|$, and $2|m_1-m_2|$. In the particular case when $|m_1|=|m_2|$, there are two situations: $m_1=m_2$ and $m_1=-m_2$. When charges have the same sign, waves add coherently producing circular interference fringes [32], and the only m-charged vortex is located at the beam center. When vortices have opposite charges, they compete with each other for the central position. For small ρ, we may neglect the Gaussian envelope of the beams, and write the amplitude at the core as

$$E(\rho,\varphi)\propto\rho^{|m|}[A_{s1}e^{im\varphi}+A_{s2}e^{-im\varphi}], \tag{23}$$

which determines the phase near the core as $\arctan[(A_{s1}-A_{s2})/(A_{s1}+A_{s2})\tan(m\varphi)]$. This means that the vortex with larger host wave amplitude will win, but becomes anisotropic [30]. Finally, if $A_{s1}=A_{s2}$, vortices annihilate each other. The interference pattern displays $2|m|$ fringes radiating from the center.

EXPERIMENTAL OBSERVATION OF VORTICES IN COMBINED BEAM

The setup for our experiment is shown in Fig. 7. The linearly polarized output beam of a 10-mW He-Ne laser operating on the TEM_{00} mode is split first at beamsplitter BS1. The directly transmitted beam is diffracted at the holographic grating, and the first diffracted order (singular beam) is selected with the iris aperture. The holographic gratings are phase holograms restoring charge -1 and 3 singular beams, both blazed for greatest efficiency into first order [25]. The reflected beam from beamsplitter BS1 is further split at BS2 into the "background" Gaussian beam (reflected beam) and reference wave (transmitted beam). Lenses $L1-L4$ are used to control the sizes and radii of curvature of the Gaussian and reference wave. Lens $L2$ is finely controlled with a translation stage to allow fine adjustments of the wave-front curvature. Polarizing beamsplitters (PBS's) are used to control the relative intensities of the singular and Gaussian beams. BS3 recombines the singular and Gaussian beams, and BS4 interferes with the combined beam and the reference wave. (Double reflection of the singular beam from BS3 and mirror M does not change the sign of its topological charge.) Finally, lens $L5$ creates a magnified image of the beam on the screen, recorded with a charge-coupled device (CCD) camera.

The experimental technique consisted, first, of an alignment of the singular and Gaussian beams to make them coaxial. The resulting interference pattern, observed in the far field, was then used to determine the relative difference in curvature of wave fronts between the two beams. This was

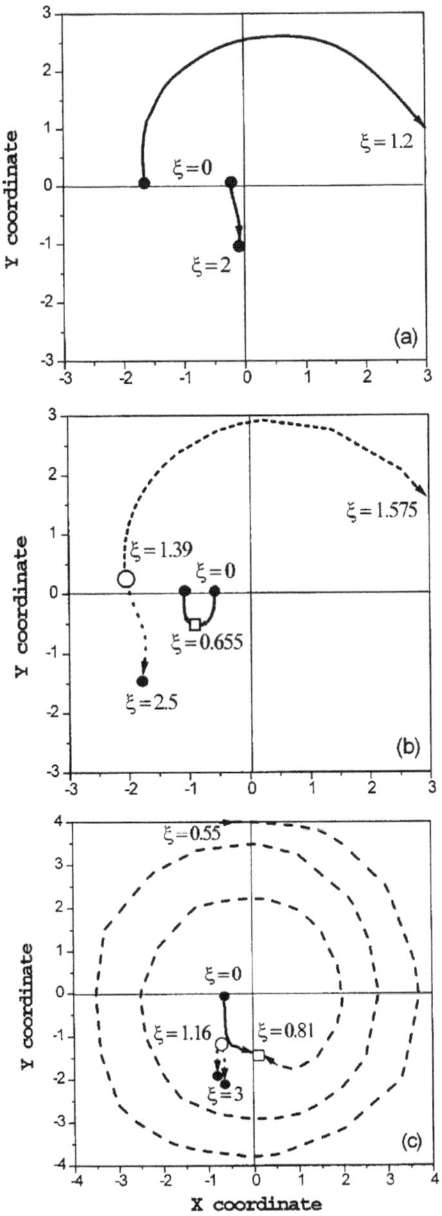

FIG. 6. Vortex trajectories in a combined beam cross section. (a) For $m=1$, $\kappa=2$, and $\eta=0.5$. The combined beam starts with two oppositely charged vortices shown by dots. The negative vortex leaves the beam at $\xi=2$. (b) Two vortices in a beam with $\eta=1.05$ annihilate in collision at $\xi\approx0.655$. A pair of two new vortices with opposite charges appear at $\xi\approx1.39$. They repel each other and finally one (negatively charged, dashed line) disappears at infinity ($r\rightarrow\infty$) at $\xi=2$. (c) Combined beam ($m=1$, $\kappa=0.5$, $\eta=4.05$) starts with only the primary vortex (shifted from the center), and an additional negative vortex enters the beam at $\xi=0.5$ (dashed line) coming from infinity. After several round trips it meets the prime vortex, and both annihilate in collision at $\xi\approx0.81$. New vortices' pair originates at $\xi\approx1.16$ (dotted lines).

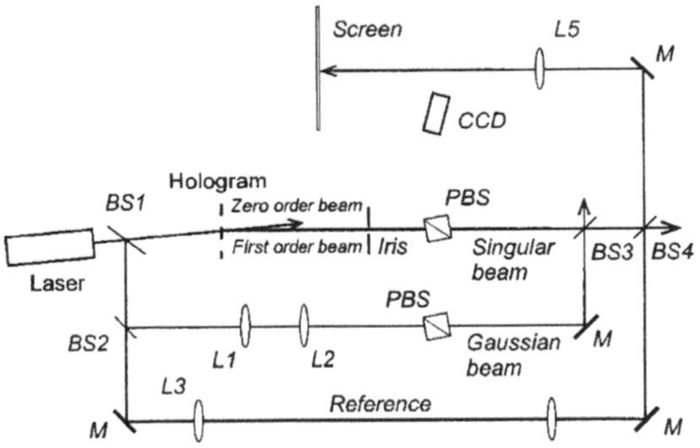

FIG. 7. Sketch of the experimental setup. The He-Ne laser beam splits into three channels. In the first channel a singular beam is created by the hologram, the second channel carries the Gaussian wave, and the reference wave is formed in the third channel. The interference patterns are observed on a screen.

adjusted using the position of lens $L2$, so that there was at most one round of dark fringe in the pattern. At this point, the wavefronts of two beams were effectively matched at the observation screen.

The profiles of energy distribution in the Gaussian and singular beams were obtained from separate CCD images, and the relative adjustment in beam intensities using PBS's could then be carried out to obtain the necessary number of amplitude graph intersection points. The corresponding intensity distributions are shown in Figs. 8 and 9 for Gaussian and singular beams (charges -1 and 3). Once the two beams were roughly the required relative intensity and size for additional vortices to be observable, the reference wave was used to interfere the intensity pattern to observe the presence of fringe dislocations, and hence the vortices in the original pattern.

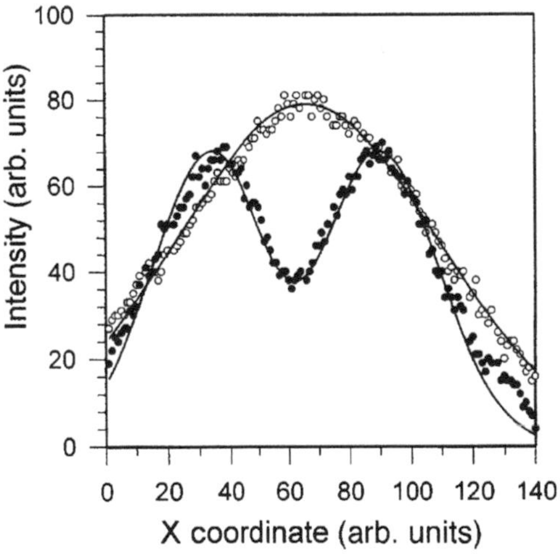

FIG. 8. Experimental intensity distributions for Gaussian ($m=0$, open circles) and singular ($m=-1$, closed circles) beams at the observation plane. Solid lines are numerical fits according to formulas (24).

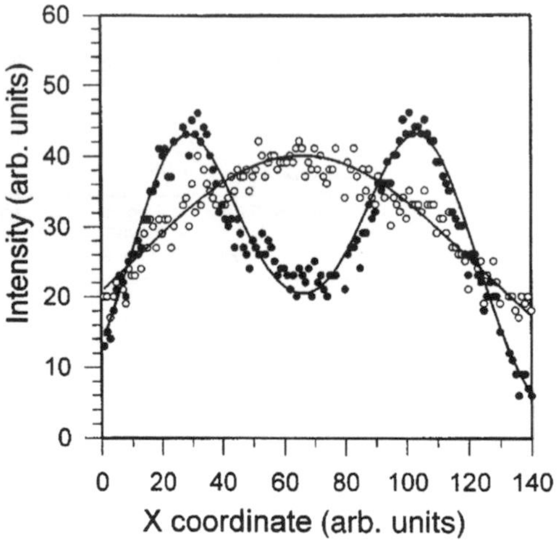

FIG. 9. The same distributions as in Fig. 8, for Gaussian and charge 3 singular beams.

The reference beam was simply adjusted through the choice of lenses $L3$ and $L4$ to be of much larger size than the combined beam cross section. The interference picture produced by the superposition of the reference beam and the combined beam at BS4 could be roughly controlled to give a reasonable number of fringes for sufficient resolution of any fringe dislocations present. To avoid reduction of fringe visibility due to a loss of coherence through multilongitudinal mode operation of the laser, all path differences were made integer multiples of the laser cavity length.

The parameters obtained from a numerical fit of the data were used for a comparison of the experimental results with theoretical predictions. The following functions were used to generate the intensity profiles of Gaussian and singular beams numerically:

$$P_g(x)=\left\{A_g\exp\left[-\frac{(x-x_0)^2}{w_g^2}\right]\right\}^2,$$

$$P_{-1}(x)=\left\{\frac{A_{-1}}{w_{-1}}[(x-x_0)^2+y_0^2]^{1/2}\exp\left[-\frac{(x-x_0)^2+y_0^2}{w_{-1}^2}\right]\right\}^2,$$

$$P_3(x)=\left\{\frac{A_3}{w_3^3}[(x-x_0)^2+y_0^2]^{3/2}\exp\left[-\frac{(x-x_0)^2+y_0^2}{w_3^2}\right]\right\}^2, \tag{24}$$

where A_g, A_{-1}, and A_3 are the amplitudes of the respective waves (Gaussian, singular charges -1 and 3) and w_g, w_{-1}, and w_3 are the beam sizes. The function $P_g(x)$ is the intensity profile of a Gaussian beam defined in the x,y plane, where the profile is taken along the x direction with $y=0$. $P_{-1,3}$ are the intensity profiles along x axis of singular beams, charges -1 and 3, at $y=y_0$, in order to take into account a small misalignment of the beam centers, which prevents the central minimum of the singular beam from going to zero. Parameters x_0 and y_0 determine the beam center. The data obtained from the CCD (in relative units) were fitted to functions (24) with fitting parameters: $A_g=8.86$, $w_g=85.22$, and $x_0=65.76$ for the Gaussian beam, and $A_{-1}=19.23$, $(A_{-1m}=8.25)$, $w_{-1}=45.81$, $x_0=62.25$, and $y_0=16.73$ for the singular (charge -1) beam. The calculated curves are shown in Fig. 8 (solid lines). The ratio of the beam sizes is $w_g/w_s=1.86$, i.e., the Gaussian beam is wider than the singular. The amplitude ratio is $A_g/A_{-1m}=1.07$, and according to the theoretical predictions (see the diagram in Fig. 3) an additional vortex should appear in the combined beam with a sign opposite to the original vortex. The presence of intersection points of the curves also indicates two vortices existence in the combined beam.

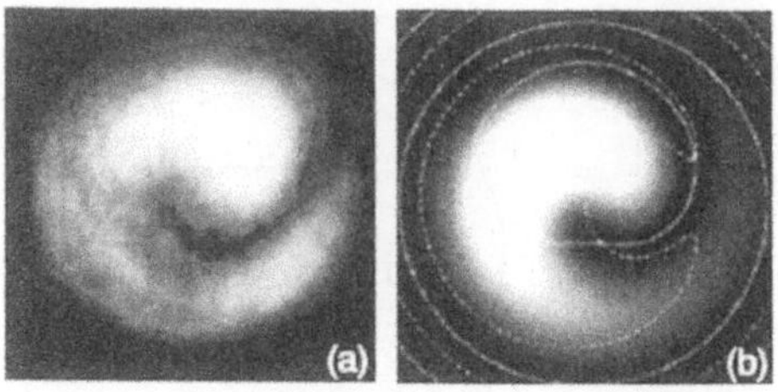

FIG. 10. Experimental (a) and numerically generated (b) intensity distributions in a combined (Gaussian plus $m=-1$ singular) beam cross section. Zero-amplitude lines are shown in (b), when the solid line represents the real part of the complex amplitude, and the dashed line the imaginary part.

For the profiles of the Gaussian and charge 3 singular beam, the fitting parameters were $A_g=6.29$, $w_g=115.92$, and $x_0=66.30$ for the Gaussian beam, and $A_3=15.82$ $(A_{3m}=6.49)$, $w_3=41.23$, $x_0=66.05$, and $y_0=33.91$ for the singular beam (Fig. 9). The Gaussian beam is again wider than the singular, and the amplitude ratio corresponds to the case of three additional vortices appearing, with the total topological charge of the combined beam being zero.

For the two superpositions investigated, namely, Gaussian beam with charge -1 singular and Gaussian with charge 3 singular, the intensity patterns were calculated in gray scale and are shown in Figs. 10 and 11. To generate the two-dimensional intensity distribution $I(x,y)$ in a beam combined from Gaussian and charge -1 singular beams numerically, the expression for $I(x,y)$ is represented as follows:

$$I(x,y)=P_g(x,y)+P_{-1}(x,y)+2\sqrt{P_g(x,y)P_{-1}(x,y)}\times\cos\left(\frac{x^2+y^2}{R^2}-\arctan\frac{y}{x}+\delta\right), \tag{25}$$

where R is the radius of relative curvature of the Gaussian wave front with respect to the singular wave, δ is an adjustable phase factor, and $\arctan(y/x)$ is the azimuth angle. The intensity distribution for the combined beam with Gaussian and charge 3 singular wave has the same form as Eq. (25), only the azimuth factor becomes $3\arctan(y/x)$. Figures 10 and 11 also show the experimental patterns of combined beam intensity in the far field and calculated intensity distributions with lines of zeros for the functions $\mathrm{Re}[E(x,y)]=0$ (solid line) and $\mathrm{Im}[E(x,y)]=0$ (dashed line). The points of intersection of the zero lines correspond to the positions of vortices. For the case of the combination of a Gaussian and a

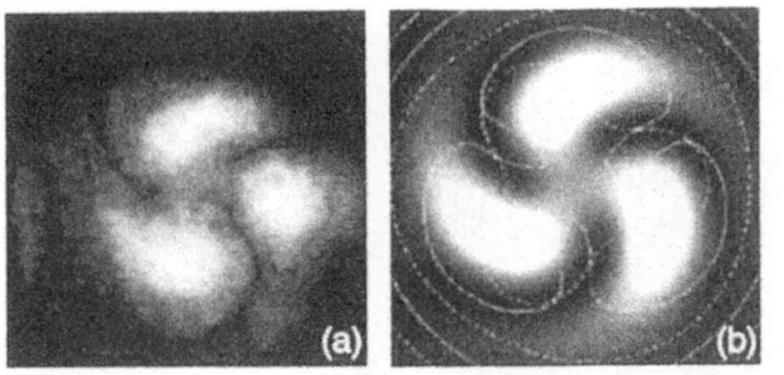

FIG. 11. The same intensity distributions as in Fig. 10, but for Gaussian plus $m=3$ singular combined beam.

charge -1 singular beam, a single dark spiral fringe is seen [Fig. 10(a)]. The curved shape of the fringe is due to slight difference in wave-front curvature. The zero-amplitude lines have two points of intersection, corresponding to two vortices existing within the dark fringe [Fig. 10(b)]. For the Gaussian and charge 3 singular beam, three dark fringes radiating from the center of the pattern are present [Fig. 11(a)]. There are six intersection points of the zero lines in Fig. 11(b), indicating two vortices with opposite charges located in each dark fringe.

To demonstrate the existence of vortices in the combined beams clearly, the interference patterns of the combined beam with the reference wave are shown in Figs. 12 and 13. The results of experimental observations are compared with calculated intensity distributions. For the Gaussian and charge -1 combined beam, both theory and experiment show a dark spiral fringe which begins at the center of the pattern and ends after nearly two turns [Figs. 12(a) and 12 (b)]. Thereafter, the interference fringes form rings corresponding well with the theory: the phase has no rotating component at this area. Thus two oppositely charged vortices make the total topological charge zero.

For the Gaussian and charge 3 combined beam, there are three inner interference "forks," equally separated by 120° and directed clockwise, and three outer "forks" directed anticlockwise (Fig. 13). Hence there are six single vortices altogether, or three pairs of oppositely charged vortices. At the area outside, the fringes again form rings.

The experimental observations clearly demonstrate the appearance of additional vortices in combined beams, changing the total topological charge of a beam. We have thus shown in both theory and experiment that the topological charge of a beam containing phase singularities is not a constant while propagating in free space.

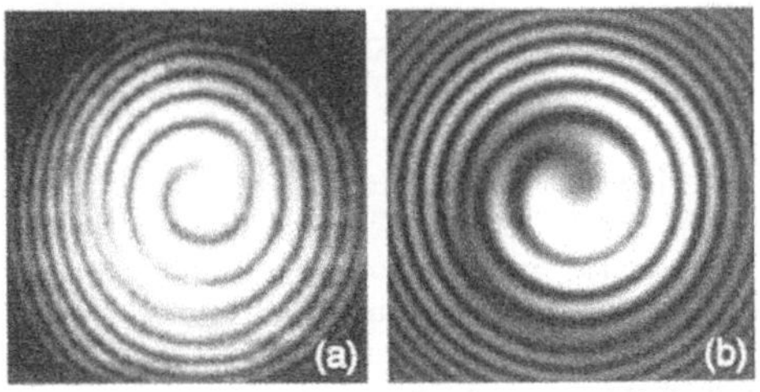

FIG. 12. The interference pattern of the combined beam (Gaussian plus $m=-1$ singular) and reference wave. The primary vortex is located near the center, and an additional vortex is seen with positive charge: (a) experimental picture; (b) numerical simulation.

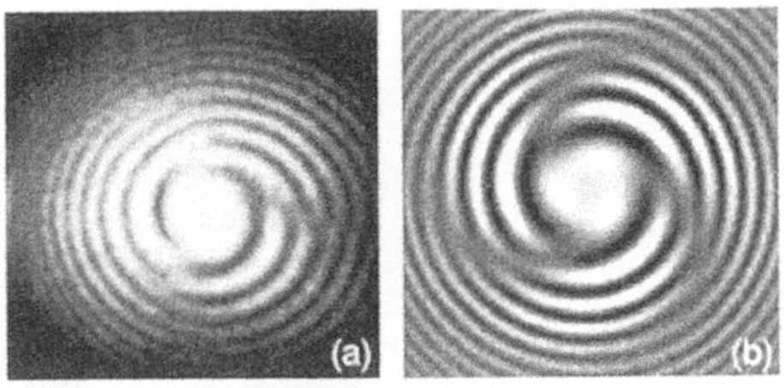

FIG. 13. The interference pattern of the combined beam (Gaussian plus $m=3$ singular) and reference wave. The primary vortex is split into three vortices, and an additional three vortices with negative charge produce, with them, three pairs: (a) experimental picture; (b) numerical simulation.

ANGULAR MOMENTUM OF LIGHT BEAM CARRYING OPTICAL VORTICES

The rotation of phase around the vortex axis causes a nonzero value of angular momentum of a beam [23,33]. The origin of the angular momentum may be easy explained from simple evaluations. As the wave front of a singular beam has a helicoidal shape, the Poynting vector $P(\rho,\varphi,z)$ which is perpendicular to the wave front surface has at each point a nonzero tangential component. In the paraxial approximation this component equals $P_{\_}(\rho,\varphi,z)=-mP/k\rho$, and the angular-momentum density in the z-axis direction is $M_z(\rho,\varphi,z)=(\rho/c^2)P_{\perp}(\rho,\varphi,z)$. As the Poynting vector is proportional to a light wave intensity, $P\propto|E_s(\rho,\varphi,z)|^2$, we may obtain the expression for angular-momentum density of a singular beam:

$$M_z(\rho,\varphi,z)\propto -m|E_s(\rho,\varphi,z)|^2. \tag{26}$$

The time-averaged density of angular momentum directed along the z axis in a combined beam cross section may be calculated in a general form [23,33] as

$$M_z=\frac{i}{2}\,\omega\epsilon_0\left[x\left(E^*\frac{\partial E}{\partial y}-E\frac{\partial E^*}{\partial y}\right)-y\left(E^*\frac{\partial E}{\partial x}-E\frac{\partial E^*}{\partial x}\right)\right], \tag{27}$$

where ϵ_0 is the permittivity of free space. The total angular momentum L_z of the beam is an integral over the beam cross section

$$L_z=\int_{-\infty}^{\infty}\int_{-\infty}^{\infty}M_z dx\,dy. \tag{28}$$

The simplest combined beam is a sum of singular [Eq. (4)] and Gaussian [Eq. (10)] beams, and its angular momentum calculated according to Eq. (27) is

$$\begin{aligned}M_z(\rho,\varphi,z)=&-\omega\epsilon_0 m\{|E_s(\rho,\varphi,z)|^2\\&+|E_s(\rho,\varphi,z)E_g(\rho,z)|\cos[\Phi_s(\rho,\varphi,z)\\&-\Phi_g(\rho,z)]\},\end{aligned} \tag{29}$$

which attains a simple form for a pure singular beam ($E_g=0$):

$$M_z(\rho,\varphi,z)=-\omega\epsilon_0 m|E_s(\rho,\varphi,z)|^2, \tag{30}$$

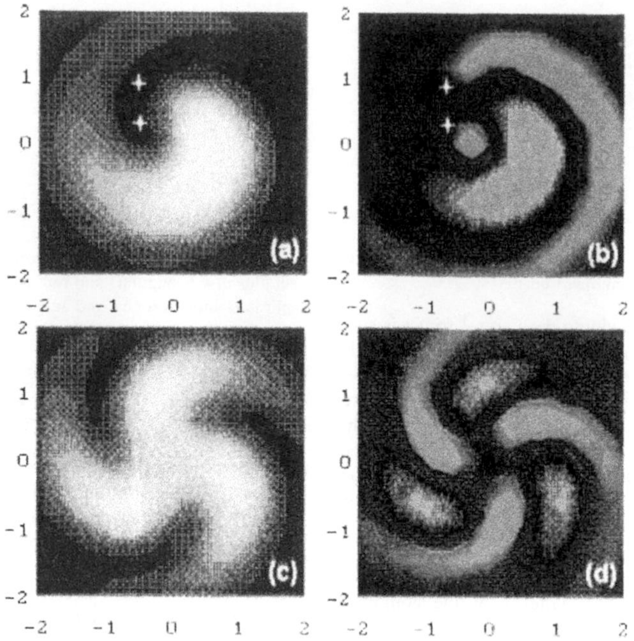

FIG. 14. (Color) Distribution of energy and angular momentum in a combined beam. (a) and (b) $m=1$, $E_g/E_{sm}=1.05$, $\kappa=2$, and $\xi=0.3$. Green is for energy distribution, red for positive angular momentum density $M_z>0$, and blue for negative angular-momentum density $M_z<0$. The positions of two oppositely charged vortices are shown by white crosses. (c) and (d) The same distributions for $m=3$, $E_g/E_{sm}=2.5$, and $\kappa=2$. No vortices exist in the combined beam with this amplitude ratio, but a modulation of the angular-momentum density is clearly seen. The transverse coordinates are x and y, normalized on ρ_s.

coinciding with Eq. (26), and

$$L_z=-2\pi\omega\epsilon_0 m E_s^2\left(\frac{\rho_s}{w_s}\right)^2\int_0^\infty \rho d\rho\left(\frac{\rho}{w_s}\right)^{2|m|}\exp\left(-\frac{2\rho^2}{w_s^2}\right)$$

$$=-\frac{\pi m|m|!}{2^{|m|+1}}\,\omega\epsilon_0\rho_s^2E_s^2. \tag{31}$$

The density of the angular momentum of a singular beam M_z is therefore proportional to the beam intensity, and the total angular momentum L_z to the beam energy. In a medium without losses, the total momentum L_z is conserved, as well as the beam energy.

The relation obtained for the angular momentum of a combined beam (29) may be generalized as a law of coaxial addition of N beams carrying optical vortices with topological charges m_i:

$$M_z(\rho,\varphi,z)=-\frac{\omega\epsilon_0}{2}\sum_{i,j}^{N}(m_i+m_j)|E_i(\rho,\varphi,z)E_j(\rho,\varphi,z)|$$

$$\times\cos[\Phi_i(\rho,\varphi,z)-\Phi_j(\rho,\varphi,z)]. \tag{32}$$

This law is valid not only for singular waves with a Gaussian envelope, but for all kinds of waves with axially symmetric amplitude distributions.

An interesting consequence following from Eq. (29) is that even a small amount of a singular wave combined with a strong Gaussian wave produces a substantial modulation of the angular-momentum density. Figure 14 represents the calculation of angular-momentum density of combined beams with parameters $m=1$, $\kappa=2$, and $\eta=1.05$ (a) and $m=3$, $\kappa=2$, and $\eta=2.5$ (b).

For a combined beam with a density of angular-momentum distribution given by Eq. (29), the total angular momentum is equal to that of the singular beam alone, because an interference term gives a zero amount in integral (28). This demonstrates that coherent addition of a wave with zero angular momentum does not affect the resulting angular momentum of the combined beam. Using the general form of the angular-momentum distribution in a combined beam (32), we are able now to calculate the result of coaxial interference of N arbitrary singular beams,

$$L_z=\sum_i^N L_{zi}, \tag{33}$$

which establishes a rule: In a beam combined from N coaxial beams, the angular momenta L_{zi} add arithmetically.

This rule gives interesting consequences for the addition of singular beams with equal and opposite topological charges. Two identical singular beams with equal energy but opposite charges form a vortex-free beam with zero angular momentum, independent of the phase relation between them. Identical beams with equal topological charges of vortices do not obey the general rule (33), because the interference term is a constant in this case, and the energy of the summed beam is strongly dependent on relative phases of components. Further, adding a vortex-free wave with zero angular momentum cannot change the total angular momentum L_z of a beam, even though it may suppress all the vortices in the combined beam.

To understand the transformations of angular momentum associated with an m-charged vortex in a combined beam where the number of vortices may change and even become zero, we need to imagine the wave front of a combined beam which is not a symmetrical m-pitched helicoid as for a pure singular beam. When all the vortices are suppressed by the strong background wave, the wave front of the combined beam still has folds, but is a smooth surface without defects. The inclination of local "rays" which are perpendicular to the surface produces both positive and negative components of the angular momentum. With an increase of Gaussian background wave amplitude, the folds become smaller, but the beam amplitude grows proportionally, and resulting negative and positive parts of angular momentum have nearly the same value. The difference between the negative and positive components of the angular momentum remains exactly equal to the initial angular momentum of the singular beam, but is now due to small residual amplitude modulation over the combined beam cross section. In the case of multiple vortices localized within a combined beam, the distribution of the angular momentum becomes more complicated, but the resulting total angular momentum remains constant.

CONCLUSIONS

Our analysis of coherent coaxial addition of optical waves carrying (at least one) optical vortices has revealed the general properties of combined beams. We have studied in detail the behavior of vortices in a beam combined from singular and "background" Gaussian waves in order to check the principle of topological charge conservation. In brief, we obtained the following results.

(1) Addition of a coherent coaxial vortex-free wave to a singular wave with a m-charged vortex may change the number of vortices and the total topological charge in the combined beam. Depending on background wave parameters, the number of vortices in the combined beam may vary between zero, $|m|$, and $2|m|$. The total topological charge is m or zero.

(2) In free-space propagation from $z=0$ to infinity, the total topological charge and number of vortices in a combined beam do change for any choice of background wave parameters, except the case $\kappa=1$ (equal transverse sizes).

(3) We demonstrate for the first time to our knowledge, the possibility, in free-space propagation, of additional vortices appearing from the far periphery of a beam (from infinity), and their disappearance at a beam periphery.

(4) The obtained results are also applicable to the case of combinations of singular beams.

(5) We show that the total angular momentum of a beam is conserved in all cases in free-space propagation, in contrast with the total topological charge. We establish the main rules for addition and subtraction of angular momenta of light beams.

(6) We analyzed the transverse distribution of the angular momentum in a combined beam cross section, and found a strong spatial modulation of the angular-momentum density even in the case of absence of vortices in a combined beam.

ACKNOWLEDGMENTS

This research was supported, in part, by the International Science Foundation, and by the Australian Department of Industry, Science and Tourism.

[1] J. F. Nye and M. V. Berry, Proc. R. Soc. London, Ser. A **336**, 165 (1974).

[2] M. V. Berry, in *Les Houches Lecture Series Session XXXV*, edited by R. Balian, M. Klaéman, and J.-P. Poirier (North-Holland, Amsterdam, 1981), p. 453.

[3] N. B. Baranova, A. V. Mamaev, N. F. Pilipetskii, V. V. Shkunov, and B. Ya. Zel'dovich, J. Opt. Soc. Am. **73**, 525 (1983).

[4] P. Coullet, L. Gil, and F. Rocca, Opt. Commun. **73**, 403 (1989).

[5] V. Yu. Bazhenov, M. V. Vasnetsov, and M. S. Soskin, Pis'ma Zh. Eksp. Teor. Fiz. **52**, 1037 (1990) [JETP Lett. **52**, 429 (1990)].

[6] N. R. Heckenberg, R. McDuff, C. P. Smith, H. Rubinsztein-Dunlop, and M. J. Wegener, Opt. Quantum Electron. **24**, S951 (1992).

[7] S. N. Khonina, V. V. Kotlyar, M. V. Shinkarev, V. A. Soifer, and G. V. Uspleniev, J. Mod. Opt. **39**, 1147 (1992).

[8] S. Tidwell, G. Kim, and W. Kimura, Appl. Opt. **32**, 5222 (1993).

[9] M. W. Beijersbergen, R. P. C. Coerwinkel, M. Kristiensen, and J. P. Woerdman, Opt. Commun. **112**, 321 (1994).

[10] J. M. Vaughan and D. V. Willetts, J. Opt. Soc. Am. **73**, 1018 (1983).

[11] C. Tamm and C. O. Weiss, J. Opt. Soc. Am. B **7**, 1034 (1990).

[12] F. T. Arecchi, G. Giacomelli, P. L. Ramazza, and S. Residori, Phys. Rev. Lett. **67**, 3749 (1991).

[13] S. R. Liu and G. Indebetouw, J. Opt. Soc. Am. B **9**, 1507 (1992).

[14] M. Ya. Darsht, I. V. Kataevskaya, N. D. Kundikova, and B. Ya. Zel'dovich, Zh. Eksp. Teor. Fiz. **107**, 1 (1995) [Sov. Phys. JETP **80**, 817 (1995)].

[15] A. V. Ilyenkov, A. I. Khizniak, L. V. Kreminskaya, M. S. Soskin, and M. V. Vasnetsov, Appl. Phys. B **62**, 465 (1996).

[16] T. Ackemann, E. Kriege, and W. Lange. Opt. Commun. **115**, 339 (1995).

[17] V. Yu. Bazhenov, M. S. Soskin, and M. V. Vasnetsov, J. Mod. Opt. **39**, 985 (1992).

[18] A. G. White, C. P. Smith, N. R. Heckenberg, H. Rubinsztein-Dunlop, R. McDuff, C. O. Weiss, and Chr. Tamm, J. Mod. Opt. **38**, 2531 (1991).

[19] O. Bryngdahl, J. Opt. Soc. Am. **63**, 1098 (1973).

[20] G. Indebetouw, J. Mod. Opt. **40**, 73 (1993).

[21] I. V. Basistiy, V. Yu. Bazhenov, M. S. Soskin, and M. V.

Vasnetsov, Opt. Commun. **103**, 422 (1993).
[22] I. V. Basistiy, M. S. Soskin, and M. V. Vasnetsov, Opt. Commun. **119**, 604 (1995).
[23] M. W. Beijersbergen, L. Allen, H. E. L. O. van der Ween, and J. P. Woerdman, Opt. Commun. **96**, 122 (1993).
[24] E. Abramochkin and V. Volostnikov, Opt. Commun. **83**, 123 (1991).
[25] H. He, N. R. Heckenberg, and H. Rubinsztein-Dunlop, J. Mod. Opt. **42**, 217 (1995).
[26] G. Swartzlander, Jr. and C. Law, Phys. Rev. Lett. **69**, 2503 (1992).
[27] B. Luther-Davies, R. Powles, and V. Tikhonenko, Opt. Lett. **19**, 1816 (1994).
[28] J. Durnin, J. Opt. Soc. Am. A **4**, 651 (1987).
[29] F. Gori, G. Guatari, and C. Padovani, Opt. Commun. **64**, 491 (1987).
[30] I. Freund, N. Shvartsman, and V. Freilikher, Opt. Commun. **101**, 247 (1993).
[31] N. R. Heckenberg, M. Vaupel, J. T. Malos, and C. O. Weiss, Phys. Rev. A **54**, 2369 (1996).
[32] M. Harris, C. A. Hill, and J. M. Vaughan, Opt. Commun. **106**, 161 (1994).
[33] L. Allen, M. W. Beijersbergen, R. J. C. Spreeuw, and J. P. Woerdman, Phys. Rev. A **45**, 8185 (1992).

Helical-wavefront laser beams produced with a spiral phaseplate

M.W. Beijersbergen, R.P.C. Coerwinkel, M. Kristensen [1], J.P. Woerdman
Huygens Laboratory, University of Leiden, P.O. Box 9504, 2300 RA Leiden, The Netherlands

Received 30 August 1994

Abstract

We demonstrate experimentally that a spiral phaseplate can convert a TEM_{00} laser beam into a helical-wavefront beam with a phase singularity at its axis. The diffractive-optical effect of the spiral phaseplate is implemented by index matching a macroscopic structure in an optical immersion. We discuss the optical properties of a helical wavefront beam produced this way by means of a mode analysis and by Fraunhofer diffraction calculations.

1. Introduction

The special properties of helical-wavefront or spiral laser beams have recently aroused appreciable interest [1–8]. The complex amplitude u of such a beam has the property

$$u(\rho, \phi, z) \propto \exp(-il\phi), \tag{1}$$

where ρ, ϕ and z are cylindrical coordinates and l is an integer. At the center the phase has a screw dislocation [9], also called a phase singularity, optical vortex, or topological defect of charge l. The equiphase surface is a spiral sheet with a pitch equal to l times the wavelength. Our interest into these beams is due to the fact that fields with such a phase distribution are orbital angular momentum eigenmodes if their intensity distribution is rotationally symmetric [10]. Our ultimate aim is to determine the mechanical consequences of this orbital angular momentum of light [11].

Note that circular polarization of the light beam gives rise to a different type of helicity. The helicity we discuss here is entirely caused by the shape of the wavefront, not by the polarization. A separation in this sense is only possible in the paraxial approximation [12].

One special set of paraxial spiral beams are the Laguerre-Gaussian (LG) modes. The mode character of these beams expresses itself in the fact that, apart from a change in scale in the transverse plane, the intensity distribution does not change upon propagation in free space.

A helical-wavefront laser beam can be produced with a laser that operates in a LG mode. This can be accomplished by locking the phases of a set of Hermite-Gaussian (HG) transverse modes of a laser [8]. Alternatively, the phase and amplitude distribution of a non-helical laser beam can be changed outside the laser cavity such that a spiral beam is obtained. A cylindrical-lens mode converter may do this if the input beam is a higher-order HG mode. The output is then a pure LG mode. As previously shown, this mode conversion can be fully described in terms of a finite set of Gaussian modes [6,13].

Holography is an alternative and more general way to influence the phase and/or amplitude distribution

[1] Present address: Mikroelektronik Centret, DTU, DK-2800 Lyngby, Denmark.

of a field. It has been used to produce beams with a single [14,15] or multiple phase singularities [16]. Depending on the type of holography (phase or amplitude, binary or sinusoidal) a larger or smaller fraction of the incident beam is diffracted into a helical-wavefront beam, that could be a pure LG mode. A modal description of this conversion strongly depends upon the properties of the hologram.

In this paper we introduce a novel way to produce helical wavefront laser beams, using a spiral phaseplate. Phaseplates in general have often been used to influence the amplitude distribution of a beam [17]. Such a thin transparent plate typically has strips or radial sectors that are half a wavelength thicker, obtained by coating or etching, than the rest of the plate. The spatially dependent 180° phase shift can be used to homogenize the phase distribution of a beam that has lobes that are in anti-phase, such as a higher order Hermite-Gaussian beam. Obviously, a phaseplate without spiral symmetry cannot produce a helical wavefront beam. We report here a spiral phaseplate that *can* cause a single screw dislocation to appear in the center of the beam, resulting in a helical wavefront laser beam.

Such a plate has been previously been used to produce a radially polarized laser beam [18]. For that application the function of the plate was interferometric in nature; in the analysis diffraction was neglected, so that the change of the intensity distribution in the far field is ignored. We instead introduce the spiral phaseplate as a means of producing helical-wavefront laser beams. We theoretically analyze the propagation properties of the beam thus produced in two ways. First, we give a paraxial mode analysis with the LG modes as a basis, and next, we discuss Fraunhofer diffraction of these beams. Finally, we show experimentally that it is possible to produce such beams using a spiral phaseplate. For that purpose we developed an immersion variety of the spiral phaseplate that is simple to produce and can be adjusted for a wide range of applications.

2. The spiral phaseplate

A spiral phaseplate is a transparent plate whose thickness increases proportional to the azimuthal angle ϕ around a point in the middle of the plate. The surface resembles one turn of a staircase, the step height being h (Fig. 1). When a beam is passed through such a plate, the helical surface can be expected to give a helical character to the beam. A rigorous calculation of the operation of the plate would require vector-diffraction theory. However, for beams with small divergence and with a height of the step that is sufficiently small we remain in the paraxial regime, so that the operation of the plate can be considered to be an operation on the phase only. If the incident beam is described by the complex amplitude $u(\rho, \phi, z)$, the amplitude u' directly after the plate is then given by

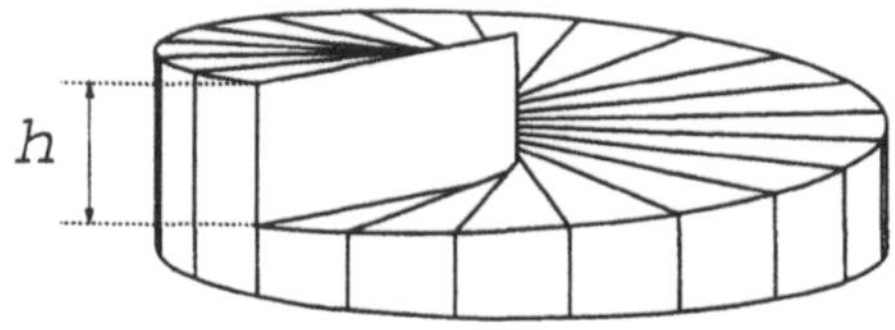

Fig. 1. Sketch of the spiral phaseplate.

$$u' = u \exp(-\mathrm{i}\Delta l \phi). \qquad (2)$$

Here Δl is the height of the step in wavelengths given by

$$\Delta l = \Delta n \; h/\lambda, \qquad (3)$$

where Δn is the difference in refractive index between the plate and its surroundings and λ the vacuum wavelength. Note that the paraxial approximation was not valid in a previously performed microwave experiment in a waveguide, where a similar spiral retarder was used [19]. This violation leads to a coupling between the amplitude and the local polarization, similar to spin-orbit coupling.

The spiral phaseplate may transform a non-helical beam into a helical one, but also change the helicity of a helical beam. When the helicity of a laser beam is modified according to Eq. (2), the phase distribution is changed while the amplitude distribution is not. In most cases the resulting beam is no longer a pure mode that propagates with only rescaling of the intensity distribution, but rather a superposition of modes. The spiral phaseplate is therefore in general not a pure mode converter.

In discussing the operation of the phaseplate we will limit ourselves to the case of an incident LG mode. For these modes we use different indices than used normally. Our preferred indices n, m make the relation between HG and LG modes more evident [6]. The amplitude $u_{nm}^{\rm LG}$ is defined by

$$u_{nm}^{\rm LG}(\rho,\phi,z) = C_{nm}^{\rm LG} \times (1/w)\exp(-ik\rho^2/2R)\exp(-\rho^2/w^2) \times \exp[-i(n+m+1)\psi]\exp[-i(n-m)\phi] \times (-1)^{\min(n,m)}(\rho\sqrt{2}/w)^{|n-m|} L_{\min(n,m)}^{|n-m|}(2\rho^2/w^2), \quad (4)$$

with

$$R(z) = (z_R^2 + z^2)/z, \quad (5)$$

$$\tfrac{1}{2}kw^2(z) = (z_R^2 + z^2)/z_R, \quad (6)$$

$$\psi(z) = \arctan(z/z_R), \quad (7)$$

where $C_{nm}^{\rm LG}$ is a normalization constant, $k = 2\pi/\lambda$ the wave number, z_R the Rayleigh range, and $L_p^l(x)$ the generalized Laguerre polynomial [6]. The more common indices p, l are related to n, m by $p = \min(n, m)$ and $l = n - m$.

3. Mode analysis

As pointed out, a LG mode with modified helicity is in general no longer a pure mode. Since the LG modes themselves form a complete set of helical modes this is still a useful basis for expanding the modified mode. There is a twofold infinite number of LG basis sets, since there are two free Gaussian beam parameters (e.g. Rayleigh range and z coordinate of the waist). If these are chosen equal to those of the incident beam, the mode decomposition is unique. The mode decomposition of a mode $u_{nm}^{\rm LG}$ whose helicity has been modified by Δl is then defined by the expansion coefficients

$$a_{nm,st} = \langle u_{st}^{\rm LG}|\ \exp(-i\Delta l\phi)\ |u_{nm}^{\rm LG}\rangle, \quad (8)$$

where the brackets denote integration in the transverse plane (ρ, ϕ). The relative weight of the modes is given by

$$I_{nm,st} = |a_{nm,st}|^2. \quad (9)$$

As can be seen from the ϕ integral, the only modes present for integer Δl are those that differ by Δl in helicity with the input mode, that is

$$s - t = n - m + \Delta l. \quad (10)$$

Table 1
Mode contents of a LG_{00} mode that has acquired a phase $\exp(-i\phi)$ due to a spiral phaseplate with $\Delta l = 1$. Indicated is the relative intensity in percentages of the LG_{st} modes; the mode index s is on the horizontal and t on the vertical axis

6							
5							0.79
4						1.17	
3					1.92		
2				3.68			
1			9.82				
0		78.5					
(00)	0	1	2	3	4	5	6

The coefficients $I_{nm,st}$ have been calculated numerically by calculating the integrals on a 100×100 mesh with a width and height equal to 8 times the Gaussian radius of the input beam. Table 1 gives the results for a LG_{00} (=TEM_{00}) beam passed through a spiral phaseplate with $\Delta l = 1$; the values in Table 1 represent $I_{00,st}$ as percentage of the incoming TEM_{00} intensity. Even though the input beam has a maximum at the axis, more than 78% ends up in the LG_{10} mode which is dark at the center. Increasing the order of the input mode gives even higher purity of the output. As an example, the LG_{30} mode is converted for 94% into the LG_{40} mode, leaving only 6% for the other modes.

Table 2 shows a few decompositions for modes passed through a plate with $\Delta l = 2$. A LG_{01} mode is converted into a pure LG_{10} mode, which is obvious since they have the same intensity distribution but inverted phase distributions. This pure mode conversion occurs for all transformations where the phase is inverted, which is the case if

$$\Delta l = -2(n - m). \quad (11)$$

In the other cases the decomposition can contain either a finite or an infinite number of modes.

The coefficients of Tables 1 and 2 can also be deduced analytically. Comparison of some of the coefficients shows that the errors in the numerical calculation are below 10^{-4}. Furthermore, the analytical results show that there is a distinction between even and odd Δl. For odd Δl, the number of modes in the decomposition is always infinite, all individual modes having the expected helicity (Eq. (10)). For even Δl, there are three situations. If Eq. (11) holds, there is only one mode in the output. If Δl is smaller than the

Table 2
Mode contents of a LG_{00}, LG_{01} and LG_{02} mode passed through a spiral phaseplate with $\Delta l = 2$ (notation as in Table 1)

(00)	0	1	2	3	4	5	6
6							
5							
4							3.33
3						5.00	
2					8.33		
1				16.67			
0			50.00				

(01)	0	1	2	3	4
2					
1					
0		100			

(02)	0	1	2	3	4
3					
2					
1		50.00			
0	50.00				

Table 3
Mode contents of a LG_{00} mode passed through a spiral phaseplate with $\Delta l = \frac{1}{2}$ (notation as in Table 1)

(00)	0	1	2	3	4	5	6
6	0.01	0.03	0.05	0.06	0.06	0.04	
5	0.03	0.07	0.09	0.08	0.05		0.32
4	0.08	0.14	0.14	0.09		0.48	0.15
3	0.25	0.27	0.17		0.78	0.23	0.12
2	0.81	0.45		1.49	0.38	0.17	0.08
1	3.54		3.98	0.75	0.27	0.11	0.05
0	40.5	31.8	2.25	0.48	0.14	0.05	0.02

right-hand side of Eq. (11), there is a finite number of modes present; if it is larger, the number of modes is infinite.

In Table 3 we show, as an illustration of the effect of non-integer Δl, the modal decomposition of a TEM_{00} laser beam after passing a spiral phaseplate with $\Delta l = \frac{1}{2}$. Here modes with all values of $(s - t)$ are present in the output. Only higher-order zero-helicity modes are forbidden by symmetry.

4. Diffraction theory

Since a mode whose helicity is modified is no longer a pure mode, its far-field intensity distribution will differ from that in the near-field. While the intensity distribution in the near-field is the same as that of the input mode, the far-field intensity distribution reflects the presence of a phase dislocation.

The propagation properties of the paraxial beam are decribed by scalar diffraction theory; the far-field is given by the Fraunhofer integral. Consider the central spot of the far field of a "spiralized" TEM_{00} beam. As long as Δl is integer, there is for each source point (ρ, ϕ) an accompanying source point $(\rho, \phi + \pi/\Delta l)$ which gives an equal but opposite contribution to the Fraunhofer integral at a point on the axis in the far field. The result is a dark spot at the center of the far-field. A similar argument predicts non-zero intensity at the center of the far-field of a helical beam that has been "despiralized".

For non-integer values of Δl there is a discontinuity of the phase along the step of the plate. The intensity distibution in this case is less simple to determine. Therefore we have calculated the far-field intensity distribution on a 128×128 mesh by taking the two-dimensional Fourrier transform of the input mode multiplied by $\exp(-i\Delta l\phi)$. The top row of Fig. 2 shows the result for a LG_{00} input mode at different values of Δl. At $\Delta l = \frac{1}{2}$, a phase singularity appears next to the center, visible as a zero point in the intensity. At $\Delta l = 1$, the singularity has moved to the center and the far-field resembles a LG_{10} mode. For higher half-integer Δl, the patterns shows several phase dislocations. At integer Δl they merge into one single phase dislocation of higher charge, giving a helical wavefront beam with a larger pitch. The bottom row of Fig. 2 shows the results for a LG_{01} input mode. For $\Delta l = 1$, intensity appears at the center which indicates that the beam has been despiralized. At $\Delta l = 2\frac{1}{2}$, two off-axis dislocations appear.

Note that the number of phase dislocations has to be integer, even for non-integer Δl. Thus, the number of dislocations changes discontinuously with step height. The non-integer nature of Δl is reflected in the fact that the intensity is low around a dislocation that appears for Δl just above integer, and rises when Δl is increased towards the next integer.

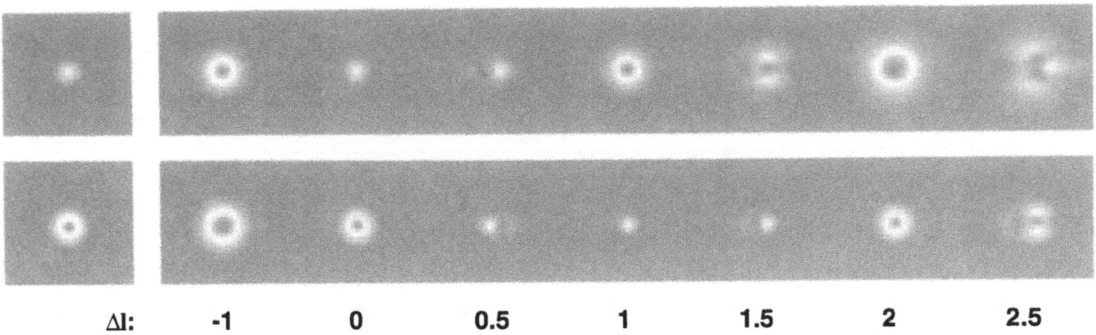

Fig. 2. Far-field calculations of (de)spiralized beams. The top row shows a LG_{00} (=TEM_{00}) beam passed through a spiral phaseplate with different heights of the step. The bottom row shows the same for an input LG_{01} mode.

5. Experiments

In Ref. [18] a spiral phaseplate has been described that was made using a dielectric coating, made for operation at λ = 10.6 μm. The coating thickness was constant in the radial direction, but increased by 6.3 μm over a full turn. In order to avoid the complexity of producing a micron-scale structure with a continuously changing thickness, especially so since we aim at a visible wavelength (λ = 633 nm), we have chosen to index-match a plate from acrylic (PMMA, $n = 1.49$) machined with normal tools. A milling tool of 1 mm diameter was used, removing material along radial lines from the edge to the center. This was done repetitively, each time rotating the plate by 5 degrees and moving the tool 10 μm down. This results in a step $h = 0.72$ mm or $\Delta l = 577$ at a wavelength of 633 nm. To bring the step down to about one wavelength, the plate was immersed into a liquid with nearly the same index of refraction. At λ = 633 nm, a difference in index of refraction $\Delta n = 8.7 \times 10^{-4}$ will result in a stepsize of one wavelength ($\Delta l = 1$). By tuning the temperature the effective step size could be tuned, due to the difference in temperature coefficient of the refractive index of the immersion liquid and the acrylic.

With this method the phaseplate, and in fact many other diffractive elements, can be produced with ordinary machining. Immersion has often been used to adjust the strength of optical elements, but to our knowledge this is the first use of immersion to produce diffractive optics. Of course it is impossible to also scale down the *transverse* structure to microscopic dimensions using immersion. In our plate this is apparent at the center. There should ideally be a infinitely small singular point, but the 1 mm diameter of the tool causes an area of about 1.5 mm diameter to be distorted.

To study the operation of the spiral phaseplate we used a setup as sketched in Fig. 3. The spiral phaseplate is put into a gold-coated brass cell with two BK7 windows, anti-reflection coated on the outside. The cell is filled with index-matching fluid of $n_D = 1.4922$ at 25°C [20]. The temperature of the cell is controlled with a Peltier element. It took about ten minutes for the liquid to adjust from one equilibrium temperature to a 1°C higher temperature.

The input beam is generated by a 633 nm HeNe laser with a Brewster-windowed discharge tube and external mirrors, separated by 83 cm. Laser polariza-

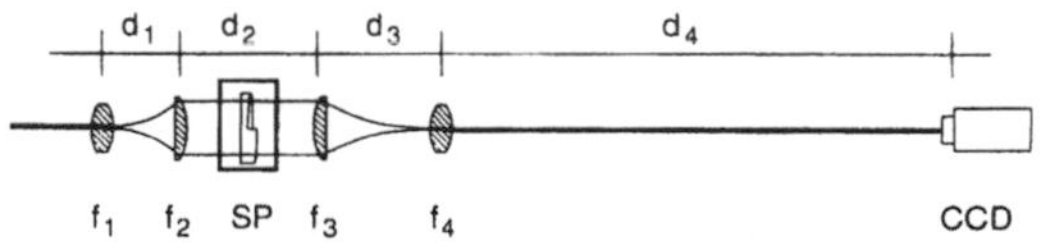

Fig. 3. Experimental setup used to convert a Laguerre-Gaussian mode into a mode with a different helicity using a spiral phaseplate. Input is a LG_{00} or LG_{01} mode, produced with a HeNe laser and a cylindrical-lens mode converter (not shown). A telescope (f_1, f_2) expands the beam to about 10 mm diameter. This beam passes through the cell in which the spiral phaseplate is immersed. Lens f_3 produces a waist about 20 mm in front of lens f_4, which projects the far-field image in the waist onto a CCD camera. The focal lengths of the lenses are f_1 = 20 mm, f_2 = 200 mm, f_3 = 300 mm and f_4 = 25 mm; the distances are d_1 = 0.22 m, d_2 = 0.40 m, d_3 = 0.34 m and d_4 = 1.53 m.

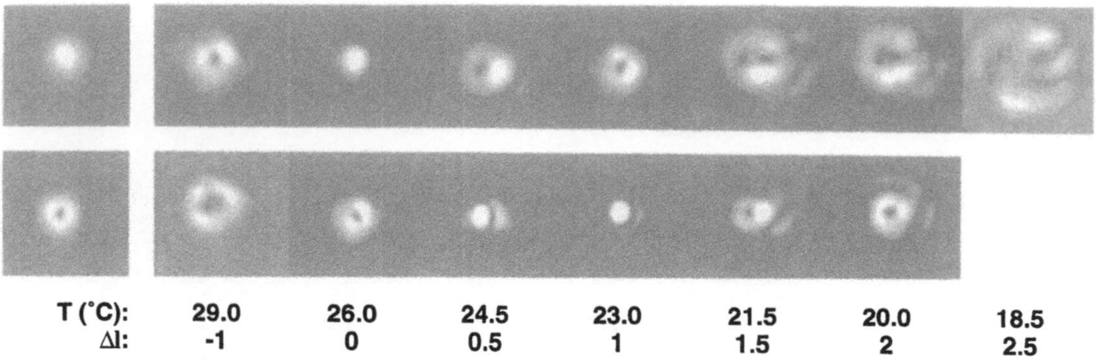

Fig. 4. Experimentally observed far-field patterns of a LG_{00} and LG_{01} laser beam passed through a spiral phaseplate. On the left the far-field image is shown when the phaseplate is removed from the setup; next to it are the far-field images at different temperatures of the cell.

tion is linear and several longitudinal modes may oscillate. Transverse-mode oscillation is normally in a TEM_{00} mode, but can be forced to a HG_{01} mode by means of a thin wire (20 μm diameter) in the cavity. This HG mode is converted into a LG mode with a cylindrical-lens mode converter outside the laser cavity. This part of the setup is identical to that used in Ref. [6], and produces either a LG_{00} mode ($l = 0$) or a LG_{01} mode ($l = -1$), selected by insertion or removal of the intra-cavity wire. This beam is enlarged to a diameter of about 10 mm and collimated, and is passed through the spiral phaseplate in the cell. The large diameter reduces the influence of the imperfection at the center of the plate. The output beam was focussed with a lens, forming the far-field image in the focal plane. A second lens projects this image enlarged onto a CCD camera. The resulting far-field image was recorded as a function of the temperature of the cell, both for an incident LG_{00} and LG_{01} mode.

In Fig. 4, top row, the results for the LG_{00} mode are shown. The leftmost picture is the far-field image of the beam before inserting the plate into the cell. It is followed by the images with the plate inserted into the cell, at different cell temperatures. At 26°C index-matching is perfect ($\Delta l = 0$). At 24.5°C a dark spot appears next to the maximum, which has moved to the center at 23°C, where it looks like a LG_{01} mode ($\Delta l = 1$). At larger values of Δl two dark spots appear (21.5°C), which merge to a larger, central hole at 20°C ($\Delta l = 2$). At 18.5°C a third dark spot appears.

The bottom row of Fig. 4 shows the results for the LG_{01} mode, which itself is already a helical mode. At 23°C the dark spot in the center has disappeared, so that the output resembles a LG_{00} mode, which indicates that the helicity of the wavefront has been removed by the spiral phase plate. At 20°C the mode looks like the input mode again, but will now have the opposite helicity ($\Delta l = 2$).

Note that, especially for large Δl, the output shows severe distortion. In an earlier experiment, using a home-made mixture of benzyl alcohol and ethylene glycol as index-matching fluid, this was much more pronounced. The distortion was partly caused by schlieren and other kinds of inhomogeneities. This is more severe for the home-made mixture due to its high viscosity and chemical instability. In the experiments of Fig. 4 the distortion is due to imperfections of the phaseplate (proportional to Δl), temperature inhomogeneities of the liquid (independent of Δl) and spherical abberations of the imaging optics, which is more severe for large beams. These aspects were not optimized; improvement of the performance should therefore be possible.

6. Conclusion

We have shown that a spiral phaseplate can be used to produce a helical wavefront laser beam. The mode decomposition of a transformed input mode has been determined, and the Fraunhofer diffraction pattern has been calculated. We have shown theoretically and experimentally that the plate transforms between beams with different helicity. Its design and fabrication has been described. The plate was produced using conventional machining and converted into a diffractive element by immersion. For producing a *solid* device, one could cover the helical surface with a nearly index-matched optical quality epoxy or thermoplastic and flatten the outer surfaces. Finally, since the plate directly transforms the orbital angular momentum of a beam, it might be interesting to study the mechanical consequences of this transformation by suspending the plate by a thin wire so as to form a torsional pendulum [11].

Acknowledgement

We thank K. Benning for producing the spiral waveplate. This work is part of the research program of the Foundation for Fundamental Research on Matter (FOM) and was made possible by financial support from the Netherlands Organization for Scientific Research (NWO).

References

[1] F.T. Arecchi, G. Giacomelli, P.L. Ramazza and S. Residori, Phys. Rev. Lett. 67 (1991) 3749.
[2] M. Brambilla, F. Battipede, L.A. Lugiato, V. Penna, F. Prati, C. Tamm and C.O. Weiss, Phys. Rev. A 43 (1991) 5090.
[3] V.I. Kruglov, Yu.A. Logvin and V.M. Volkov, J. Mod. Optics 39 (1992) 2277.
[4] G.A. Swartzlander, Jr. and C.T. Law, Phys. Rev.Lett. 69 (1992) 2503.
[5] E. Abramochkin and V. Volostnikov, Optics Comm. 102 (1993) 336.
[6] M.W. Beijersbergen, L. Allen, H.E.L.O. van der Veen and J.P. Woerdman, Optics Comm. 96 (1993) 123.
[7] I.V. Basistiy, V.Yu. Bazhenov, M.S. Soskin and M.V. Vasnetsov, Optics Comm. 102 (1993) 422.
[8] M. Harris, C.A. Hill and J.M. Vaughan, Optics Comm. 106 (1994) 161.
[9] M. Berry, in: Physics of Defects, eds. R. Balian, M. Kléman and J.-P. Poirier (North Holland, Amsterdam, 1981).
[10] S.J. van Enk and G. Nienhuis, Optics Comm. 94 (1992) 147.
[11] L. Allen, M.W. Beijersbergen, R.J.C. Spreeuw and J.P. Woerdman, Phys. Rev. A 45 (1992) 8185.
[12] S.J. van Enk and G. Nienhuis, J. Mod. Optics 41 (1994) 963.
[13] E. Abramochkin and V. Volostnikov, Optics Comm. 83 (1991) 123.
[14] N.R. Heckenberg, R. McDuff, C.P. Smith and A.G. White, Optics Lett. 17 (1992) 221.
[15] F.S. Roux, Applied Optics 32 (1993) 4191.
[16] C. Paterson, J. Mod. Optics 41 (1994) 757.
[17] L.W. Casperson, Laser Focus World 30(5) (1994) 223, and references therein.
[18] S.C. Tidwell, G.H. Kim and W.D. Kimura, Appl. Optics 32 (1993) 5222.
[19] M. Kristensen, M.W. Beijersbergen and J.P. Woerdman, Optics Comm. 104 (1994) 229.
[20] R.P. Cargille Laboratories Inc., Standard Immersion Liquid nr. 5040, $n_D = 1.4922$ @ 25°C.

PAPER 3.12

The generation of free-space Laguerre-Gaussian modes at millimetre-wave frequencies by use of a spiral phaseplate

G.A. Turnbull, D.A. Robertson, G.M. Smith, L. Allen, M.J. Padgett

School of Physics and Astronomy, North Haugh, The University of St. Andrews, St. Andrews, Fife, KY16 9SS, Scotland, UK

Received 20 September 1995; revised 16 November 1995; accepted 18 December 1995

Abstract

A spiral phaseplate is used at millimetre-wave frequencies to transform a free-space, fundamental Hermite-Gaussian mode into a Laguerre-Gaussian mode with an azimuthal phase component. A ray optics analysis confirms that the Laguerre-Gaussian beam produced has a total angular momentum equivalent to $\ell\hbar$ per photon.

1. Introduction

Laguerre-Gaussian (LG) modes, like Hermite-Gaussian (HG) modes, form a complete basis set for paraxial light beams. The former exhibit circular symmetry, the latter rectangular [1]. Two indices identify a given mode, and the modes are normally denoted LG_p^ℓ and HG_{mn}. For a Hermite-Gaussian mode m and n are the numbers of nodes in the x and y directions respectively. For a Laguerre-Gaussian mode, ℓ is the number of 2π cycles in phase around the circumference and $(p+1)$ the number of radial nodes. The amplitude, u_p^ℓ of the LG_p^ℓ mode in cylindrical co-ordinates is

$$u_p^\ell(r,\phi,z) \propto e^{-ikr^2/2R} e^{-r^2/w^2} e^{-i(2p+\ell+1)\psi} e^{-i\ell\phi} \times(-1)^p(r\sqrt{2}/w)^\ell L_p^\ell(2r^2/w^2), \quad (1)$$

where R is the wavefront radius of curvature, w is the radius for which the Gaussian term falls to $1/e$ of its on-axis value, ψ is the Gouy phase and L_p^ℓ a generalised Laguerre polynomial. The azimuthal phase term, $e^{i\ell\phi}$, distinguishes the Laguerre-Gaussian modes from the Hermite-Gaussian modes. This phase term creates helical wavefronts for the Laguerre-Gaussian modes in contrast to the planar wavefronts of the Hermite-Gaussian modes [2]. Laguerre-Gaussian modes possess an orbital angular momentum of $\ell\hbar$ per photon [3], quite distinct from the spin angular momentum associated with the polarisation state. This orbital angular momentum has been recently demonstrated through an optical interaction with microscopic particles [4].

In recent years, the production of Laguerre-Gaussian modes at optical frequencies has attracted considerable interest. Laguerre-Gaussian laser beams may be produced directly [5] or by the conversion of Hermite-Gaussian modes. To date, three different classes of mode converter have been demonstrated. Two of these, spiral phaseplates [6] and computer generated holographic converters [7,8], introduce the azimuthal phase term to a HG_{00} beam. The spiral phaseplate technique has also been employed within a waveguide configuration incorporating a 10 GHz source [9]. In all these devices a screw phase-dislocation, produced on-axis, causes destructive interference leading to the characteristic ring intensity pattern in the far field. The spiral phaseplate may also be used to convert between any two LG_p^{ℓ} modes separated by $e^{i\ell\phi}$ phase terms. In general, the purity of Laguerre-Gaussian modes produced by these methods is limited by the co-production of higher order modes. The other class of converter is the cylindrical-lens mode converter [10] which converts higher order Hermite-Gaussian modes to the corresponding Laguerre-Gaussian mode. Unlike the spiral phaseplate and the holographic converter, this method can produce pure Laguerre-Gaussian modes.

For a Laguerre-Gaussian mode, the orbital angular momentum in the beam is equivalent to $\ell\hbar$ per photon. Consequently, for a fixed power, the angular momentum in the beam is proportional to the wavelength; unlike linear momentum, h/λ per photon, where for a fixed power the linear momentum in the beam is wavelength independent.

In this paper we extend the production of free-space, Laguerre-Gaussian modes to millimetre-wave frequencies ($\sim$ 100 GHz), where the wavelength is $\sim 10^4$ times that at optical frequencies. Hence, for the same power, the orbital angular momentum is also $\sim 10^4$ time larger, which opens the possibility for observing the transfer of angular momentum to a macroscopic object. We demonstrate the use of a spiral phaseplate to convert the fundamental Hermite-Gaussian to higher order Laguerre-Gaussian modes. The phaseplate is preferable to the cylindrical lens converter because of the relative difficulty of producing high-order, free-space, Hermite-Gaussian beams at millimetre-wave frequencies.

The total angular momentum, J_z, of a Laguerre-Gaussian beam is the sum of orbital and spin angular momenta [3]. Thus for left-hand or right-hand circularly polarised beams $J_z = l \pm 1$. The Hermite-Gaussian mode converted in this work has a well-defined linear polarisation and consequently the total angular momentum in the beam is due entirely to orbital angular momentum.

2. Ray optics analysis of a spiral phaseplate mode converter

The spiral phaseplate (Fig. 1) has one planar and one spiral surface. The spiral surface forms one

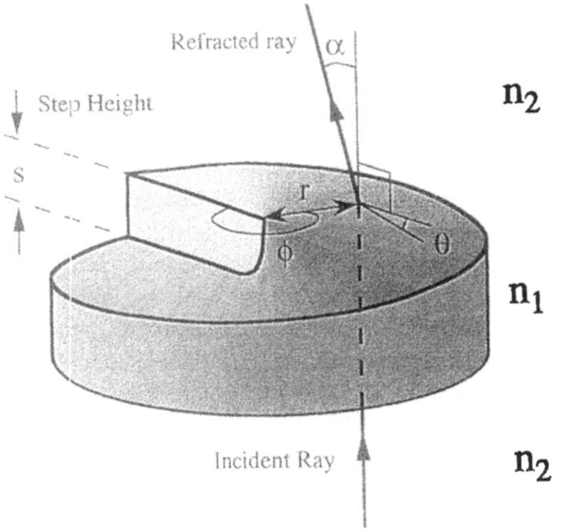

Fig. 1. The spiral phaseplate, showing the refraction of a single ray upon transmission.

period of a helix, with a step discontinuity. Upon transmission through the phaseplate, a beam of wavelength λ is subject to a phase delay, ψ, which depends on the azimuthal angle, ϕ, where

$$\psi = \frac{(n_1 - n_2)s}{\lambda}\phi, \tag{2}$$

n_1 and n_2 are the refractive indices of the phaseplate and surrounding media respectively and s is the physical step height at $\phi = 0$. For a Laguerre-Gaussian mode, the total phase delay around the phaseplate must be an integer multiple of 2π, i.e. $2\pi\ell$. Thus, to produce a Laguerre-Gaussian mode, the physical height of the step in the spiral phaseplate is given by

$$s = \frac{l\lambda}{(n_1 - n_2)}. \tag{3}$$

When the step height is not an integer number of wavelengths, the phase of the beam is discontinuous at the step and this is observed as a break in the ring intensity pattern. Beijersbergen et al. [6] have modelled the detuning of the step height through the transition from one Laguerre-Gaussian mode to another. In their small-angle approximation, the converter only changes the phase and not the intensity of the beam. The annular intensity pattern arises from the far field diffraction of the beam's screw dislocation. However, the beam produced is not a pure mode, but an infinite superposition of Laguerre-Gaussian modes. The conversion from the HG_{00} to the LG_0^1 mode was calculated to be 78% efficient.

A rigorous analysis of the mechanical interaction between the beam and the spiral phaseplate requires the solution of Maxwell's equations at the boundaries of the phaseplate. However, as with many optical and quasi-optical systems, a ray optics analysis gives useful insight into the nature of the interaction. In this paper, we use a ray optics model (Fig. 1) to gain insight into how the orbital angular momentum content of the beam arises from the mode converter. Although the orbital angular momentum is a property of the beam as a whole, it is useful to consider this in terms of the equivalent angular momentum per photon.

Consider a ring of radius r, projected on the spiral surface. The angle, θ, of the local azimuthal slope of the spiral surface is given by

$$\tan\theta = \frac{s}{2\pi r}. \tag{4}$$

A ray parallel to, but a distance r from, the optical axis will be refracted as it emerges from the spiral surface. The deflection angle, α, may be found using Snell's Law:

$$n_2 \sin(\theta + \alpha) = n_1 \sin(\theta). \tag{5}$$

Before refraction, the beam has a linear momentum of $n_2 h/\lambda$ per photon. After refraction, there is a component of linear momentum in the azimuthal direction, p_ϕ, given by

$$p_\phi = n_2 \frac{h}{\lambda} \sin\alpha. \tag{6}$$

To achieve this there is a transfer of angular momentum, L, between the spiral phaseplate and the beam of light of

$$L = r p_\phi = n_2 \frac{h}{\lambda} r \sin\alpha \tag{7}$$

per photon in the beam.

Let us consider the small-angle case where (4), (5) and (7) reduce to

$$\theta \approx \frac{2}{2\pi r}, \tag{8}$$

$$n_2(\theta + \alpha) \approx n_1 \theta, \tag{9}$$

$$L \approx n_2 \frac{h}{\lambda} r\alpha. \tag{10}$$

Combining Eqs. (8), (9) and (10) with s set by Eq. (3) (the condition for a Laguerre-Gaussian mode), the angular momentum exchanged, L, between the light beam and the phaseplate is

$$L \approx \hbar \frac{s(n_1 - n_2)}{\lambda} \approx \ell\hbar \tag{11}$$

per photon in the beam.

This agrees with the result for Laguerre-Gaussian beams derived from the analysis of Maxwell's equations [3].

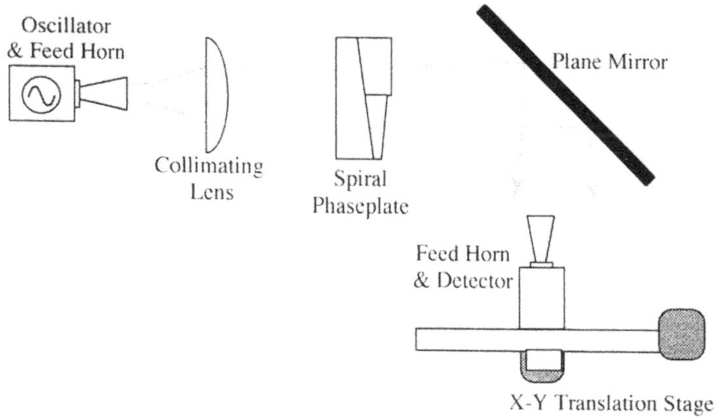

Fig. 2. The experimental configuration for obtaining Laguerre-Gaussian modes at millimetre-wave frequencies.

As r decreases it is unreasonable to approximate tan $\theta \approx \theta$ and the small-angle calculation for the angular momentum diverges from the exact result. When $r = \ell\lambda$ the difference is approximately 10%. Therefore, we require most of the energy in the incident beam to fall in the region $r > \ell\lambda$. This is easily satisfied in the optical regime, where $\ell\lambda$ is small compared with the typical beam radius. The criterion becomes more important when the wavelength is increased to the millimetre-wave range. Although it is impractical to scale the beam diameter by the same amount as wavelength, it is not to difficult to ensure that most of the beam power satisfies the small-angle approximation; this occurs when $w > \ell\lambda$.

The intensity distribution of our experimental results is in good agreement with theory and indicates that most of the energy is being transformed into a single Laguerre-Gaussian mode. It therefore seems reasonable to assume that the prediction of the ray optics model for the orbital angular momentum content of the beam of $\ell\hbar$ per photon is itself valid providing $w > \ell\lambda$.

It follows from (7) that for a fixed step height and constant power, the angular momentum transferred to the beam is independent of the wavelength. Consequently, a spiral phaseplate converter with a step height of a few millimetres may be used to produce an optical beam with a large angular momentum. However, the efficiency of coupling into a single Laguerre-Gaussian mode falls significantly as l increases [6], hence, the generated beam is not a pure mode.

3. Experimental configuration

Fig. 2 shows the experimental configuration to produce millimetre-wave, free-space, Laguerre-Gaussian modes. The source was an InP Gunn diode oscillator with a peak output power of 10–20 mW. Adjusting the dimensions of the resonant cavity tuned the linearly polarised output from 72 to 95 GHz [11]. The circular-aperture, corrugated feed-horn produced a ~ 98% pure HG_{00} beam with Rayleigh range of 50 mm [12]. A polyethylene lens of focal length 120 mm collimated the beam with $w \approx 25$ mm. We see that $w \gg \ell\lambda$ and therefore the small-angle approximation is valid in this case. The phaseplate was also made of polyethylene, which has a refractive index of 1.52 at millimetre-wave frequencies [13]. Two different phaseplates were used, one to generate the LG_0^1 mode and the other to generate the LG_0^2 mode. The step heights were 6.7 mm and 13.4 mm respectively to give a single and a double wavelength step at 86 GHz. The planar surface of the phaseplate and both surfaces of the collimating lens were cut with an anti-reflection texture of quarter-wavelength deep concentric grooves.

An aluminium mirror reflected the output from the phaseplate onto a detector mounted on an x-y scanning stage placed in the far field of the converter. The detector used was an Anritsu MP81B/ML83A with an identical feed horn to that on the oscillator. The antenna pattern of the horn is Gaussian in form, and so the measured intensity profile is the convolution of the true far field diffraction pattern and a Gaussian point spread function. The x-y scanning stage and detector were computer controlled to measure a 50×50 grid over a square area with a side of 100 mm. The readings were transferred to *Mathematica* [14], in which they were interpolated and displayed as density plots.

4. Results

Fig. 3(a) shows the result of the conversion from HG_{00} to LG_0^1. The central minimum, a characteristic of the Laguerre-Gaussian mode, is well defined. Fig. 3(b) shows the corresponding result for the LG_0^2 mode. As expected, the radius of maximum intensity [15] of the LG_0^2 is $\sqrt{2}$ times that of the LG_0^1. The linear polarisation state of the Laguerre-Gaussian beams was demonstrated using a wire-grid polariser, with which the beam could be completely attenuated. This confirms that, unlike previous waveguide-based work in the microwave regime [9] there is no combination of the spin polarisation and the orbital angular momentum in our case. For both of the generated Laguerre-Gaussian modes, "hot-spots" were observed in the ring with an increased intensity of about 10%. Two possible sources of these have been modelled. One is an imperfection at the centre of the phaseplate, caused by the finite size of the machining

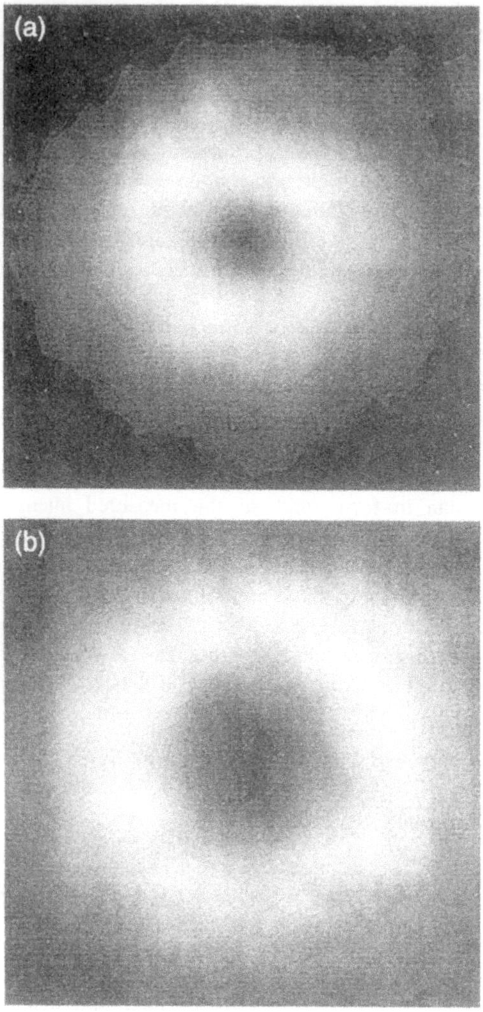

Fig. 3. Far-field intensity distribution for the observed Laguerre-Gaussian modes (a) LG_0^1 and (b) LG_0^2.

tool; the other is slight misalignment of the axis of the HG_{00} beam and the axis of the phaseplate. The magnitude of the observed ''hot-spots'' are consistent with the precision of our experimental configuration.

5. Conclusions

In this paper we have used a spiral phaseplate at millimetre-wave frequencies, to transform a free-space, fundamental Hermite-Gaussian mode into Laguerre-Gaussian modes with an azimuthal phase component. As with earlier work, in the optical regime [6], most of the energy of the incident mode can be converted into a single Laguerre-Gaussian mode, the order of which is set by the step height of the spiral phaseplate. A ray optics analysis of the spiral phaseplate confirms that, in the small-angle approximation, the Laguerre-Gaussian mode produced has a total angular momentum equivalent to $\ell\hbar$ per photon in the beam.

We believe that the comparatively large angular momentum associated with free-space Laguerre-Gaussian modes in the millimetre-wave range, opens the possibility of observing the transfer of orbital angular momentum from an electro-magnetic wave to a macroscopic object. Previously, angular momentum has been transferred, within a waveguide, from a 10 GHz beam to a suitable antenna. However, complications associated with near-field diffraction and the apparent interchange of spin and orbital angular momentum prevented quantitative agreement between theory and observations [9].

For the free space modes generated in this work we calculate that for a 20 mW Laguerre-Gaussian $(l = 2)$ beam, at 86 GHz, the total angular momentum flow per second is 7.4×10^{-14} kg m^2s^{-2}. Total absorption of such a beam and its associated angular momentum by a 4 mm diameter homogeneous sphere of density 1000 kg m^{-3} would result in an angular acceleration of the particle of $0.03° \, s^{-2}$. A suitable material for such a sphere is ferrite loaded epoxy resin, which has a high absorption and can be easily set into any required shape.

Acknowledgements

We would like to acknowledge Willie Smith for his skill and enthusiasm in producing a number of the spiral phaseplates. Miles Padgett acknowledges the support of the Royal Society of Edinburgh and the Royal Society. Les Allen was working at JILA, University of Colorado during the preparation of this manuscript.

References

[1] A.E. Siegman, Lasers (University Science Books, Mill Valley, 1986) sec. 17.5, pp. 685–695.
[2] J.M. Vaughan and D.V. Willetts, Optics Comm. 30 (1979) 263.
[3] L. Allen, M.W. Beijersbergen, R.J.C. Spreeuw and J.P. Woerdman, Phys. Rev. A 45 (1992) 8185.
[4] H. He, M.E.J. Friese, N.R. Heckenberg and H. Rubinsztein-Dunlop, Phys. Rev. Lett. 75 (1995) 826.
[5] M. Harris, C.A. Hill and J.M. Vaughan, Optics Comm. 106 (1994) 161.
[6] M.W. Beijersbergen, R.P.C. Coerwinkel, M. Kristensen and J.P Woerdman, Optics Comm. 112 (1994) 321.
[7] N R Heckenberg, R McDuff, C P Smith and A G White, Optics Lett. 17 (1992) 221.
[8] N.R. Heckenberg, R. McDuff, C.P. Smith, H. Rubinsztein-Dunlop and M.J. Wegener, Opt. Quant. Electron. 24 (1992) S951.
[9] M.Kristensen, M.W. Beijersbergen and J. P. Woerdman, Optics Comm. 104 (1994) 229.
[10] M.W. Beijersbergen, L. Allen, H.E.L.O. van der Veen and J.P. Woerdman, Optics Comm. 96 (1993) 123.
[11] G.M. Smith, TEO's at mm-wave frequencies and their characterisation using quasi-optical techniques. Ph.D. Thesis, St. Andrews (1990).
[12] R.J.Wylde, Proc. IEE, part H, 13 (1984) 258.
[13] J.C.G. Lesurf, Millimetre-wave Optics, Devices and Systems (Adam Hilger /IOP, 1990).
[14] Wolfram Research, Inc., Mathematica, Version 2.2, Champaign, Illinois, USA (1994).
[15] M.J Padgett and L. Allen, Optics Comm. 121 (1995) 36.

Section 4

OPTICAL FORCES AND TORQUES ON PARTICLES

In 1986 Ashkin *et al.* (4.1, **Paper 4.1**) published their seminal paper on optical tweezers which describes the way in which a tightly focused laser beam may be used to trap and manipulate micron-sized particles in three dimensions. The paper has received over 400 scientific citations and, apart from its great importance in optical and atomic physics, has spawned a multitude of biological experiments including measurements of individual muscle forces, stretching of DNA and, when combined with other lasers, the cutting and manipulation of biological materials (4.2).

When placed in an electric field gradient, dielectric materials experience a force directed towards the region of highest field. The direction of the force does not depend upon the polarity of the field and so the AC field gradient associated with a tightly focused laser produces a force on a dielectric particle which moves it in the direction of the highest intensity of the beam. An additional force arises from light scattering which causes a re-direction of the input light's linear momentum into all directions. The goal in well-designed optical tweezers is to obtain a gradient force sufficiently large to overcome the scattering force and allow the formation of a 3D trap. The biggest contribution to the achievement of this goal comes from an extremely tight focusing of the trapping laser beam. Additional forces arise from the absorption of the light, which results in the transfer of both linear and angular momentum from the light beam to the particle. All of these forces have been observed in optical tweezers.

Ashkin showed by use of a ray-optical picture, that it was the refraction of the off-axis rays which were equivalent to the gradient force and that these could lift a transparent object towards the beam focus. The on-axis rays did not give rise to a gradient force, only a scattering force. He suggested that, in optical tweezers, an annular intensity profile should be preferable to an $HG_{0,0}$ mode (4.3) as this would result in a greater overall trapping force. Laguerre–Gaussian (LG) beams have been used for this purpose and a number of groups have demonstrated measurable improvements (4.4, 4.5). The on-axis intensity null associated with LG beams has also been used in optical tweezers to trap particles with a refractive index lower than that of the surrounding fluid (4.6). However, it should be emphasised that in all these applications the orbital angular momentum of the LG mode plays no role; it is the consequences of the annular intensity distribution which are being investigated.

The first use of a particle trapped in a Laguerre–Gaussian mode, and its associated orbital angular momentum, was reported in 1995 by He *et al.* (4.7, 4.8, **Paper 4.2**). They observed that when a tightly focused LG mode ($l = 3$) was directed at ceramic, micron-sized particles, absorption of the light and its associated angular momentum set the particles into rotation about their axis. In this experiment, the particles were confined to the beam axis in two dimensions; the base of the sample cell provided containment in the third dimension. Subsequently, the same group showed that the additional spin angular momentum obtained by making the beam circularly polarised could cause the rotation of the particle to speed up or slow down (4.9, **Paper 4.3**).

In 1997, Simpson *et al.* (4.10, **Paper 4.4**) trapped a slightly absorbing Teflon particle at the focus of a circularly polarised LG mode. The circular polarisation corresponded to a spin angular momentum of $\hbar$ per photon and the $l = 1$ mode possessed an orbital angular momentum of $\hbar$ per photon. The particle was trapped in three dimensions in true optical tweezers and the absorption of the light gave a transfer mechanism for both the spin and orbital angular momentum from the beam to the particle. The insertion of a half waveplate into the beam changed the sign of the spin angular momentum, such that spin and orbital AM acted additively to give $2\hbar$ per photon or in opposite directions to give zero total angular momentum. The observed stop-start rotation of the particle confirmed the magnitude and sign of the orbital angular momentum term and, for $l = 1$, demonstrated the mechanical equivalence of spin and orbital angular momentum.

Optical tweezers have also been used to perform the microscopic equivalent of Beth's 1936 experiment (4.11, **Paper 1.2**), where the spin angular momentum associated with circular polarisation has been transferred to a birefringent element. Friese *et al.* (4.12) trapped microscopic fragments of calcite in circularly polarised optical tweezers. For calcite, the birefringence is high enough so that even microscopic particles change the polarisation state of the transmitted light. As with the Beth experiment, transformation of the light polarisation from circular to linear results in an angular momentum transfer of $\hbar$ per photon. They also demonstrated that a change to elliptically polarised light reduced the torque on the particle accordingly (4.13).

The interaction of orbital angular momentum with a cylindrical lens mode converter is the direct analogue to the transfer of spin angular momentum to a birefringent particle. To date no one has demonstrated the transfer of orbital angular momentum by the direct re-phasing of the component modes. However, light scattering offers a possible transfer mechanism for orbital angular momentum. While observing the behaviour of a non-absorbing metallic particle near the focus of a circularly polarised LG mode, O'Neil and Padgett (4.14) noted the particle rotated around the beam axis. It was found that the sense of circular polarisation made no difference to the motion of the particle so, clearly, spin angular momentum was not being transferred. In any case, such a transfer would have led to the particle being set into rotation about its own axis rather than that of the beam rather than about its own. They concluded that the scattering of light had caused the rotation. Even at the beam waist, the diameter of the LG mode exceeds that of a particle and the gradient force confines a dielectric particle to the annulus of maximum intensity which is at a uniform radius from the beam axis. If the particle were birefringent, then the change in the polarisation of the transmitted light would result in a subsequent transfer of spin angular momentum and the production of a torque acting on the particle about its own axis. If scattering is the predominant transfer mechanism, then the orbital angular momentum produces a torque on the particle about the beam axis. This spinning and orbiting of the particle is, therefore, identifiable with the spin and orbital angular momentum components respectively. A circularly polarised LG mode ($l = 8$) has been used to demonstrate that spin and orbital angular momentum may be observed to act independently within the same tweezers configuration (4.15, **Paper 2.7**).

REFERENCES

4.1 A Ashkin, JM Dziedzic, J E Bjorkholm and S Chu, 1986, *Opt. Lett.* **11** 288.
4.2 JE Molloy and MJ Padgett, 2002, *Contemp. Phys.* **43** 241.
4.3 A Ashkin, 1992, *Biophys. J.* **61** 569.
4.4 MEJ Friese, H Rubinsztein-Dunlop, NR Heckenberg and EW Dearden, 1998, *Appl. Opt.* **35** 7112.
4.5 AT O'Neil and MJ Padgett, 2001, *Opt. Commun.* **193** 45.
4.6 KT Gahagan and GA Swartzlander, 1996, *Opt. Lett.* **21** 827.
4.7 H He, NR Heckenberg and H Rubinsztein-Dunlop, 1995, *J. Mod. Opt.* **42** 217.
4.8 H He, MEJ Friese, NR Heckenberg and H Rubinsztein-Dunlop, 1995, *Phys. Rev. Lett.* **75** 826.
4.9 MEJ Friese, J Enger, H Rubinsztein-Dunlop and NR Heckenberg, 1996, *Phys. Rev. A* **54** 1593.
4.10 NB Simpson, K Dholakia, L Allen and MJ Padgett, 1997, *Opt. Lett.* **22** 52.
4.11 RA Beth, 1936, *Phys. Rev.* **50** 115.
4.12 MEJ Friese, TA Nieminen, NR Heckenberg and H Rubinsztein-Dunlop, 1998, *Nature* **394** 348.
4.13 MEJ Friese, TA Nieminen, NR Heckenberg and H Rubinsztein-Dunlop, 1998, *Opt. Lett.* **23** 1.
4.14 AT O'Neil and MJ Padgett, 2000, *Opt. Commun.* **185** 139.
4.15 AT O'Neil, I MacVicar, L Allen and MJ Padgett, 2001, *Phys. Rev. Lett.* **88** 053601.

Observation of a single-beam gradient force optical trap for dielectric particles

A. Ashkin, J. M. Dziedzic, J. E. Bjorkholm, and Steven Chu

AT&T Bell Laboratories, Holmdel, New Jersey 07733

Received December 23, 1985; accepted March 4, 1986

Optical trapping of dielectric particles by a single-beam gradient force trap was demonstrated for the first reported time. This confirms the concept of negative light pressure due to the gradient force. Trapping was observed over the entire range of particle size from 10 μm to ~25 nm in water. Use of the new trap extends the size range of macroscopic particles accessible to optical trapping and manipulation well into the Rayleigh size regime. Application of this trapping principle to atom trapping is considered.

We report the first experimental observation to our knowledge of a single-beam gradient force radiation-pressure particle trap.[1] With such traps dielectric particles in the size range from 10 μm down to ~25 nm were stably trapped in water solution. These results confirm the principles of the single-beam gradient force trap and in essence demonstrate the existence of negative radiation pressure, or a backward force component, that is due to an axial intensity gradient. They also open a new size regime to optical trapping encompassing macromolecules, colloids, small aerosols, and possibly biological particles. The results are of relevance to proposals for the trapping and cooling of atoms by resonance radiation pressure.

A wide variety of optical traps based on the basic scattering and gradient forces of radiation pressure have been demonstrated or proposed for the trapping of neutral dielectric particles and atoms.[2–4] The scattering force is proportional to the optical intensity and points in the direction of the incident light. The gradient force is proportional to the gradient of intensity and points in the direction of the intensity gradient. The single-beam gradient force trap is conceptually and practically one of the simplest radiation-pressure traps. Although it was originally proposed as an atom trap,[1] we show that its uses also cover the full spectrum of Mie and Rayleigh particles.

It is distinguished by the feature that it is the only all-optical single-beam trap. It uses only a single strongly focused beam in which the axial gradient force is so large that it dominates the axial stability. In the only previous single-beam trap, the so-called optical levitation trap, the axial stability relies on the balance of the scattering force and gravity.[5] In that trap the axial gradient force is small, and if one turns off or reverses the direction of gravity the particle is driven out of the trap by the axial scattering force.

There were also relevant experiments using gradient forces on Rayleigh particles that did not strictly involve traps, in which liquid suspensions of submicrometer particles acted as an artificial nonlinear optical Kerr medium.[6]

The physical origin of the backward gradient force in single-beam gradient force traps is most obvious for particles in the Mie size regime, where the diameter is large compared with λ. Here one can use simply ray optics to describe the scattering and optical momentum transfer to the particle.[7,8] In Fig. 1a) we show the scattering of a typical pair of rays A of a highly focused beam incident upon a 10-μm lossless dielectric sphere, for example. The principal part of the momentum transfer from the incident light to the particle is due to the emergent rays A′, which are refracted by the particle. Successive surface reflections, such as R_1 and R_2, contribute a lesser scattering. For a glass particle in water the effective index m, equal to the index of the particle divided by the index of the medium, is about 1.1 to 1.2, and the sphere acts as a weak positive lens. If we consider the direction of the resulting forces F_A on the particle that are due to refraction of rays A in the weak-lens regime, we see as shown in Fig. 1a) that there is a substantial net backward trapping-force component toward the beam focus.

Figure 2 sketches the apparatus used for trapping Mie or Rayleigh particles. Spatially filtered argon-laser light at 514.5 nm is incident upon a high-numerical-aperture (N.A. 1.25) water-immersion microscope objective, which focuses a strongly convergent downward-directed beam into a water-filled glass cell. Glass Mie particles are introduced into the trap by an auxiliary vertically directed holding beam,[5] which lifts particles off the bottom of the cell and manipulates them to the focus. Rayleigh particles are simply dispersed in water solution at reasonable concentrations and enter the trapping volume by Brownian diffusion. A microscope M is used to view the trapped particles visually off a beam splitter S or by recording the 90° scatter with a detector D.

Figure 1b) is a photograph of a 10-μm glass sphere of index about 1.6 trapped and levitated just below the beam focus of a ~100-mW beam. The picture was taken through a green-blocking filter using the red fluorescence of the argon laser beam in water in order to make the trajectories of the incident and scattered

a)

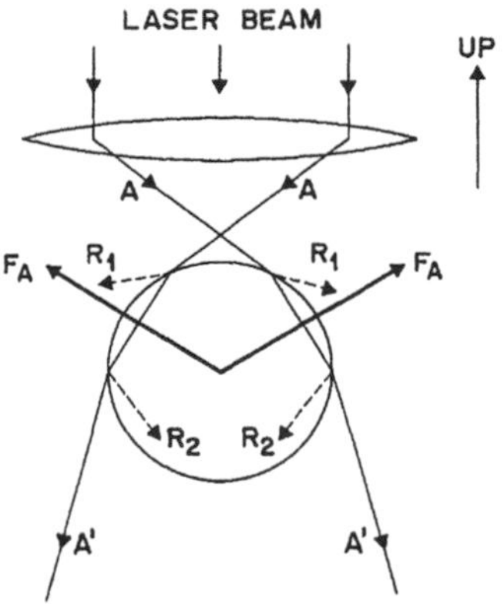

b)

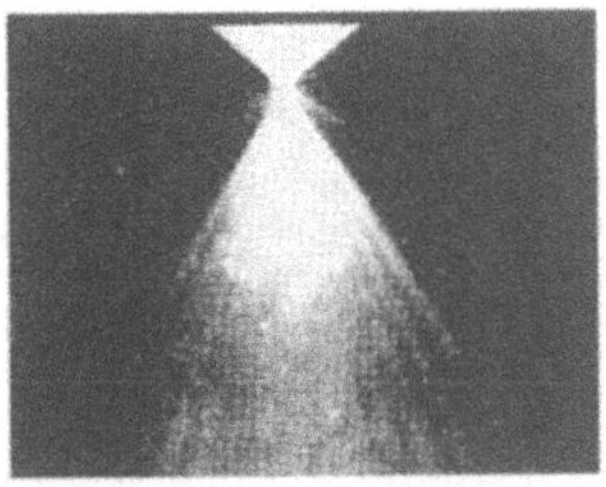

Fig. 1. a) Diagram showing the ray optics of a spherical Mie particle trapped in water by the highly convergent light of a single-beam gradient force trap. b) Photograph, taken in fluorescence, of a 10-μm sphere trapped in water, showing the paths of the incident and scattered light rays.

beams visible. The sizable decrease in beam angle of the scattered light, which gives rise to the backward force, is clearly seen. The stria in the forward-scattered light arise from the usual Mie-scattering ring pattern.

Next consider the possibility of single-beam trapping of submicrometer Rayleigh particles whose diameter $2r$ is much less that λ. Although we are now in the wave-optic regime, we will again see the role of the strong axial gradient in producing a net backward axial force component. For Rayleigh particles in a medium of index n_b the scattering force in the direction of the incident power is $F_{scat} = n_b P_{scat}/c$, where P_{scat} is the power scattered.[9] In terms of the intensity I_0 and effective index m

$$F_{scat} = \frac{I_0}{c}\frac{128\pi^5 r^6}{3\lambda^4}\left(\frac{m^2-1}{m^2+2}\right)^2 n_b. \qquad (1)$$

The gradient force F_{grad} in the direction of the intensity gradient for a spherical Rayleigh particle of polarizability α is[6]

$$F_{grad} = -\frac{n_b}{2}\alpha\nabla E^2 = -\frac{n_b^3 r^3}{2}\left(\frac{m^2-1}{m^2-2}\right)\nabla E^2. \qquad (2)$$

This Rayleigh force component, in analogy with the gradient force for Mie particles, can be related to the lenslike properties of the scatterer.

As for atoms,[1] the criterion for axial stability of a single-beam trap is that R, the ratio of the backward axial gradient force to the forward-scattering force, be greater than unity at the position of maximum axial intensity gradient. For a Gaussian beam of focal spot size w_0 this occurs at an axial position $z = \pi w_0^2/\sqrt{3}\ \lambda$, and we find that

$$R = \frac{F_{grad}}{F_{scat}} = \frac{3\sqrt{3}}{64\pi^5}\frac{n_b^2}{\left(\frac{m^2-1}{m^2+2}\right)}\frac{\lambda^5}{r^3 w_0^2} \geq 1, \qquad (3)$$

where λ is the wavelength in the medium. This condition applies only in the Rayleigh regime where the particle diameter $2r \lesssim 0.2\lambda \cong 80$ nm. In practice we require R to be larger than unity. For example, for polystyrene latex spheres in water with $m = 1.65/1.33 = 1.24$ and $2w_0 = 1.5\lambda = 0.58\ \mu$m we find for $R \geq 3$ that $2r \leq 95$ nm. Thus with this choice of spot size we meet the stability criterion over the full Rayleigh regime. The fact that $R < 3$ for $2r > 95$ nm does not necessarily imply a lack of stability for such larger particles since we are beyond the range of validity of the formula. Indeed, as we enter the transition region to Mie scattering we expect the ray-optic forward-scattering picture to be increasingly valid. As will be seen experimentally we have stability from the Rayleigh regime, through the transition region, into the full Mie regime. For silica particles in water with $m = 1.46/1.33 = 1.10$ and $2w_0 = 0.58\ \mu$m we find for $R \geq 3$ that $2r \leq 126$ nm. For high-index particles with $m \equiv 3.0/1.33 = 2.3$ we find that $2r \lesssim 61$ nm.

The stability condition on the dominance of the backward axial gradient force is independent of power and is therefore a necessary but not sufficient condition for Rayleigh trapping. As an additional sufficient trapping condition we have the requirement[1] that the Boltzmann factor $\exp(-U/kT) \ll 1$, where $U = n_b \alpha E^2/2$ is the potential of the gradient force. As was previously pointed out,[6] this is equivalent to requiring that the time to pull a particle into the trap be much less than the time for the particle to diffuse out of the trap by Brownian motion. If we set $U/kT \geq 10$,

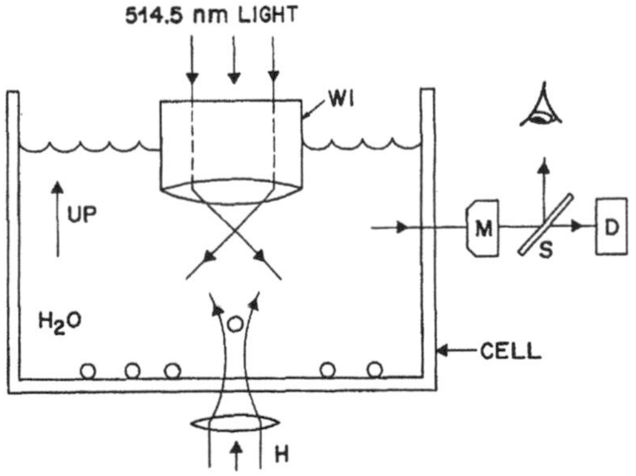

Fig. 2. Sketch of the basic apparatus used for the optical trapping of Mie and Rayleigh particles in water by means of a single-beam gradient force radiation-pressure trap.

for example, and use a power of ~1.5 W focused close to the limiting spot diameter of 0.58 μm ≅ 1.5λ, we find for silica that the minimum theoretical particle size that satisfies this condition is $2r = 19$ nm. For polystyrene latex, the minimum particle size that can be trapped under these conditions is $2r = 14$ nm. With a high-index particle of $m = 3/1.33 = 2.3$ the theoretical minimum size is $2r \cong 9$ nm.

Additional experiments were performed on individual colloidal polystyrene latex particles in water. Unfortunately the particles exhibit a form of optical damage at high optical intensities. For 1.0-μm spheres with a trapping power of a fraction of a milliwatt, particles survived for tens of minutes and then shrank in size and disappeared. Spheres of 0.173 μm were trapped for several minutes with a power of a milliwatt before being lost. Particles of 0.109-μm diameter required about 12–15 mW and survived about 25 sec. With 85- and 38-nm latex particles the damage was so rapid that it was difficult to observe the scattering reliably. It was nevertheless clear that trapping occurred over full size range from Mie to Rayleigh particles.

The remarkable uniformity of latex particles was evident from the small variation of ±15% in the 90° scatter of 0.109-μm particles. Since the scattering is closely Rayleigh this corresponds to a diameter variation of ±2.4%. Subsequently we determined the size of unknown silica Rayleigh particles by comparing their scatter with the scatter from the 0.109-μm particles taken as a standard, using Eq. (1). Although the 0.109-μm particles are not strictly Rayleigh, one can make a modest theoretical correction[10] of ~1.06 to the effective particle size.

Trapping of nominally spherical colloidal silica particles was observed by using commercially available Nalco and Ludox samples[11] diluted with distilled water. With a high concentration we quickly collect many particles in the trap and observe a correspondingly large scattering. At reduced concentration we can observe single particles trapped for extended time. Once a particle is captured at the beam focus we observe an apparent cessation of all Brownian motion and a large increase in particle scattering.

With silica samples we always observe a wide distribution of particle sizes as evidenced by the more than an order-of-magnitude difference in scattering from particles trapped with a given laser power. Particle damage by the light was not a serious problem with silica particles. The smaller particles of the distribution showed only slight changes of scattering over times of minutes. The larger particles would often decay by factors up to 3 in comparable times.

Measurements were made on a Nalco 1060 sample with a nominal size of about 60 nm and an initial concentration of silica of 50% by weight diluted to one part in 10^5–10^6 by volume. Trapping powers of 100–400 mW were used. The absolute size of the Nalco 1060 particles as determined by comparison with the 0.109-μm latex standard varied from ~50 to 90 nm with many at ~75 nm. We also studied smaller silica particles using a Ludox TM sample with nominal particle diameter of ~21 nm and a Nalco 1030 sample with nominal distribution of 11 to 16 nm. Dilutions of ~10^6–10^7 and powers of ~500 mW to 1.4 W were used. With both samples we were limited by laser power in the minimum-size particle that could be trapped. With 1.4 W of power the smallest particle trapped had a scattering that was a factor of ~3×10^4 less than from the 0.109-μm standard. This gives a minimum particle size of 26 nm assuming a single spherical scatterer. The measured minimum size of 26 nm compares with the theoretically estimated minimum size of ~19 nm for this power, based on $U/kT = 10$ and spot size $2w_0 = 0.58$ μm. This difference could be resolved by assuming a spot size $2w_0 = 0.74$ μm $= 1.28$ (0.58 μm) $\cong 1.9\ \lambda$.

Experimentally we found that we could introduce a significant drift of the fluid relative to the trapped particle by moving the entire cell transversely relative to the fixed microscope objective. This technique gives a direct method of measuring the maximum trapping force. It also implies the ability to separate a single trapped particle from surrounding untrapped particles by a simple flushing technique.

Our observation of trapping of a 26-nm silica particle with 1.4 W implies, by simple scaling, the ability to trap a 19.5-nm particle with $m = 1.6/1.33 = 1.20$ and a 12.5-nm particle of $m = 3.0/1.33 = 2.26$ at the same power. These results suggest the use of the single-beam gradient force traps for other colloidal systems, macromolecules, polymers, and biological particles such as viruses. In addition to lossless particles with real m there is the possibility of trapping Rayleigh particles with complex m for which one can in principle achieve resonantly large values of the polarizability α. Finally, we expect that these single-beam traps will work for trapping atoms[1] as well as for macroscopic Rayleigh particles since atoms can be viewed as Rayleigh particles with a different polarizability.

References

1. A. Ashkin, Phys. Rev. Lett. **40,** 729 (1978).
2. A. Ashkin, Science **210,** 1081 (1980); V. S. Letokhov and V. G. Minogin, Phys. Rep. **73,** 1 (1981).
3. A. Ashkin and J. P. Gordon, Opt. Lett. **8,** 511 (1983).
4. A. Ashkin and J. M. Dziedzic, Phys. Rev. Lett. **54,** 1245 (1985).
5. A. Ashkin and J.M. Dziedzic, Appl. Phys. Lett. **19,** 283 (1971).
6. P. W. Smith, A. Ashkin, and W. J. Tomlinson, Opt. Lett. **6,** 284 (1981); and A. Ashkin, J. M. Dziedzic, and P. W. Smith, Opt. Lett. **7,** 276 (1982).
7. A. Ashkin, Phys. Rev. Lett. **24,** 146 (1970).
8. G. Roosen, Can. J. Phys. **57,** 1260 (1979).
9. See, for example, M. Kerker, *The Scattering of Light* (Academic, New York, 1969), p. 37.
10. W. Heller, J. Chem. Phys. **42,** 1609 (1965).
11. Nalco Chemical Company, Chicago, Illinois; Ludox colloidal silica by DuPont Corporation, Wilmington, Delaware.

PAPER 4.2

Direct Observation of Transfer of Angular Momentum to Absorptive Particles from a Laser Beam with a Phase Singularity

H. He, M. E. J. Friese, N. R. Heckenberg, and H. Rubinsztein-Dunlop

Department of Physics, The University of Queensland, Brisbane, Queensland, Australia Q4072
(Received 28 November 1994; revised manuscript received 4 April 1995)

Black or reflective particles can be trapped in the dark central minimum of a doughnut laser beam produced using a high efficiency computer generated hologram. Such beams carry angular momentum due to the helical wave-front structure associated with the central phase singularity even when linearly polarized. Trapped absorptive particles spin due to absorption of this angular momentum transferred from the singularity beam. The direction of spin can be reversed by changing the sign of the singularity.

PACS numbers: 42.25.Md, 42.40.My

It is well known that a circularly polarized beam carries angular momentum. Each photon of such a beam has an angular momentum of $\hbar$. The effects produced by the optical angular momentum are hard to observe in most circumstances as they represent extremely small quantities. The angular momentum flux carried by a circularly polarized 10 mW He-Ne laser beam is of the order of 10^{-18} mN. The first attempt to measure the torque produced by the optical angular momentum was made by Beth [1] 59 years ago. Beth reported that his results agreed with theory in sign and magnitude. More recently Santamato *et al.* [2] observed the light induced rotation of liquid crystal molecules, Chang and Lee [3] calculated the optical torque acting on a weakly absorbing optically levitated sphere, and Ashkin and Dziedzic [4] observed rotation of particles optically levitated in air.

A linearly polarized wave containing a central phase singularity also carries angular momentum associated with its helical structure. Figure 1 shows this structure for a TEM_{01}^{*} beam [5]. Each photon carries $l\hbar$ angular momentum where l is called the topological charge of the singularity. This contribution to the angular momentum is sometimes referred to as "orbital angular momentum" to distinguish it from "spin" angular momentum associated with circular polarization [6,7].

Allen *et al.* [6] have proposed to measure the angular momentum carried by a doughnut laser beam by measuring the torque acting on an optical device which reverses the chirality of the beam.

We have demonstrated in a previous paper [8] that black or reflective particles of sizes of 1–2 μm can be trapped optically in a liquid by higher order doughnuts produced by high efficiency computer generated holograms. We also mentioned that some bigger absorptive particles were set into rotation by slightly defocused doughnuts. In this paper we report on further experiments concerned with the transfer of angular momentum to absorptive particles and their subsequent rotation. The detection is performed using a video camera. A large number of small particles have been trapped and have been recorded. Our results clearly show that the rotation directions of all these trapped particles agree with the sign of the doughnut.

In this Letter we analyze the results of the rotating particles in terms of possible torques acting on a trapped absorptive particle.

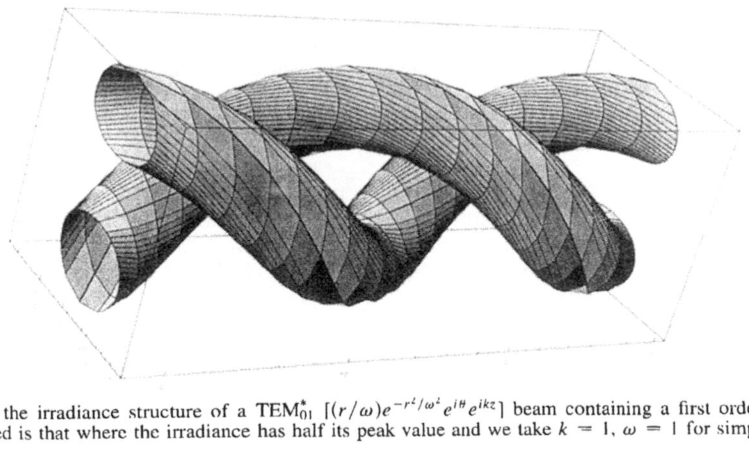

FIG. 1. Snapshot of the irradiance structure of a TEM_{01}^{*} $[(r/\omega)e^{-r^2/\omega^2}e^{i\theta}e^{ikz}]$ beam containing a first order phase singularity. The surface represented is that where the irradiance has half its peak value and we take $k = 1$, $\omega = 1$ for simplicity.

The angular momentum carried by light can be understood in two ways. Classically, electromagnetic radiation carries momentum, which can be both linear and angular. Allen *et al.* [6] have shown that for a Laguerre-Gaussian beam the angular momentum density is given as

$$M_z = \frac{l}{\omega}|u|^2 + \frac{\delta_z r}{2\omega}\frac{\partial |u|^2}{\partial r}, \quad (1)$$

where u is the amplitude of the light field, l is the azimuthal index number of the Laguerre polynomial, $\delta_z = \mp 1$ for right-handed or left-handed circularly polarized light and $\delta_z = 0$ for linearly polarized light.

Quantum mechanically, one would say that each photon carries $(l \mp \delta_z)\hbar$ of angular momentum because of the well-known analogy between paraxial theory and quantum mechanics [9].

Although the above simple relationship is valid only within the paraxial approximation as shown recently by Barnett and Allen [7], for a linearly polarized light ($\delta_z = 0$), even when tightly focused as in an optical trapping experiment, the total angular momentum per second is still given by

$$\Gamma_z = \frac{P}{\omega}l, \quad (2)$$

where P is the laser power.

A linearly polarized charge 3 singularity beam corresponds to $l = 3, \delta_z = 0$.

This suggests that an absorptive particle illuminated by such a singularity beam should be set into rotation in the same sense as the helical beam. It should reverse its rotation direction when the direction of rotation of the helical wave is reversed.

The experimental setup for the optical trap used in this work is similar to the one used in [10]. A linearly polarized TEM_{00} laser beam from a 15 mW He-Ne laser (632.8 nm) illuminates a blazed charge 3 phase hologram. The hologram then produces a phase singular laser doughnut beam equivalent to a linearly polarized Gauss-Laguerre LG_{03} mode with a power of approximately 7 mW. As the hologram used here is blazed, the sign of the doughnut can be simply reversed by turning the hologram around. We can also switch the diffracted beam between doughnut mode and Gaussian mode by moving the hologram sideways. The laser beam is now introduced into a microscope (Olympus CHT) through the aperture for the vertical illuminator. A dichroic mirror reflects the beam to fill the back aperture of an oil-immersion, high numerical aperture ($NA = 1.30$), 100× objective. The tightly focused doughnut beam had a diffraction limited beam waist within micrometers of the object plane of the objective.

The particles were illuminated from below by a lamp and green filters. A video camera placed vertically on the top of the microscope was used to record the motion of the trapped particles. We normally used black high-T_c superconductor ceramic powder as absorptive particles, dispersed in kerosene. Sizes of the particles that can be trapped are around 1–2 μm. We have performed similar experiments using CuO particles dispersed in water.

Using the charge 3 doughnut, we can trap a particle in the dark central spot. It can then be moved around relative to its surroundings by moving the microscope slide. The maximum measured trapping speed is around 5 μm/s. Particles adhering to the slide can often be set free by switching from the doughnut to the Gaussian mode and "kicking" them with the strong repulsive force.

Using a video camera, we clearly see a trapped particle rotating always in one direction determined by the helicity of the beam. The particle also keeps rotating in the same direction while being moved relative to its surroundings. The rotation of a trapped particle can be easily maintained for long periods of time. The particle can be set free and trapped again with the same rotation direction. In a session, more than 30 particles have been trapped, moved, set free, and trapped again. They all rotate in the same direction as the helicity of the beam. A few particles with very irregular shapes tumble wildly and occasionally rotate in the opposite direction for one or two turns but much more slowly than they rotate the other way. However, most of the time they rotate in the same direction as other particles. Those particles with close to spherical shapes always rotate in the same direction with high constant speed.

Turning the hologram back to front, that is reversing the charge of the singularity, we repeat the above procedures and we observe the same effects except that the rotation direction of all particles is reversed.

The trapped particle rotation speed varies from 1 to 10 Hz depending on shape and size.

Figure 2 shows six successive frames of a video recording. A fairly asymmetric particle, about 2 μm across, tumbling and rotating at about 1 Hz, has been chosen for illustration. It lies near the top of the frame and over a period of 0.24 s rotates through an angle of a little over $\pi/2$. Surrounding objects are stationary. More rapidly rotating symmetrical particles show up poorly in individual frames. However, the rotation is clear in continuous playback, in spite of motional blurring, very limited depth of field, and the fact that the microscope is operating near its limit of resolution.

When trapping CuO particles in water, if a little detergent is added, all the particles move very freely, and when one is trapped, almost immediately surrounding particles begin to move in radially and join a general circulation about it, in the same direction as the rotation of the trapped one.

Taking losses into consideration, the power incident on the focal plane is around 4 mW. According to Eq. (2), the angular momentum flux of such a doughnut beam produced by a charge 3 blazed hologram illuminated by a 633 nm HeNe laser is about 4×10^{-18} mN.

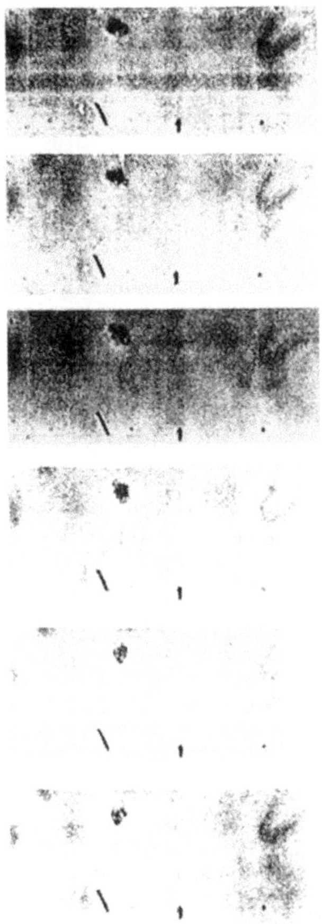

FIG. 2. Six successive frames of a video recording of a particle of black ceramic trapped in a charge three doughnut beam. The particle is near the top; other objects in the field are stationary. A rotation of just over $\pi/2$ occurs in the period shown.

Assume a particle absorbs 25% of the beam and hence 25% of the angular momentum. The rotation speed will become constant when the torque produced by the doughnut laser beam is balanced by the drag torque exerted by the surrounding liquid.

The drag torque acting on a spherical particle rotating with angular velocity ω_a is given by [11]

$$\tau_v = -8\pi\eta r^3\omega_a , \qquad (3)$$

where r is the radius of the particle, η is the viscosity of the liquid, taken as 1.58×10^{-3} N s m^{-2} for kerosene.

For our particles, with radius about 1 μm, this leads to a rotation speed of around 4 Hz, consistent with our observations, bearing in mind the actual form of our particles, and the fact that we have neglected the effects of the nearby slide surface. Our assumption that the particles absorb 25% of the incident power is based on the case where the sizes of particles and the central dark spot are the same. We define the size of the dark spot to be the diameter of the maximum intensity ring of the doughnut beam. This estimate does not affect the relative sizes of the torques hypothetically present as they all would be proportional to the absorbed power.

Although absorbtion of angular momentum therefore accounts satisfactorily for the observed rotation, we briefly consider other explanations that might be advanced. It is possible that an unbalanced force, such as a thermal force, or a scattering force, may produce a torque acting upon a nonsymmetric particle. Among these forces, the scattering force is the biggest. However, since the direction of the scattering force is predominantly downward, the direction of the torque produced by the scattering force is therefore perpendicular to the beam axis. We believe that such torques are responsible for the irregular tumbling exhibited by asymmetric particles.

However, it is conceivable that reflections on the surface of an asymmetric particle may produce a torque in the same direction as that of the angular momentum from the beam.

Such a torque will be, at most,

$$\tau_a = k\gamma F r , \qquad (4)$$

where F is the force acting on the particle, r is an effective radius, γ is the reflection coefficient, and k is an asymmetry index, expressing the ratio of the area of a typical irregularity to the total area.

The scattering force (radiation pressure) cannot exceed

$$F_s = \frac{P}{c} = 1.3 \times 10^{-11}\ \text{N} .$$

The value of k can vary from 0 to 1. However, only if the particle has a perfect "propeller" shape will k have a value approaching 1. Since our particles are mostly spheroidal and the effect of the trap is to center them in the beam, as the main asymmetry is due to the bumps on the surface of the particle, the value of k is approximately equal to the ratio of the bump area to the particle cross sectional area. Using a scanning electron microscopy image of these particles, we estimate that the value of k will normally be less than 1×10^{-2} as the irregularities are smaller than 10%. It is safe to assume the reflection coefficient to be smaller than 0.1 as our particles are highly absorptive. Hence, the torque due to asymmetric scattering forces is less than 10^{-20} N m, which is much smaller than the angular momentum from the doughnut laser beam. Obviously, a particle with an irregular shape may have a large value of k for a short period of time during which the scattering torque may dominate. However, the scattering torque is not constant, and hence the rotation direction changes. Furthermore, such a scattering force would be independent of the sign of the doughnut and therefore not able to explain our experimental results.

It is very difficult to estimate the magnitude of possible thermal effects like the photophoretic forces known to act on illuminated particles in air [12], but they could be expected to be smaller in our experiments because of the much higher thermal conductivities of the liquid media and the particles used. We have been unable to find any reports of such forces in liquids. Even if such forces are acting, considering the high thermal conductivity of the material and the fact that irregularities are small relative to the particle size and the wavelength, it is unlikely that they would exert torques leading to regular rotation with direction depending on the helicity of the wave.

Finally, if the rotation we observe was due to some thermally mediated torque, in order that angular momentum be conserved, it would be necessary that the surrounding liquid circulate in the opposite direction to the particle. However, our observations of the motion of nearby particles clearly eliminates this possibility. We see nearby particles swept around in the same direction as the rotating particle, consistent with a picture of an externally driven trapped particle stirring the surrounding liquid.

In conclusion, it has been demonstrated that absorptive particles trapped in the dark central minimum of a doughnut laser beam are set into rotation. The rotational motion of the particles is caused by the transfer of angular momentum carried by the photons. Since the laser beam is linearly polarized, this must originate in the "orbital" angular momentum associated with the helical wave-front structure and central phase singularity. We have shown that the direction of the rotational motion is determined by the chirality of the helical wave front. With a laser power of a few milliwatts, the rotation speed of the particles lies between 1 and 10 Hz depending on their sizes and shapes. This is in agreement with a simple model of absorption of the angular momentum of the radiation field.

We would like to thank Professor P. Drummond for useful discussions. We also want to thank R. McDuff and A. Noskoff for assistance with the images. This work was supported by the Australian Research Council.

[1] R. A. Beth, Phys. Rev. **50**, 115 (1936).
[2] E. Santamato, B. Daino, M. Romagnoli, M. Settemlre, and Y. R. Shen, Phys. Rev. Lett. **57**, 2433 (1986).
[3] S. Chang and S. S. Lee, J. Opt. Soc. Am. B **2**, 1853 (1985).
[4] A. Ashkin, Science **210**, 4474 (1980); **210**, 1081 (1980).
[5] N. R. Heckenberg, R. G. McDuff, C. P. Smith, H. Rubinsztein-Dunlop, and M. J. Wegerner, Opt. Quant. Electron. **24**, S951 (1992).
[6] L. Allen, M. W. Beijersbergen, R. J. C. Spreeuw, and J. P. Woerdman, Phys. Rev. A **45**, 8185 (1992).
[7] S. M. Barnett and L. Allen, Opt. Commun. **110**, 670 (1994).
[8] H. He, N. R. Heckenberg, and H. Rubinsztein-Dunlop, J. Mod. Opt. **42**, 217 (1995).
[9] D. Marcuse, *Light Transmission Optics* (Van Nostrand, New York, 1972).
[10] C. D'Helon, E. W. Dearden, H. Rubinsztein-Dunlop, and N. R. Heckenberg, J. Mod. Opt. **41**, 595 (1994).
[11] S. Oka, in *Rheology,* edited by F. R. Eirich (Academic Press, New York, 1960), Vol. 3.
[12] L. R. Eaton and S. L. Neste, AIAA J. **17**, 261 (1979).

PAPER 4.3

Optical angular-momentum transfer to trapped absorbing particles

M. E. J. Friese,[1] J. Enger,[2] H. Rubinsztein-Dunlop,[1] and N. R. Heckenberg[1]
[1]*Department of Physics, The University of Queensland, Queensland, Australia 4072*
[2]*Department of Physics, Chalmers University of Technology, S-41296 Göteborg, Sweden*
(Received 21 November 1995)

Particle rotation resulting from the absorption of light carrying angular momentum has been measured. When absorbing CuO particles ($1-5\mu$m) were trapped in a focused "donut" laser beam, they rotated, due to the helical phase structure of the beam. Changing the polarization of the light from plane to circular caused the rotation frequency to increase or decrease, depending on the sense of the polarization with respect to the helicity of the beam. Rotation frequencies were obtained by Fourier analysis of amplitude fluctuations in the backscattered light from the particles. [S1050-2947(96)08908-1]

PACS number(s): 42.50.Vk, 42.25.Ja, 42.40.My

In this article, we report on the transfer of both orbital and spin angular momentum from a polarized Laguerre-Gaussian (LG) mode laser beam to macroscopic particles. The angular momentum carried by light can be characterized by the "spin" angular momentum associated with circular polarization [1] and the "orbital" angular momentum associated with the spatial distribution of the wave [2].

To our knowledge, first experimental observation of the torque on a macroscopic object resulting from interaction with light was by Beth [1] in 1936, who observed the deflection of a quartz wave plate suspended from a thin quartz fiber when circularly polarized light passed through it. An experiment was proposed in 1957 [3] to measure radiation torque using microwave radiation, and measurements of light torques were made in 1966 [4]. The experiments of Allen [4] showed that the torque on a suspended dipole due to circularly polarized radiation increased linearly with the intensity of the light. Recently, Allen *et al.* [2] have shown that a Laguerre-Gaussian laser mode has a well defined orbital angular momentum, and proposed an experiment analogous to the experiment of Beth, to observe the torque on suspended cylindrical lenses arising from the reversal of the helicity of a LG mode. In 1995, He *et al.* [5] observed transfer of angular momentum from a linearly polarized LG mode laser beam to absorbing particles. Micrometer-sized CuO particles were trapped in a focused "donut" beam, and observed to rotate when viewed through an optical microscope. The direction of the particle rotation was found to be determined by the direction of the helicity of the beam.

We present measurements of rotation frequency of micron-sized black CuO particles trapped in a donut laser beam, for different states of polarization of the incident beam. The measurements clearly show that an absorbing particle trapped and rotating in a focused plane-polarized donut laser beam rotates faster if the beam is changed to circularly polarized with spin of the same sense as the helicity of the donut, and slower if the beam is changed to circularly polarized with spin of the opposite sense to that of the helicity. The magnitude of the change in rotation frequency agrees with that expected from theory [6], within the accuracy of the experiment.

The transfer of angular momentum from light to an absorbing particle can be understood using classical electromagnetic theory. The torque due to the polarization of the light on a particle that absorbs power P_{abs} is $\Gamma_{\sigma_z}=P_{abs}\sigma_z/\omega$, where σ_z is ± 1 for circularly polarized light and 0 for plane-polarized light and ω is the frequency of the light. Our experiments were conducted using what is essentially a LG mode, which has an angular dependence of $e^{-il\phi}$, where l is the *azimuthal mode index*, or *charge* [7]. Such beams have a donut shape with a central zero. Even when plane polarized, the field has a helical structure due to the $e^{-il\phi}$ term. The torque on an absorbing particle, which absorbs power P_{abs}, from a plane-polarized LG mode beam, determined using electromagnetic theory [2], is $\Gamma_l=P_{abs}l/\omega$. In the paraxial limit, the torque on the particle due to both the polarization and the helical Poynting vector of the LG mode is simply [2]

$$\Gamma=\frac{P_{abs}}{\omega}(l+\sigma_z). \tag{1}$$

More intuitively, the light torque can be seen to arise from the angular momentum of photons. Each photon of energy $\hbar\omega$ can be assigned a "spin" of $\sigma_z\hbar$. The LG modes can be seen as eigenmodes of the angular-momentum operator L_z [2] and thus carry an "orbital" angular momentum of $l\hbar$ per photon. The angular momentum carried by a photon of a polarized Laguerre-Gaussian mode is then $(l+\sigma_z)\hbar$. When this is multiplied by the number of photons absorbed per second, the same resulting torque is obtained as that found using electromagnetic theory.

However, this very simple result is restricted to the paraxial approximation. Barnett and Allen [6] have developed a general nonparaxial theory, using a mode which is a general LG mode in the paraxial limit. The simple dichotomy between orbital and spin angular momentum that exists in the paraxial limit is no longer found. Instead, the results of Barnett and Allen imply that the torque will be

$$\Gamma=\frac{P_{abs}}{\omega}\left\{(l+\sigma_z)+\sigma_z\left(\frac{2kz_R}{2p+l+1}+1\right)^{-1}\right\}. \tag{2}$$

Here p and l are the mode indices, k is the wave number, and z_R is a length term, which in the paraxial limit is associated with the Rayleigh range [6]. As the beam becomes

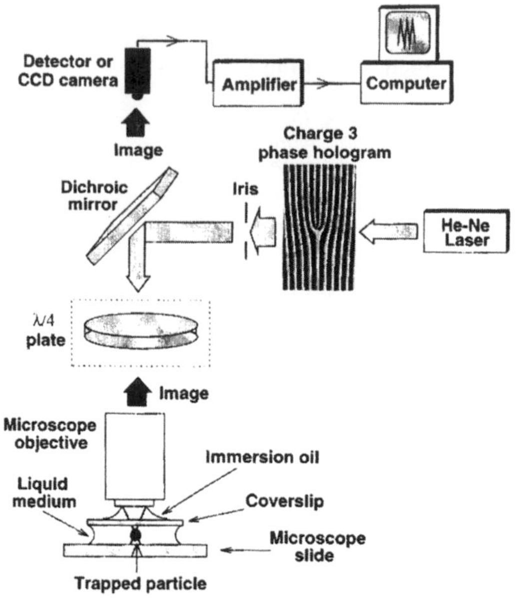

FIG. 1. Experimental setup for observation of optical angular-momentum transfer.

more strongly focused, the mixed term in Eq. (2) becomes more significant. However, if the laser beam were focused to one wavelength across, the most this term could contribute would be 20%, and substituting values appropriate to our experiments, the contribution for our laser beam (which closely approximates a Laguerre-Gaussian LG_{03} mode in the far field) focused to a 2-μm waist is only 4%. Thus we expect to observe the torques on a particle trapped in a LG_{03} mode laser beam of left-circular, plane, and right-circular polarization to be approximately in the ratio 2:3:4.

The quantity P_{abs} can, in principle, be determined from the surface area and absorptivity α of the particles and the beam profile [5]. A body rotating in a viscous fluid also experiences a viscous drag torque which is proportional to the angular velocity $\omega_{particle}$. An equilibrium angular velocity will be reached when the two torques balance, so the rotation angular frequency can be used as a measure of the optical torque.

The essential result to be tested in our experiments is thus that the angular momentum carried by photons of a left-circularly polarized, plane polarized, and right-circularly polarized charge-3 donut beam is in the ratio 2:3:4 (for a right-helical beam). This translates to the angular velocity of a particle trapped in such a beam changing in the same 2:3:4 ratio as the polarization is changed.

Our experiments were performed using an *optical tweezers* setup, as shown in Fig. 1. A Gaussian (TEM_{00}) beam from a 17-mW He-Ne laser was passed through a computer-generated phase hologram [8], which produced a beam closely approximating a Laguerre-Gaussian LG_{03} mode in the far field [7]. This donut beam was then introduced into an optical microscope. A quarter-wave plate was introduced into the beam path directly before a 100× microscope objective (N.A. 1.3), which focused the donut beam to a waist approximately 2 μm in diameter.

As explained above, the objective of this experiment was to determine if transfer of spin angular momentum from light to absorbing particles is observable on a macroscopic scale. The $\lambda/4$ plate, (as shown in Fig. 1) was used to change the polarization of the donut beam. Since the reflectivity of the beamsplitter used to direct the laser beam into the microscope objective is polarization dependent, the $\lambda/4$ plate was placed directly before the objective, after the beamsplitter, in the beam path. The polarization was checked after the objective lens using an analyzer, and the power of the beam was measured for each polarization (left- and right-circularly polarized and plane polarized.) The laser power varied less than 1% between different polarizations. The $\lambda/4$ plate was also carefully positioned so that the beam alignment into the objective did not change when varying the state of polarization of the incoming beam, since poor alignment can cause unstable trapping and thus affect the rotation speed of trapped particles. The samples of absorbing particles (CuO of sizes up to 20 μm, in kerosene mixed with oil to optimize the stability of the trap) were placed between a glass microscope slide and coverslip.

Absorbing particles of size 1–5 μm can be trapped above the focus of the laser beam, where the spot size is rapidly changing. This can be easily understood by considering that Poynting vector and hence the force due to radiation pressure has a component toward the center of the beam as well as one in the direction of propagation. Thus absorbing particles in a converging donut beam should experience a force trapping them in the center and pushing them in the direction of propagation, so they are pressed to the glass surface and trapped in the transverse plane.

The helical structure of the beam means that the Poynting vector also has a tangential component that causes the CuO particles to begin to rotate as soon as they are trapped [5]. The rotation frequency of particles trapped in a linearly polarized helical donut beam has previously been measured by viewing video recordings of rotating particles frame by frame. However, this method is difficult and of limited accuracy as the particles move out of focus when they are trapped and thus their images are not well defined on video. In the present experiments, the rotation frequency was measured using a photodetector positioned off center to intercept a portion of the light reflected from the rotating particle. The particles are irregularly shaped, and protruding parts of the particle reflect a "flash" of light onto the photodetector. The signal detected by the photodetector is then processed using Fourier transform methods, and the transform yields the rotation frequency of the particle. If the particle has two protruding parts positioned approximately opposite one another, a peak at double the rotation frequency shows up in the spectrum as well as the peak due to the actual rotation frequency. An example of such a signal is shown in Fig. 2, and the resulting spectrum is shown in Fig. 3. That the main peak shown in Fig. 3 corresponds to the rotation frequency of the particle was verified by a frame-by-frame study of a simultaneous video recording of the same particle rotating. However, the sensitivity of the reflected signal method is such that measurements may be made on less asymmetric particles than can be easily followed in video recordings. This

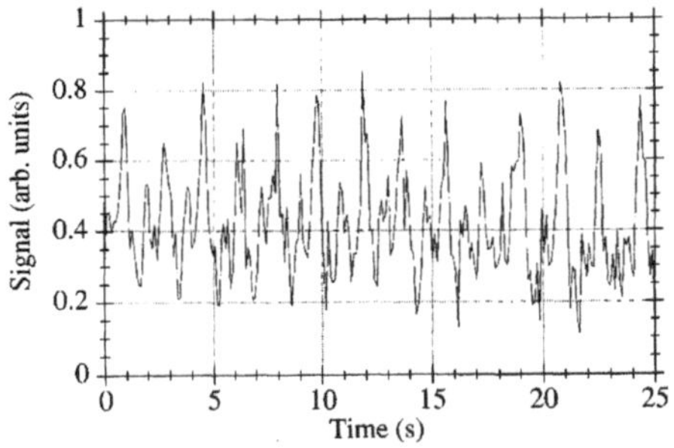

FIG. 2. Sample of raw signal detected from a rotating CuO particle. This particle has two comparatively large protrusions approximately opposite each other.

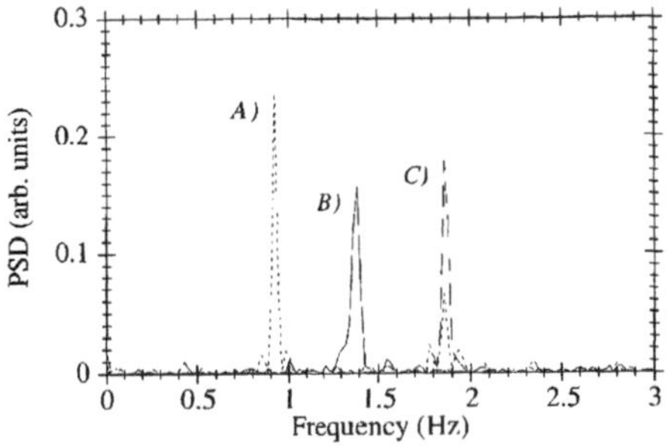

FIG. 4. Typical spectra showing the change in rotation frequency of a trapped particle as the polarization of the beam is changed. (*A*) is the peak corresponding to rotation in a left-circularly-polarized right-helical LG_{03} beam, (*B*) corresponds to a plane-polarized right-helical LG_{03} beam, and (*C*) corresponds to rotation in a right-circularly-polarized right-helical LG_{03} beam.

results in smoother rotation, monitored over longer periods, with a consequent great improvement in the accuracy of the rotational velocity estimates.

The sequence of the experiment was as follows: CuO particles were trapped using a plane-polarized donut beam. The photodetector was used to measure the reflected light, and the signal was sampled at 20 Hz for a period of 100 sec. The $\lambda/4$ plate was then rotated to give left-circularly polarized light while the particle remained trapped, and the signal was again sampled. This procedure was repeated for right-circularly-polarized light and again for plane-polarized light, in that order. In this way, the rotation frequency for a particular particle was obtained for each polarization. Variations of the order in which the polarization was changed were also used. For each experiment, the first and final data sets were taken with the laser beam in the same polarization state, to ensure that any systematic increase or decrease in rotation speed would be evident. The spectra obtained (Fig. 4) clearly show an increase or decrease in rotation speed of the particle due to the circular polarization of the light. The rotation frequency increases when the helicity of the electric field vector (due to the circular polarization) has the same direction as the helicity of the Poynting vector (due to the donut mode), and the decrease in rotation frequency corresponds to the case where the torque due to polarization opposes the torque due to the Laguerre-Gaussian mode. To verify this observation, the hologram which is used to produce the Laguerre-Gaussian mode was turned around, reversing the sense of the helicity of the Poynting vector. The trapped particles now all rotate in the opposite direction to the previous situation, and are slowed down by the polarization which previously caused an increase in rotation frequency and sped up by the polarization which previously slowed them down. Spectra obtained for this orientation of the donut are of the same quality as that shown in Fig. 4.

We have plotted the fractional increase or decrease in rotation frequency due to circular polarization against the rotation frequency for plane-polarized light in Fig. 5. From this graph we see that the proportional change in rotation speed in going from plane-polarized light to left- or right-circularly-polarized light is between 20% and 45%.

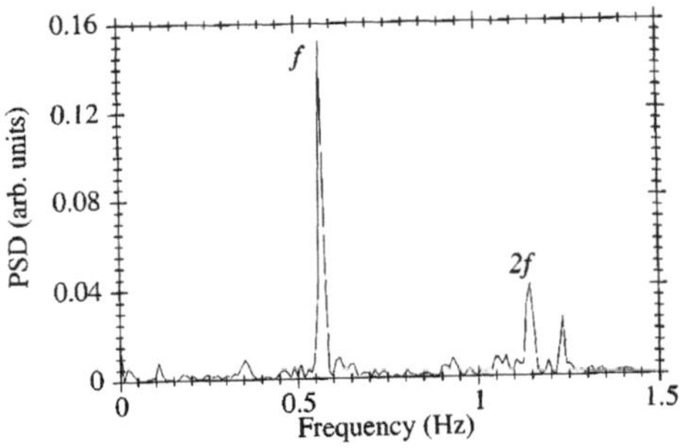

FIG. 3. Spectrum obtained from a CuO particle with two protrusions resulting in strong peaks at both f and $2f$, where f is the actual rotation frequency of the particle. That f is indeed the true rotation frequency of the particle was verified by observation of a video recording made simultaneously with the taking of the data shown in Fig. 2.

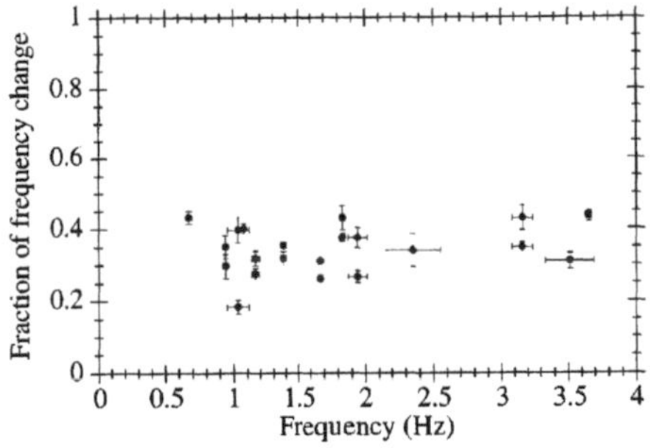

FIG. 5. Plot of fractional change in rotation frequency vs rotation frequency for CuO particles. The change in rotation frequency is due to rotation of a $\lambda/4$ plate by 45°, changing the polarization of the beam from plane polarized to circularly polarized. The error bars reflect the uncertainties in frequency measurements on the individual particles.

According to theory for the donut beam of charge 3, we expect to observe the torques on an absorbing particle trapped in a donut beam with left-circularly, plane-, and right-circularly-polarized light to be in the ratio 2:3:4 approximately. However, static friction between the particle and the slide can affect the ratio. As the particle size increases, we expect the torque required to initiate rotation to increase, causing all rotation frequencies to be reduced by approximately the same amount. This will have the effect of enlarging the proportional frequency change, which was indeed observed in our data.

In spite of this, the results show that, independent of frequency over a 4:1 range, the fractional frequency changes observed cluster about the expected 0.33 figure. Although the spread in values is too great to allow an accurate estimate of the correction term in Eq. (2) it can certainly be seen not to be dominant. Since, in practical terms, ours is a very strongly focused beam, this suggests that the simple paraxial result that the total angular momentum of the field is a simple sum of orbital and spin components will be adequate in the great majority of cases.

The results also provide further proof that the rotation previously reported [5] is the result of optical angular-momentum transfer rather than, say, some thermal effect, as such effects certainly could not be polarization sensitive in this way.

Our experiments show that the rotational speed of small absorbing particles trapped in a focused laser beam depends on the polarization of the light. The dependence is consistent with the transfer of angular momentum to the particles from a field exhibiting two forms of angular momentum— an orbital component associated with the helical phase structure of the donut beam, and a spin component associated with circular polarization of the light— as expected for a paraxial Gauss-Laguerre mode.

[1] R. Beth, Phys. Rev. **50**, 115 (1936).

[2] L. Allen, M.W. Beijersbergen, R.J.C. Speeuw, and J.P. Woerdman, Phys. Rev. A **45**, 8185 (1992).

[3] G.T. di Francia, Nuovo Cimento 1, 150 (1957).

[4] P. Allen, Am. J. Phys. **74**, 1185 (1966).

[5] H. He, M.E.J. Friese, N.R. Heckenberg, and H. Rubinsztein-Dunlop, Phys. Rev. Lett. **75**, 826 (1995).

[6] S.M. Barnett and L. Allen, Opt. Commun. **110**, 670 (1994).

[7] N. R. Heckenberg *et al.*, Opt. Quantum Electron. **24**, S951 (1992).

[8] H. He, N.R. Heckenberg, and H. Rubinsztein-Dunlop, J. Mod. Opt. **42**, 217 (1995).

PAPER 4.4

Mechanical equivalence of spin and orbital angular momentum of light: an optical spanner

N. B. Simpson, K. Dholakia, L. Allen, and M. J. Padgett

J. F. Allen Physics Research Laboratories, Department of Physics and Astronomy, University of St. Andrews, North Haugh, St. Andrews, Fife KY16 9SS, Scotland

Received August 27, 1996

We use a Laguerre–Gaussian laser mode within an optical tweezers arrangement to demonstrate the transfer of the orbital angular momentum of a laser mode to a trapped particle. The particle is optically confined in three dimensions and can be made to rotate; thus the apparatus is an optical spanner. We show that the spin angular momentum of $\pm\hbar$ per photon associated with circularly polarized light can add to, or subtract from, the orbital angular momentum to give a total angular momentum. The observed cancellation of the spin and orbital angular momentum shows that, as predicted, a Laguerre–Gaussian mode with an azimuthal mode index $l = 1$ has a well-defined orbital angular momentum corresponding to $\hbar$ per photon.

© 1997 Optical Society of America

The circularly symmetric Laguerre–Gaussian (LG) modes form a complete basis set for paraxial light beams.[1] Two indices identify a given mode, and the modes are normally denoted $LG_p{}^l$, where l is the number of 2π cycles in phase around the circumference and $(p + 1)$ is the number of radial nodes. In contrast to the planar wave fronts of the Hermite–Gaussian (HG) mode, for $l \neq 0$ the azimuthal phase term $\exp(il\phi)$ in the LG modes results in helical wave fronts.[2] These helical wave fronts are predicted by Allen *et al.* to give rise to an orbital angular momentum for linearly polarized light of $l\hbar$ per photon.[3] This orbital angular momentum is distinct from the angular momentum associated with the photon spin manifested in circularly polarized light. For a collimated beam of circularly polarized light the photon spin is well known to be $\pm\hbar$; however, for a tightly focused beam the polarization state is no longer well defined.[4] The total angular momentum of a beam, found by a rigorous solution of Maxwell's equations that reduces to a LG beam in the paraxial approximation, is given by[5]

$$\left[l + \sigma_z + \sigma_z\left(\frac{2kz_r}{2p + l + 1} + 1\right)^{-1}\right]\hbar \qquad (1)$$

per photon, where $\sigma_z = 0, \pm1$ for linearly and circularly polarized light, respectively, and where z_r is the Rayleigh range and k is the wave number of the light. Note that for a collimated beam $kz_r \gg 1$, and expression (2) reduces to

$$(l + \sigma_z)\hbar \qquad (2)$$

per photon, whereas for linearly polarized light $l\hbar$ per photon is strictly correct even in the absence of any approximation.

Lasers have been made that operate in LG modes,[6] but it is easier to obtain them from the conversion of HG modes. Three different classes of mode converter have been demonstrated. Two of these, spiral phase plates[7] and computer-generated holographic converters,[8,9] introduce the azimuthal phase term to a $HG_{0,0}$ Gaussian beam to produce a LG mode with $p = 0$ and specific values of l. In all these devices a screw phase dislocation, produced on axis, causes destructive interference, leading to the characteristic ring intensity pattern in the far field. The other class of converter is the cylindrical-lens mode converter,[10] which employs the change in Gouy phase in the region of an elliptically focused beam to convert higher-order HG modes, of indices m and n, into the corresponding LG modes (or vice versa) with 100% efficiency. The transformation is such that $l = |m - n|$ and $p = \min(m, n)$.

Consider the various macroscopic mechanisms whereby angular momentum can be transferred between a well-defined transverse laser mode and matter. One can transfer spin angular momentum by using a birefringent optical component, resulting in a change in the polarization state of the light. One can transfer orbital angular momentum by using a component that possesses an azimuthal dependence of its optical thickness, as with a cylindrical lens or a spiral phase plate. In contrast, absorption of the light allows both spin and orbital angular momentum to be transferred.

The prediction that LG laser modes possess well-defined orbital angular momentum has led to considerable research activity.[3,6–14] The transfer of orbital angular momentum from a light beam to small particles held at the focus has been modeled,[12] but to date only one group of researchers has demonstrated that these modes do indeed possess orbital angular momentum. He *et al.*[13] demonstrated that orbital angular momentum could be transferred from the laser mode to micrometer-sized ceramic and metal-oxide particles near the focus of a LG beam with an azimuthal mode index $l = 3$. More recently the same authors reported the use of a circularly polarized mode and the corresponding speeding up or slowing down of the rotation, depending on the relative sign of the spin and orbital angular-momentum terms.[14] In each case the particles were trapped with radiation pressure, first forcing the totally absorbing particle into the dark region on the beam axis (x–y trapping) and then forcing the particle against the microscope slide (z trapping). LG modes were recently also used in an optical tweezers geometry for trapping hollow glass spheres[15]; however, no rotation was observed because these spheres were totally transparent.

Here we use weakly absorbing dielectric particles, larger than the dimensions of the tightly focused LG mode. The particle experiences a net force within the electric field gradient toward the focus of the beam and is thus held in an all-optical $x-y-z$ trap. This clearly distinguishes our experiments from previous ones,[13,14] which relied on mechanical restraint in the z direction. Such all-optical traps, using zero-order Gaussian modes, were demonstrated by Ashkin *et al.*[16] and are commonly referred to as optical tweezers. The transfer of angular momentum and the subsequent rotation of the trapped particle through the use of a LG mode leads us to refer to our modified optical geometry as an optical spanner.

Figure 1 shows our experimental arrangement. We use a diode-pumped Nd:YLF laser at 1047 nm operating in a high-order HG mode. This mode is converted into the corresponding LG mode by the use of the previously mentioned cylindrical-lens mode converter; a quarter-wave plate sets the polarization state. A telescope arrangement optimizes the mode size to fill the back aperture of the objective lens, and an adjustable mirror enables the focused beam to be translated within the field of view of the objective lens. A dichroic beam splitter couples the beam into the optical spanner. The optical spanner is based on a 1.3-N.A., 100× oil-immersion microscope objective. The associated tube lens forms an image of the trapped particle on a CCD array. The sample cell comprises a microscope slide, an ≈60-μm-thick vinyl spacer, and a cover slip and is backlit with a standard tungsten bulb. The LG mode is focused such that the trapped object is held in the focal plane of the objective lens, and a colored glass filter can be inserted before the CCD array to attenuate the reflected laser light. We independently confirmed the polarization state of the light in the plane of the trapped particle by monitoring the power transmitted through a linear polarizer at various orientations.

When the $l = 1$ LG mode was focused by the objective lens, its beam waist was approximately 0.8 μm, which corresponds to a high-intensity ring 1.1 μm in diameter. For dielectric particles smaller than ~1 μm we observe that the gradient force results in the particles' being trapped off axis, centered within the high-intensity region of the ring itself. However, particles 1 μm or larger are trapped on the beam axis and for incident power above a few milliwatts are held stably in three dimensions. If the particle is partially absorbing, the transfer of angular momentum from the laser beam to the particle causes the particle to rotate. The strong z-trapping force enables us to lift the particle optically off the button of the sample cell, allowing it to rotate more freely. We have successfully rotated a number of a different particle types, including particles of absorbing glass (e.g., Schott BG38 glass) and Teflon spheres. The particles were dispersed in various fluids, including water, methanol, and ethanol.

For the linearly polarized $l = 1$ LG mode and an incident power of ≈25 mW, typical rotation speeds for the Teflon particles were ~1 Hz. The relationship between applied torque τ and limiting angular velocity ω_{lim} of a sphere of radius r in a viscous medium of viscosity η is[17]

$$\omega_{\text{lim}} = \tau/(8\pi\eta r^3). \quad (3)$$

This implies that the absorption of the Teflon particle is of the order of 2%.

The principal motivation of this study was to demonstrate that, with respect to absorption and the resulting mechanical rotation of a particle, the spin and the orbital angular momenta of light are equivalent. As discussed above, for the spin and orbital angular-momentum content of the laser mode to interact in an equivalent manner with the trapped particle it is essential that the dominant mechanism for transfer of the angular momentum be absorption. The birefringence or astigmatism of the trapped particle will preferentially transfer spin or orbital angular momentum, respectively. For comparison of spin and orbital angular momentum we selected Teflon particles, or amalgamated groups of Teflon particles, larger than the focused beam waist. Trapped on the beam axis, these particles interact uniformly with the whole of the beam, can be optically levitated off the bottom of the sample cell, and are observed to rotate smoothly with a constant angular velocity. This ability to lift the particles optically away from the cell boundary differentiates our research from that previously reported. Less regular particles could be lifted and made to rotate in the same sense as the more regular particles but often failed to stop when the spin and orbital angular-momentum terms were subtractive. This imperfect cancellation is due to additional transfer of orbital angular momentum owing to the asymmetry of the particle.

For our experimental configuration and an $l = 1$ mode, expression (1) shows that the total angular

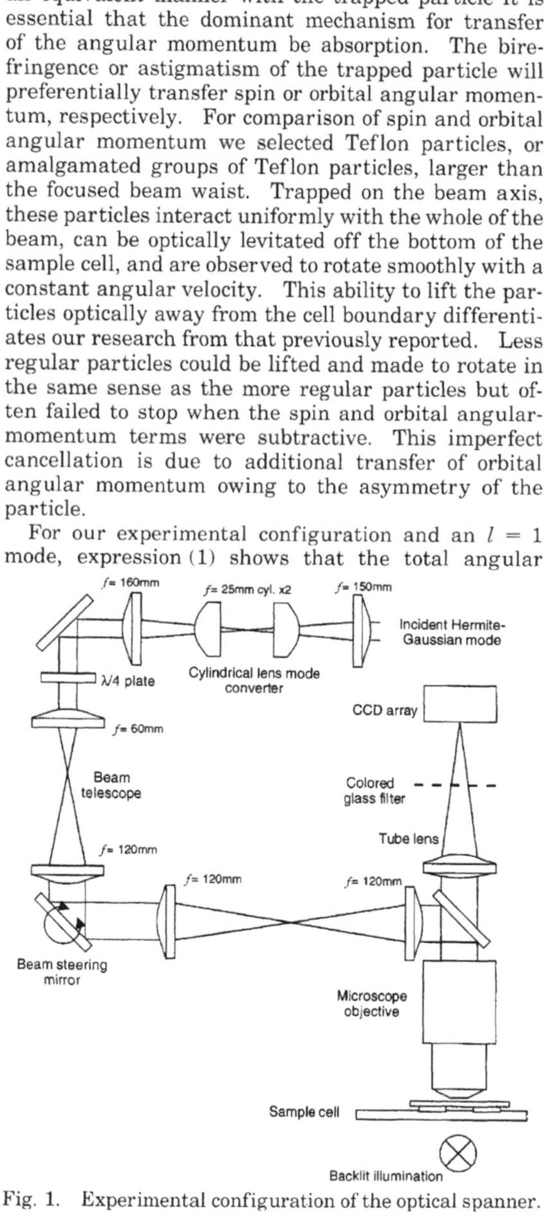

Fig. 1. Experimental configuration of the optical spanner.

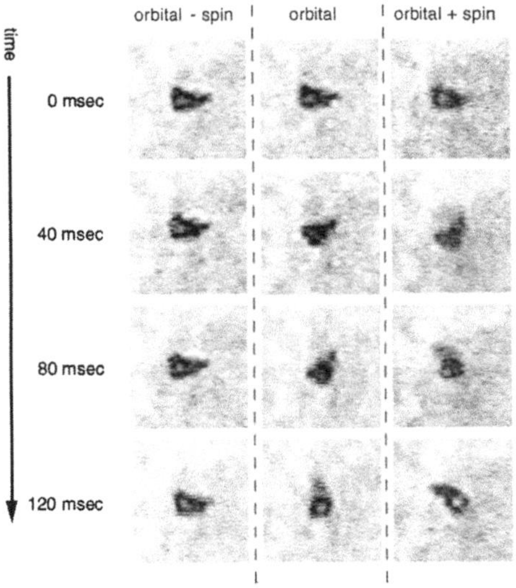

Fig. 2. Successive frames of the video image showing the stop–start behavior of a 2-μm-diameter Teflon particle held with the optical spanner.

momentum is $\sim 2.06\hbar$ per photon when the spin and orbital terms are additive and $\sim 0.06\hbar$ per photon when the spin and orbital terms are subtractive.

Figure 2 shows successive frames of the video image of a 2-μm-diameter Teflon particle held within the optical spanner, trapped with an $l = 1$ mode of various polarization states. The slight asymmetry in the particle geometry allows the particle rotation to be assessed. In more than 80% of cases, once a smoothly rotating particle had been selected, orienting the wave plate to circularly polarize the light would cause the particle to speed up or stop completely, depending on whether the spin and orbital angular-momentum terms were added or subtracted. In the other 20% of cases, although the particles would speed up or slow down, they could not be stopped completely. We believe that this was because these particles were insufficiently uniform, resulting in an unwanted mode transformation and giving an additional exchange of orbital angular momentum. With the well-behaved particles, when the spin and orbital terms are subtracted the low value of the torque, owing to a total angular momentum of no more than $0.06\hbar$ per absorbed photon, is insufficient to overcome the stiction present within the system, and the particle ceases to rotate. Stiction can arise from slight particle asymmetry coupled with residual astigmatism in the laser mode, which will favor the particles' being trapped in particular orientations. When the spin and orbital terms are additive the rotation speed increases significantly, but in only few cases is it seen to double. We attribute this to a nonlinear relationship between the applied torque and the terminal rotation speed, and the particle asymmetry. We confirmed this nonlinearity in the relationship between torque and rotation speed by deliberately changing the optical power while maintaining the same laser mode.

We observe, in agreement with our interpretation, that reversing the sense of the cylindrical-lens mode converter caused the particles to rotate in the opposite direction; similarly, by use of a $HG_{0,0}$ mode, particles could be rotated in either direction when circularly polarized light of the appropriate handedness was used.

Our experiment uses a LG mode within an all-optical x–y–z trap to form an optical spanner for rotating micrometer-sized particles. By controlling the polarization state of a LG mode with $l = 1$ we can arrange for the angular momentum associated with the photon spin to add to the orbital angular momentum, giving a total angular momentum of $2\hbar$ per photon, in which case the Teflon particle spins more quickly. Alternatively, that momentum can subtract, resulting in no overall angular momentum, and the particle comes to a halt. We observe the cancellation of the spin and orbital angular momentum in a macroscopic system, which verifies that for an $l = 1$ LG mode the orbital angular momentum is indeed well defined and corresponds to $\hbar$ per photon.

This research is supported by Engineering and Physical Sciences Research Council grant GR/K11536. M. I. Padgett is a Royal Society Research Fellow.

References

1. A. E. Siegman, *Lasers* (University Science, Mill Valley, Calif., 1986), Sec. 17.5, pp. 685–695.
2. J. M. Vaughan and D. V. Willetts, Opt. Commun. **30,** 263 (1979).
3. L. Allen, M. W. Beijersbergen, R. J. C. Spreeuw, and J. P. Woerdman, Phys. Rev. A **45,** 8185 (1992).
4. D. N. Pattanayak and G. P. Agrawal, Phys. Rev. A **22,** 1159 (1980).
5. S. M. Barnett and L. Allen, Opt. Commun. **110,** 670 (1994).
6. M. Harris, C. A. Hill, and J. M. Vaughan, Opt. Commun. **106,** 161 (1994).
7. M. W. Beijersbergen, R. P. C. Coerwinkel, M. Kristensen, and J. P. Woerdman, Opt. Commun. **112,** 321 (1994).
8. N. R. Heckenberg, R. McDuff, C. P. Smith, and A. G. White, Opt. Lett. **17,** 221 (1992).
9. N. R. Heckenberg, R. McDuff, C. P. Smith, H. Rubinsztein-Dunlop, and M. J. Wegener, Opt. Quantum Electron. **24,** S951 (1992).
10. M. W. Beijersbergen, L. Allen, H. E. L. O. van der Veen, and J. P. Woerdman, Opt. Commun. **96,** 123 (1993).
11. M. Babiker, W. L. Power, and L. Allen, Phys. Rev. Lett. **73,** 1239 (1994); S. J. van Enk and G. Nienhuis, Opt. Commun. **94,** 147 (1992).
12. N. B. Simpson, L. Allen, and M. J. Padgett, J. Mod. Opt. **43,** 2485 (1996).
13. H. He, M. E. J. Friese, N. R. Heckenberg, and H. Rubinsztein-Dunlop, Phys. Rev. Lett. **75,** 826 (1995).
14. M. E. J. Friese, J. Enger, H. Rubinsztein-Dunlop, and N. R. Heckenberg, Phys. Rev. A **54,** 1593 (1996).
15. K. T. Gahagan and G. A. Swartzlander, Jr., Opt. Lett. **21,** 827 (1996).
16. A. Ashkin, J. M. Dziedzic, J. E. Bjorkholm, and S. Chu, Opt. Lett. **11,** 288 (1986).
17. S. Oka, in *Rheology,* F. R. Eirich, ed. (Academic, New York, 1960), Vol. 3.

Section 5

OPTICAL FORCES AND TORQUES ON ATOMS

It has been known for some time that the interaction of laser light with a free atom can give rise to electromagnetic pressure forces that act on its centre of mass. These forces have been extensively examined both theoretically and experimentally: see, for example Metcalf and van der Straten (5.1) or Adams and Riis (5.2). A simple model of a two-level atom subject to a plane electromagnetic wave may be shown to give rise to two kinds of force acting on the atom centre of mass. These are, a dissipative force arising from the absorption of the light by the atom and its subsequent spontaneous emission in a random direction and a dipole force arising from the non-uniformity of the field distribution which can attract the atom to the regions of intense field. These forces underpin the manipulation of atoms by lasers in a variety of beam configurations. The dissipative force has been exploited in the Doppler cooling of the atomic motion (5.3, 5.4) and the dipole force used for trapping (5.5), while a combination of these effects has led to the realisation of Bose–Einstein condensation (5.6). These forces are, of course, closely related to the ones discussed in Section 4.

It appeared probable that the interaction of atoms with beams possessing orbital angular momentum should lead to new effects. Consequently, theoretical studies of the interaction of light with atoms have been conducted to examine how the main features of Doppler cooling and trapping are modified when a plane wave or a fundamental Gaussian beam is replaced with Laguerre–Gaussian light.

Allen *et al.* (5.7, **Paper 5.1**), studied the effects of the orbital angular momentum of light on atoms by considering the theory of forces due to Laguerre–Gaussian light and their effects on a two-level atom. In order to explore more fully the effects of orbital angular momentum on atomic motion, they extended their investigations to more than one beam. The theory was developed in terms of the optical Bloch equations following Dalibard and Cohen-Tannoudji (5.8), which allows the *ab initio* inclusion of relaxation effects and incorporates saturation phenomena. The solution of the optical Bloch equations in the adiabatic, or constant velocity, approximation gave insight into the time evolution of angular momentum effects for an atom in a coherent light beam assumed to have a complex amplitude and a Laguerre–Gaussian spatial distribution. In this paper, use is made of the semi-classical approximation by replacing the position and momentum operators and by their expectation values, while maintaining a quantum treatment for the internal dynamics of the atom. The validity of the semi-classical approximation requires that the spatial extent of the atomic wave-packet be much smaller than the wavelength of the radiation field and that the uncertainty in the

Doppler shift be much smaller than the upper state linewidth of the atom. This is the case for most atoms if the recoil energy of the atom is much smaller than the upper state linewidth. Within the semi-classical approximation, the atomic density matrix can be written readily and its time evolution determined in accordance with the Heisenberg equation of motion. The paper specifically investigates the motion of a magnesium ion for a number of different beam configurations. Perhaps the most interesting aspect of the motion is that which arises from the reciprocal interplay between motions in orthogonal directions.

Related work in this area includes that of Babiker *et al.* (5.9) who show that a two-level atom moving in a Laguerre–Gaussian beam is subject to a light induced torque about the beam axis directly proportional to the orbital angular momentum. This torque reduces in the saturation limit to the simple form $T = \hbar\ell\Gamma$, where 2Γ is the decay rate of the excited state. A proposed experiment to observe the torque on the atom has been published by Power and Thompson (5.10). Allen *et al.* (5.11, **Paper 6.5**) showed that an atom moving in a Laguerre–Gaussian beam experiences an azimuthal shift in resonant frequency of $\ell V_\phi/r$ where r is the radial atomic position and V_ϕ the azimuthal component of velocity. Lembessis (5.12) has considered an atom moving in a linearly polarised beam and included the Roentgen term which arises from the motion of the electric dipole interacting with the magnetic part of the optical field. He found terms which couple the photon angular momentum with that of the atomic particle. The other interesting aspect of this type of investigation is that of Allen, Lembessis and Babiker (5.13), who recognised that the azimuthal component of the dissipative force contains a term which is proportional to $\sigma\ell$. Clearly, just as reversing the sign of the orbital angular momentum changes the direction of the force, so too does changing the handedness of the circular polarisation of the light. Normally the handedness of circularly polarised light would not be expected to determine the gross motion of the atom, only its internal state. Although the term is small and is comparable with other small terms ignored in trapping calculations, it is none the less an example of spin–orbit coupling in light. We are used to the spin–orbit coupling of electrons but not of light. Liu and Milburn (5.14) investigated the classical 2-D nonlinear dynamics of cold atoms in far-off-resonant Laguerre–Gaussian beams. They showed that chaotic developments exist provided $\ell > 1$, when the beam is periodically modulated. The atoms are predicted to accumulate on several ring regions when the system enters a regime of global chaos.

It has been recognised for a very long time that although dipole radiation dominates the physics of the interaction of light and matter, other higher order processes are also possible. This is particularly important as the selection rules of quadrupole and higher order radiation explicitly have both spin and orbital angular momentum contributions. Until the use of Laguerre–Gaussian beams the only higher order processes observed were ones where it was not possible, even in principle, to observe a change in the orbital angular momentum of the light. In a study of the selection rules and centre-of-mass motion of ultra-cold atoms, van Enk (5.15, **Paper 5.2**) considered how spin and orbital angular momentum might be transferred to internal and external angular momentum of an atom in dipole and quadrupole transitions.

For a cold atom the recoil effect on the atom of even a single photon cannot be neglected and the state of the atom may become changed. When the atom is sufficiently cold the external atomic motion can no longer be described classically. Indeed, when its de Broglie wavelength becomes comparable to the wavelength of the light, the external motion must also be quantised. This is the situation analysed by van Enk who shows that the spin and orbital angular momenta are distributed over the internal and external angular momenta of the atom. The key feature of this work, as with other work on the torque and forces on atoms, is that the transition rate depends on the specific spatial dependence of the field, while the internal selection rules are independent of the mode structure and do not depend on the external atomic motion.

The most successful, and most interesting, experiment concerning the interaction of orbital angular momentum and atoms is a fascinating one on cold caesium atoms conducted by Tabosa and Petrov (5.16, **Paper 5.3**). They demonstrated, using a non-degenerate four-wave mixing process, that orbital angular momentum can be transferred from the optical beam to the cold caesium atoms. They point out that this process is similar to the phenomenon of optical pumping where orientation and alignment is induced in atomic systems by means of polarised light, or spin angular momentum. Their observation, therefore, is the first example of optical pumping of the orbital angular momentum of light. The remarkable aspect of their experiment is that they observed the interchange of orbital angular momentum between two waves of different frequency as a result of pumping the caesium atoms. They argued that the low temperature of laser cooled atoms, as well as diminishing the Doppler broadening of the radiation, could also give access to the direct observation of mechanical effects induced by an atom–field interaction. However, as yet, experiments to observe any net transfer of angular momentum to the atomic system during the lifetime of the process remain to be carried out. At the time of writing it would appear that, as the orbital angular momentum is conserved, radiation processes of higher order than dipole should demonstrate mode conversion when both orbital and spin selection rules become invoked.

The characteristic shape of the intensity distribution of Laguerre–Gaussian modes, irrespective of their orbital angular momentum, make them ideal for interactions with atoms and for the creation of traps with dark zones in them. It is not surprising therefore that this topic has a fast growing literature. It is not, however, an area of work relevant to the interests of this volume.

Just as the intensity distribution of Laguerre–Gaussian beams suggest annular trapping schemes, it follows that such beams can also be used for the creation of Bose–Einstein condensates; see for example (5.17, 5.18). The effect of what occurs when the orbital angular momentum aspects of the beams are exploited, is yet to be studied.

REFERENCES

5.1 H Metcalf and P van der Straten, 1994, *Phys. Rep.* **244** 203.

5.2 CS Adams and E Riis, 1997, *Prog. Quant. Elect.* **21** 1.

5.3 D Wineland and H Dehmelt, 1975, *Bull. Am. Phys. Soc.* **20** 637.

5.4 T Hansch and A Schawlow, 1975, *Opt. Commun.* **13** 68.

5.5 S Chu, JE Bjorkholm, A Ashkin and A Cable, 1986, *Phys. Rev. Lett.* **57** 314.

5.6 MH Anderson, JR Ensher, MR Matthews, CE Wieman and EA Cornell, 1995, *Science* **269** 198.

5.7 L Allen, M Babiker, WK Lai and VE Lembessis, 1996, *Phys. Rev. A* **54** 4259.

5.8 J Dalibard and C Cohen-Tannoudji, 1985, *J. Phys. B* **18** 1661.

5.9 M Babiker, WL Power and L Allen, 1994, *Phys. Rev. Lett.* **73** 1239.

5.10 WL Power and RC Thompson, 1996, *Opt. Commun.* **132** 371.

5.11 L Allen, M Babiker and WL Power, 1994, *Opt. Commun.* **112** 144.

5.12 VE Lembessis, 1999, *Opt. Commun.* **159** 243.
5.13 L Allen, VE Lembessis and M Babiker, 1996, *Phys. Rev. A* **53** R2937.
5.14 XM Liu and G Milburn, 1999, *Phys. Rev. E* **59** 2842.
5.15 SJ van Enk, 1994, *Quantum Opt.* **6** 445.
5.16 JWR Tabosa and DV Petrov, 1999, *Phys. Rev. Lett.* **83** 4967.
5.17 EL Bolda and DF Walls, 1998, *Phys. Lett. A* **246** 32.
5.18 EM Wright, J Arlt and K Dholakia, 2000, *Phys. Rev. A* **63** 013608.

Atom dynamics in multiple Laguerre-Gaussian beams

L. Allen, M. Babiker, W. K. Lai, and V. E. Lembessis
Department of Physics, University of Essex, Colchester, Essex C04 3SQ, United Kingdom
(Received 10 April 1996)

The leading radiation forces acting on an atom or ion subject to linearly polarized Laguerre-Gaussian (LG) light are studied. Particular emphasis is laid on the orbital angular momentum effects associated with LG light. The optical Bloch equations appropriate for the adiabatic approximation are derived and used to evaluate the forces and associated torque governing the atomic motion. The steady-state dynamics of the atom are explored for atoms subject to a single beam and multiple independent counterpropagating beams. The main features responsible for the dynamics of the atom, together with the dipole potentials characteristic of Laguerre-Gaussian light, are identified and discussed. The theory is illustrated by the numerical integration of the equation of motion for Mg^- ions in various beam configurations. This yields information on trajectories, velocity evolution, and vibrational frequencies at potential minima. Interesting effects involving a reciprocal interplay between motions in orthogonal directions are demonstrated. Such features are purely dependent on the orbital angular momentum property of the light. Their possible use in controlling atomic motion is investigated. [S1050-2947(96)07210-1]

PACS number(s): 32.80.Pj, 42.50.Vk

I. INTRODUCTION

The radiation forces associated with the near-resonant interaction of laser light with atoms and ions have been the subject of intensive theoretical and experimental study [1–4] since the basic mechanisms were first recognized [5]. The simplest features can be described with reference to a two-level atom subject to an electromagnetic wave. Near resonance, such an atom experiences two distinct forces: a dissipative force that arises from the absorption of the light by the atom followed by its spontaneous emission and a dipole force that arises from the nonuniformity of the field distribution. These basic forces underpin many of the applications involving the manipulation of atoms by lasers in a variety of beam configurations. The dissipative force has been exploited in cooling the atomic motion [6] and the dipole force used for trapping [7].

Much of the previous theoretical work in this context has assumed plane-wave modes. However, the demonstration that Laguerre-Gaussian (LG) laser beams possess well-defined orbital angular momentum $l\hbar$ [8,9] that originates in the azimuthal phase dependence of the field distribution has aroused new interest in the basic physics. The orbital angular momentum of LG beams is quite distinct from the spin angular momentum associated with circularly polarized light and can occur in linearly polarized LG modes. A circularly polarized LG beam possesses spin angular momentum as well as orbital angular momentum and can exhibit features involving spin-orbit coupling [10]. In our recent work [11] we presented a theory for the motion of a two-level atom in a Laguerre-Gaussian beam with spontaneous emission and saturation effects taken into account heuristically. The results found were in the form of an azimuthal shift in the atomic resonance and a torque about the beam axis.

The purpose of this paper is twofold: first, to present a more rigorous theory for the forces due to LG light and their effects on a two-level atom and, second, we extend our investigations on the orbital angular momentum effects beyond the one beam case in order to explore more fully the effects of the orbital angular momentum on atomic motion. The theory is developed in terms of the optical Bloch equations (OBEs) [1,12,13] that allow the *ab initio* inclusion of relaxation effects and naturally incorporate saturation phenomena. The solution of the OBEs in the adiabatic, or constant velocity, approximation lends insight into the time evolution of angular momentum effects for an atom in a LG beam. In the long-time limit, these solutions lead directly to the steady-state results for the dissipative and dipole forces due to a LG beam.

In Sec. II we set up the density-matrix formalism appropriate for a two-level atom interacting with Laguerre-Gaussian light. In the adiabatic approximation this leads to the optical Bloch equations that formally enable the calculation of the average mechanical force (defined as the rate of change of the atomic momentum). We argue that as the concept of force is a classical one, the rate of change of momentum can only be interpreted as a force for elapsed times greater than the spontaneous decay time Γ^{-1}, where time is measured from the instant the light beam is switched on [1]. With this restriction, we solve the OBEs to determine the evolution of the average "force" components from the instant the LG beam is switched on. It is possible to examine the time evolution in a number of limits, but we discuss primarily the steady state. Further insight into the nature of the forces is gained by the numerical solution of the optical Bloch equations for a typical set of parameters. In Sec. III we give an analysis of the steady-state dissipative and dipole forces for various beam configurations and identify the features directly attributable to the angular momentum of the LG beams. The motion of a Mg^+ ion in counterpropagating LG fields is described in Sec. IV after solving the equation of motion numerically. The results demonstrate the effects of a characteristic torque and of reciprocating forces between axial and azimuthal motions. Section V contains conclusions and further comment.

II. FORMALISM

We wish to examine the evolution of the average force acting on a two-level atom or ion, henceforth referred to as the atom, due to its interaction with light. The light is in the form of a coherent beam with a complex amplitude α and has a LG distribution [8]. An appropriate Hamiltonian is given by

$$H=H_A+H_F+H_{\text{int}}, \tag{1}$$

where H_A and H_F are the unperturbed Hamiltonians for the atom and field, respectively, and are

$$H_A=\frac{\mathbf{P}^2}{2M}+\hbar\omega_0\pi^\dagger\pi, \tag{2}$$

$$H_F=\hbar\omega a^\dagger a. \tag{3}$$

Here $\mathbf{P}$ is the center-of-mass momentum operator and π and $\pi^\dagger$ are the atomic lowering and raising operators; M is the mass of the atom and ω_0 is the atomic transition frequency. In Eq. (3) a and $a^\dagger$ are the annihilation and creation operators and ω is the frequency of the field.

The interaction Hamiltonian H_{int} in Eq. (1) describes the coupling of the atom to the electromagnetic field and is given in the electric dipole approximation by

$$H_{\text{int}}=-\mathbf{d}\cdot\mathbf{E}(\mathbf{R}), \tag{4}$$

where $\mathbf{d}$ is the atomic dipole moment operator and $\mathbf{E}(\mathbf{R})$ is the electric field evaluated at the position $\mathbf{R}$ of the atom. The atomic dipole moment operator may be written as

$$\mathbf{d}=\mathbf{D}_{12}(\pi+\pi^\dagger), \tag{5}$$

where $\mathbf{D}_{12}$ is the dipole matrix element of the atomic transition. The electric-field vector associated with a Laguerre-Gaussian mode propagating along the z axis is given by

$$\mathbf{E}(\mathbf{R})=i[a\hat{\boldsymbol{\epsilon}}\mathcal{E}_{klp}(\mathbf{R})e^{i\Theta_{klp}(\mathbf{R})}-\text{H.c.}], \tag{6}$$

where $\hat{\boldsymbol{\epsilon}}$ is a polarization vector in the x-y plane. The electric field of a LG beam has a small vector component along the z axis [11,14], which we have ignored. It can easily be shown that the ignored term is of the the order λ/w_0 relative to the principal component along $\hat{\boldsymbol{\epsilon}}$ [11]. In Eq. (6) $\mathcal{E}_{klp}(\mathbf{R})$ and $\Theta_{klp}(\mathbf{R})$ are, respectively, the mode amplitude and phase of the electric field, which may be written as [11,14]

$$\mathcal{E}_{klp}(\mathbf{R})=\mathcal{E}_{k00}\frac{C_{lp}}{(1+z^2/z_R^2)^{1/2}}\left(\frac{\sqrt{2}r}{w(z)}\right)^{|l|}L_p^{|l|}\left(\frac{2r^2}{w^2(z)}\right)e^{-r^2/w^2(z)}, \tag{7}$$

$$\Theta_{klp}(\mathbf{R})=\frac{kr^2z}{2(z^2+z_R^2)}+l\phi+(2p+l+1)\tan^{-1}(z/z_R)+kz. \tag{8}$$

Here $C_{lp}=\sqrt{p!/(|l|+p)!}$ is a normalization constant; $w(z)$ is given by $w^2(z)=2(z^2+z_R^2)/kz_R$, where z_R is the Rayleigh range. The integers l and p are indices characterizing the LG mode. It has been shown [8] that $l\hbar$ represents the orbital angular momentum of each quantum in the mode. Finally, in Eq. (7) $\mathcal{E}_{k00}$ corresponds to the plane-wave amplitude for an axial wave vector k. The plane-wave amplitude and phase emerge directly from Eqs. (7) and (8) by setting $l=0$, $p=0$, and $z_R\to\infty$.

We now transform to an interaction picture with respect to the unperturbed field Hamiltonian $\hbar\omega a^\dagger a$. The field annihilation and creation operators then acquire the time dependences

$$a(t)=e^{i\omega a^\dagger at}ae^{-i\omega a^\dagger at}=ae^{-i\omega t}, \tag{9}$$

$$a^\dagger(t)=e^{i\omega a^\dagger at}a^\dagger e^{-i\omega a^\dagger at}=a^\dagger e^{i\omega t}. \tag{10}$$

In the classical limit in which the field forms a coherent beam, we may replace the field operators by c numbers

$$a(t)\to\alpha e^{-i\omega t}, \tag{11}$$

$$a^\dagger(t)\to\alpha^* e^{i\omega t}. \tag{12}$$

The coupling between the atom and field may then be written as

$$H_{\text{int}}=-\mathbf{d}\cdot\mathbf{E}(\mathbf{R})=-i\hbar[\tilde{\pi}^\dagger\alpha f(\mathbf{R})-\text{H.c.}], \tag{13}$$

where in writing Eq. (13) we have made use of the rotating-wave approximation and have defined

$$\tilde{\pi}=\pi e^{i\omega t}, \tag{14}$$

$$f(\mathbf{R})=(\mathbf{D}_{12}\cdot\hat{\boldsymbol{\epsilon}})\mathcal{E}(\mathbf{R})e^{i\Theta(\mathbf{R})}/\hbar. \tag{15}$$

For convenience, we have not explicitly shown the LG labels klp associated with $\mathcal{E}_{klp}(\mathbf{R})$ and $\Theta_{klp}(\mathbf{R})$ by virtue of Eqs. (7) and (8). In the rest of this section we continue to use this simple notation but resort to the full notation subsequently.

To derive the optical Bloch equations for the atomic density-matrix elements we make the assumption that the position and momentum operators $\mathbf{R}$ and $\mathbf{P}$ may be replaced by their expectation values $\mathbf{R}_0$ and $\mathbf{P}_0$, respectively. This approximation allows the gross motion of the atom to be treated classically, while maintaining a quantum treatment for the internal dynamics of the atom. The validity of the semiclassical approximation requires that the spatial extent of the atomic wave packet be much smaller than the wavelength of the radiation field and that the uncertainty in the Doppler shift be much smaller than the upper-state linewidth of the atom. This is the case for most atoms [1] if the recoil energy of the atom is much smaller than the upper-state linewidth.

Within the semiclassical approximation, the atomic density matrix can be written as

$$\rho=\delta(\mathbf{R}-\mathbf{R}_0)\delta(\mathbf{P}-\mathbf{P}_0)\rho(t), \tag{16}$$

where the internal dynamics of the atom are now contained in $\rho(t)$. The evolution of $\rho(t)$ is given by the well-known relation

$$\frac{d\rho}{dt}=-\frac{i}{\hbar}[H,\rho]+\mathcal{R}\rho, \tag{17}$$

where $\mathcal{R}\rho$ accounts for the relaxation dynamics of the atomic system. By substitution of H and use of the coupling given in

Eq. (13), we obtain the following optical Bloch equations for the atomic density-matrix elements:

$$\frac{d\rho_{22}}{dt} = -2\Gamma\rho_{22} - \alpha f(\mathbf{R}_0)\tilde{\rho}_{12} - \alpha^* f^*(\mathbf{R}_0)\tilde{\rho}_{21}, \quad (18)$$

$$\frac{d\tilde{\rho}_{21}}{dt} = -(\Gamma - i\Delta_0)\tilde{\rho}_{21} + \alpha f(\mathbf{R}_0)(\rho_{22} - \rho_{11}), \quad (19)$$

where $\Delta_0 = \omega - \omega_0$ is the detuning of the field frequency from atomic resonance and $\tilde{\rho}_{21} = \langle \tilde{\pi} \rangle$.

The average radiation force acting on the atom is defined as the average rate of change of the atomic momentum. We may write

$$\langle \mathbf{F} \rangle = -\langle \nabla H_{\text{int}} \rangle. \quad (20)$$

Substitution of Eq. (13) into Eq. (20) and use of Eq. (15) allows the force to be written as $\langle \mathbf{F} \rangle = \langle \mathbf{F}_{\text{diss}} \rangle + \langle \mathbf{F}_{\text{dipole}} \rangle$. Here $\langle \mathbf{F}_{\text{diss}} \rangle$ is the dissipative force given by

$$\langle \mathbf{F}_{\text{diss}} \rangle = -\hbar \nabla \Theta(\mathbf{R}_0)\{\tilde{\rho}_{12}(t)\alpha f(\mathbf{R}_0) + \tilde{\rho}_{21}(t)\alpha^* f^*(\mathbf{R}_0)\} \quad (21)$$

and $\langle \mathbf{F}_{\text{dipole}} \rangle$ is the dipole force given by

$$\langle \mathbf{F}_{\text{dipole}} \rangle = i\hbar \frac{\nabla \Omega(\mathbf{R}_0)}{\Omega(\mathbf{R}_0)} \{\tilde{\rho}_{12}(t)\alpha f(\mathbf{R}_0) - \tilde{\rho}_{21}(t)\alpha^* f^*(\mathbf{R}_0)\}, \quad (22)$$

where we have introduced a position-dependent Rabi frequency as $\Omega(\mathbf{R}_0) = |\alpha(\mathbf{D}_{12}\cdot\hat{\boldsymbol{\epsilon}})\mathcal{E}(\mathbf{R}_0)|/\hbar$.

In the adiabatic approximation [15], the atomic velocity $\mathbf{V} = \mathbf{P}_0/M$ is assumed to be constant during the time taken for the dipole moment to relax to its steady-state value. The position $\mathbf{R}_0$ of the atom at time t is then given by

$$\mathbf{R}_0 = \mathbf{r}_0 + \mathbf{V}t, \quad (23)$$

where $\mathbf{r}_0$ is the (initial) position of the atom when the beam was switched on. Thus we can write

$$f(\mathbf{R}_0) = f(\mathbf{r}_0 + \mathbf{V}t) \quad (24)$$

$$\simeq f(\mathbf{r}_0)e^{i\nabla\Theta(\mathbf{r}_0)\cdot\mathbf{V}t}, \quad (25)$$

where we have assumed that the change in the field amplitude is negligible during the time taken for the dipole moment to relax to its steady-state value.

Within the adiabatic approximation, the optical Bloch equations take the form

$$\frac{d\rho_{22}}{dt} = -2\Gamma\rho_{22} - \alpha f(\mathbf{r}_0)\hat{\rho}_{12} - \alpha^* f^*(\mathbf{r}_0)\rho_{22}, \quad (26)$$

$$\frac{d\hat{\rho}_{21}}{dt} = -[\Gamma - i\Delta(\mathbf{r}_0,\mathbf{V})]\hat{\rho}_{21} + \alpha f(\mathbf{r}_0)(\rho_{22} - \rho_{11}), \quad (27)$$

where the total detuning $\Delta(\mathbf{r}_0,\mathbf{V}) = \Delta_0 - \nabla\Theta(\mathbf{r}_0)\cdot\mathbf{V}$ and $\hat{\rho}_{21} = \tilde{\rho}_{21}e^{-i\mathbf{V}\cdot\nabla\Theta(\mathbf{r}_0)}$. The forces can now be written as

$$\langle \mathbf{F}_{\text{diss}} \rangle = -\hbar \nabla\Theta(\mathbf{r}_0)\{\hat{\rho}_{12}(t)\alpha f(\mathbf{r}_0) + \hat{\rho}_{21}\alpha^* f^*(\mathbf{r}_0)\}, \quad (28)$$

$$\langle \mathbf{F}_{\text{dipole}} \rangle = i\hbar \frac{\nabla \Omega(\mathbf{r}_0)}{\Omega(\mathbf{r}_0)} \{\hat{\rho}_{12}(t)\alpha f(\mathbf{r}_0) - \hat{\rho}_{21}(t)\alpha^* f^*(\mathbf{r}_0)\}. \quad (29)$$

For given initial conditions the solution of the optical Bloch equations (26) and (27) leads formally to the determination of the forces by direct substitution in Eqs. (28) and (29).

A. Steady state

The steady state occurs when all time derivatives in the optical Bloch equations are set equal to zero and corresponds to the long-time limit. It is not difficult to show that the steady-state solutions to the optical Bloch equations (26) and (27) yield the following expressions for the steady-state forces:

$$\langle \mathbf{F} \rangle = \langle \mathbf{F}_{\text{diss}} \rangle + \langle \mathbf{F}_{\text{dipole}} \rangle, \quad (30)$$

where

$$\langle \mathbf{F}_{\text{diss}}(\mathbf{R},\mathbf{V}) \rangle = 2\hbar\Gamma\Omega^2(\mathbf{R})\left(\frac{\nabla\Theta(\mathbf{R})}{\Delta^2(\mathbf{R},\mathbf{V}) + 2\Omega^2(\mathbf{R}) + \Gamma^2}\right), \quad (31)$$

$$\langle \mathbf{F}_{\text{dipole}}(\mathbf{R},\mathbf{V}) \rangle = -2\hbar\Omega(\mathbf{R})\nabla\Omega(\mathbf{R}) \times \left(\frac{\Delta(\mathbf{R},\mathbf{V})}{\Delta^2(\mathbf{R},\mathbf{V}) + 2\Omega^2(\mathbf{R}) + \Gamma^2}\right), \quad (32)$$

where we have redefined the notation such that $\mathbf{R}$ now stands for the position of the atom (instead of $\mathbf{r}_0$). The above results are the same as those presented in our previous work where perturbation techniques for time-dependent Heisenberg operators have been used [11]. The dependence on the decay constant and on saturation are in agreement with our earlier heuristic approach.

B. Transients

Torrey [16] gave detailed solutions of the original optical Bloch equations. He also recognized that there were three special cases of interest that have relatively simple solutions. These were for strong collisions when the natural lifetime of the state may be replaced by the collision shortened lifetime, exact resonance and for intense external fields. His approach was applied by Allen and Eberly [17] to the optical Bloch equations. Consequently, in a similar way, the evolution of the forces from the instant the light beam is switched on can also be examined for a number of special cases. In fact, radiation effects have been examined in detail for atoms excited by plane-wave light [18]; the cases considered were (i) an atom with all relaxation constants equal to zero, (ii) a weak beam, (iii) exact resonance, and (iv) steady state achieved by an intense field. This treatment may be readily generalized for Laguerre-Gaussian light.

We shall settle simply for the steady-state case already considered in Sec. II A because the general time dependence of the density-matrix elements can be determined more readily for arbitrary parameter values by the numerical solution of the optical Bloch equations (26) and (27). This enables the evolution of the corresponding forces to be displayed. We display the results for a Laguerre-Gaussian mode

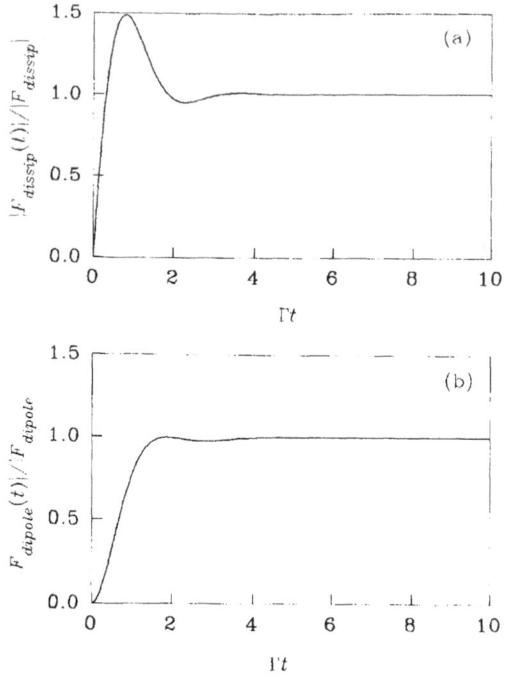

FIG. 1. Variation with time (in units of Γ^{-1}) of (a) the average dissipative force and (b) the average dipole force in units of the corresponding steady-state values for a stationary atom in a single LG beam. [The time variation of the corresponding torque would be the same as in (a), but in units of the steady-state torque.] See the text for the values assumed for the parameters.

with $l=1$ and $p=0$ such that $\Omega(\mathbf{r}_0)=\Gamma$; $\Delta=-\Gamma$ and $\omega(0)=35\lambda$, where $\lambda=280$ nm is the atomic transition wavelength. The results shown in Figs. 1(a) and 1(b) depict the evolutions of the dissipative and dipole forces [given by Eqs. (28) and (29), respectively]. These figures show clearly that the dipole moment, and hence force components, relax to their steady-state values within a time of the order of Γ^{-1}.

III. STEADY-STATE DYNAMICS

A. Single LG beam

From the results of the preceding section it is clear that, for elapsed times greater than the inverse relaxation parameters, the total force on a two-level atom has a steady-state value, exhibits position dependence, and is naturally divisible into two terms. Restoring the explicit reference to a specific Laguerre-Gaussian mode, the steady-state force on a moving atom due to a single Laguerre-Gaussian beam propagating along the positive z axis is written

$$\langle \mathbf{F}\rangle_{klp}=\langle \mathbf{F}_{\mathrm{diss}}\rangle_{klp}+\langle \mathbf{F}_{\mathrm{dipole}}\rangle_{klp}, \tag{33}$$

where

$$\langle \mathbf{F}_{\mathrm{diss}}(\mathbf{R},\mathbf{V})\rangle_{klp} = 2\hbar\Gamma\Omega^2_{klp}(\mathbf{R})\left(\frac{\nabla\Theta_{klp}(\mathbf{R})}{\Delta^2_{klp}(\mathbf{R},\mathbf{V})+2\Omega^2_{klp}(\mathbf{R})+\Gamma^2}\right) \tag{34}$$

and

$$\langle \mathbf{F}_{\mathrm{dipole}}(\mathbf{R},\mathbf{V})\rangle_{klp} = -2\hbar\Omega_{klp}(\mathbf{R})\nabla\Omega_{klp}\left(\frac{\Delta_{klp}(\mathbf{R},\mathbf{V})}{\Delta^2_{klp}(\mathbf{R},\mathbf{V})+2\Omega^2_{klp}(\mathbf{R})+\Gamma^2}\right), \tag{35}$$

where $\mathbf{R}(t)$ now denotes the current position vector of the atom and $\mathbf{V}=\dot{\mathbf{R}}$. The effective detuning $\Delta_{klp}(\mathbf{R},\mathbf{V})$ is now both position and velocity dependent

$$\Delta_{klp}(\mathbf{R},\mathbf{V})=\Delta_0-\mathbf{V}\cdot\nabla\Theta_{klp}(\mathbf{R},\mathbf{V}). \tag{36}$$

The dissipative force, proportional to the phase gradient, is given by Eq. (34). This force can be visualized as arising locally from the absorption followed by spontaneous emission of light by the atom. The dipole force, which is proportional to the gradient of the field intensity, subsumed in the position-dependent Rabi frequency, is given by Eq. (35). Both forces play important roles in the cooling and trapping of the atom. The dissipative component is responsible for the existence of a frictional force in a configuration involving two counterpropagating waves, while the dipole force confines the atom to the high-intensity regions of the field when the detuning is below resonance [7].

B. Low-velocity limit

In order to elucidate the nature of the interaction between the LG beam and the atom we consider the low-velocity limit of the dissipative and dipole forces. However, in the computational evaluation of the full extent of the interaction to study the dynamics of the atom described later in this paper, this approximation will not be made. The assumption involved in the low-velocity limit is that the Doppler shift induced by the motion of the atom is smaller than the atomic width $\mathbf{V}\cdot\nabla\Theta\ll\Gamma$. In this case we may expand the denominators of Eqs. (34) and (35) retaining terms up to those linear in the velocity. We can thus write each force as the sum of a static (velocity-independent) and dynamic (velocity-dependent) components. The static components are given by

$$\langle \mathbf{F}^0_{\mathrm{diss}}(\mathbf{R})\rangle_{klp}=\frac{2\hbar\Gamma\Omega^2_{klp}(\mathbf{R})}{\Delta_0^2+2\Omega^2_{klp}(\mathbf{R})+\Gamma^2}\left[\{\eta_{klp}(\mathbf{R})+k\}\hat{\mathbf{z}}+\frac{l}{r}\hat{\boldsymbol{\phi}} + \xi_k(\mathbf{R})\hat{\mathbf{r}}\right], \tag{37}$$

$$\langle \mathbf{F}^0_{\mathrm{react}}(\mathbf{R})\rangle_{klp}=-\frac{2\hbar\Delta_0\Omega_{klp}(\mathbf{R})\nabla\Omega_{klp}(\mathbf{R})}{\Delta_0^2+2\Omega^2_{klp}(\mathbf{R})+\Gamma^2} \tag{38}$$

and the dynamic components by

$$\langle \mathbf{F}^V_{\rm diss}(\mathbf{R},\mathbf{V})\rangle = \frac{4\hbar\Gamma\Delta_0\Omega^2_{klp}(\mathbf{R})}{[\Delta_0^2+2\Omega^2_{klp}(\mathbf{R})+\Gamma^2]^2}\left[\{\eta_{klp}(\mathbf{R})+k\}\hat{\mathbf{z}} + \frac{l}{r}\hat{\phi}+\xi_k(\mathbf{R})\hat{\mathbf{r}}\right]\left[\{\eta_{klp}(\mathbf{R})+k\}V_z+\frac{l}{r}V_\phi + \xi_k(\mathbf{R})V_r\right], \tag{39}$$

$$\langle \mathbf{F}^V_{\rm dipole}(\mathbf{R},\mathbf{V})\rangle = \frac{2\hbar\Omega_{klp}(\mathbf{R})\nabla\Omega(\mathbf{R})}{[\Delta_0^2+2\Omega^2_{klp}(\mathbf{R})+\Gamma^2]}\times\left(1-\frac{2\Delta_0^2}{[\Delta_0^2+2\Omega^2_{klp}(\mathbf{R})+\Gamma^2]}\right)\times\left[\{\eta_{klp}(\mathbf{R})+k\}V_z+\frac{l}{r}V_\phi+\xi_k(\mathbf{R})V_r\right]. \tag{40}$$

In the equations above V_z, V_ϕ, and V_r are, respectively, the axial, azimuthal, and radial components of the velocity and the functions $\eta_{klp}(\mathbf{R})$ and $\xi_k(\mathbf{R})$ are defined by

$$\eta_{\pm klp} = \pm\frac{kr^2}{2(z^2+z_R^2)}\left[1-\frac{2z^2}{z^2+z_R^2}\right]\pm(2p\pm l+1)\frac{z_R}{z^2+z_R^2}, \tag{41}$$

$$\xi_{\pm k}(\mathbf{R}) = \pm\frac{krz}{z^2+z_R^2}. \tag{42}$$

We make the additional assumption that the atom moves in a region of the beam for which $z \ll z_R$ and we can then ignore the z dependence in $\Omega_{klp}(\mathbf{R})$ and set $\eta_{klp}(\mathbf{R})=0$ and $\xi_k(\mathbf{R})=0$. We may also write to a good approximation

$$\nabla\Omega_{klp}(\mathbf{R}) \approx \left\{\left[\frac{|l|}{r}-\frac{2r}{w_0^2}\right]\Omega_{klp}(\mathbf{R}) - \frac{2\sqrt{2p}}{w_0}\Omega_{k|l|+1\,p-1}(\mathbf{R})\right\}\hat{\mathbf{r}}. \tag{43}$$

In the low-velocity limit with $z \ll z_R$, the static dissipative and dipole forces become

$$\langle \mathbf{F}^0_{\rm diss}(\mathbf{R})\rangle_{klp} \approx \frac{2\hbar\Gamma\Omega^2_{klp}(\mathbf{R})}{\Delta_0^2+2\Omega^2_{klp}(\mathbf{R})+\Gamma^2}\left[k\hat{\mathbf{z}}+\frac{l}{r}\hat{\phi}\right], \tag{44}$$

$$\langle \mathbf{F}^0_{\rm dipole}(\mathbf{R})\rangle_{klp} \approx -\frac{2\hbar\Delta_0\Omega_{klp}(\mathbf{R})}{\Delta_0^2+2\Omega^2_{klp}(\mathbf{R})+\Gamma^2}\left(\left[\frac{|l|}{r}-\frac{2r}{w_0^2}\right]\Omega_{klp}(\mathbf{R}) - \frac{2\sqrt{2p}}{w_0}\Omega_{k|l|+1\,p-1}(\mathbf{R})\right)\hat{\mathbf{r}}, \tag{45}$$

while the dynamic dissipative and dipole forces become

$$\langle \mathbf{F}^V_{\rm diss}(\mathbf{R},\mathbf{V})\rangle \approx \frac{4\hbar\Gamma\Delta_0\Omega^2_{klp}(\mathbf{R})}{[\Delta_0^2+2\Omega^2_{klp}(\mathbf{R})+\Gamma^2]^2}\left[\left\{k^2V_z+\frac{kl}{r}V_\phi\right\}\hat{\mathbf{z}} + \frac{kl}{r}V_z\hat{\phi}\right], \tag{46}$$

$$\langle \mathbf{F}^V_{\rm dipole}(\mathbf{R},\mathbf{V})\rangle \approx \frac{2\hbar\Omega_{klp}(\mathbf{R})}{[\Delta_0^2+2\Omega^2_{klp}(\mathbf{R})+\Gamma^2]^2}\times\left(1-\frac{2\Delta_0^2}{[\Delta_0^2+2\Omega^2_{klp}(\mathbf{R})+\Gamma^2]}\right)\times\left[kV_z+\frac{l}{r}V_\phi\right]\left(\left[\frac{|l|}{r}-\frac{2r}{w_0^2}\right]\Omega_{klp}(\mathbf{R}) - \frac{2\sqrt{2p}}{w_0}\Omega_{k|l|+1\,p-1}(\mathbf{R})\right)\hat{\mathbf{r}}. \tag{47}$$

Equations (44) and (46) show that the dissipative force has static components in both the axial and azimuthal directions; the latter is equivalent to a torque about the beam axis. These forces combine with dynamic components in the axial and azimuthal directions. Note that within this approximation, Eq. (46) shows that there is a reciprocal relationship between the axial and azimuthal motions. An atom moving initially in the z direction will induce a force in the azimuthal direction and vice versa. It may be seen from Eqs. (45) and (47) that the dipole force consists of static and dynamic components, both of which are in the radial direction.

The static component of the dipole force, given by Eq. (38), attracts the atom to the high-intensity regions of the field when the detuning is below resonance. This force can be derived from a potential [7]

$$\langle U(\mathbf{R})\rangle_{klp} = \frac{\hbar\Delta_0}{2}\ln\left[1+\frac{2\Omega^2_{klp}(\mathbf{R})}{\Delta_0^2+\Gamma^2}\right] \tag{48}$$

such that $\langle \mathbf{F}^0_{klp}\rangle = -\nabla\langle U(\mathbf{R})\rangle_{klp}$. This potential exhibits minima in the high-intensity regions of the beam for an atom tuned below resonance where $\Delta_0<0$. For $\Delta_0>0$, we have trapping in the low-intensity (dark) regions of the field. As an illustration, we consider the LG mode for which $l=1$, $p=0$. The potential is

$$\langle U\rangle_{k10} = \frac{\hbar\Delta_0}{2}\ln\left[1+\frac{2\Omega^2_{k10}(\mathbf{R})}{\Delta_0^2+\Gamma^2}\right]. \tag{49}$$

At the beam waist $z=0$, the minimum occurs at $r=r_0$ where

$$r_0 = w_0/\sqrt{2}. \tag{50}$$

For a beam propagating along the z axis it is easy to verify that the locus of the potential minimum in the xy plane is a circle given by

$$x^2+y^2=r_0^2. \tag{51}$$

Expanding the potential in powers of $(r-r_0)$ we have the parabolic approximation

$$\langle U\rangle_{k10} \approx U_0+\tfrac{1}{2}\Lambda_{k10}(r-r_0)^2, \tag{52}$$

where U_0 is the potential depth given by

$$U_0=\frac{1}{2}\hbar\Delta_0 \ln\left[1+\frac{2\Omega_{k10}^2(r_0)}{\Delta_0^2+\Gamma^2}\right] \tag{53}$$

and Λ_{k10} is an effective elastic constant given by

$$\Lambda_{k10}=\frac{4\hbar|\Delta_0|}{\Delta_0^2+2e^{-1}\Omega_{k00}^2+\Gamma^2}\left(\frac{e^{-1}\Omega_{k00}^2}{w_0^2}\right). \tag{54}$$

The atom is considered trapped if its kinetic energy is less than U_0 and will exhibit quasiharmonic vibrational motion about $r=r_0$. The characteristic angular frequency is equal to $\sqrt{\Lambda_{k10}/M}$, where M is the atomic mass.

C. Counterpropagating LG beams

1. One-dimensional case

We have seen above that an atom immersed in a Laguerre-Gaussian beam will experience a dissipative force that is predominantly in the direction of propagation and a dipole force in the radial direction. If a second beam is added propagating in the opposite direction, we have a configuration that can be referred to as the one-dimensional (1D) counterpropagating beam configuration. In this paper the beams are assumed to be independent of each other in that their phases are not locked. The case in which the beams are phase locked to form a standing wave will be considered elsewhere. For independent counterpropagating LG beams we can write the mean force on the atom as a sum of forces due to individual beams

$$\langle \mathbf{F}_{\text{diss}}\rangle_{kl_1p_1,-kl_2p_2}=2\hbar\Gamma\Omega_{klp}^2(\mathbf{R})\times\left[\frac{\nabla\Theta_{kl_1p_1}(\mathbf{R})}{\Delta_{kl_1p_1}^2(\mathbf{R},\mathbf{V})+2\Omega_{kl_1p_1}^2(\mathbf{R})+\Gamma^2}+\frac{\nabla\Theta_{-kl_2p_2}(\mathbf{R})}{\Delta_{-kl_2p_2}^2(\mathbf{R},\mathbf{V})+2\Omega_{-kl_2p_2}^2(\mathbf{R})+\Gamma^2}\right], \tag{55}$$

$$\langle \mathbf{F}_{\text{dipole}}\rangle_{kl_1p_1,-kl_2p_2}=-2\hbar\Omega_{klp}(\mathbf{R})\nabla\Omega_{klp}\times\left[\frac{\Delta_{kl_1p_1}(\mathbf{R},\mathbf{V})}{\Delta_{kl_1p_1}^2(\mathbf{R},\mathbf{V})+2\Omega_{kl_1p_1}^2(\mathbf{R})+\Gamma^2}+\frac{\Delta_{-kl_2p_2}(\mathbf{R},\mathbf{V})}{\Delta_{-kl_2p_2}^2(\mathbf{R},\mathbf{V})+2\Omega_{-kl_2p_2}^2(\mathbf{R})+\Gamma^2}\right], \tag{56}$$

where we have assumed that $p_1=p_2=p$ and either $(l_1=-l_2=l)$ or $(l_1=l_2=l)$.

In the low-velocity regime, for an atom close to the beam waist, we may make use of Eqs. (44)–(47). The total static dissipative and dipole forces are then given by

$$\langle \mathbf{F}_{\text{diss}}^0(\mathbf{R})\rangle_{kl_1p_1-kl_2p_2}\approx\frac{2\hbar\Gamma\Omega_{klp}^2(\mathbf{R})}{\Delta_0^2+2\Omega_{klp}^2(\mathbf{R})+\Gamma^2}\left[\frac{l_1}{r}+\frac{l_2}{r}\right]\hat{\phi}, \tag{57}$$

$$\langle \mathbf{F}_{\text{dipole}}^0(\mathbf{R})\rangle_{kl_1p_1-kl_2p_2}\approx-\frac{4\hbar\Delta_0\Omega_{klp}(\mathbf{R})}{\Delta_0^2+2\Omega_{klp}^2(\mathbf{R})+\Gamma^2}\times\left(\left[\frac{|l|}{r}-\frac{2r}{w_0^2}\right]\Omega_{klp}(\mathbf{R})-\frac{2\sqrt{2p}}{w_0}\Omega_{k|l|+1\,p-1}(\mathbf{R})\right)\hat{\mathbf{r}} \tag{58}$$

and the total dynamic dissipative and dipole forces are

$$\langle \mathbf{F}_{\text{diss}}^V(\mathbf{R},\mathbf{V})\rangle_{kl_1p_1-kl_2p_2}\approx\frac{4\hbar\Gamma\Delta_0\Omega_{klp}^2(\mathbf{R})}{[\Delta_0^2+2\Omega_{klp}^2(\mathbf{R})+\Gamma^2]^2}\times\left[\left\{2k^2V_z+\frac{k}{r}(l_1-l_2)V_\phi\right\}\times\hat{\mathbf{z}}+\frac{k}{r}(l_1-l_2)V_z\hat{\phi}\right], \tag{59}$$

$$\langle \mathbf{F}_{\text{dipole}}^V(\mathbf{R},\mathbf{V})\rangle_{kl_1p_1-kl_2p_2}\approx\frac{2\hbar\Omega_{klp}(\mathbf{R})}{[\Delta_0^2+2\Omega_{klp}^2(\mathbf{R})+\Gamma^2]}\times\left(1-\frac{2\Delta_0^2}{[\Delta_0^2+2\Omega_{klp}^2(\mathbf{R})+\Gamma^2]}\right)\times\left(\left[\frac{|l|}{r}-\frac{2r}{w_0^2}\right]\Omega_{klp}(\mathbf{R})-\frac{2\sqrt{2p}}{w_0}\Omega_{k|l|+1\,p-1}(\mathbf{R})\right)\times(l_1+l_2)\frac{V_\phi}{r}\hat{\mathbf{r}}. \tag{60}$$

From Eq. (58) we see that the velocity-independent dipole force is simply double that of a single-beam case. The dissipative force, however, depends on the relative signs of l_1 and l_2. For $l_1=l_2=l$ we have

$$\langle \mathbf{F}_{\text{dipole}}^0(\mathbf{R})\rangle_{klp-klp}\approx\frac{4\hbar\Gamma\Omega_{klp}^2(\mathbf{R})}{\Delta_0^2+2\Omega_{klp}^2(\mathbf{R})+\Gamma^2}\left(\frac{l}{r}\right)\hat{\phi}, \tag{61}$$

$$\langle \mathbf{F}_{\text{diss}}^V(\mathbf{R},\mathbf{V})\rangle_{klp-klp}\approx\frac{8\hbar\Gamma\Delta_0\Omega_{klp}^2(\mathbf{R})}{[\Delta_0^2+2\Omega_{klp}^2(\mathbf{R})+\Gamma^2]^2}k^2V_z\hat{\mathbf{z}}. \tag{62}$$

Thus, for $l_1=l_2=l$ we have a torque about the beam axis and an axial cooling or heating force, depending on the sign of Δ_0.

For the case $l_1=-l_2=l$, we have

$$\langle \mathbf{F}_{\text{diss}}^0(\mathbf{R})\rangle_{klp-k-lp}=0, \tag{63}$$

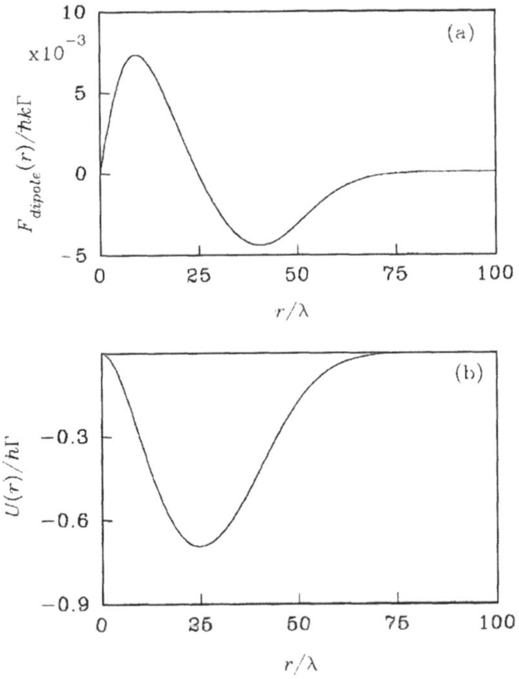

FIG. 2. (a) Radial distribution of the dipole force due to 1D counterpropagating Laguerre-Gaussian beams at $z=0$. Here $l_1=-l_2=1$, $p_1=p_2=0$, and the parameters are $\Delta_0=-\Gamma$, $\Omega_{k00}=1.648\Gamma$, and $\omega_0=35\lambda$. (b) Radial potential distribution corresponding to (a).

$$\langle \mathbf{F}^V_{\text{diss}}(\mathbf{R},\mathbf{V})\rangle_{klp-k-lp} \approx \frac{8\hbar\Gamma\Delta_0\Omega^2_{klp}(\mathbf{R})}{[\Delta_0^2+2\Omega^2_{klp}(\mathbf{R})+\Gamma^2]^2} \times\left\{\left[k^2V_z+\frac{kl}{r}V_\phi\right]\hat{\mathbf{z}}+\frac{kl}{r}V_z\hat{\phi}\right\}. \tag{64}$$

The static force in this case is zero, while the velocity-dependent force contains extra terms that arise from the orbital angular momentum of the counterpropagating Laguerre-Gaussian beams. As with the one-beam case, we again see force components arising from the reciprocating interchange between the axial and azimuthal motions.

2. Two-dimensional, three-dimensional, and three coplanar beams

The 2D case arises when a second pair of counterpropagating LG beams is arranged orthogonal to the first pair. The total force can again be written as a sum of forces from each of the beams. However, in addition to the reciprocal action between the azimuthal and axial motions in each pair of beams, there is also the fact that the azimuthal atomic motion associated with one beam is part of the axial motion in the other. In other words, there is an additional level of reciprocity between the components of the motion arising from the presence of two pairs of counterpropagating beams.

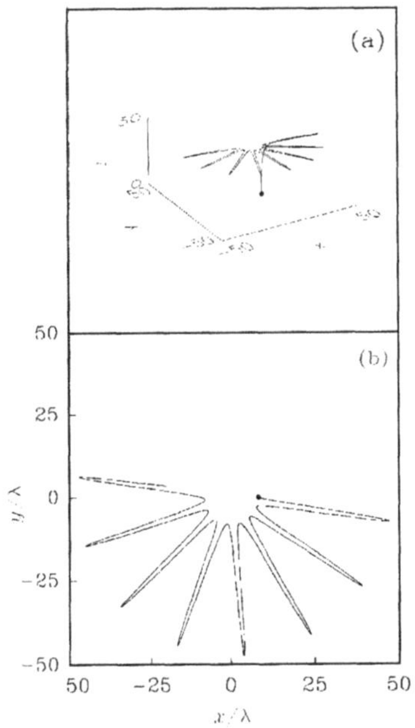

FIG. 3. (a) Trajectory of a Mg^+ ion in 1D counterpropagating LG beams for $l_1=-l_2=1$ and $p_1=p_2=0$. All distances are in units of the wavelength of the light λ. The initial position is $\mathbf{R}_0=8\lambda\hat{\mathbf{x}}$ and the initial velocity components are $V_z=5.0$ ms^{-1} and $V_r=0=V_\phi$. (b) Projection in the xy plane of the ion trajectory shown in (a). In this and subsequent figures, the initial position is indicated by a full circle. See the text for the values assumed for the other parameters.

There are also two overlapping dipole potential distributions arising from the orthogonal beams. It is easy to see for beam pairs for which $l_1=-l_2=1$ and $p_1=p_2=0$ and where the axes are such that one pair is along the z axis and the second along the x axis, the potential minima are four times as deep as that of a single beam. The minima are situated at the space points defined by the two equations

$$x^2+y^2=w_0^2/2, \tag{65}$$

$$y^2+z^2=w_0^2/2. \tag{66}$$

These equations apply the additional constraint $x=\pm z$. Atoms subject to such 2D counterpropagating beams will congregate at points lying on the curve defined by two intersecting circles, one on the plane $x+z=0$ and the other on the plane $x-z=0$.

When a third set of beams is arranged orthogonal to the other two orthogonal pairs we have 3D counterpropagating LG beams. The common potential minima in this case occur at eight distinct points defined by

$$x=\pm w_0/\sqrt{2}, \quad y=\pm w_0/\sqrt{2}, \quad z=\pm w_0/\sqrt{2} \tag{67}$$

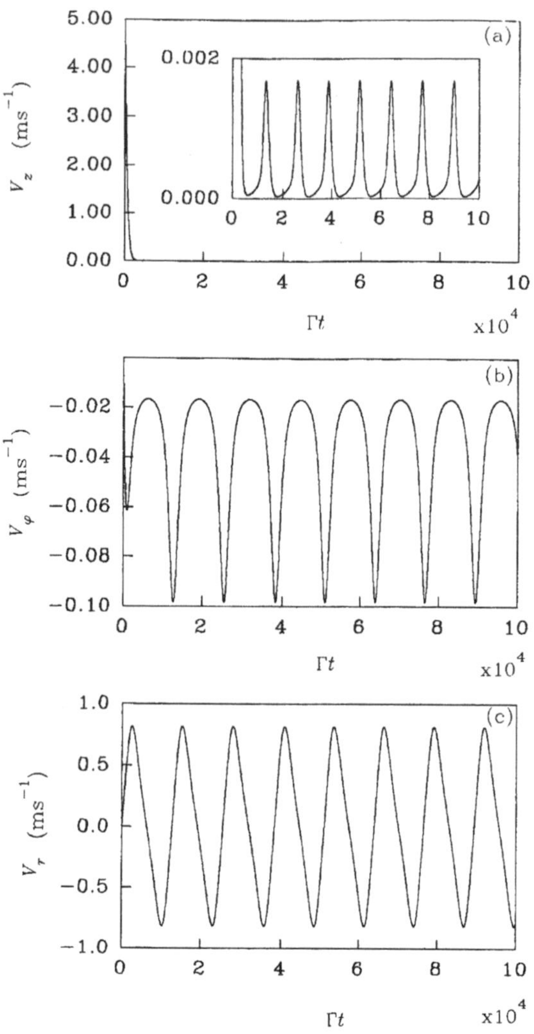

FIG. 4. Variations of the velocity components for the case in Fig. 3. (a) Evolution of V_z indicating axial cooling. The inset to this figure shows small oscillations of V_z due to reciprocating effects. (b) Evolution of V_ϕ and (c) evolution of V_r. Both (b) and (c) indicate the rapid onset of oscillatory motions of the same period.

and are six times as deep as the potential due to one beam. However, in this case the detailed polarization gradients are such as to make further study of this configuration nontrivial.

Finally, we consider the case of three coplanar beams [19] in the x-y plane in a symmetric configuration in which the angle between adjacent beams is $2\pi/3$. This leads to three overlapping circles that meet at two distinct points at

$$x=0, \quad y=0, \quad z=\pm w_0/\sqrt{2} \tag{68}$$

and the potential well is three times as deep as for a single beam.

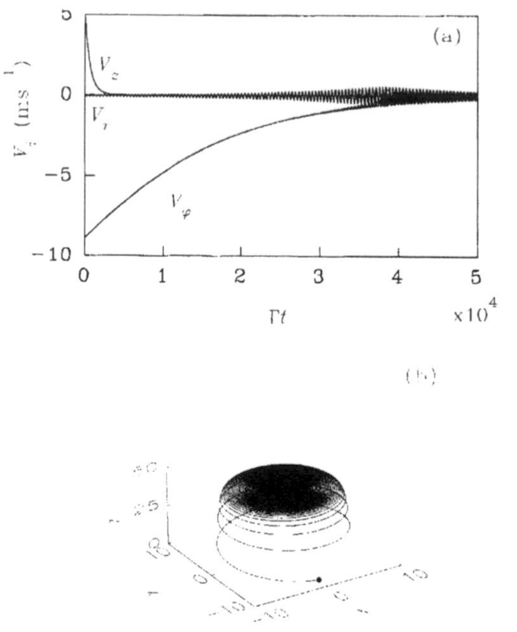

FIG. 5. (a) Evolution of the velocity components of a Mg^+ ion subject to Laguerre-Gaussian 1D counterpropagating beams with $l_1=l_2=1$, $p_1=p_2=0$, and $|\mathbf{B}|=1$ T. Initially the ion possesses both azimuthal and axial velocity components $V_z=5.0$ ms^{-1} and $V_\phi=-8.9$ ms^{-1}. (b) Trajectory of the Mg^+ in (a). All distances are in units of the wavelength of the light λ.

IV. Mg^+ IN MULTIPLE BEAMS

The emphasis throughout this paper is on the physics introduced by the orbital angular momentum aspects of the interaction of atoms with Laguerre-Gaussian light. In the theoretical analysis presented in the preceding section we were able to infer that an atom in a configuration of such beams is subject to axial forces and various forms of static and dynamic rotational forces and that axial and rotational motions influence each other in a rather intricate way. Furthermore, a system of multiple Laguerre-Gaussian beams presents an atom with well-defined potential landscapes that depend on the angular momentum quantum numbers and beam configuration. For example, in the 1D case with a given set of parameters, a given atom should have well-defined quasiharmonic vibrational states associated with the potential profiles.

To illustrate these features we consider the case of Mg^+ in Laguerre-Gaussian light. The Mg^+ mass is $M=4.0\times10^{-26}$ kg; the transition wavelength is $\lambda=280.1$ nm and its half-width is $\Gamma=2.7\times10^8$ s^{-1}. To illustrate the theory typical beam parameters are exemplified by the choices $\Delta_0=-\Gamma$, $\Omega_{k00}=1.648\Gamma$, and $w_0=35\lambda$. The equation of motion of a Mg^+ ion in multiple LG beams is written as

$$M\frac{\partial^2\mathbf{R}(t)}{\partial t^2}=\sum_i \{\langle\mathbf{F}_{\mathrm{diss}}(\mathbf{R},\mathbf{V})\rangle_{k_il_ip_i}+\langle\mathbf{F}_{\mathrm{dipole}}(\mathbf{R},\mathbf{V})\rangle_{k_il_ip_i}\} + Q\mathbf{V}\times\mathbf{B}, \tag{69}$$

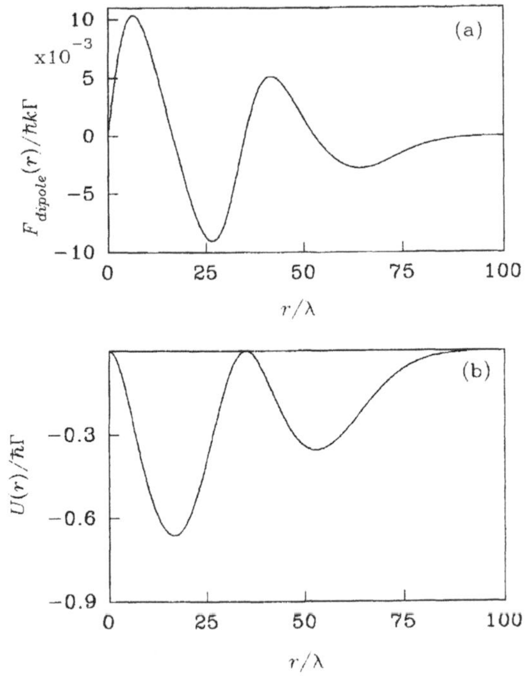

FIG. 6. (a) Radial distribution of the dipole force due to 1D counterpropagating Laguerre-Gaussian beams at $z=0$. Here $l_1=-l_2=1$, $p_1=p_2=1$, and the parameters are $\Delta_0=-\Gamma$, $\Omega_{k00}=1.648\Gamma$, and $\omega_0=35\lambda$. (b) Radial potential distribution corresponding to (a).

where Q is the ionic charge and we have included the last term on the right-hand side to allow for the possibility of an applied external magnetic field. The summation indicates the vector addition over force contributions arising from individual beams. The forces from each beam are taken in their unapproximated forms given by Eqs. (34) and (35).

We begin by considering 1D counterpropagating beams in the absence of the magnetic field. Figure 2(a) displays the dipole force as given by Eq. (56) as a function of radial distance r at $y=0$ and $z=0$. The beam quantum numbers are such that $l_1=-l_2=1$ and $p_1=p_2=0$ and the parameters are $\Delta_0=-\Gamma$, $\Omega_{k00}=1.648\Gamma$, and $w_0=35\lambda$. Figure 2(b) displays the corresponding radial potential distribution. The maximum intensity is located at points where $r=w_0/\sqrt{2}=24.75\lambda$. As expected, we see that for $\Delta_0<0$ the dipole potential exhibits a minimum at points where the intensity is maximum. The vibrational states in the parabolic approximation have an elastic constant that is twice that for the one-beam case as given by Eq. (54). The vibrational frequency corresponding to the above parameters is

$$\nu \approx \frac{1}{2\pi}\left(\frac{8\hbar|\Delta_0|e^{-1}\Omega_{k00}^2}{Mw_0^2[\Delta_0^2+2e^{-1}\Omega_{k00}^2+\Gamma^2]}\right)^{1/2}. \qquad (70)$$

For the parameter values specified above this yields $\nu\approx 2.0\times10^4\Gamma$.

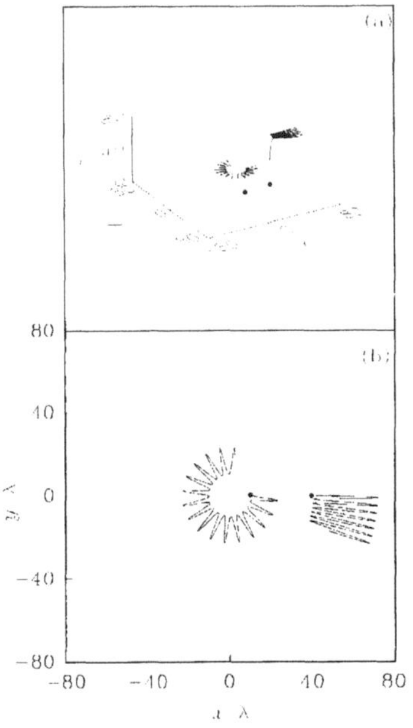

FIG. 7. (a) Trajectories of a Mg^+ ion in 1D counterpropagating LG beams for $l_1=-l_2=1$ and $p_1=p_2=1$ for two different initial positions: one is at $\mathbf{R}_0=10\lambda\hat{\mathbf{x}}$ and the second is at $\mathbf{R}_0=40\lambda\hat{\mathbf{x}}$. All distances are in units of the wavelength of the light λ. The initial velocity components in both cases are $V_z=5.0$ ms^{-1} and $V_r=0=V_\phi$. (b) Projections in the xy plane of the ion trajectories shown in (a). See the text for the assumed values of the other parameters.

Figure 3(a) displays the trajectory of the ion as a function of time and Fig. 3(b) depicts its projection onto the xy plane. The initial position is at $\mathbf{R}_0=8\lambda\hat{\mathbf{x}}$ and the initial velocity components are $V_z=5.0$ ms^{-1}, $V_\phi=0=V_r$. It is clear from the figure that the atom, subject to an axial friction force, has been slowed axially. Once the atom is moving sufficiently slowly, it starts a vibrational motion about the radial coordinate $r=w_0/\sqrt{2}$, accompanied by a slow rotational motion. The latter, according to Eq. (46), is attributed to the azimuthal component of the dissipative force induced by the axial motion.

Figure 4 displays the evolution of the velocity components. The axial velocity is seen to decay almost to zero. However, closer inspection, as shown by the inset to Fig. 4(a), reveals that the axial motion exhibits periodic oscillations that are attributable to a reciprocating force arising from the periodic azimuthal motion, depicted in Fig. 4(b). The period associated with these figures is indeed about $2.0\times10^4\Gamma$ as in the estimate based on Eq. (70). An important feature displayed by the results depicted in Figs. 3 and 4 is that changing the sign of the angular momentum quantum number l from $+1$ to -1, which is readily achievable [8,9], causes the change in the rotational motion from clockwise to the opposite (counterclockwise) sense.

Figure 5 is concerned with the case $l_1=l_2=+1$ and $p_1=p_2=0$ in the presence of a magnetic field $|\mathbf{B}|=1$ T directed along the positive z axis. From Eqs. (61) and (62) we deduce that, besides the ion cyclotron motion due the magnetic field, the main effects are in the form of an axial friction force provided that $\Delta_0<0$ and a static torque about the beam axis that acts upon the ion azimuthally. Figure 5(a) displays the evolution of velocity components for the case $\Delta_0=-\Gamma$. The initial ion position is $\mathbf{R}(0)=-8\lambda\hat{\mathbf{y}}$ and the initial velocity components are $V_r=0$, $V_\phi=-8.9$ ms^{-1}, and $V_z=5.0$ ms^{-1}. We see that the torque due to the LG beams generates a braking effect on the cyclotron motion, while the axial motion is gradually cooled by the axial friction force. All these features can be inferred from the trajectory shown in Fig. 5(b). If the sign of l in both beams were to be changed, but $\mathbf{B}$ kept in the same direction, we would have heating of the azimuthal motion, while the axial motion would still be cooled. Clearly the former case amounts to a decrease in angular motion due to the LG beam, while the latter is equivalent to the enhancement of the angular motion. These phenomena are attributable only to the angular momentum properties of the LG beams [20].

Figure 6 is concerned with counterpropagating beams with the next-higher-order LG modes $l_1=-l_2=1$ and $p_1=p_2=1$ and no external magnetic field. Other parameter values are $\Omega_{k00}=1.648$ and $\Delta_0=-\Gamma$. Figures 3(a) and 3(b) display the radial distribution of the dipole force and corresponding potential, respectively. We now have two potential wells with minima at $r=0.468w_0\equiv16.38\lambda$ and $r=1.5w_0\equiv52.86\lambda$. The ion is destined to oscillate about one of these points, depending on the initial conditions. This can be seen in Fig. 7 for an atom with $V_x=0=V_y$ and $V_z=5.0$ ms^{-1}. The inner curve depicts the trajectory when the atom begins at $\mathbf{R}_0=10\lambda\hat{\mathbf{x}}$ and the outer curve when it begins $\mathbf{R}_0=40\lambda\hat{\mathbf{y}}$.

Figure 8(a) shows the trajectory in a 2D counterpropagating beam case with $l_1=-l_2=1$ and $p_1=p_2=0$ and Fig. 8(b) shows the corresponding projection in the xy plane. The initial position is at $\mathbf{R}_0=10\lambda\hat{\mathbf{x}}$ and the initial velocity components are (0.02,0.02,0.05) ms^{-1}. Figure 9 shows the evolution of the velocity components. From Figs. 8 and 9 it can be seen that the atom is subject to friction forces from all directions, which result in it coming to rest at a point within the potential profile. We have shown earlier that the locus of the dipole potential minimum for the 2D case with (1,0) beams is in the form of two intersecting circles, satisfying Eqs. (65) and (66). This is shown in Fig. 10 for the case $w_0=35\lambda$. The trajectory end point for the case depicted in Figs. 8 and 9 lies on the curve shown in Fig. 10. Thus this theory assigns predetermined end points for a given ion under given initial conditions. That this is clearly the case can be seen from Table I, where the coordinates of the trajectory end point recorded for various starting points satisfy Eqs. (65) and (66) for the two intersecting circles shown in Fig. 10.

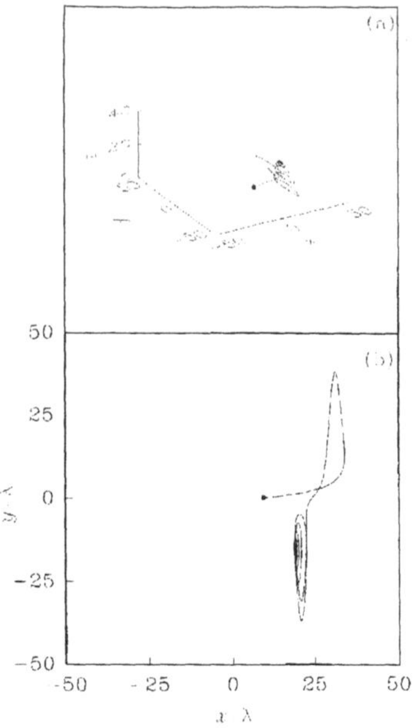

FIG. 8. (a) Trajectory of a Mg^+ ion in 2D counterpropagating LG beams involving two orthogonal pairs: one pair has the z axis as a common axis with $l_1=-l_2=1$ and $p_1=p_2=0$ and the second pair has the y axis as a common axis and $l_3=-l_4=1$ and $p_3=p_4=0$. All distances are in units of the wavelength of the light λ. The initial position of the Mg^+ ion is at $\mathbf{R}_0=20\lambda\hat{\mathbf{x}}$ and the initial velocity components are (0.02, 0.02, 0.05) ms^{-1}. (b) Projection in the xy plane of the trajectory in (a).

V. COMMENTS AND CONCLUSIONS

In this paper we have explored the nature of the radiation forces and their influence on atomic motion for a specific type of laser light, namely, Laguerre-Gaussian laser light in the form of a single beam and for multiple beams in various configurations. We have emphasized from the outset that the orbital angular momentum effects characterizing these modes give rise to different physical phenomena when such light is made to interact with atoms at near resonance. We have shown that a variety of forces come into play when the LG light is arranged in well-defined multiple beams particularly linear, orthogonal 2D, and symmetric coplanar three-beam configurations. We have, for simplicity, considered only coaxial multiple LG beams of the same kind whose beam waists coincide and have assumed that all counterpropagating pairs have the same magnitude of orbital angular momentum quantum numbers l and p. Notwithstanding the simplification inherent in these symmetric configurations, the physics has been intricate, but has given rise to effects associated with the orbital angular momentum of LG beams.

The results show that LG light generates a potential arising from the dipole force, while the dissipative force provides a mechanism to cool the atom axially and that there is a torque that can be utilized to cool or heat the azimuthal motion. Furthermore, there are reciprocating forces involving an interplay between motions in orthogonal directions that can generate oscillatory and precessional motions.

The model we have adopted involves linearly polarized light. We have also assumed that the beams are independent and possess no fixed phase relationship, thus excluding interference or multiphoton processes [1], for example, absorption from one beam followed by emission into the other beam. The atom responds therefore to the sum of the individual forces acting upon it. This is distinct from the case in which the two counterpropagating beams form a standing

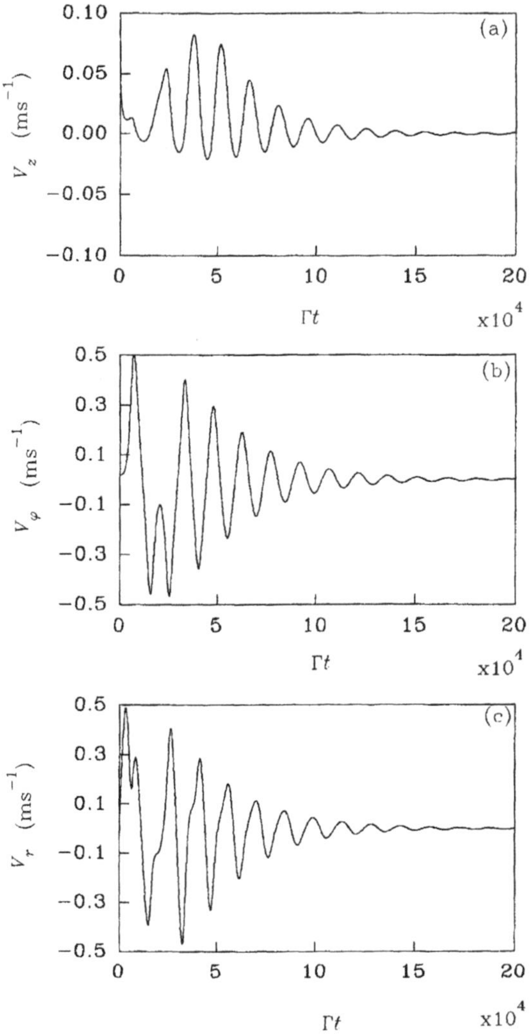

FIG. 9. Evolution of the velocity components corresponding to case with initial position at $\mathbf{R}_0 = 10\lambda\hat{\mathbf{x}}$ (a) $V_z(t)$, (b) $V_\phi(t)$, and (c) $V_r(t)$. Note that all components of velocity go to zero after a sufficiently long time.

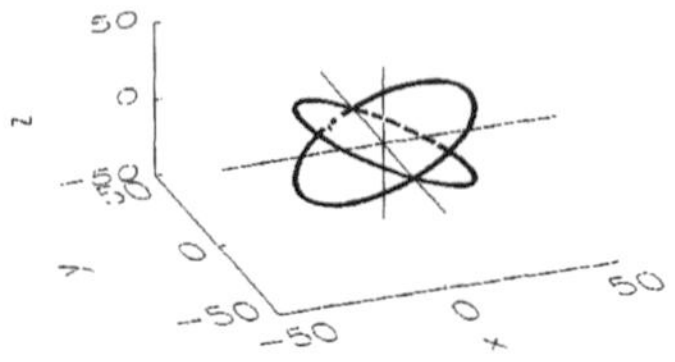

FIG. 10. Locus of spatial points where the dipole potential profile due to a system of two orthogonal pairs of counterpropagating LG beams possesses the lowest minimum. All distances are in units of the wavelength of the light λ.

wave and the possible Doppleron effects [1] that can arise under such circumstances. Work on the case of a standing LG wave is planned to be reported elsewhere.

In summary, this paper has dealt with the basic features that can arise when an atom interacts with multiple Laguerre-Gaussian beams possessing orbital angular momentum. The main effects of LG light on atomic behavior are elucidated for the 1D counterpropagating beams where we find reciprocity between axial and azimuthal motions and the existence of a static torque and a characteristic dipole potential. In the case of 2D counterpropagating beams the potential profile indicates that cooled atoms are forced to congregate into well-defined loci depending on the size of the light beams. Initial cooling need not be effected by the same LG beams; the primary aim of the LG beams at the late cooling stages could be the installation of the dipole potential; artificially generated dark field beams have been experimentally exploited [21]. We have illustrated the results by considering the case of beams of order of tens of wavelengths diameter. This results in atoms sitting on loci separated by distances of tens of wavelengths. However, the generation of results for diameters of order of millimeters is straightforward and would lead to atomic loci separated by distances in the mms scale. We have also briefly considered the loci for the 3D case and for the three coplanar converging LG beams. For these cases no solutions of the dynamical equation were presented and we have only pointed out the characteristic potential profiles and the points at which cooled atoms would congregate.

The effects of orbital angular momentum have been discussed here in connection with Laguerre-Gaussian modes. It

TABLE I. Coordinates x_f, y_f, z_f of the trajectory end points against initial coordinates x_0, y_0, z_0. All distances are in units of the wavelength of the light λ. The last three columns demonstrate that the ion end points always lie on the two intersecting circles shown in Fig. 10.

x_0	y_0	z_0	x_f	y_f	z_f	$\sqrt{x_f^2+y_f^2}$	$\sqrt{y_f^2+z_f^2}$	$w_0/\sqrt{2}$
20	0	0	15.96	18.92	15.96	24.75	24.75	24.75
20	20	0	2.84	24.61	2.84	24.75	24.75	24.75
−20	20	10	−14.27	20.22	14.27	24.75	24.75	24.75
−20	20	−10	−13.81	20.54	−13.81	24.75	24.75	24.75

appears likely, however, that the dynamically induced stabilization of the atomic motion, the so-called supermolasses configuration, which arises from a small displacement of the molasses fields [22], can be related to the azimuthal forces arising from orbital angular momentum, as probably can the macroscopic vortex force due to the offset beams in the spin-polarized spontaneous force-atom trap of Walker *et al.* [23]. As we have shown, the orbital angular momentum of the Laguerre-Gaussian modes is explicit and their influence on atomic motion is more straightforward to interpret.

ACKNOWLEDGMENTS

The authors are grateful to Dr. E. Riis and Dr. R. C. Thompson for useful discussions. This work was carried out under EPSRC Grant No. GR/J64009, which also provided support for W.K.L. and V.E.L. L.A. was working at JILA, University of Colorado during the preparation of the paper.

---

[1] V. S. Letokhov and V. G. Minogin, *Laser Light Pressure on Atoms* (Gordon and Breach, New York, 1987).

[2] A. P. Kazantsev, G. I. Surdutovich, and V. P. Yakovlev, *Mechanical Action of Light on Atoms* (World Scientific, Singapore, 1990).

[3] W. D. Phillips, in *Fundamental Systems in Quantum Optics*, Proceedings of the International School of Physics, Les Houches, Session LIII, 1990, edited by J. Dalibard, J.-M. Raimond, and J. Zinn-Justin (Elsevier, Amsterdam, 1992).

[4] H. Metcalf and P. van der Straten, Phys. Rep. **244**, 203 (1994).

[5] A. Ashkin, Phys. Rev. Lett. **24**, 156 (1970); **25**, 1321 (1970).

[6] D. Wineland and H. Dehmelt, Bull. Am. Phys. Soc. **20**, 637 (1975); T. Hansch and A. Schawlow, Opt. Commun. **13**, 68 (1975).

[7] S. Chu, J. E. Bjorkholm, A. Ashkin, and A. Cable, Phys. Rev. Lett. **57**, 314 (1986).

[8] L. Allen, M. W. Beijersbergen, R. J. C. Spreeuw, and J. P. Woerdman, Phys. Rev. A **45**, 8185 (1992).

[9] M. W. Beijersbergen, L. Allen, H. E. L. O. van der Veen, and J. P. Woerdman, Opt. Commun. **96**, 123 (1993); S. M. Barnett and L. Allen, *ibid.* **110**, 670 (1994); S. J. van Enk and G. Nienhaus, *ibid.* **94**, 147 (1992).

[10] L. Allen, V. E. Lembessis, and M. Babiker, Phys. Rev. A **53**, R2937 (1996).

[11] W. L. Power, L. Allen, M. Babiker, and V. E. Lembessis, Phys. Rev. A **52**, 479 (1995); M. Babiker, L. Allen, and W. L. Power, Phys. Rev. Lett. **73**, 1239 (1994); L. Allen, M. Babiker, and W. L. Power, Opt. Commun. **112**, 141 (1994).

[12] R. J. Cook, Phys. Rev. A **20**, 224 (1979).

[13] J. Dalibard and C. Cohen-Tannoudji, J. Phys. B **18**, 1661 (1985).

[14] H. A. Haus, *Waves and Fields in Optoelectronics* (Prentice-Hall, Englewood Cliffs, NJ, 1984).

[15] C. Cohen-Tannoudji, J. Dupont-Roc, and G. Grynberg, *Atom-Photon Interactions: Basic Processes and Applications* (Wiley Interscience, New York, 1975).

[16] H. C. Torrey, Phys. Rev. **76**, 1059 (1949).

[17] L. Allen and J. H. Eberly, *Optical Resonance and Two-Level Atoms* (Wiley Interscience, New York, 1975).

[18] H. Al-Hilfy and R. Loudon, Opt. Acta **32**, 995 (1985).

[19] G. Grynberg, B. Lunis, P. Verkerk, J.-Y. Courtis, and C. Salomon, Phys. Rev. Lett. **70**, 2249 (1993).

[20] M. Babiker, V. E. Lembessis, W. K. Lai, and L. Allen, Opt. Commun. **123**, 523 (1996).

[21] M. H. Anderson, W. Petrich, J. R. Ensher, and E. A. Cornell, Phys. Rev. A **50**, R3597 (1994).

[22] S. Chu, M. G. Prentiss, A. Cable, and J. Bjorkholm, in *Laser Spectroscopy VII*, edited by W. P. Person and S. Svanberg (Springer-Verlag, Berlin, 1987); V. S. Banato, N. Bigelow, G. I. Surdutovitch, and S. Zilio, Opt. Lett. **19**, 1568 (1994).

[23] T. Walker, P. Feng, D. Hoffman, and R. S. Williamson III, Phys. Rev. Lett. **69**, 2168 (1992).

Selection rules and centre-of-mass motion of ultracold atoms

S J van Enk
Max-Planck Institut für Quantenoptik, Hans-Kopfermann-Strasse 1, D-85748 Garching, Germany
and
Foundation for Research and Technology-Hellas, Institute of Electronic Structure and Laser, PO Box 1527, Heraklion 71110, Crete, Greece

Received 3 March 1994, in final form 18 April 1994

Abstract. In recent years much attention has been paid to the quantized evolution of the centre-of-mass momentum and position of ultracold atoms in light fields. We consider the effects resulting from the quantization of the external angular momentum variables. We investigate how spin and orbital angular momentum of light are transferred to internal and external angular momentum of an atom in dipole and quadrupole transitions.

1. Introduction

When a light field interacts with an atom, one can usually neglect the centre-of-mass motion of the atom, and consider the effects on the internal electronic motion only. It suffices, e.g., to specify the change in energy and angular momentum of the electron in the centre-of-mass system.

For a cold atom, however, the recoil effect of even a single photon can no longer be neglected: a single photon can change the external state of the atom. Even in this case the external atomic motion can in general still be described classically. Only when the atom is cooled down further to the recoil temperature, and its de Broglie wavelength λ_B becomes comparable to the wavelength λ of the light, must the external motion be quantized [1].

The prime example of a pure quantum effect arising from the quantized external motion of an ultracold atom is the effect of velocity-selective coherent population trapping [2]. An atom in a light field gets trapped in a state that is insensitive to the light, and which is a superposition of two states with different external momenta, differing by two photon momenta. A cooling scheme based on this mechanism has been shown to lead to temperatures below the recoil temperature [2]. This idea has been extended to three dimensions [3] and to different atomic transitions [4].

Another example is the quantized motion of cold atoms in optical potential wells with the size of a wavelength, produced by two counterpropagating laser beams [5]. Here atoms can be trapped in a single well and can occupy a single quantized energy level therein. Also the occurrence of tunnelling from one well to an adjacent one has been predicted [6]. The presence of the discrete energy levels and the localization of an atom in one well have been observed experimentally [7, 8]. Recently cooling of atoms and their quantized motion in optical wells has also been observed in two- and three-dimensional laser beam configurations [9–11].

Now photons also carry angular momentum [12]. Therefore, photons can exert, in addition to a force, also a torque on an atom. There is one important difference, however:

the momentum of the photon is always transferred to the *external* momentum of the atom, whereas the photon angular momentum in general is transferred to the *internal* atomic angular momentum. In fact, the well known dipole selection rules $\delta l = \pm 1$ refer exclusively to the internal electronic state, and are understood as arising from the spin angular momentum of the photon, which is equal to 1. Recently it has been shown that also orbital angular momentum of light, arising from the transverse spatial dependence of a light beam, is a meaningful concept, accessible to experimental verification [13–16]. This quantity will in general be transferred to the *external* angular momentum of an atom [17]. This implies that it is the *orbital* angular momentum of light that can make atoms rotate around a given point or axis.

Some questions remain: what effects arise for ultracold atoms, when the external angular momentum must be quantized? What happens in a quadrupole transition, where $\delta l = \pm 2$? Does the second unit of angular momentum come from the orbital angular momentum of the photon? In order to answer these questions, we investigate the selection rules concerning (internal and external) angular momenta during the absorption or emission of a photon by an atom. We wish to clarify the conservation laws underlying these rules, and in particular the role of the centre-of-mass motion of the atom. We will discuss some unfamiliar effects on cold atoms resulting from angular momentum transfer between photons and atoms.

Throughout this paper we will use the long-wavelength approximation, i.e. we assume that the wavelength λ of the light is large compared to the size a of the atomic system under consideration. This justifies making an expansion in the small parameter a/λ. It should be noted that this approximation does not imply any restriction on the de Broglie wavelength of the atom: λ_B and a are independent atomic quantities, determined by the size of the external and of the internal part of the wavefunction, respectively. This paper is organized as follows. In order to be able to discuss conservation laws of angular momentum one needs to define electromagnetic field modes with well defined angular momenta. Several possible definitions are reviewed in section 2. In section 3 we examine in some detail the selection rules for photon absorptions and emissions by an atom. Emphasis is put on the modifications that arise due to the inclusion of the centre-of-mass motion of the atom. Some explicit examples are presented in section 4. They serve to show how spin and orbital angular momenta of a photon are distributed over the internal and external angular momenta of the atom, and why the temperature plays an important role. The results are summarized and discussed in section 5. Finally, in the appendix we define the internal and external variables as used in this paper.

2. Photons and angular momentum

We first establish the notation and definitions for the well known multipole waves [12, 18], in which photons are in eigenstates of the operators for total field angular momentum J^2 and its projection J_z. In the second subsection, we define photons in eigenstates of the projected 'orbital' and 'spin' angular momenta L_z and S_z [17]. Finally, we will briefly discuss Laguerre–Gaussian modes, which have been shown to be producable by a transformation of laser beams [15], and which also possess well defined S_z and L_z.

A photon is here defined as an elementary excitation of a field mode. A field mode is represented by a mode function $\boldsymbol{F}_\lambda(\boldsymbol{r})$, which is a transverse vector solution to the wave equation with wave number $k_\lambda = \omega_\lambda / c$. Each mode function can be chosen as the eigenfunction of a prescribed set of four commuting Hermitian operators. The mode is then

Table 1. Notation for the quantum numbers related to the various angular momenta of photons and atoms.

Species	Operator	Quantum number	Eigenvalue
Photon	J^2	j	$\hbar^2 j(j+1)$
Photon	J_z	m	$m\hbar$
Photon	S_z	s	$s\hbar k_z/k$
Atom	J^2_{int}	l	$\hbar^2 l(l+1)$
Atom	$J_{\text{int},z}$	m	$m\hbar$
Atom	J^2_{ext}	L	$\hbar^2 L(L+1)$
Atom	$J_{\text{ext},z}$	M	$M\hbar$
Atom	J^2_{total}	J	$\hbar^2 J(J+1)$

specified by the four eigenvalues of these operators, denoted collectively by λ, which are at the same time the quantum numbers of the photons from that field mode.

Once we have thus defined a complete set of normalized transverse vector functions $\boldsymbol{F}_\lambda$, the operator $\boldsymbol{A}$ for the vector potential in the Coulomb gauge can be expanded in this set as [19]

$$\boldsymbol{A}(\boldsymbol{r}) = \sum_\lambda \sqrt{\frac{\hbar}{2\epsilon_0\omega_\lambda}}[a_\lambda \boldsymbol{F}_\lambda(\boldsymbol{r}) + a_\lambda^\dagger \boldsymbol{F}_\lambda^*(\boldsymbol{r})] \tag{1}$$

where a_λ and $a_\lambda^\dagger$ are the annihilation and creation operators for photons in the mode λ. In the examples we will consider, two of the four quantum numbers will refer to angular momentum variables. The notation for these variables is summarized in table 1.

2.1. Multipole waves

We review here some well known results concerning the multipole waves, as described in [12, 19]. One constructs eigenstates of the four commuting Hermitian operators for total angular momentum J^2, its projection along the z axis J_z, energy, and parity. The set of quantum numbers is correspondingly given by $\lambda = (j, m, \omega, P)$, where the parity P takes the values ± 1, and where ω is the frequency. There are two types of waves: electric multipole waves with parity $P = (-1)^j$ and magnetic multipole waves with opposite parity $P = (-1)^{j+1}$. The explicit expressions for the mode functions can be found in [12], or, for the corresponding electric and magnetic fields in [19]. The only multipole fields that are non-vanishing in the origin are the electric dipole waves with $j = 1$ and $m = \pm 1, 0$.

2.2. Bessel waves

One can also construct field modes as the eigenfunctions of the commuting set of operators for the projection of the angular momentum J_z, linear momentum P_z, energy and 'spin' S_z. The corresponding quantum numbers are $\lambda = m, k_z, \omega, s$, where $\hbar k_z$ is the momentum in the z direction, and where $s = \pm 1$ denotes the polarization or helicity (right or left hand). Only for waves propagating in the z direction, i.e. for waves with $k_z = \pm k$, is the quantum operator S_z a true spin angular momentum operator [17, 20]. In that case it has discrete eigenvalues $\pm\hbar$. Explicit expressions for the corresponding mode functions $\boldsymbol{F}_\lambda$ were derived in [17]. The result takes the form (the $\pm$ refers to the polarization index s)

$$\frac{1}{\sqrt{2}}(F_x + \mathrm{i}F_y) = \frac{k_z \mp k}{2k} f(k_t, k_z, m+1)$$

$$\frac{1}{\sqrt{2}}(F_x - \mathrm{i}F_y) = \frac{k_z \pm k}{2k} f(k_t, k_z, m-1)$$

$$F_z = \frac{k_t}{\sqrt{2}k} f(k_t, k_z, m) \tag{2}$$

where the functions $f(k_t, k_z, m)$ are defined in cylindrical coordinates (ρ, z, ϕ) as

$$f(k_t, k_z, m) = J_m(k_t\rho)\exp(\mathrm{i}k_z z)\exp(\mathrm{i}m\phi)/N. \tag{3}$$

Here J_m is the mth-order Bessel function, $\hbar k_t$ is the momentum in the direction perpendicular to the z direction: $k_t^2 = k^2 - k_z^2$, and N is a normalization constant. Note that these modes do not have well defined total angular momentum J^2, nor a definite parity. The only fields that are non-vanishing in the origin are those with $m = \pm 1, 0$.

These waves generalize so-called Bessel beams, which are proportional to the zeroth-order Bessel functions, and which have been produced recently to moderately high intensity using a specific zone plate configuration [21].

2.3. Laguerre–Gaussian beams

Laguerre–Gaussian (LG) beams are special solutions to the paraxial equation [22]. They belong, therefore, to a class of exact solutions to an approximate equation describing light beams with a well defined propagation direction. The LG modes are cylindrically symmetric beams, with an azimuthal dependence given by $\exp(\mathrm{i}m\phi)$. They can be viewed as superpositions of Bessel waves with fixed value of m and with different $k_t \ll k$. These beams can be produced from (laser) Hermite–Gaussian beams by using a special configuration of astigmatic cylindrical lenses [13–15]. Thus one has been able to produce modes with up to three units of orbital angular momentum L_z per photon [15].

3. Selection rules

We review here how the selection rules for photon absorption and emission processes arise in the case that the atomic centre of mass motion is neglected. For more details see [18]. Subsequently we include this external motion, and investigate the resulting modifications of the selection rules and conservation laws.

3.1. Centre of mass is neglected

The atom is modelled by a spinless electron with coordinate $\boldsymbol{r}$ and momentum $\boldsymbol{p}$, charge $-e$ and mass μ that is bound to the origin by a given potential $V(r)$. The electron is confined by this potential to move in a region $|\boldsymbol{r}| < a$. We neglect the centre-of-mass motion, effectively assuming the atom to have an infinitely heavy nucleus. The centre-of-mass position $\boldsymbol{R}$ is not a dynamical variable, but its value is fixed and equal to $\boldsymbol{R} = 0$.

The electron interacts with an external electromagnetic field, which is represented by the vector potential $\boldsymbol{A}$ in the Coulomb gauge. The interaction part of the Hamiltonian is, neglecting the $\boldsymbol{A}^2$ term†, given by

$$H_I = \frac{e}{\mu}\boldsymbol{A}(\boldsymbol{r}) \cdot \boldsymbol{p}. \tag{4}$$

The wavelength of the light is assumed to be long, so that it is legitimate to expand in a/λ. This boils down to expanding the operator $\boldsymbol{A}(\boldsymbol{r})$ (i.e. the field modes $\boldsymbol{F}_\lambda$) around the origin,

$$\boldsymbol{A}(\boldsymbol{r}) = \boldsymbol{A}(0) + (\boldsymbol{r} \cdot \nabla)\boldsymbol{A}(0) + \ldots. \tag{5}$$

† At the low intensities used in laser cooling this is certainly a valid approximation. Furthermore, in lowest order in a/λ, this term contributes only a (small) energy shift of all states.

3.1.1. Dipole transitions. The first term of the expansion yields the dipole interaction

$$H_I^{\mathrm{ED}} = \frac{e}{\mu} \boldsymbol{p} \cdot \boldsymbol{A}(0). \tag{6}$$

If the matrix element

$$\langle \psi_{\mathrm{f}} | H_I^{\mathrm{ED}} | \psi_{\mathrm{i}} \rangle$$

of this operator between some initial state $|\psi_{\mathrm{i}}\rangle$ and some final state $|\psi_{\mathrm{f}}\rangle$ is non-zero then the corresponding transition is an electric dipole transition. Let the initial atomic state be an eigenstate of total angular momentum J_{int}^2 and of the projection $J_{\mathrm{int},z}$ with eigenvalues $l(l+1)$ and m: $|\psi_{\mathrm{i}}\rangle = |l, m\rangle$. Then a dipole transition to a final state of the form $|\psi_{\mathrm{f}}\rangle = |l', m'\rangle$ is possible if and only if at least one the matrix elements

$$\langle l', m' | \boldsymbol{p} | l, m \rangle$$

is non-zero. This leads to the well known dipole selection rules [18] for $\delta m \equiv m' - m$ and for $\delta l \equiv l' - l$,

$$\delta m = \pm 1, 0 \qquad \delta l = \pm 1. \tag{7}$$

The usual interpretation of these rules is that they express angular momentum conservation, since the photon has spin 1. Indeed, if one makes an expansion of the field in multipole waves, then only one term has a non-vanishing value at the origin, namely the electric dipole field with total angular momentum $j = 1$, and projections $m = \pm 1, 0$. Thus, angular momentum is conserved, as only the electric dipole wave contributes to the dipole interaction (6). Parity is conserved as well, since the parity of an elctric dipole photon is odd, while the parity of the electron state changes sign.

3.1.2. Higher order transitions. The second term in the expansion (5), of first order in a/λ, yields two different types of interaction terms: first the magnetic dipole Hamiltonian of the form

$$H_I^{\mathrm{MD}} = \frac{e}{\mu} \overleftrightarrow{Q}_{\mathrm{MD}} : (\boldsymbol{r}\boldsymbol{p} - \boldsymbol{p}\boldsymbol{r}) \tag{8}$$

and second the electric quadrupole interaction

$$H_I^{\mathrm{EQ}} = \frac{e}{\mu} \overleftrightarrow{Q}_{\mathrm{EQ}} : (\boldsymbol{r}\boldsymbol{p} + \boldsymbol{p}\boldsymbol{r}). \tag{9}$$

We defined here

$$\overleftrightarrow{Q}_{\mathrm{MD}} = \tfrac{1}{2}([\boldsymbol{\nabla}\boldsymbol{A}(0)]^t - \boldsymbol{\nabla}\boldsymbol{A}(0)) \qquad \overleftrightarrow{Q}_{\mathrm{EQ}} = \tfrac{1}{2}([\boldsymbol{\nabla}\boldsymbol{A}(0)]^t + \boldsymbol{\nabla}\boldsymbol{A}(0)) \tag{10}$$

where the superscript t indicates the transpose of a tensor: $T_{ij}^t = T_{ji}$. The quantity $\overleftrightarrow{Q}_{\mathrm{MD}}$ is proportional to the magnetic field, $\overleftrightarrow{Q}_{\mathrm{EQ}}$ to the gradient of the electric field, both evaluated at the origin. A magnetic dipole transition is allowed if one of the matrix elements

$$\langle \psi_{\mathrm{f}} | (\boldsymbol{r}\boldsymbol{p} - \boldsymbol{p}\boldsymbol{r}) | \psi_{\mathrm{i}} \rangle$$

is non-zero. The off-diagonal components of this tensor operator contain the cartesian components of the angular momentum operator, the diagonal part is a pure number. Neither operator can change the value of l, while only L_x and L_y can change m by one unit. Thus the selection rules are found to be [18]

$$\delta m = \pm 1, 0 \qquad \delta l = 0. \tag{11}$$

Analogously, one finds the electric quadrupole selection rules [18]

$$\delta m = \pm 2, \pm 1, 0 \qquad \delta l = \pm 2, 0. \tag{12}$$

In the multipole expansion the gradient of the field is non-vanishing at the origin for two kinds of waves: the magnetic dipole wave, with angular momentum 1, and the electric quadrupole wave with angular momentum 2, both with even parity. Hence, the selection rules (11) and (12) express, again, conservation of both angular momentum and parity. The usual comment is, that since the spin of the photon is 1, the second unit of angular momentum in a quadrupole transition must come from orbital angular momentum of the light field. In the next subsection we will show that this is not strictly true. Summarizing we note that in the multipolar expansion of the interaction

(i) in each different order of the expansion the atom interacts with a different multipole wave,

(ii) in lowest (electric dipole) order the spatial field dependence is neglected. Each higher order takes higher-order gradients of the field into account,

(iii) each expansion term of the interaction Hamiltonian is invariant under rotations around the origin. Therefore the selection rules express conservation of angular momentum of the corresponding multipole field and the electron,

(iv) external atomic angular momentum is excluded. The internal angular momentum change is determined completely by the angular momentum of the multipole wave.

3.2. Centre of mass is included

In this subsection we consider a bound system consisting of $N+1$ charges q_i, with masses μ_i, where the total mass is M and the total charge is zero. The interaction Hamiltonian is given by

$$H_I = -\sum_{i=0}^{N} \frac{q_i \boldsymbol{A}(\boldsymbol{r}_i) \cdot \boldsymbol{p}_i}{\mu_i}. \tag{13}$$

This has to be rewritten in terms of internal and external variables, by using (A3) from the appendix. The former variables are denoted by Greek symbols, the latter by capital ones. We keep the long-wavelength approximation, but allow the mass M to be finite. We thus include the centre-of-mass motion. Therefore, instead of expanding the field around the origin, we now have an expansion of $\boldsymbol{A}$ around the centre of mass (see appendix),

$$\boldsymbol{A}(\boldsymbol{r}_0) = \boldsymbol{A}(\boldsymbol{R}) - \left(\sum_{i=1}^{N} \frac{\mu_i \boldsymbol{\rho}_i}{M} \cdot \nabla\right) \boldsymbol{A}(\boldsymbol{R}) + \ldots$$

$$\boldsymbol{A}(\boldsymbol{r}_i) = \boldsymbol{A}(\boldsymbol{R}) - \left(\sum_{j=1}^{N} \frac{\mu_j \boldsymbol{\rho}_j}{M} \cdot \nabla\right) \boldsymbol{A}(\boldsymbol{R}) + (\boldsymbol{\rho}_i \cdot \nabla) \boldsymbol{A}(\boldsymbol{R}) + \ldots. \tag{14}$$

Notice that one does not obtain the multipolar Hamiltonian in this way. In order to do so, one should in addition apply a unitary transformation [19, 23]: this step is not necessary, however, for obtaining the selection rules.

The state vector of the atom is given by the direct product of an internal part $|\psi\rangle$ and an external part $|\Psi\rangle$. Now also the latter part can change during the emission or absorption of a photon, leading to selection rules for external variables.

3.2.1. Dipole transitions. To lowest order in a/λ one finds the dipole term

$$H_I^{\mathrm{ED}} = \sum_{i=1}^{N} \left(\frac{q_0}{\mu_0} - \frac{q_i}{\mu_i} \right) \boldsymbol{A}(\boldsymbol{R}) \cdot \pi_i. \tag{15}$$

Note that the *full* spatial dependence of the field has been taken into account, since $\boldsymbol{R}$ is now a dynamical variable. Each single term in the summation leads to the usual dipole selection rules for the corresponding internal variables m_i and l_i, whereas the other quantum numbers for $j \neq i$ do not change. Therefore, for the internal quantum numbers m_i, l_i and also for $m = \sum m_i$ and for $\sum l_i$ one finds the selection rules (7).

Additional selection rules now exist for the external angular momentum. They are found from the requirement that the matrix elements of $\boldsymbol{A}(\boldsymbol{R})$ between the external parts of the atomic initial and final states be non-vanishing. These selection rules depend explicitly on the spatial dependence of the field modes. In particular, they are different for multipole waves and Bessel waves. Explicit examples will be given in section 4.

Hence, the conservation law now refers to the *total* angular momentum, including the external atomic part. This fact implies, for instance, that an electric dipole transition is also possible in a field with angular momentum $j > 1$. This can be understood in two ways. First, the remaining angular momentum can now be absorbed by the external motion of the atom. Second, the field is sampled by the atomic wavefunction, not only at the origin, but around $\boldsymbol{R}$ with a width determined by the de Broglie wavelength. Therefore, the fact that a higher-order multipole field vanishes at the origin, no longer implies that the transition is forbidden.

3.2.2. Higher order transitions. The first-order terms in H_I, which are linear in the gradient of the vector potential, can be grouped according to

$$H_I^1 = \sum_{i=1}^{N} \sum_{j=1}^{N} \left(\frac{q_i}{\mu_i} - \frac{q_0}{\mu_0} \right) \left[\frac{\mu_j \rho_j}{M} \cdot \nabla \right] (\boldsymbol{A}(\boldsymbol{R}) \cdot \pi_i) - \sum_{i=1}^{N} \frac{q_i}{\mu_i} [\rho_i \cdot \nabla] (\boldsymbol{A}(\boldsymbol{R}) \cdot \pi_i)$$
$$- \sum_{i=1}^{N} \frac{q_i}{M} [\rho_i \cdot \nabla] (\boldsymbol{A}(\boldsymbol{R}) \cdot \boldsymbol{P}). \tag{16}$$

For an atom the ratio μ_i/M is small, so that the first term is negligible compared to the second. The third term describes in effect a dipole coupling of the external motion with the magnetic field: it is equivalent to the Röntgen term, which has recently been shown to be necessary for a consistent description of the centre-of-mass motion for cooled atoms [23], and which leads to the existence of a geometric phase analogous to the Aharonov–Casher effect [24]. We concentrate on the second line, which can be separated into two terms,

$$H_I^{\mathrm{MD}} = \overleftrightarrow{Q}_{\mathrm{MD}}(\boldsymbol{R}) : \sum_{i=1}^{N} \frac{q_i}{\mu_i} (\rho_i \pi_i - \pi_i \rho_i)$$
$$H_I^{\mathrm{EQ}} = \overleftrightarrow{Q}_{\mathrm{EQ}}(\boldsymbol{R}) : \sum_{i=1}^{N} \frac{q_i}{\mu_i} (\rho_i \pi_i + \pi_i \rho_i) \tag{17}$$

where $\overleftrightarrow{Q}_{\mathrm{EQ}}$ and $\overleftrightarrow{Q}_{\mathrm{MD}}$ are given by the same expressions (10) but now evaluated at $\boldsymbol{R}$. For each internal angular momentum quantum number m_i and l_i one finds the same selection rules (11) and (12) as before. For the external angular momentum one finds additional selection rules from the requirement that the gradient of $\boldsymbol{A}$ have non-zero matrix elements between initial and final external atomic states. Again, the latter rules depend on the explicit spatial dependence of the field modes, and will be different from those in a dipole transition. For details see section 4.

We note that the inclusion of the external atomic motion leads to the following modifications in comparison with the results from the preceding subsection.

(i) In each order of the expansion the atom interacts with the total field.

(ii) In each order the full spatial field dependence is included.

(iii) Each expansion term of the interaction Hamiltonian is invariant under rotations around the origin. Therefore the selection rules express conservation of angular momentum of the total field and the atom.

(iv) External atomic angular momentum is included. The change in internal angular momentum is determined not only by the angular momentum of the field but also by the change in external angular momentum.

4. Illustrations

We start this section by having a closer look at dipole transitions in a field with arbitrary angular momentum $j > 1$. Next we consider electric quadrupole transitions, being the lowest-order transitions in which the atomic angular momentum can change by two units, in a field with angular momentum 1.

4.1. Electric dipole transitions

The transition probability for going from a given initial to some final state in a dipole transition is determined by the square of the matrix element

$$\sum_{i=1}^{N} \langle \Psi_{\rm f} | \boldsymbol{A}(\boldsymbol{R}) | \Psi_{\rm i} \rangle \cdot \left(\frac{q_0}{\mu_0} - \frac{q_i}{\mu_i} \right) \langle \psi_{\rm f} | \pi_i | \psi_{\rm i} \rangle. \tag{18}$$

If one makes the usual approximation of neglecting the centre-of-mass motion, then first of all the variable $\boldsymbol{R}$ is a fixed position in space. The first part of the matrix element, then, reduces to

$$\boldsymbol{A}(\boldsymbol{R}) \langle \Psi_{\rm f} | \Psi_{\rm i} \rangle$$

which implies that, indeed, the external state of the atom does not change. This part will contribute only to the total transition rate, but does not change the angular momentum or the parity of $|\Psi_{\rm i}\rangle$. Furthermore, if the vector potential vanishes at the centre-of-mass position, then the transition is forbidden. Now, when the external motion is quantized, $\boldsymbol{R}$ becomes a dynamical variable. The matrix element

$$\langle \Psi_{\rm f} | \boldsymbol{A}(\boldsymbol{R}) | \Psi_{\rm i} \rangle = \int {\rm d}\boldsymbol{R}\, \Psi_{\rm f}^*(\boldsymbol{R}) \boldsymbol{A}(\boldsymbol{R}) \Psi_{\rm i}(\boldsymbol{R}) \tag{19}$$

is non-zero only when the external state changes; the external state can change parity and can absorb angular momentum. If $\boldsymbol{A}$ vanishes at the *average* centre-of-mass position, it is no longer implied that the corresponding matrix element vanishes. Still, the electric dipole transition probability in such a field will be negligibly small in general. For example, in an electric multipole wave with angular momentum j, the field around the origin is proportional to [12, 19]

$$|\boldsymbol{A}(\boldsymbol{r})| \propto \left(\frac{2\pi r}{\lambda}\right)^{j-1} \tag{20}$$

for $r/\lambda \ll 1$. Thus, if the width of the centre-of-mass wavefunctions of the atom is given by $\Delta x \ll \lambda$, the transition probability is proportional to $(2\pi\Delta x/\lambda)^{2j-2}$, which is negligibly small, except for $j = 1$. For ultracold atoms, however, the width of the wavefunctions of $|\Psi_i\rangle$ and $|\Psi_f\rangle$ is appreciable, as it is determined by the small momentum spread Δp, according to $\Delta x > \hbar/\Delta p$. For atoms cooled down to the recoil limit, $\Delta p \approx h/\lambda$, one obtains

$$\Delta x > \frac{\lambda}{2\pi}. \tag{21}$$

Hence, in this case the field is probed over a distance of the order of the wavelength, so that effectively all higher-order gradients of the field will contribute. Thus the electric dipole transition probability becomes finite, for any multipole wave. As an example, consider an electric quadrupole wave with $j = 2$ and projection $m = 2$. If an atom makes an electric dipole transition, then the selection rules for the internal variables are not changed:

$$\delta l = \pm 1 \qquad \delta m = -1, 0, 1. \tag{22}$$

Thus the atom absorbs only one unit of angular momentum, although the photon contains two units. The deficit is accounted for by the external angular momentum of the atom. For instance, the spherical components of $\boldsymbol{A}$ are proportional to

$$(A_x + \mathrm{i}A_y) \propto \exp(3\mathrm{i}\phi) \qquad A_z \propto \exp(2\mathrm{i}\phi) \qquad (A_x - \mathrm{i}A_y) \propto \exp(\mathrm{i}\phi) \tag{23}$$

so that the selection rules for external quantum number M are

$$\delta M = 3, 2, 1 \tag{24}$$

such that $\delta m + \delta M = 2$. The quantum number J for the length of the total angular momentum $\boldsymbol{J}_{\text{total}} = \boldsymbol{J}_{\text{int}} + \boldsymbol{J}_{\text{ext}}$, changes according to

$$\delta J = \pm 2 \tag{25}$$

which expresses angular momentum conservation. For L, the eigenvalue pertaining to J^2_{ext}, the selection rules are less strict and follow from the familiar triangle and parity rules

$$L + l \geqslant J \geqslant |L - l| \qquad L' + l' \geqslant J' \geqslant |L' - l'| \tag{26}$$

and $\delta L = L' - L$ is odd.

4.2. Electric quadrupole transitions

The transition probability for electric quadrupole transitions is determined by the square of the matrix elements

$$\sum_{i=1}^{N}\langle\Psi_{\rm f}|\overleftrightarrow{Q}_{\rm EQ}(\boldsymbol{R})|\Psi_{\rm i}\rangle : \langle\psi_{\rm f}|\frac{q_i}{\mu_i}(\rho_i\pi_i+\pi_i\rho_i)|\psi_{\rm i}\rangle. \tag{27}$$

The main conclusions from the preceding subsection can, *mutatis mutandis*, be carried over to this case. For instance, in a field where the quadrupole tensor vanishes in the origin, the probability of making an electric quadrupole transition is negligble, unless the atom is very cold. As an explicit example we consider a circularly polarized Bessel beam (or a Laguerre–Gaussian beam) propagating in the positive z direction, such that $k_z \approx k$, with $L_z = 0$ and $S_z = \hbar$. From (2) one sees that there is only one non-vanishing component of the field:

$$F_x - {\rm i}F_y = \sqrt{2}f(k_{\rm t}, k_z, 0) \tag{28}$$

where $k_{\rm t} \ll k_z$. Suppose an atom absorbs a photon from this beam by making an electric quadrupole transition. Since only one spherical component (28) of the field is non-vanishing, one is left with a reduced set of quadrupole selection rules

$$\delta m = 0, 1, 2 \tag{29}$$

for the projections of internal angular momentum on the z axis. From the explicit spatial dependence of (28) one gets the following selection rules for the external angular momentum,

$$\delta M = 1, 0, -1 \tag{30}$$

respectively, such that $\delta m + \delta M = 1$. Therefore, the magnetic quantum number m can change by 2 units of angular momentum, even though the field possesses only one unit of spin. The second unit does not come from orbital angular momentum of the photon, since $L_z = 0$. Rather the external motion of the atom loses one unit of angular momentum in that case: $\delta M = -1$. For completeness we note that in a wave with $L_z = m\hbar$ and $S_z = \hbar$, the selection rule for δm is not affected, while for δM one has now

$$\delta M = m + 1, m, m - 1 \tag{31}$$

respectively. The extra amount of orbital angular momentum of the photon is thus always transferred to the centre-of-mass motion of the atom. The same conclusion holds for dipole transitions [17]. Similarly, in an electric dipole wave with $j = 1$, the quadrupole transition matrix elements lead to the rules

$$\delta l = \pm 2 \tag{32}$$

even though the field possesses only one unit angular momentum. Again, the other unit comes from the external motion of the atom, not from orbital angular momentum of light.

5. Discussion and conclusions

If one assumes the centre of mass of an atom to be at rest at a given point, the selection rules for one-photon transitions express angular momentum and parity conservation: as is well known [12], an electric or magnetic 2^j transition with $\delta l = j$ is allowed if and only if the emitted or absorbed photon has total angular momentum j, and parity $(-1)^j$ or $(-1)^{j+1}$, respectively.

When the external motion of the atom is included, things are different: a multipole transition is possible also in fields with lower angular momentum. In particular, we showed that a quadrupole transition with $\delta l = \pm 2$ and $\delta m = \pm 2$ is allowed both in a field with total angular momentum $j = 1$, and in a field with vanishing orbital angular momentum $L_z = 0$, and spin $S_z = \pm 1$. Conversely, a multipole transition is also allowed in a field with higher angular momentum: for instance, an atom may emit a quadrupole photon in a dipole transition. In all these cases there is no violation of conservation of angular momentum or parity, since the external motion can absorb angular momentum and since the parity of the external wavefunction can change sign.

These changes originate from the fact that the external centre-of-mass position $\boldsymbol{R}$ must be considered a dynamical variable. Consequently, both the transition probability and the selection rules depend on the explicit spatial dependence of the field. The internal selection rules, on the other hand, are independent of the mode structure, and do not depend on the inclusion of external atomic motion, quantized or classical.

When is it relevant to include the centre-of-mass motion? If the atom is cooled to low temperatures, then the angular momentum transfer, i.e. the torque, of one photon is not negligible. Thus for a cooled atom one expects to see above-mentioned deviations from the usual emission and absorption behaviour. Indeed, one may note that the de Broglie wavelength of the atom becomes larger as its temperature becomes lower. Therefore the cooled atom sees a larger part of the field around its centre-of-mass position. Hence, the spatial dependence of the field becomes important, and thereby also the orbital angular momentum of the field. This part of the field angular momentum is in general absorbed by the external motion of the atom, in the sense discussed in section 4.

This also indicates how an atom can be made to rotate around a given axis. When the atom is cooled in a Doppler cooling scheme with two counter-propagating Laguerre–Gaussian laser beams with azimuthal index m (i.e. $L_z = m\hbar$ per photon), then at each stimulated absorption the atomic external angular momentum changes by $m\hbar$. The angular momentum will on average not change by spontaneous emissions. The net rate of absorption of external angular momentum is then equal to $\mathrm{d}M/\mathrm{d}t = AP_e m\hbar$, where A is the spontaneous decay rate and P_e the probability for the atom to be in the excited state.

Finally, let us note that we discussed only single-photon transitions. Multi-photon transitions can, to lowest order, be seen as a sequence of electric dipole transitions, and the foregoing conclusions can easily be extended to this case.

Acknowledgments

It is a pleasure to thank Professor P Lambropoulos for fruitful discussions. This work is supported by the Human Capital and Mobility programme under contract no ERBCHBGCT920192.

Appendix A. Internal and external variables

We consider an $N+1$-particle system, consisting of particles $i = 0 \ldots N$. Their positions and momenta are denoted by $\boldsymbol{r}_i$ and $\boldsymbol{p}_i$. Here we will define internal and external position and momentum variables for this system. The internal variables will be denoted by greek symbols, the external variables by capital symbols. The internal position variables $\boldsymbol{\rho}_i$ for $i = 1 \ldots N$ are chosen relative to the position of particle $i = 0$, which for an atom would be the nucleus. The external position vector $\boldsymbol{R}$ is the centre-of-mass position. Thus we define

$$\boldsymbol{R} = \sum_{j=0}^{N} \frac{\mu_j \boldsymbol{r}_j}{M} \qquad \boldsymbol{\rho}_i = \boldsymbol{r}_i - \boldsymbol{r}_0 \tag{A1}$$

for $i = 1 \ldots N$, where $M = \sum \mu_i$ is the total mass. The canonically conjugate momenta follow from these definitions, and are given by

$$\boldsymbol{P} = \sum_{j=0}^{N} \boldsymbol{p}_j \qquad \boldsymbol{\pi}_i = \boldsymbol{p}_i - \frac{\mu_i}{M} \sum_{j=0}^{N} \boldsymbol{p}_j. \tag{A2}$$

These expressions are valid both classically and quantum mechanically, and the quantum operators satisfy the canonical commutation relations. Note that if one defines internal coordinates with respect to the centre of mass, one has an overcomplete set of internal variables. Therefore, these variables have to satisfy a constraint condition, and do not satisfy the canonical commutation rules, which, however, does not lead to any difficulties [23].

The inverse relations read

$$\begin{aligned} \boldsymbol{r}_0 &= \boldsymbol{R} - \sum_{j=1}^{N} \frac{\mu_j \boldsymbol{\rho}_j}{M} \qquad & \boldsymbol{r}_i &= \boldsymbol{R} + \boldsymbol{\rho}_i - \sum_{j=1}^{N} \frac{\mu_j \boldsymbol{\rho}_j}{M} \\ \boldsymbol{p}_0 &= \frac{\mu_0}{M} \boldsymbol{P} - \sum_{j=1}^{N} \boldsymbol{\pi}_j \qquad & \boldsymbol{p}_i &= \frac{\mu_i}{M} \boldsymbol{P} + \boldsymbol{\pi}_i. \end{aligned} \tag{A3}$$

Finally, the atomic angular momentum can be separated as

$$\boldsymbol{J}_{\text{atom}} = \sum_{i=0}^{N} \boldsymbol{r}_i \times \boldsymbol{p}_i = \boldsymbol{R} \times \boldsymbol{P} + \sum_{i=1}^{N} \boldsymbol{\rho}_i \times \boldsymbol{\pi}_i \qquad \equiv \boldsymbol{J}_{\text{ext}} + \boldsymbol{J}_{\text{int}}. \tag{A4}$$

References

[1] Cohen-Tannoudji C 1993 *Atomic Physics* vol 13 ed H Walther, T W Hänsch and B Neizert (New York: American Institute of Physics) and references therein

[2] Aspect A, Arimondo E, Kaiser R, Vansteenkiste N and Cohen-Tannoudji C 1988 *Phys. Rev. Lett.* **61** 826 1989 *J. Opt. Soc. Am.* B **6** 2112

[3] Ol'shanii M A and Minogin V G 1991 *Quantum Opt.* **3** 317

[4] Papoff F, Mauri F and Arimondo E 1992 *J. Opt. Soc. Am.* B **9** 321

[5] Castin Y and Dalibard J 1991 *Europhys. Lett.* **14** 761

[6] Berg-Sørensen K, Castin Y, Mølmer K and Dalibard J 1993 *Europhys. Lett.* **22** 663

[7] Verkerk P *et al* 1992 *Phys. Rev. Lett.* **68** 3861
[8] Jessen P S *et al* 1992 *Phys. Rev. Lett.* **69** 94
[9] Hemmerich A and Hänsch T W 1993 *Phys. Rev. Lett.* **70** 410
[10] Grynberg G, Lounis B, Verkerk P, Courtois J-Y and Salomon C 1993 *Phys. Rev. Lett.* **70** 2249
[11] Hemmerich A, Zimmermann C and Hänsch T W 1993 *Europhys. Lett.* **22** 89
[12] Berestetskii V B, Lifshitz E M and Pitaevskii L P 1982 *Quantum Electrodynamics* (Oxford: Pergamon)
[13] Allen L, Beijersbergen M W, Spreeuw R J C and Woerdman J P 1992 *Phys. Rev.* A **45** 8185
[14] S.J. van Enk and Nienhuis G 1992 *Opt. Commun.* **94** 147
[15] Beijersbergen M W, Allen L, van der Veen H E L O and Woerdman J P 1993 *Opt. Commun.* **96** 123
[16] Kristensen M, Beijersbergen M W and Woerdman J P 1994 *Opt. Commun.* **104** 229
[17] van Enk S J and Nienhuis G 1994 *J. Mod. Opt.* **41** 963
[18] Cohen-Tannoudji C, Diu C and Laloë F 1977 *Quantum Mechanics* (Paris: Wiley and Herman)
[19] Cohen-Tannoudji C, Dupont-Roc J and Grynberg G 1989 *Photons and Atoms* (New York: Wiley)
[20] van Enk S J and Nienhuis G 1994 *Europhys. Lett.* **25** 497
[21] Wulle T and Herminghaus S 1993 *Phys. Rev. Lett.* **70** 1401
[22] Siegman A E 1986 *Lasers* (Mill Valley, CA: University Science Books)
[23] Baxter C, Babiker M and Loudon R 1993 *Phys. Rev.* A **47** 1278
[24] Wilkens M 1994 *Phys. Rev. Lett.* **72** 5

PAPER 5.3

Optical Pumping of Orbital Angular Momentum of Light in Cold Cesium Atoms

J. W. R. Tabosa
Departamento de Física, Universidade Federal de Pernambuco, 50670-901, Recife, PE, Brazil

D. V. Petrov
Departamento de Química Fundamental, Universidade Federal de Pernambuco, 50670-901, Recife, PE, Brazil
(Received 25 May 1999)

We present experimental results on the transfer of the orbital angular momentum of light to a system of cold cesium atoms. A nondegenerate four-wave mixing process was used as an indirect tool to observe this transfer. Our experiments show, in particular, that the orbital angular momentum of light can be transferred, via optical pumping in the cold atomic sample, from one beam to another oscillating at a different frequency.

PACS numbers: 32.80.Pj, 42.50.Vk, 42.65.Hw

The Laguerre-Gaussian (LG) laser modes possess well-defined angular momentum that can be decomposed into an orbital component and a spin component associated with its polarization [1]. Optical beams with orbital angular momentum (OAM) include screw topological wave front dislocations or vortices. Vortices appear as spiral phase ramps around a singularity where the phase of the wave is undefined; thus its amplitude must vanish. The order of the singularity multiplied by its sign is referred to as the topological charge of the dislocation.

There has been a great deal of interest in the mechanical and optical effects that light beams with OAM can exert on material systems. This angular momentum has been transferred to macroscopic particles trapped in an optical tweezer causing them to rotate [2,3]. An atom moving in a beam with OAM experiences a torque and an azimuthal shift in its resonant frequency in addition to the usual axial Doppler shift and recoil shift [4]. Also the LG modes are important in laser cooling and trapping experiments. For example, in Ref. [5] such beams were used for high-resolution spectroscopy in a magneto-optical trap, and a novel trapping scheme using LG modes was proposed and experimentally demonstrated in Ref. [6].

The linear propagation of beams with phase front dislocations and also their single-frequency propagation in cubic and photorefractive nonlinear bulk media in single pass and cavity configurations have been studied in [7–10]. The up-conversion by second harmonic generation of light in quadratic nonlinear media has been investigated experimentally in Refs. [11–14]. However, only very limited experimental work has been done with these beams in atomic systems, and has mainly dealt with nonlinear propagation effects [15].

A system of cold atoms as a medium for studying optical processes with beams carrying OAM possesses a number of interesting properties. First, nonlinear effects arising by individual optical transitions in the system of atomic energy levels constitute by itself basic steps that exist in any nonlinear processes in an arbitrary medium. The OAM of the input beams can be transferred to the atomic system and, hence, modify these nonlinear optical processes. Furthermore, the low temperature of laser cooled atoms, besides allowing to diminish the Doppler broadening of the atomic levels, can also give access to the direct observation of mechanical effects induced by the atom-field interaction [16].

In this Letter, we present experimental results on the nondegenerate four-wave mixing process (NDFWM) performed in cold cesium atoms with optical beams carrying OAM. We use this process as an indirect tool in order to demonstrate that the OAM can be transferred from the optical beam to the system of cold cesium atoms. This effect is similar to the phenomenon of optical pumping induced orientation and alignment in atomic systems due to the spin angular momentum of the photon [17], and to our knowledge corresponds to the first observation of the optical pumping of OAM of light.

We have employed a sample of cold atoms obtained from a vapor cell magneto-optical trap. The trapping beams are provided by a stabilized Ti-sapphire laser, and are red detuned by about 12 MHz from the resonance frequency of the cesium cycling transition $6S_{1/2}, F = 4–6P_{3/2}, F' = 5$ at $\lambda = 852$ nm, as indicated in Fig. 1(a). The necessary repumping laser is a long external cavity diode laser, which is tuned into resonance with the $6S_{1/2}, F = 3–6P_{3/2}, F' = 3$ transition. Typically the number of trapped atoms, estimated by measuring the absorption of a weak probe beam by the atomic cloud is of the order of 10^7 atoms. Recently, four-wave mixing experiments in this atomic system were done with beams without OAM [18].

The experimental setup is shown in Fig. 1(b). The forward pump beams F and P with linear polarizations have the same frequency ω_1, and are incident in the trap forming an angle of $2\theta \simeq 2°$. These beams are provided by a grating-stabilized diode laser (LASER1), and are locked in resonance with the noncycling transition $6S_{1/2}, F = 4–6P_{3/2}, F' = 4$, using an auxiliary saturated

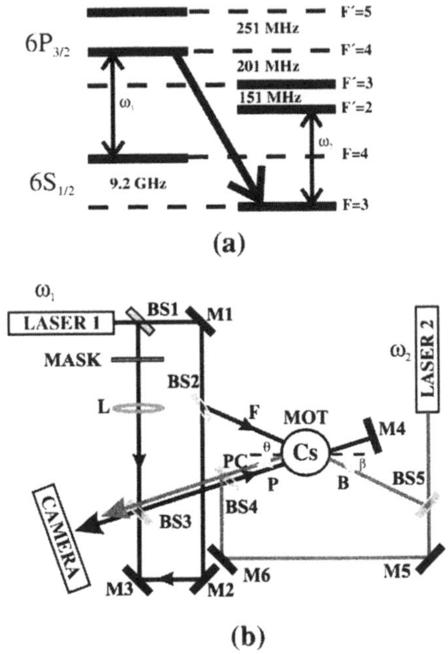

FIG. 1. (a) Hyperfine energy levels of cesium D_2 line. (b) Experimental setup.

absorption signal. Single-charge and double-charge topological screw dislocations were nested in the beam P using two different computer-generated spiral zone plates MASK [19]. A lens L allowed us to separate the beam with a nested vortex from the undiffracted light and high-order Fresnel images.

The beam P is focused so that its waist in the trap i much smaller than the trap size (about 2 mm), while th beam F is collimated to a diameter of the order of 3 mm To reveal the topological phase structure of the beam P we used a part of the beam with the frequency ω_1 a a reference beam. The reference beam is reflected fron the mirrors $M1$, $M2$, and $M3$, and after a beam splitte BS3 provided an interference pattern with the inciden beam P, back reflected from the mirror $M4$. This mirro is removed during the four-wave mixing experiments The intensity pattern was analyzed by a charge couple device camera.

Another independent, grating-stabilized diode lase (LASER2) generated the backward pump beam B. Th frequency ω_2 of the beam B is set, using other auxiliar reference absorption signal, around the frequency of th cycling transition $6S_{1/2}, F = 3\text{–}6P_{3/2}, F' = 2$. This beam, also linearly polarized, has a diameter of the order of 3 mm and is incident in the trap satisfying the phase-matching condition, i.e., $\sin\beta = (\omega_1/\omega_2)\sin\theta$. The powers of beams P, F, and B are $P_P = 5\ \mu$W, $P_F = 10\ \mu$W, and $P_B = 200\ \mu$W, respectively. The generated beam, PC, is therefore nearly counterpropagating to the beam P. Its phase structure was revealed using an interference pattern after a beam splitter BS4 obtained with the reference beam of frequency ω_2. To measure the phase structure of the incident beam P we stop the beam from the LASER2, while to measure the phase structure of the beam PC we stop the reflection from the mirrors $M2$ and $M4$. The efficiency of the NDFWM process, measured by the ratio P_{PC}/P_P, was of the order of 0.1%.

Figures 2(a) and 2(b) show, respectively, the intensity profile and interferogram of the pump beam P, when the MASK generates into the beam P a single charge phase dislocation. Figures 2(c) and 2(d) show the corresponding distributions for the generated beam PC. As seen, the beam PC includes also a phase front dislocation and hence, an OAM. Owing to the reversed sense of rotation of the interference spiral due to the reflection in mirror $M4$, the topological charge of the beams P and PC are the same. Therefore, as these beams are nearly counterpropagating, they carry opposite OAM.

If the topological charge of the pump beam P is equal to 2, the corresponding intensity profile and interferogram are shown in Figs. 3(a)–3(d). We again observe the generation of the PC beam with the same topological charge as the one of the beam P.

We have modeled our system by considering the incident FWM beams to be of the form $\vec{E}_\mu = \frac{1}{2}\vec{A}_\mu \exp[i(\omega_\mu t - \vec{k}_\mu \cdot \vec{r})] + \text{c.c.}$, with $\mu = F, P, B$; $\omega_F = \omega_P = \omega_1$; $\omega_B = \omega_2$, and $\vec{A}_\mu$ being the corresponding complex field amplitudes. The fields $\vec{E}_F$ and $\vec{E}_B$ are taken as plane waves, so that $\vec{A}_F$ and $\vec{A}_B$ do

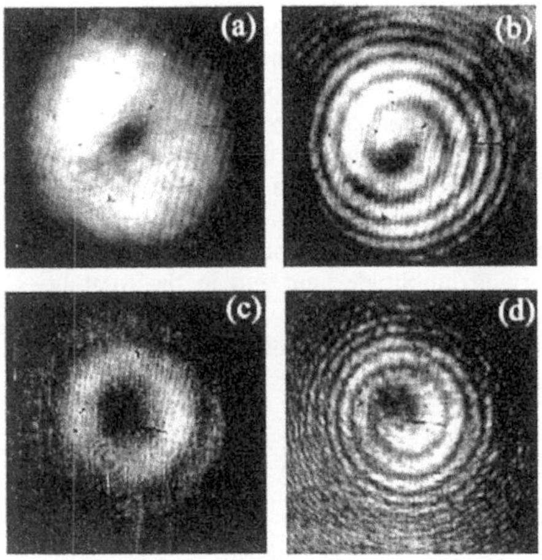

FIG. 2. Observed light distributions and interferograms of the pump beam P [(a) and (b)] and the diffracted beam PC [(c) and (d)]. The topological charge of phase dislocation in the pump beam P is equal to 1 and is the same as that of the beam PC since mirror $M4$ reverses the sense of rotation of the spiral as discussed in the text.

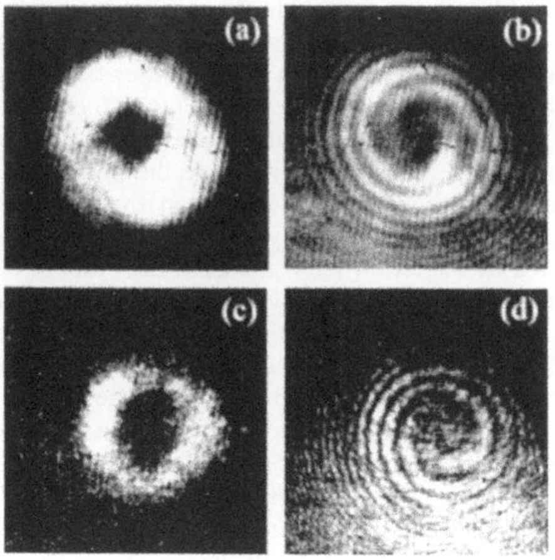

FIG. 3. Same as in Fig. 2 but with the double topological charge of phase dislocation in the pump beam P.

not depend on the coordinates, while for $\vec{E}_P$, the field amplitude of the LG mode at the beam waist, located at $z = 0$ and coincident with the trap position, is given by $\vec{A}_P = \vec{A}_0(\frac{\sqrt{2}r}{w})^l \exp(-il\phi)\exp[-(\frac{r^2}{w^2})]L_p^l(2r^2/w^2)$. Here w is the half-beam width, $\vec{A}_0$ is the beam amplitude, L_p^l is the associated Laguerre polynomial, and r and ϕ are, respectively, the radial and angular coordinates in cylindrical polar coordinate system with its z axis being along the beam propagation direction. The beam P employed in the experiments corresponds to $l = 1$; $p = 0$ (charge 1), and $l = 2$; $p = 0$ (charge 2).

The effect of the trapping and the repumping beams is taken into account through incoherent optical pumping rates specified by γ_T and γ_R, respectively. We have supposed the same relaxation rate γ for all the excited states. We also have assumed that the residual Doppler shift is much smaller than all the relaxation rates, i.e., $k_\mu \cdot \vec{v} \ll \gamma, \gamma_R$, where $\vec{v}$ is the atomic velocity. A straightforward calculation using the density matrix in the slowly varying amplitude limit and in the rotating-wave approximation yields for the on-resonance third-order induced coherence at the transition $F = 3 - F' = 2$ the following result [18]:

$$\rho^{(3)}_{32'} = \frac{iN\mu_{32'}|\mu_{44'}|^2}{\hbar^3\gamma^2}\left(\frac{\gamma_T + \gamma}{2\gamma_T + \gamma}\right)\left(\frac{1}{\gamma} + \frac{1}{\gamma_R}\right)A_B A_F A_P^* \times \exp\{i[\omega_2 t - (\vec{k}_F + \vec{k}_B - \vec{k}_P)\cdot\vec{r}]\}.$$

In this expression $\mu_{44'}$ and $\mu_{32'}$ are the dipole matrix elements of the transitions involved, and N is the total number of atoms in the trap. This induced polarization generates a field oscillating at the frequency ω_2 and propagating along the direction $\vec{k}_{PC} = \vec{k}_F + \vec{k}_B - \vec{k}_P$, under the phase-matching condition. The complex field amplitude of the generated beam is proportional to the complex conjugated amplitude of the beam P. This can lead to the observation of wave-front rectification and especially in this case to the transfer of the OAM from one beam to another beam with different frequency. Furthermore, the complex amplitude of the generated phase conjugated beam PC has the topological phase defect of the same sign as the one in the beam P, since the induced polarization is proportional to A_P^*, but the propagation directions of the beams P and PC are opposite.

The NDFWM signal can be interpreted in terms of the Bragg diffraction into a transferred population grating. The grating is produced by excitation of the noncycling $6S_{1/2}, F = 4$–$6P_{3/2}, F' = 4$ transition by the beams F and P. The excited-state population grating created by these beams is transferred through spontaneous emission into the $6S_{1/2}, F = 3$ ground-state hyperfine level. The transferred grating is then monitored by the third laser beam B.

In the writing process of the excited-state population grating by absorption of each photon of the beam P, the OAM of this beam is transferred to the atomic system. The angular momentum is stored by the atomic system as the rotational movement of the atoms and is conserved during the spontaneous emission into the $6S_{1/2}, F = 3$ ground-state level. In the reading process by absorption of each photon from the beam B the atomic system generates a photon with OAM in the beam PC. The net OAM transferred to the atomic system depends on the number of photons absorbed in the beam P and the number of photons emitted in the beam PC. Since the atoms in the lower ground state $6S_{1/2}, F = 3$ do not experience any confining force, there is no external torque acting on the atoms to compensate the corresponding change of OAM due to the interaction with the light. Therefore, the atomic sample as a whole should rotate. By taking into account the measured Bragg diffraction efficiency and the absorption of beam P, and the fact that each absorbed and generated photon carries $\pm l\hbar$ of OAM, we can roughly estimate the flux of OAM transferred to the atomic system. For the beam P with $l = 1$ this value is of the order of 1×10^{-21} Nm/s. Further investigations to detect directly this atomic motion are currently under way. As we have mentioned before, the transfer of OAM to the cold atomic sample presents a close analogy with the well-known phenomenon of optical pumping induced orientation in atomic systems, a phenomenon which has been used to the production of cold sample of spin-polarized atoms [20].

The present effect could in principle be observed in other nonlinear media, as, for example, in a hot atomic vapor. However, cold atoms with low velocities seem to be essential if one wants to observe the mechanical effects induced on the atomic system due to the OAM exchange with light. Moreover, as light beams carrying

OAM have been suggested to excite vortex states in a Bose-Einstein condensate (BEC) [21], our results, in fact, present a first step towards this goal. Also, as has been demonstrated recently [22], a BEC can be used to perform coherent four-wave mixing of matter waves prepared in well-defined momentum states. The extension of this phenomenon to include the OAM or vortex in the atomic wave packets certainly is of great interest and should provide a better understanding for the coherent matter-wave interaction.

In conclusion, we have observed experimentally the interchange of the OAM between two waves of different frequencies as a result of the optical pumping of the OAM into the system of cold cesium atoms. We hope that this demonstration will trigger further investigation on the coupling between the internal and external degrees of freedom of an atomic system interacting with light, where, in addition to the well-known effect of linear momentum transfer, one also will need to consider its angular counterpart for light carrying OAM.

We are grateful to R. Y. Chiao and J. Rios Leite for stimulating discussions. The work of J. W. R. T. was supported by FINEP, CNPq-Pronex, Brazilian Agencies.

[1] L. Allen, M. W. Beijersbergen, R. J. C. Spreeuw, and J. P. Woerdman, Phys. Rev. A **45**, 8185 (1992).

[2] H. He, N. R. Heckenberg, and H. Rubinsztein-Dunlop, J. Mod. Opt. **42**, 217 (1995); H. He, M. E. J. Friese, N. R. Heckenberg, and H. Rubinsztein-Dunlop, Phys. Rev. Lett. **75**, 826 (1995).

[3] N. B. Simpson, K. Dholakia, L. Allen, and M. J. Padgett, Opt. Lett. **22**, 52 (1997).

[4] M. Babiker, W. L. Power, and L. Allen, Phys. Rev. Lett. **73**, 1239 (1994); L. Allen, M. Babiker, W. K. Lai, and V. E. Lembessis, Phys. Rev. A **54**, 4259 (1996); W. K. Lai, M. Babiker, and L. Allen, Opt. Commun. **133**, 487 (1997); J. Courtial, D. A. Robertson, K. Dholakia, L. Allen, and M. J. Padgett, Phys. Rev. Lett. **81**, 4828 (1998).

[5] M. J. Snadden, A. S. Bell, R. B. M. Clarke, E. Riis, and D. H. McIntyre, J. Opt. Soc. Am. B **14**, 544 (1997).

[6] T. Kuga, Y. Torii, N. Shiokawa, T. Hirano, Y. Shimizu, and H. Sasada, Phys. Rev. Lett. **78**, 4713 (1997).

[7] E. M. Wright, R. Y. Chiao, and J. C. Garrison, Chaos Solitons Fractals **4**, 1797 (1994).

[8] A. V. Mamaev, M. Saffman, and A. A. Zozulya, Phys. Rev. Lett. **77**, 4544 (1996); **78**, 2108 (1997).

[9] Z. Chen, M. Segev, D. W. Wilson, R. E. Muller, and P. D. Maker, Phys. Rev. Lett. **78**, 2948 (1997).

[10] G. A. Swartzlander, Jr. and C. T. Law, Phys. Rev. Lett. **69**, 2503 (1992).

[11] I. V. Basistiy, V. Y. Bazhenov, M. S. Soskin, and M. V. Vasnetsov, Opt. Commun. **103**, 422 (1993).

[12] K. Dholakia, N. B. Simpson, M. J. Padgett, and L. Allen, Phys. Rev. A **54**, R3742 (1996).

[13] A. Berzanskis, A. Matijosius, A. Piskarskas, V. Smilgevicius, and A. Stabinis, Opt. Commun. **140**, 273 (1997).

[14] D. V. Petrov and L. Torner, Phys. Rev. E **58**, 7903 (1998); D. V. Petrov, L. Torner, J. Martorell, R. Vilaseca, J. P. Torres, and C. Cojocaru, Opt. Lett. **23**, 1444 (1998).

[15] B. Luther-Davies, R. Powels, and V. Tikhonenko, Opt. Lett. **19**, 1816 (1994); V. Tikhonenko, J. Christou, and B. Luther-Davies, J. Opt. Soc. Am. B **12**, 2046 (1995); B. Luther-Davies, J. Christou, V. Tikhonenko, and Y. S. Kivshar, J. Opt. Soc. Am. B **14**, 3045 (1997); A. G. Truscott, M. E. J. Friese, N. R. Heckenberg, and H. Rubinsztein-Dunlop, Phys. Rev. Lett. **82**, 1438 (1999).

[16] J. W. R. Tabosa, G. Chen, Z. Hu, R. B. Lee, and H. J. Kimble, Phys. Rev. Lett. **66**, 3245 (1991).

[17] W. Happer, Rev. Mod. Phys. **44**, 169 (1972).

[18] G. C. Cardoso, V. R. de Carvalho, S. S. Vianna, and J. W. R. Tabosa, Phys. Rev. A **59**, 1408 (1999).

[19] N. R. Heckenberg, R. McDuff, C. P. Smith, and A. G. White, Opt. Lett. **17**, 221 (1992).

[20] T. Walker, P. Feng, D. Hoffmann, and R. S. Williamson III, Phys. Rev. Lett. **69**, 2168 (1992).

[21] K.-P. Marzlin, W. Zhang, and E. M. Wright, Phys. Rev. Lett. **79**, 4728 (1997).

[22] L. Deng, E. W. Hagley, J. Wen, M. Trippenbach, Y. Band, P. S. Julienne, J. E. Simsarian, K. Helmerson, S. L. Rolston, and W. D. Phillips, Nature (London) **398**, 218 (1999).

Section 6

ROTATIONAL FREQUENCY SHIFTS

In the late 1970s Garetz and Arnold demonstrated that when a circularly polarised ($\sigma = \pm 1$) wave was transmitted through a half-wave plate rotating at $\Omega/2$, its frequency was shifted by $\sigma\Omega$ (6.1). Garetz recognised that the frequency shift was an example of a more general angular Doppler shift (6.2, **Paper 6.1**), which was manifest whenever light was emitted, absorbed or scattered by rotating bodies. The frequency shift can be understood in terms of energy conservation or, more sophisticatedly, by a Jones matrix analysis. Subsequently, the same type of experiment was described in terms of a dynamically evolving Berry phase (6.3). The shift is analogous to the speeding up or slowing down of a clock hand when the clock is placed on a rotating turntable.

In 1993 van Enk argued that when LG modes were transformed to HG modes where $n + m = 1$ and back again, there should be a geometric phase equivalent to the Pancharatnam phase obtained for polarisation (6.4). The way in which astigmatic optical elements transformed the phase of helical phasefronts, analogous to the way birefringent components transformed the phase of polarised waves, had already been demonstrated by Beijersbergen *et al.* (6.5, **Paper 3.5**). Nienhuis predicted that the rotation of a cylindrical-lens π-mode converter would impart a frequency shift to the transmitted LG mode (6.6, **Paper 6.2**). For a mode converter rotation frequency of $\Omega/2$, the predicted frequency shift of the light beam should be $l\Omega$. The waveplate and the mode converter perform a mirror inversion of the electric field and phase structure respectively. It follows that a rotation of the optical component by α rotates the polarisation, or phase structure, by 2α and so a $\Omega/2$ rotation frequency of the optical component rotates the beam at Ω. In each case, the frequency shift is equal to the rotation rate of the beam multiplied by the angular momentum per photon. Both shifts are readily explicable in terms of transformations on the Poincaré sphere or by Jones matrices or by their equivalent for helically phased modes (6.7, **Paper 6.3** and 6.8, **Paper 2.4**).

In 1998, Courtial *et al.* (6.9) generated a variety of LG modes in the millimetre-wave region of the spectrum and passed them through a rotating π-mode converter based on a Dove prism: conveniently, this device does not rotate the polarisation state (6.10). They confirmed the $l\Omega$ frequency shift and explained it both in terms of earlier work, a quantum mechanical rotational frequency shift identified by Bialynicki-Birula and Bialynicka-Birula (6.11, **Paper 6.4**) and an azimuthal Doppler shift applied to atomic transitions (6.12, **Paper 6.5**). This frequency shift is simply explicable in terms of the electric field. Whereas the phasefront of a planar wave is invariant under rotation, the phasefront of a helical beam is not. For helical phase fronts, a rotation of the beam is indistinguishable from its temporal evolution and one rotation of the beam introduces a phase change of l cycles. A beam rotation frequency of Ω thus produces a frequency shift of $l\Omega$.

Courtial *et al.* (6.13, **Paper 6.6**) also experimented with circularly polarised LG modes, rotating the whole beam by means of a Dove prism and half-waveplate combined. They confirmed that the result was not two individual frequency components relating to the shifts arising from the spin and orbital angular momentum, but a single frequency shift that was proportional to the total angular momentum, that is $(l+\sigma)\Omega$, as predicted by the Birulas (6.11). This equivalence of action between spin and orbital angular momentum is by no means general. Again the effect can be understood in terms of the electric field distribution. A cross-section through the beam shows that the effect of circular polarisation and the helical structure is to produce an $(l+\sigma)$-fold rotational symmetry, such that upon one rotation the phase change is $(l+\sigma)$ cycles.

This general form of the angular Doppler shift should not be confused with the Doppler shift associated with rotating galaxies and other bodies. The latter effect is simply a manifestation of the translation Doppler effect where the rotation of an object results in a component of velocity in the line of sight. The angular Doppler shift is a maximum in the direction of the angular momentum vector where the linear Doppler shift is zero. Whereas the translation Doppler shift can be expressed in terms of the linear momentum multiplied by the linear velocity per photon, $k_z v_z$, the angular or rotational Doppler effect is the angular momentum per photon multiplied by the annular velocity, that is $[k_\varphi(r)/r][rv_\varphi]$, or $k_\varphi(r)v_\varphi$.

REFERENCES

6.1 BA Garetz and S Arnold, 1979, *Opt. Commun.* **31** 1.
6.2 BA Garetz, 1981, *J. Opt. Soc. Am.* **71** 609.
6.3 F Bretenaker and A Le Flock, 1990, *Phys. Rev. Lett.* **65** 2316.
6.4 SJ van Enk, 1993, *Opt. Commun.* **102** 59.
6.5 MW Beijersbergen, L Allen, HELO van der Veen and JP Woerdman, 1993, *Opt. Commun.* **96** 123.
6.6 G Nienhuis, 1996, *Opt. Commun.* **132** 8.
6.7 MJ Padgett and J Courtial, 1999, *Opt. Lett.* **24** 430.
6.8 L Allen, J Courtial and MJ Padgett, 1999, *Phys. Rev. E* **60** 7497.
6.9 J Courtial, K Dholakia, DA Robertson, L Allen and MJ Padgett, 1998, *Phys. Rev. Lett.* **80** 3217.
6.10 MJ Padgett and JP Lesso, 1999, *J. Mod. Opt.* **46** 175.
6.11 I Bialynicki-Birula and Z Bialynicka-Birula, 1997, *Phys. Rev. Lett.* **78** 2539.
6.12 L Allen, M Babiker and WL Power, 1994, *Opt. Commun.* **112** 144.
6.13 J Courtial, DA Robertson, K Dholakia, L Allen and MJ Padgett, 1998, *Phys. Rev. Lett.* **81** 4828.

PAPER 6.1

Angular Doppler effect

Bruce A. Garetz

Department of Chemistry, Polytechnic Institute of New York, Brooklyn, New York 11201

Received August 15, 1980; revised manuscript received November 3, 1980

An angular analog of the Doppler effect arising from the quantum of angular momentum carried by circularly polarized photons is presented and developed. Applications to rotational Raman scattering, fluorescence doublets, controlled frequency shifting of light, rotation-induced optical activity, and the measurement of rotational motion of small particles are discussed.

INTRODUCTION

That a photon carries a quantum of angular momentum has some interesting consequences concerning the way it interacts with matter. It is responsible for frequency shifts in circularly polarized light interacting with bodies in angular motion. Early mention of frequency shifts that are due to light interacting with rotating bodies dates back to Airy's conception of revolving plane-polarized light as the superposition of right and left circularly polarized components of different frequencies and Righi's experimental tests of this idea using a rotating Nicol.[1]

In the course of this Letter, we show that phenomena as diverse as rotational Raman spectra,[2] fluorescence doublets, and certain frequency-shifting devices[3,4] are manifestations of the angular Doppler effect (ADE). Other applications include the measurement of small-particle rotational motion and rotation-induced optical activity.

We are concerned with angular motion with an axis of rotation that is parallel to the direction of propagation. Such angular motion is not to be confused with angular motion with an axis of rotation that is normal to the propagation vector; the latter is responsible for frequency shifts that are fully described by the linear Doppler effect.[5] We are also not concerned with the relativistic transverse Doppler effect, which is due to linear motion perpendicular to the propagation vector.[6]

Although quantum mechanics is used as a convenient device for simplifying the discussion of the ADE, the ADE is essentially a classical effect.

1. ANGULAR MOMENTUM OF LIGHT

Circularly polarized (CP) light is of particular interest because it is that electromagnetic radiation that is an eigenstate of the angular momentum operator L_z, with eigenvalues $\pm\hbar = \pm h/2\pi$, where h is Planck's constant, corresponding to right and left circular polarization (RCP and LCP, respectively). This connection between circular polarization and angular momentum predates quantum theory. Poynting first suggested that circularly polarized electromagnetic waves should carry angular momentum of magnitude $\lambda/2\pi$ times the linear momentum of the wave,[7] where λ is the wavelength. This is also the quantal result since the linear momentum p has the value $h/\lambda = h\nu/c$, where ν is the frequency and c is the speed of light. Thus a photon carries linear momentum proportional to ν/c, whereas its angular momentum is *independent* of frequency. These facts dictate some important differences in the details of the linear and angular Doppler effects.

2. ANGULAR DOPPLER EFFECT

The linear Doppler effect has two aspects: (1) that light emitted from translating bodies suffers a shift $\Delta\nu = \nu v/c$, where v is the relative velocity between observer and body; and (2) that light interacting with (specifically, reflected from) translating bodies suffers a shift $\Delta\nu = 2\nu v/c$.

The angular Doppler effect displays two analogous aspects, which we treat separately.

A. Light Emitted from Rotating Bodies

Our development of the ADE closely parallels standard derivations of the linear Doppler effect, as given by Sommerfeld and others. A full treatment requires relativity. Since $v \ll c$ for the phenomena with which we are concerned, computations can be carried out classically. We follow a corpuscular description similar to one used by Sommerfeld.[8]

Consider a particle rotating with angular frequency $\nu_{\rm rot}$. If a CP photon is emitted normal to the plane of rotation, conservation of angular momentum requires that $\Delta L = \pm\hbar$. (Whether the sign is plus or minus depends on the relative senses of the circular polarization of the light and the rotational motion of the particle.) Since the rotational kinetic energy is given by $E = L^2/2I$, where I is the moment of inertia of the particle, conservation of energy dictates that

$$\frac{L_1^2}{2I} + E_1 = \frac{L_2^2}{2I} + E_2 + h(\nu + \Delta\nu), \qquad (1)$$

where L_1 and L_2 are particle angular momenta before and after emission, respectively, E_1 and E_2 represent the internal electronic energies of the emitting atom or molecule before and after emission, respectively, ν is the frequency of the emission when the particle is at rest, and $\Delta\nu$ is the angular Doppler shift in the frequency that is due to particle rotation.

The rotating particle may be some macroscopic body on which an emitting molecule or light source is fixed, or it may be the emitting molecule itself. In the latter case, $L^2/2I$ is the nuclear-rotational contribution to the molecular kinetic energy. Interpreting Eq. (1) in terms of a single molecule, we are thus assuming that the electronic Hamiltonian is separable from the nuclear-rotational Hamiltonian and that the rotational motion may be treated classically. The first assumption is essentially the Born–Oppenheimer approximation[9]; the second is appropriate in the limit that $L \gg \hbar$, or equivalently, that the rotational quantum number $J \gg 1$. This situation is frequently encountered with gas-phase molecules of moderate molecular weight at room temperature. [For iodine (I_2) at room temperature, the most probable J is approximately 70.] The practice of treating some degrees of freedom classically and others quantum mechanically is not a new one; the translational energy of atoms and molecules is commonly treated classically. We have simply shifted nuclear rotation into this classical realm.

Realizing that $E_1 - E_2 = h\nu$ yields

$$h\Delta\nu = \frac{L_1^2 - L_2^2}{2I} = \frac{(L_1 - L_2)(L_1 + L_2)}{2I} = \frac{\Delta L}{I}\left(\frac{L_1 + L_2}{2}\right), \quad (2)$$

which is $\approx(L\Delta L/I)$ to first order in L. However, $\Delta L = \pm\hbar$ and $L = I\omega_{\text{rot}} = 2\pi I\nu_{\text{rot}}$, where ω_{rot} is the particle's angular velocity in rad/sec, so that

$$h\Delta\nu = \pm 2\pi\hbar\nu_{\text{rot}} = \pm h\nu_{\text{rot}} \text{ or } \Delta\nu = \pm\nu_{\text{rot}}. \quad (3)$$

We have assumed here that, in the rotating molecule case, the molecular electronic angular momentum remains unchanged on emission. Thus we are limiting that portion of our treatment to transitions of the sort $\Sigma \rightarrow \Sigma$ in diatomic molecules.

An easy way to visualize the shift just described is to consider that CP light of frequency ν is represented by a rotating electric field vector, which makes ν rev/sec. An observer in a reference frame rotating at angular frequency ν_{rot} would count $(\nu - \nu_{\text{rot}})$ or $(\nu + \nu_{\text{rot}})$ revolutions in 1 sec, depending on whether the electric field rotation was in the same or the opposite sense as the frame rotation, respectively.

B. Light Interacting with Rotating Bodies

The next step is to determine the angular analog of light reflected from a translating mirror. A mirror reverses the sign of the linear momentum vector of the light. The angular equivalent is an interaction in which the sign of the angular momentum vector is reversed, i.e., in which RCP is converted to LCP or vice versa. A half-wave retardation plate serves this function.[3] The change in angular momentum of a CP beam suggests a torque applied to the wave plate. Such torques have been experimentally measured by Beth.[10] If the wave plate is rotating, this torque is applied through an angular displacement, meaning that work is done on or by the wave plate:

$$W = \int \tau d\theta, \quad (4)$$

where W is the work, τ is the applied torque, and θ is the angular displacement.

For a single CP photon, the change in angular momentum $\Delta L = \pm 2\hbar$ (cf. the emissive case, where $\Delta L = \pm\hbar$). For a wave plate rotating at angular velocity ω_{rot}, $L = I\omega_{\text{rot}}$ and $E_{\text{rot}} = L^2/2I$. To first order, the change in energy of the plate is

Table 1. Comparison of Linear and Angular Doppler Frequency Shifts

	Emissive Case	Interactive Case
Linear Doppler	$\pm\nu_0(v/c)$	$\pm 2\nu_0(v/c)$
Angular Doppler	$\pm\nu_{\text{rot}}$	$\pm 2\nu_{\text{rot}}$

$$\Delta E = \frac{L\Delta L}{I} = \pm 2\hbar\omega_{\text{rot}} = \pm 2h\nu_{\text{rot}}. \quad (5)$$

To conserve energy, this requires the photon to gain or lose this energy, which is manifested in a frequency change

$$\Delta\nu = \pm 2\nu_{\text{rot}}, \quad (6)$$

i.e., the frequency of an incident CP beam whose electric field is rotating in the same sense as the wave plate is downshifted by $2\nu_{\text{rot}}$, whereas a CP beam of the opposite sense is upshifted by $2\nu_{\text{rot}}$. This is analogous to the linear momentum transfer that occurs in the linear Doppler effect, which causes a frequency shift $\Delta\nu = \pm 2\nu_0 v/c$.

Interactions involving the scattering of light from rotating bodies can yield an unshifted component to the scattered light, in addition to the two shifted components described above.[2] This arises from photons scattered with their angular momentum vectors unchanged. In this case, $\Delta L = 0$ and $\Delta\nu = 0$.

Table 1 summarizes the frequency shifts associated with the linear and angular Doppler effects for both emissive and interactive cases. The factor of 2 in the interactive cases arises because the linear or angular momentum vector is reversed, causing a change in momentum of *twice* the magnitude of the momentum of the photon. The factor ν_0 for the linear Doppler effect arises from the fact that the photon linear momentum is proportional to the frequency of the light, whereas the angular momentum is independent of frequency.

3. APPLICATIONS

A. Rotational Structure of Emission Spectra: Fluorescence Doublets

In terms of the ADE, the rotational structure of vibronic transitions becomes illuminated. Consider the $B \rightarrow X$ emission of the diatomic molecule, iodine. A molecule in rotational quantum level J rotates with $L = \hbar J = I\omega$ or with a classical frequency $\nu_{\text{rot}} = \pm hJ/4\pi^2 I$. If the I_2 molecule at rest would emit light at ν_0, the ADE would predict RCP and LCP emission doublets observable at $\nu = \nu_0 \pm hJ/4\pi^2 I$. This is observed quantum mechanically as a result of the selection rule $\Delta J = \pm 1$, which, in the classical limit (large J), yields the same result. This is obviously no coincidence, since the rotational selection rule is simply a consequence of conservation of angular momentum coupled with the fact that photons carry a quantum of angular momentum. In a sense, the existence of fluorescence doublets is a classical phenomenon.

B. Rotational Raman Scattering

Rotational Raman scattering represents a case of interactive ADE. As such, the scattering of light of frequency ν_0 from a rotating diatomic molecule, such as I_2 in rotational level J (classical rotational frequency of $\nu_{\text{rot}} = hJ/4\pi^2 I$), yields scattered components at frequencies ν_0 and $\nu_0 \pm 2hJ/4\pi^2 I$. Because of quantum mechanical selection rules $\Delta J = 0, \pm 2$, this same result is obtained quantum mechanically in the classical limit. Subtle differences in the two cases are discussed by Newburgh and Borgiotti.[2]

C. Variable Frequency Shifting of CP Light

The ADE is the basis for a device capable of shifting the frequency of a CP beam of light by an arbitrary amount.[3] It involves sending CP light through a rotating quartz half-wave plate. The resulting light displays a frequency shift of $\pm 2\nu_{rot}$, where ν_{rot} is the rotational frequency of the wave plate. Because of physical limitations in rotating macroscopic quartz plates without distortion, achievable shifts are limited to the kilohertz regime, although megahertz shifts are possible if the retardation is carried out electro optically.

D. Other Applications

As mentioned at the end of Section 2.A, incident RCP and LCP waves of the same frequency are observed to have different frequencies in a rotating reference frame. If the rotating medium exhibits dispersion in absorption or refractive index, an apparent optical activity should be observed, since the right and left CP components are absorbed or refracted by different amounts by the medium. Such induced circular dichroism should be substantial in the vicinity of sharp absorption peaks or band edges.

The ADE can be used to measure the rotational velocity of small particles through the detection of frequency-shifted components of CP light scattered from such particles, the sign of the shift yielding the sense of the particle rotation. Finally, shifted frequency components in light scattering measurements in liquids that are due to processes such as rotational diffusion and rotational Brownian motion can be interpreted in terms of the ADE.

In summary, the angular Doppler effect is a novel optical principle that characterizes frequency shifts in CP light, which is emitted from or interacts with rotating bodies. The principle arises from simple conservation of energy and angular momentum considerations and has application to a variety of spectroscopic, light scattering, and optical processing phenomena.

Acknowledgment is made to the Research Corporation and the Petroleum Research Fund, administered by the American Chemical Society, for partial support of this research.

REFERENCES

1. R. W. Wood, *Physical Optics* (Dover, New York, 1961), Chap. 9, p. 363.
2. R. G. Newburgh and G. V. Borgiotti, "Short wire as a microwave analogue to molecular Raman scatterers," Appl. Opt. **14,** 2727–2730 (1975).
3. B. A. Garetz and S. Arnold, "Variable frequency shifting of circularly polarized laser radiation via a rotating half-wave plate," Opt. Commun. **31,** 1–3 (1979).
4. P. J. Allen, "A radiation torque experiment," Am. J. Phys. **34,** 1185–1192 (1966).
5. A. Sommerfeld, *Optics* (Academic, New York, 1954), Chap. 2, Sec. 15, pp. 79–82.
6. C. Møller, *The Theory of Relativity* (Clarendon, Oxford, 1972), Sec. 2.11, p. 59.
7. J. H. Poynting, "The wave motion of a revolving shaft, and a suggestion as to the angular momentum in a beam of circularly polarised light," Proc. R. Soc. London **A82,** 560–567 (1909).
8. Ref. 5, Chap. 2, Sec. 16, pp. 84–86.
9. M. Born and R. Oppenheimer, "Zur Quantentheorie der Molekeln," Ann. Phys. **84,** 457–484 (1927).
10. R. A. Beth, "Mechanical detection and measurement of the angular momentum of light," Phys. Rev. **50,** 115–125 (1936).

PAPER 6.2

Doppler effect induced by rotating lenses

Gerard Nienhuis

Huygens Laboratorium, Rijksuniversiteit Leiden, Postbus 9504, 2300 RA Leiden, The Netherlands

Received 8 February 1996; accepted 24 April 1996

Abstract

The orbital angular momentum inherent to light beams with a helical wavefront can be transferred to non-isotropic lenses. A system of three cylindrical lenses suffices to transform an arbitrary paraxial input beam by inverting one transverse direction of the mode function. This also inverts the orbital angular momentum of a Laguerre–Gaussian beam with mode index m. The corresponding transfer of angular momentum shows up directly as a frequency shift $2m\Omega$ when the lens system is set in rotation at frequency Ω around the axis.

1. Introduction

An optical element that rotates about its axis at a uniform angular velocity Ω will usually generate sidebands in the spectrum of a light beam. When the element affects only the light polarization, and the monochromatic input beam has frequency ω, the sidebands can only occur at the frequencies $\omega \pm 2\Omega$. In special cases, the frequency is simply shifted by 2Ω. An example is found in Ref. [1], where the observed frequency shift of a linearly polarized beam after a double pass through a rotating quarter-wave plate has been interpreted as a time-dependent manifestation of a geometric phase. The closed trajectory on the Poincaré sphere describing the polarization encloses a solid angle that varies linearly with time, and the resulting linear variation of the Pancharatnam phase corresponds to a frequency shift. From a less fancy point of view, frequency shifts due to a rotating birefringent plate may also be regarded as resulting from the energy exchange between the light beam and the plate. The origin of the work that the light exerts on the rotating plate lies in the torque that accompanies the exchange of angular momentum, for instance when a half-wave plate inverts the handedness of circular polarization [2]. In this picture, the frequency change has a dynamical rather than a geometrical nature. The mechanical action of this torque has been demonstrated in an early experiment by Beth [3]. The physical origin of the geometric Pancharatnam phase in transfer of angular momentum has been suggested before [4,5].

In addition to the angular momentum inherent to circular polarization, a light beam can also have an angular momentum along the axis that results from the helical nature of the wavefronts [6]. This orbital angular momentum results from the density of transverse momentum, which in turn corresponds to the transverse phase gradients that generally exist in a beam of finite extent. Therefore, one may expect that sidebands generated by rotating lenses can likewise be viewed as resulting from the exchange of orbital angular momentum between beam and lens. This idea will be explored in the present paper. Whereas the polarization-dependent angular momentum is limited to a single unit $\pm\hbar$ per photon, the orbital angular momentum can in principle amount to a large number of units. As a result, one predicts that the frequency change can be many times the rotation frequency Ω. In some cases, a cyclic transformation of the state of the orbital angular momentum of a paraxial beam can be accompanied by a geometric phase change [7]. A frequency shift will then arise when the cyclic transformation is performed by rotating optical elements, so that the geometric phase varies linearly with time. We describe an inverter consisting of three astigmatic lenses, which transforms an arbitrary paraxial beam with orbital angular momentum $\hbar m$ per photon into a beam with the opposite value $-\hbar m$. In contrast to other mode converters, the present lens system need not be adapted to the specific properties of the input beam, such as its Rayleigh range or the location of its focus. Then the frequency shift $2m\Omega$ induced by rotation of the lenses can be directly interpreted as arising from the torque on the lenses that accompanies the transfer of orbital angular momentum.

2. State-vector description of time-dependent paraxial beams

2.1. Free propagation

A moving optical element, such as a receding mirror or a rotating lens, will in general introduce non-monochromaticity into a light beam. Therefore, we use a time-dependent version of the paraxial wave equation. A light beam propagating in free space is described by Maxwell's wave equation for the electric field,

$$\nabla^2 \boldsymbol{E} = \frac{1}{c^2}\frac{\partial^2 \boldsymbol{E}}{\partial t^2}. \tag{1}$$

For a nearly monochromatic beam with polarization $\boldsymbol{\varepsilon}$ and propagating along the z-axis the electric field can be factorized as

$$\boldsymbol{E}(\boldsymbol{r}, t) = \boldsymbol{\varepsilon} u(\boldsymbol{r}, t) e^{ikz - i\omega t}, \tag{2}$$

where the field amplitude u obeys the time-dependent paraxial wave equation

$$\left(\frac{\partial}{\partial z} + \frac{1}{c}\frac{\partial}{\partial t}\right) u = \frac{i}{2k}\left(\frac{\partial^2}{\partial x^2} + \frac{\partial^2}{\partial y^2}\right) u. \tag{3}$$

This equation results after substituting Eq. (2) in the wave equation (1), while neglecting second derivatives of u with respect to z and t. This is justified when the amplitude u varies little with z and t over a wavelength or an optical cycle. Eq. (3) generalizes the standard paraxial wave equation for monochromatic beams, in which u is independent of time [8]. Then the time derivative in Eq. (3) may be omitted, and the resulting equation has the same form as Schrödinger's equation for a free particle in two dimensions, where the z coordinate plays the role of time [9]. This suggests describing the amplitude of a monochromatic paraxial beam for a given value of z in terms of a state vector $|u(z)\rangle$, in analogy to a quantum-mechanical time-dependent state vector $|\psi(t)\rangle$ [10]. The evolution of a quantum-mechanical state as described in terms of linear operators then corresponds to the propagation of the light beam. In the present case of non-monochromatic paraxial beams, the state vector depends both on z and t, and the propagation equation takes the linear form

$$\left(\frac{\partial}{\partial z} + \frac{1}{c}\frac{\partial}{\partial t}\right)|u(z, t)\rangle = -\frac{i}{2k}\left(\hat{P}_x^2 + \hat{P}_y^2\right)|u(z, t)\rangle, \tag{4}$$

where we introduced the components of a momentum operator

$$\hat{P}_x = -i\frac{\partial}{\partial x}, \quad \hat{P}_y = -i\frac{\partial}{\partial y}. \tag{5}$$

The free propagation of a beam over a distance Z can be expressed in terms of a propagation operator

$$\hat{U}(Z) = \exp\left(-\frac{iZ}{2k}\left(\hat{P}_x^2 + \hat{P}_y^2\right)\right), \tag{6}$$

in analogy to the quantum-mechanical evolution operator. However, the effect of free propagation constitutes a slight generalization with respect to quantummechanical evolution, in that the effect of retardation must be included. The solution of Eq. (4) obeys the equality

$$|u(z, t)\rangle = \hat{U}(Z)\left|u\left(z_0, t - \frac{Z}{c}\right)\right\rangle, \tag{7}$$

with $Z = z - z_0$ the traversed distance. Eq. (7) relates the field amplitude at the position z at time t to the amplitude at position z_0 at an earlier time. The time difference is simply the travel time of light between the two positions.

2.2. *Ideal lenses*

The effect of a thin dielectric object placed in the beam is described by a phase change that depends on x and y. The layer must be so thin that transverse propagation in the layer is negligible, so that for each value of x and y the output amplitude is equal to the input amplitude multiplied by the phase factor $\exp(\mathrm{i}\eta)$ with

$$\eta(x, y) = k(n-1)a(x, y). \tag{8}$$

Here n is the refractive index of the dielectric, and a is its thickness. A purely cylindrical lens with axis oriented at an angle α with respect to the x-axis can then be described by the operator $\hat{S}$, that is equal to [10,11]

$$\hat{S}(\alpha) = \exp\left(-\frac{\mathrm{i}k}{2f}(\hat{X}\cos\alpha + \hat{Y}\sin\alpha)^2\right), \tag{9}$$

where f is the focal length, and where by definition the operators $\hat{X}$ and $\hat{Y}$ multiply the amplitude function $u(x,y)$ with x or y. The orientation dependence of the lens operator is conveniently expressed as

$$\hat{S}(\alpha) = \hat{R}(\alpha)\hat{S}(O)\hat{R}^{+}(\alpha), \tag{10}$$

i.e. in terms of the rotation operator

$$\hat{R}(\alpha) = \exp(-\mathrm{i}\alpha\hat{J}), \tag{11}$$

with

$$\hat{J} = \hat{X}\hat{P}_y - \hat{Y}\hat{P}_x = -\mathrm{i}\frac{\partial}{\partial\phi} \tag{12}$$

the angular momentum operator. When the lens is rotating at a constant rotation frequency Ω, the operator $\hat{S}$ attains a time dependence imposed by the time-dependent angle $\alpha = \Omega t$. For an arbitrary incident monochromatic beam, the output amplitude is obtained simply by multiplying the input amplitude by a phase factor. For the lens indicated by Eq. (9), the factor can be expressed in polar coordinates as

$$\exp\left[-\frac{\mathrm{i}k}{2f}\rho^2\cos^2(\Omega t - \phi)\right]$$

$$= \exp\left(-\frac{\mathrm{i}k\rho^2}{4f}\right)\sum_{n=-\infty}^{\infty} J_n\left(\frac{k\rho^2}{4f}\right) \times \exp[2\mathrm{i}n(\Omega t - \phi - \pi/4)], \tag{13}$$

with J_n the ordinary Bessel functions. This demonstrates that for a monochromatic input beam, the output of the rotating lens is highly non-monochromatic. It contains sidebands at all frequencies $\omega + 2n\Omega$.

When a time-dependent input beam passes a rotating lens at the position z_0, the output state vector at position z is can be expressed as

$$|u_{\mathrm{out}}(z, t)\rangle = \hat{U}(Z)\hat{S}(t - Z/c)|u_{\mathrm{in}}(z_0, t - Z/c)\rangle, \tag{14}$$

with $Z = z - z_0$. This is easily verified by noting that the output beam obeys the time-dependent paraxial equation (4), while at $z = z_0$ it is correctly related to the input by the time-dependent lens operator.

3. Conversion of paraxial beams

3.1. *HG–LG conversion*

A Hermite–Gaussian (HG) beam can be converted into a Laguerre–Gaussian (LG) beam by using two cylindrical lenses. This has been demonstrated by Tamm and Weiss for the lowest-order Hermite–Gaussian mode [12]. In fact, the technique can be applied to modes of any order [13]. The setting of the pair of lenses must be such that they impose a difference equal to $\pi/2$ in the Gouy phase of the x-dependence and the y-dependence of the mode function. This can be done by orientating the axis of two identical cylindrical lenses by an angle of 45° with respect to the symmetry axis of the incident HG mode. The two lenses must be symmetrically positioned at opposite sides of the focus, at a distance

$$D = \beta\tan\frac{\pi}{8} = \beta(-1 + \sqrt{2}) \tag{15}$$

from the focus, where β is the Rayleigh range of the input beam, which is half its confocal parameter. The focal length of the cylindrical lenses must be equal to

$$f = D\sqrt{2} = \beta(2 - \sqrt{2}). \tag{16}$$

The Rayleigh range β' of the transformed beam factor in the astigmatic region between the two lenses is then equal to $D^2/\beta = \beta(3 - 2\sqrt{2})$.

3.2. *Harmonic-oscillator picture*

It has been shown [14] that the connection between various higher-order Gaussian modes is mathematically identical to the eigenstates of a two-dimensional quantum harmonic oscillator, and that the change of the Gouy phase during propagation has the same effect as the corresponding phase evolution of the two components of the harmonic oscillator. Higher-order modes can always be described as arising from the action of a raising operator on the fundamental mode. This fundamental mode has a Gaussian shape, and for each value of z it is fully determined by the radius of curvature s of the wavefronts and by the spot size γ. For an astigmatic beam, these parameters are different for two orthogonal transverse directions. The local values of s and γ determine both the Rayleigh range β and the distance z from the focus, by the relation

$$\frac{1}{\gamma^2} - \frac{\mathrm{i}k}{s} = \frac{k}{\beta + \mathrm{i}z}. \tag{17}$$

This equation determines the z-dependence of γ and s during free propagation. The corresponding change of the Gouy phase χ can be expressed by the differential

$$\mathrm{d} \tan \chi = \mathrm{d}z/\beta. \tag{18}$$

A thin lens with focal length f does not affect the spot size γ, but it affects the curvature radius s, according to the identity

$$\frac{1}{s_+} = \frac{1}{s_-} - \frac{1}{f}. \tag{19}$$

Of course, a cylindrical lens only affects the curvature radius in the plane in which the lens is focusing. An attractive feature of the harmonic-oscillator analogy is that the effect of propagation and of ideal lenses on a Gaussian beam of arbitrary order can be expressed solely in terms of the beam parameters γ and s, which determine the local Gaussian envelope of the mode function.

A HG beam is analogous to a factorized state of the harmonic oscillator, with a well-defined excitation number for the oscillation in two orthogonal directions. An LG beam corresponds to a state of the two-dimensional oscillator that is an eigenstate of both energy and angular momentum. When the azimuthal mode number of the LG beam is equal to m, corresponding to the dependence $\sim \exp(\mathrm{i}m\phi)$ of the mode function u, the beam carries an orbital angular momentum in the propagation direction that is equal to $\hbar m$ per photon [6]. The conversion of a HG into an LG beam is analogous to the transformation of a factorized oscillator state with symmetry axis rotated over 45° with respect to the x-axis to an angular momentum eigenstate, simply by imposing a $\pi/2$ phase difference between the oscillation in the two orthogonal directions. The resulting angular momentum quantum number m is then equal to the difference of the two one-dimensional excitation numbers.

3.3. *LG inversion*

From the harmonic-oscillator analogy it follows that an LG beam $\sim \exp(\mathrm{i}m\phi)$ is converted to an LG beam $\sim \exp(-\mathrm{i}m\phi)$ by a lens system that imposes a phase difference π for the x dependence compared with the y dependence, in the same way as a harmonic oscillator eigenstate of angular momentum is transformed into the state with opposite eigenvalue by imposing a phase difference π for the oscillation in two orthogonal directions. It is impossible to make such a converter between two opposite LG beams with just two cylindrical lenses [11]. The reason is basically that the phase difference can only build up in the region between the two lenses, where the beam is astigmatic. Since the phase change of a Gaussian beam over a focus is asymptotically equal to π, this difference of two phase changes is necessarily smaller than π.

On the other hand, one would expect that such an LG inverter can be constructed by a combination of three parallel astigmatic lenses. To be specific, we consider a symmetric configuration, with three parallel cylindrical lenses that are focusing in the xz-plane, while leaving the y dependence of the mode function unaffected. Two identical cylindrical lenses are located at the positions $z = \pm D$, and a third cylindrical lens is positioned at $z = 0$. The two outer lenses have a focal length f, and f_0 is the focal length of the central lens. The parameters D, f and f_0 must

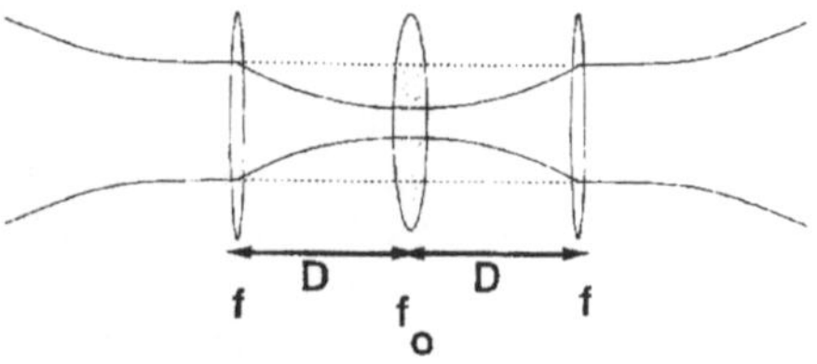

Fig. 1. Sketch of the beam inverter. The dotted line indicates the beam envelope in the yz-plane, which is not affected by the cylindrical lenses. The solid line gives the beam envelope in the xz-plane. Exact inversion occurs when the focal lengths are $f = D$, and $f_0 = D/4$.

be selected in such a way that the total difference in the Gouy phase for the x- and the y-direction amounts to π. Again, this phase difference must build up in the astigmatic region between the lenses. A sketch of the set-up and of a possible beam envelope is given in Fig. 1.

As one notices from Eqs. (15) and (16), the choice of the separation D and the focal length f in the case of the HG–LG converter depends on the specific properties of the incident beam, such as its Rayleigh range and the location of the focus. One might expect that this would also be true in the present case of an LG inverter. However, the following argument indicates that this is not so. Suppose that the LG inverter is constructed such that an incoming LG beam $\sim \exp(im\phi)$ with some specific Rayleigh range and focus position is properly converted to an output beam $\sim \exp(-im\phi)$. Since the lenses do not affect the y-dependence of the beam, the net result is a simple inversion of the x-dependence as compared with the output beam in the absence of the lenses. The relation between the LG output field u leaving the lens at $z = D$ and the unperturbed LG field u_0 at the same position in the absence of the lenses can therefore be expressed as

$$u(x, y, D) = u_0(-x, y, D) \tag{20}$$

for any LG beam mode with the same Rayleigh range and focus position. Since these LG modes form a complete set, we can expand an arbitrary input beam $u(x, y, -D)$ arriving at the first lens as a linear combination of these LG modes. The conclusion is that the effect of the lens system can be expressed as in Eq. (20) for an arbitrary paraxial input beam. This implies that the lens system is a universal beam inverter for paraxial beams.

In order to determine the parameters needed for LG–LG inversion, we express this conclusion in operator form. The net effect of this lens system on an arbitrary paraxial input beam is given by the ordered product of operators corresponding to the lenses and the intermittent free propagation, of the abbreviated form

$$\hat{C} = \hat{S}\hat{U}(D)\hat{S}_0\hat{U}(D)\hat{S}, \tag{21}$$

where $\hat{U}$ is the free-propagation operator (6), and $\hat{S}$ and $\hat{S}_0$ are operators for cylindrical lenses, with focal lengths f and f_0, respectively. Since the lenses are focussing in the x-direction, one should use Eq. (9) with $\alpha = 0$. We introduce the x-inversion operator $\hat{M}_x$ by the identity

$$\hat{M}_x u(x, y) = u(-x, y). \tag{22}$$

Since $|u_0(D)\rangle$ is related to $|u(-D)\rangle$ by the free-propagation operator $\hat{U}(2D)$, the general condition (20) for the LG inverter is equivalent to the identity

$$|u(D)\rangle = \hat{C}|u(-D)\rangle = \hat{M}_x\hat{U}(2D)|u(-D)\rangle \tag{23}$$

for an arbitrary input beam $|u(-D)\rangle$. Hence, we must select the parameters D, f and f_0 of the lens system so that the operator identity

$$\hat{C} = \hat{M}_x\hat{U}(2D) \tag{24}$$

holds. The validity conditions for this identity can be obtained from the requirement that both sides lead to the same transformation of any operator $\hat{A}$, so that

$$\hat{C}^\dagger\hat{A}\hat{C} = \hat{U}^\dagger(2D)\hat{M}_x^\dagger\hat{A}\hat{M}_x\hat{U}(2D). \tag{25}$$

It is sufficient to substitute for the operator $\hat{A}$ the complete set of operators $\hat{X}$, $\hat{P}_x$, $\hat{Y}$ and $\hat{P}_y$. Since $\hat{M}_x$, $\hat{S}$ and $\hat{S}_0$ do not operate on the y-dependence of a mode function, the identity (25) trivially holds for $\hat{A} = \hat{Y}$ and $\hat{A} = \hat{P}_y$. For the other two cases $\hat{A} = \hat{X}$ and $\hat{A} = \hat{P}_x$, we use the transformation rules [14]

$$\hat{U}^\dagger(Z)\hat{X}\hat{U}(Z) = \hat{X} + \frac{Z}{k}\hat{P}_x, \tag{26a}$$

$$\hat{S}^\dagger\hat{P}_x\hat{S} = \hat{P}_x - \frac{k}{f}\hat{X}. \tag{26b}$$

When we take $\hat{A}=\hat{X}$ in Eq. (25), repeated application of Eq. (26) leads to the condition

$$\left(1-\frac{D}{f_0}-\frac{2D}{f}+\frac{D^2}{ff_0}\right)\hat{X}+\frac{D}{k}\left(2-\frac{D}{f_0}\right)\hat{P}_x$$
$$=-\hat{X}-\frac{2D}{k}\hat{P}_x. \quad (27)$$

The resulting condition for the parameters of the lens system is given by

$$f_0=D/4, \quad f=D. \quad (28)$$

In the same way one may verify that these relations (28) are also sufficient to ensure the validity of Eq. (25) for $\hat{A}=\hat{P}_x$. This serves as a check on the consistency of the arguments given above for the universality of the LG inverter.

The conclusion is that a symmetric system of three lenses characterized by the relations (28) has the effect of a simple inversion of the x-dependence of any incident paraxial beam. For Gaussian beams, this system serves as an exact π converter, regardless the values of the Rayleigh range and the location of the focus of the input beam. In the special case of an incident LG beam, it inverts the azimuthal dependence $\exp(im\phi)\rightarrow\exp(-im\phi)$, without affecting he radial dependence.

4. Frequency shift by a rotating LG inverter

We consider the case of an LG inverter that rotates about the z-axis at an angular velocity Ω. Strictly speaking, the axes of the three lenses should not be exactly parallel, in order to account for the retardation due to the finite travel time of the light. As one may conclude from Eq. (14), the orientation of the first lens at a given instant t must be parallel to the axis of the central lens a time D/c later, so that the output of the lens system has been affected by the three lenses with parallel orientations. For realistic rotation frequencies, this retardation effect can be safely ignored.

The input–output relation of the rotating lens system is expressed by

$$|u(D,t)\rangle=\hat{R}(\Omega t)\hat{C}\hat{R}^{\dagger}(\Omega t)\,|u(-D)\rangle. \quad (29)$$

When the parameters of the lens system obey Eqs. (28), the identity (24) is valid, and the effect of the rotating lenses is that the output beam is replaced by its mirror image with respect to a line that rotates in the xy plane at an angular velocity Ω. The effect of the rotation operator $\hat{R}(\Omega t)$ is that the azimuthal angle ϕ is replaced by $\phi-\Omega t$, whereas according to Eq. (24) the operator $\hat{C}$ leads to the simple inversion $\phi\rightarrow-\phi$, in addition to the free propagation over $2D$. By applying Eq. (29), we find that the only effect of the lenses is that in the output beam the azimuthal angle ϕ is substituted by $-\phi+2\Omega t$.

We consider the case that the input beam is an LG beam with an azimuthal dependence $\sim\exp(im\phi)$. One notices from this substitution that the effect of the rotating inverter is twofold. First, the output beam has the azimuthal dependence $\sim\exp(-im\phi)$, as in the case of a stationary inverter. Second, the frequency of the output beam has the shifted value

$$\omega'=\omega-2m\Omega. \quad (30)$$

In this special case of an LG input beam, the output is monochromatic. The field in the astigmatic region between the lenses, however, is strongly non-monochromatic, as illustrated by Eq. (13).

The frequency shift can be compared to the frequency shift -2Ω that occurs when a beam of circularly polarized light is converted by a rotating half-wave plate into the opposite circular polarization [2]. Frequency shifts arising from rotating birefringent plates have been interpreted as dynamical manifestations of the Pancharatnam phase [1]. The most straightforward interpretation results when one views the light beam in a co-rotating reference frame. In such a frame, the electric-field vector of circularly polarized light rotates at a frequency that is shifted by the angular velocity. This is just a rotational analogue of the (translational) Doppler effect, which also plays a role in the interpretation of the mechanical Faraday effect [15]. The double frequency shift occurs when the circular polarization is inverted by a rotating plate, in the same way as reflection from a moving mirror produces a double Doppler shift.

In the present case, it is the helicity of the wave fronts that is inverted rather than the light polarization. An LG beam with azimuthal quantum number m appears at a frequency that is shifted by $-m\Omega$ when it is viewed from a rotating frame, simply because a single full turn around the axis corresponds to m oscillations. This frequency down-shift is doubled when the inverted LG beam is transformed back to the non-rotating frame. Hence the frequency shift $-2m\Omega$ allows a simple geometrical interpretation as a rotational Doppler shift.

The frequency shift of the beam by the rotating LG inverter corresponds to an exchange of energy from the light beam to the lens system. Suppose that the beam intensity in number of photons per unit time is N. Photons in the input beam have an (orbital) angular momentum $\hbar m$ per photon, whereas the photons in the output beam have angular momentum $-\hbar m$. This demonstrates that the system of lenses gains an angular momentum per unit time that is equal to

$$T = 2N\hbar m, \tag{31}$$

which therefore equals the combined torque on the lenses. Since the lens system is rotating at an angular velocity Ω, the work that the light torque exerts on the lenses corresponds to a gain of rotational energy per unit time

$$W = T\Omega. \tag{32}$$

The resulting power loss per photon $2\hbar m\Omega$ corresponds precisely to the frequency shift $\omega - \omega'$, that was obtained in Eq. (30). Of course, a similar dynamical interpretation can also be given for the frequency shift accompanying a rotating half-wave plate that inverses circular polarization, or, for that matter, for the Doppler shift of light reflected from a receding mirror.

5. Conclusion

With a system consisting of three parallel astigmatic lenses, one can construct an exact beam inverter, so that the output beam is the full mirror image of what the output beam would have been in the absence of the lenses. The system consists of two identical lenses in a confocal configuration, and a third lens located at their common focus. The focal length of this central lens is one fourth of the focal length of the two outer lenses. The system is insensitive to the specific properties of the input beam. The only limitation is the validity of the paraxial approximation. It had been noted before that an inverter for LG beams can be realized only approximately with two lenses [11,13]. When the system is used to invert a Laguerre–Gaussian mode with azimuthal mode index m, the orbital angular momentum that is transferred to the lens system amounts to $2\hbar m$ per photon. This transfer of angular momentum shows up when the lens system is set into rotation with angular velocity Ω, in that the output frequency is downshifted by the amount $2m\Omega$. The corresponding power loss of the light beam corresponds to the mechanical work exerted by the torque on the lenses. Since both the input and the output beam have a circular nature, there can be no net torque on the two outer lenses, which implies that the full angular momentum transfer is picked up by the central lens. This effective rotational Doppler effect allows both a geometrical and a dynamical interpretation. When the input beam is not an exact Laguerre–Gaussian beam, the output beam acquires a series of sidebands. This means that the monochromaticity of the output may serve as a check on the azimuthal dependence of the input beam.

References

[1] R. Simon, H.J. Kimble and E.C.G Sudarshan, Phys. Rev. Lett. 61 (1988) 19.
[2] F. Bretenaker and A. Le Floch, Phys. Rev. Lett. 65 (1990) 2316.
[3] R.A. Beth, Phys. Rev. 50 (1936) 115.
[4] H. Jiao, S.R. Wilkinson, R.Y. Chiao and H. Nathel, Phys. Rev. A 39 (1989) 3475.
[5] S.C. Tiwari, J. Mod. Optics 39 (1992) 1097.
[6] L. Allen, M.W. Beijersbergen, R.J.C. Spreeuw and J.P. Woerdman, Phys. Rev. A 45 (1992) 8185.
[7] S.J. van Enk, Optics Comm. 102 (1993) 59.
[8] H.A. Haus, Waves and Fields in Optoelectronics (Prentice–Hall, Englewood Cliffs, NJ, 1984).
[9] D. Gloge and D. Marcuse, J. Opt. Soc. Am. 59 (1969) 1629.
[10] D. Stoler, J. Opt. Soc. Am. 71 (1981) 334.
[11] S.J. van Enk and G. Nienhuis, Optics Comm. 94 (1992) 147.
[12] C. Tamm and C.O. Weiss, J. Opt. Soc. Am. B 7 (1990) 1034.
[13] M.W. Beijersbergen, L. Allen, H.E.L.O. van der Veen and J.P. Woerdman, Optics Comm. 96 (1993) 123.
[14] G. Nienhuis and L. Allen, Phys. Rev. A 48 (1993) 656.
[15] G. Nienhuis, J.P. Woerdman and I. Kuščer, Phys. Rev. A 46 (1992) 7079.

PAPER 6.3

Poincaré-sphere equivalent for light beams containing orbital angular momentum

M. J. Padgett and J. Courtial

School of Physics & Astronomy, University of St Andrews, St Andrews, Fife, KY16 9SS Scotland, UK

Received December 10, 1998

The polarization state of a light beam is related to its spin angular momentum and can be represented on the Poincaré sphere. We propose a sphere for light beams in analogous orbital angular momentum states. Using the Poincaré-sphere equivalent, we interpret the rotational frequency shift for light beams with orbital angular momentum [Phys. Rev. Lett. **80**, 3217 (1998)] as a dynamically evolving geometric phase. © 1999 Optical Society of America

OCIS code: 350.1370.

Light beams possess a spin angular momentum that is associated with their polarization state. It is less well known that light beams also possess an orbital angular momentum that is associated with their azimuthal phase structure. This orbital angular momentum is independent of the polarization state of the light and therefore independent of the spin. Both spin and orbital angular momentum have been experimentally transferred to matter, causing rotation.[1-3]

Right- and left-handed circularly polarized light, the spin eigenstates, have a spin angular momentum of $+\hbar$ and $-\hbar$ per photon, respectively. In 1992, Allen *et al.* found that Laguerre–Gaussian modes, which have an azimuthal phase term of $\exp(il\phi)$, are orbital angular momentum eigenstates and carry an orbital angular momentum of $l\hbar$ per photon.[4] Laguerre–Gaussian modes are denoted $LG_p{}^l$, where l and p are the azimuthal and the radial mode indices, respectively. The quantity $N = 2p + |l|$ is referred to as the order of the mode.[5]

The polarization state of a monochromatic light beam can be completely characterized in terms of the Stokes parameters[6]

$$p_1 = \frac{I_{0°} - I_{90°}}{I_{0°} + I_{90°}},$$

$$p_2 = \frac{I_{45°} - I_{135°}}{I_{45°} + I_{135°}},$$

$$p_3 = \frac{I_{\text{right}} - I_{\text{left}}}{I_{\text{right}} + I_{\text{left}}}, \tag{1}$$

where $I_{0°}$, $I_{45°}$, $I_{90°}$, and $I_{135°}$ are the intensities of the light recorded through various orientations of linear polarizers and I_{right} and I_{left} are the intensities of the circularly polarized components in the beam. For a completely polarized light beam the squares of the Stokes parameters add up to unity, i.e.,

$$p_1{}^2 + p_2{}^2 + p_3{}^2 = 1. \tag{2}$$

Consequently, the Stokes parameters are the Cartesian coordinates of a space in which any completely polarized light beam is represented by a point on a sphere with unit radius around the origin. This sphere is known as the Poincaré sphere [Fig. 1(a)] and has proved to be a useful tool in dealing with transformations of the polarization state.[6]

The north and south poles of the Poincaré sphere represent the spin eigenstates, left- and right-handed circularly polarized light, respectively. Any state of complete polarization, and therefore any point on the Poincaré sphere, can be described as a superposition of left- and right-handed circular polarizations. For example, linearly polarized light is a superposition of equal intensities of left- and right-handed circularly polarized light; the relative phase of the superposition determines the orientation of the linear polarization.

We construct an analogous sphere for superpositions of left- and right-handed Laguerre–Gaussian modes with azimuthal phase terms of $\exp(\pm i\phi)$, respectively, which are orbital angular momentum eigenstates. These Laguerre–Gaussian modes (Fig. 2) are denoted $LG_0{}^{+1}$ and $LG_0{}^{-1}$ and possess an orbital angular momentum of $+\hbar$ and $-\hbar$ per photon, respectively. Their superpositions form structurally stable beams (modes) of mode order[7] $N = 1$. Therefore, as every point on the Poincaré-sphere analog corresponds to a mode of order 1, and also every mode of order 1 corresponds to a point on the Poincaré-sphere analog, we refer to this sphere as the sphere of first-order modes. For example, a superposition of equal intensities of left- and right-handed Laguerre–Gaussian modes forms a Hermite–Gaussian mode with indices $m = 1$ and $n = 0$, denoted $HG_{1,0}$, which is of order[5] $N = m + n = 1$. The

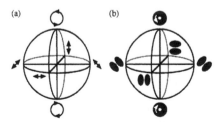

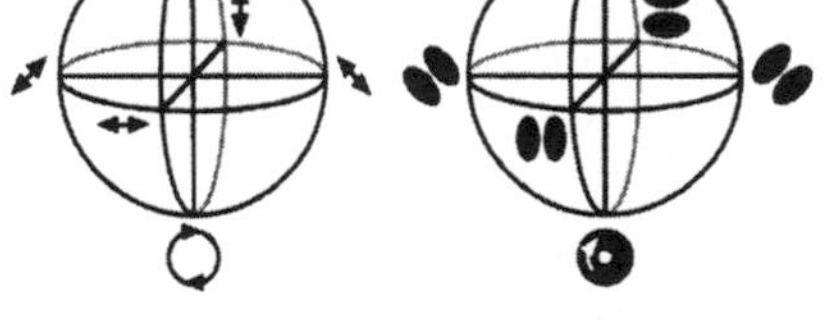

Fig. 1. (a) Poincaré sphere and (b) sphere of first-order modes. Some commonly encountered polarization states and modes are indicated.

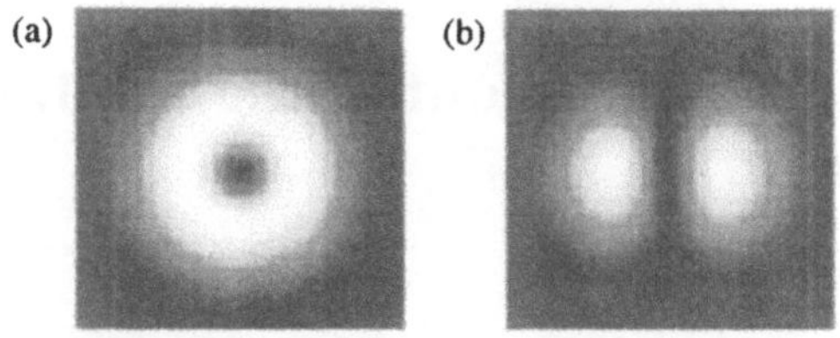

Fig. 2. Intensity cross section of (a) $LG_0^{\pm 1}$ modes and (b) a $HG_{1,0}$ mode. The modes LG_0^{-1} and LG_0^{-1} differ only in the handedness of their azimuthal mode structure.

relative phase of the superposition determines the orientation of the mode.

In analogy to the Stokes parameters, we define a set of parameters, o_1, o_2, o_3, for the newly proposed sphere:

$$o_1 = \frac{I_{HG_{1,0}^{0°}} - I_{HG_{1,0}^{90°}}}{I_{HG_{1,0}^{0°}} + I_{HG_{1,0}^{90°}}},$$

$$o_2 = \frac{I_{HG_{1,0}^{45°}} - I_{HG_{1,0}^{135°}}}{I_{HG_{1,0}^{45°}} + I_{HG_{1,0}^{135°}}},$$

$$o_3 = \frac{I_{LG_0^1} - I_{LG_0^{-1}}}{I_{LG_0^1} + I_{LG_0^{-1}}}, \quad (3)$$

where $I_{HG_{1,0}^{\alpha}}$ stands for the intensity of the Hermite–Gaussian $m = 1, n = 0$ mode at angle α that is present in the beam and $I_{LG_0^{\pm 1}}$ stands for the intensities of the Laguerre–Gaussian $l = \pm 1$, $p = 0$ modes in the beam. Writing all the intensities as the square of the moduli of the corresponding mode functions, and expanding all the mode functions in terms of Laguerre–Gaussian $l = \pm 1$, $p = 0$ mode functions,[5] we can show that

$$o_1^2 + o_2^2 + o_3^2 = 1. \quad (4)$$

Starting with left-handed circularly polarized light, a quarter-wave plate transforms the light to a linear polarization at 45° to the axis of the wave plate. On the Poincaré sphere, this transformation is represented by a move from the north pole to a point on the equator, the longitude of which depends on the orientation of the linear polarization.[8] Rotation of the wave plate through an angle α advances the longitude of the transformation by an angle 2α. Similarly, a half-wave plate transforms from left-handed to right-handed circular polarization. This transformation is presented by a move from pole to pole along a great circle, the longitude of which depends on the rotation angle of the wave plate.[8] This dependence on the rotation angle can be demonstrated if we consider the half-wave plate as two identical quarter-wave plates: As before, the state after the first wave plate is represented on the equator, and its longitude depends on the rotational alignment of the fast axis of the wave plate. Rotation of the wave plate through an angle α advances the longitude of the trajectory taken during the transformation by 2α.

In a fashion similar to that for a birefringent wave plate, which controls the relative phase between two linear polarizations at 90° to each other, the relative phase of two Hermite–Gaussian $m = 1$, $n = 0$ modes at 90° to each other (Fig. 3) can be controlled by an arrangement of cylindrical lenses. Beijersbergen *et al.*[5] detailed the design and operation of cylindrical-lens mode converters that utilize changes in relative Gouy phase to transform Hermite–Gaussian modes into Laguerre–Gaussian modes and vice versa. Both π converters and $\pi/2$ converters exist that perform tasks analogous to those performed by half-wave and quarter-wave plates, respectively. One may best appreciate the analogy by considering both the wave plates and the cylindrical lenses as means of introducing phase shifts between two orthogonal components of the beam.

In 1979 Garetz and Arnold[9] demonstrated that when circularly polarized light is transmitted through a rotating half-wave plate a frequency shift results. The shift is equal to twice the rotation frequency of the wave plate. Although the shift can be simply explained by Jones matrix polarization analysis, a similar experiment was also performed and explained in terms of a dynamic geometric (or Berry) phase by Simon *et al.*[8] In their experiment a combination of wave plates was used to transform the polarization state of the light around a closed loop on the Poincaré sphere. The resulting geometric phase shift was equal to half the solid angle enclosed by the loop and, in addition, to the constant phase shift that was due to the optical thickness of the components. Uniform rotation of one of the wave plates gave a constant rate of change of the enclosed solid angle, resulting in a frequency shift of the light beam.

Recently a similar experiment was performed with light beams containing orbital angular momentum.[10] When it is transmitted through a rotating π converter, the frequency of a $LG_0^{\pm 1}$ mode is shifted by twice the rotation frequency of the π converter. The mode transformation in the π converter can be interpreted as a geometric phase effect[7]; we interpret the frequency shift that occurs with transmission of the mode through a rotating π converter as evolution of this geometric phase in time.

Figure 4 illustrates the equivalence of the geometric phase shift for light beams containing either spin or orbital angular momentum after transmission through two half-wave plates or two π converters, respectively. In both cases the geometric phase shift is equal to half

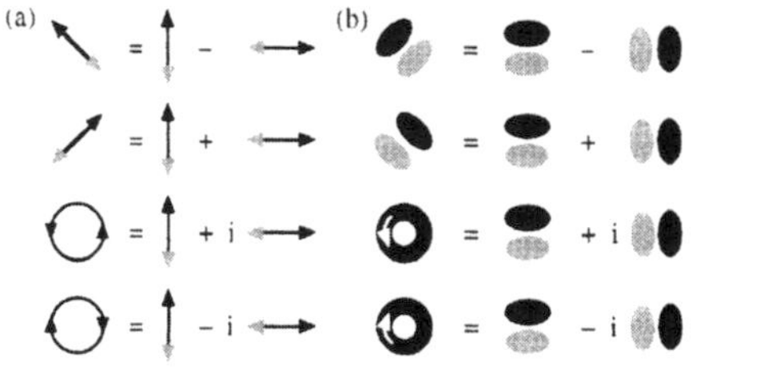

Fig. 3. Decompositions of (a) various polarization states and (b) first-order modes in terms of orthogonal linear polarization states and orthogonal Hermite–Gaussian $m = 1$, $n = 0$ modes, respectively. The black and gray shading indicates the relative phase of the components.

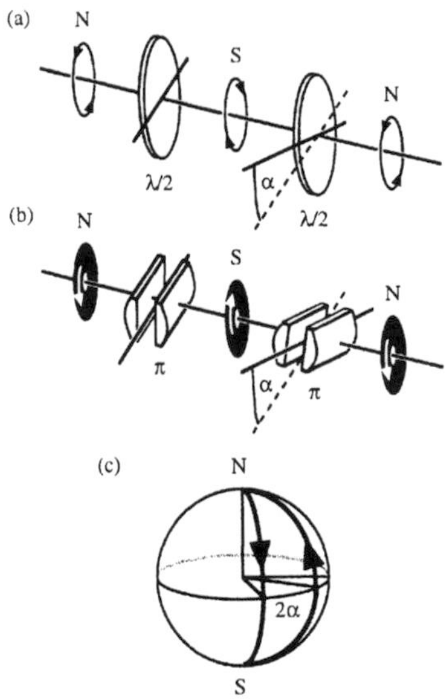

Fig. 4. Generation of evolving geometric phases for light beams containing (a) spin and (b) orbital angular momentum. (c) Path taken on the Poincaré sphere or the sphere of first-order modes.

the solid angle enclosed by the loop on the surfaces of the respective spheres. Therefore, for a change in the angular orientation between the two optical components of α, the resulting geometric phase shift is 2α. Thus, in both cases, the frequency of the light beam is shifted by twice the rotation frequency of the optical component.

We have proposed a Poincaré-sphere equivalent for light beams containing orbital angular momentum, called the sphere of first-order modes. On this sphere cylindrical-lens mode converters perform transformations analogous to those performed by wave plates on the Poincaré sphere. For the geometric phase accompanying transformations of first-order modes, the sphere of first-order modes is what the Poincaré sphere is for the geometric phase accompanying transformations of the polarization state.

We thank L. Allen, S. M. Barnett, and K. K. Wan for very helpful discussions. M. J. Padgett is a Royal Society research fellow. J. Courtial's e-mail address is jc4@st-and.ac.uk.

References

1. H. He, M. E. J. Friese, N. R. Heckenberg, and H. Rubinsztein-Dunlop, Phys. Rev. Lett. **75,** 826 (1995).
2. M. E. J. Friese, J. Enger, H. Rubinsztein-Dunlop, and N. R. Heckenberg, Phys. Rev. A **54,** 1593 (1996).
3. N. B. Simpson, K. Dholakia, L. Allen, and M. J. Padgett, Opt. Lett. **22,** 52 (1997).
4. L. Allen, M. W. Beijersbergen, R. J. C. Spreeuw, and J. P. Woerdman, Phys. Rev. A **45,** 8185 (1992).
5. M. W. Beijersbergen, L. Allen, H. E. L. O. van der Veen, and J. P. Woerdman, Opt. Commun. **96,** 123 (1993).
6. M. Born and E. Wolf, *Principles of Optics* (Pergamon, New York, 1980).
7. S. J. van Enk, Opt. Commun. **102,** 59 (1993).
8. R. Simon, H. J. Kimble, and E. C. G. Sundarshan, Phys. Rev. Lett. **61,** 19 (1988).
9. B. A. Garetz and S. Arnold, Opt. Commun. **31,** 1 (1979).
10. J. Courtial, K. Dholakia, D. A. Robertson, L. Allen, and M. J. Padgett, Phys. Rev. Lett. **80,** 3217 (1998).

PAPER 6.4

Rotational Frequency Shift

Iwo Bialynicki-Birula
Center for Theoretical Physics, Lotników 32/46, 02-668 Warsaw, Poland

Zofia Bialynicka-Birula
Institute of Physics, Lotników 32/46, 02-668 Warsaw, Poland
(Received 5 November 1996)

The notion of the rotational frequency shift, an analog of the Doppler shift, is introduced. This new frequency shift occurs for atomic systems that lack rotational invariance, but have stationary states in a rotating frame. The rotational frequency shift is given by the scalar product of the angular velocity and the angular momentum of the emitted photon in full analogy with the standard Doppler shift which is given by the scalar product of the linear velocity of the source and the linear momentum of the photon. The rotational frequency shift can be observed only in a Mössbauer-like regime when the angular recoil is negligible. [S0031-9007(97)02782-8]

PACS numbers: 32.70.Jz, 06.30.Ft, 32.90.+a, 41.60.-m

The purpose of this Letter is to describe a new effect: the frequency shift of emitted (or absorbed) photons that is due to the rotation of the radiating system. This rotational frequency shift (RFS) is a close analog of the standard, first-order Doppler shift (the latter might be called a translational frequency shift) but there are two important differences.

First, the dynamical laws are invariant under a uniform translation, but they are not invariant under a uniform rotation. Therefore, the differences between the energy levels are not changed by a uniform translation but they are dynamically modified by a uniform rotation, owing to the centrifugal and Coriolis forces. Hence, the frequency of a photon emitted (absorbed) by a system moving with constant velocity is modified *only* by the Doppler shift, while in the case of a rotating system the photon frequency is modified *both* by the changes in the energy levels and by the RFS. For rotationally invariant systems, these two effects completely cancel each other as a result of the conservation of angular momentum, and the observed photon frequency is unchanged, as one would have anticipated.

Second, under normal conditions the recoil corrections are much more significant for the RFS than for the Doppler shift. In order to observe the RFS one must work in a Mössbauer-like regime. The atomic system must be embedded in a larger structure that will provide the angular momentum of the emitted photon.

The RFS should not be confused with the ordinary linear Doppler shift observed for rotating objects (for example, stars or galaxies) that is due to the instantaneous linear motion of the emitter. This linear Doppler shift is maximal in the plane of rotation while the RFS is maximal along the angular velocity, that is in the direction perpendicular to the instantaneous velocity. Thus the RFS competes with the quadratic Doppler shift rather than with the linear Doppler shift.

Our discussion will be based on the nonrelativistic Schrödinger equation for the atomic system with a time dependent Hamiltonian. The photon emission will be treated in the first order of perturbation theory. Our derivation of the RFS will be done in parallel with the calculation of the standard Doppler shift in order to exhibit their similarities and differences.

As a generic model of a radiating system we shall consider an electron bound by a potential $V(t)$ and interacting via minimal coupling with the quantized electromagnetic field. We assume that the time dependence of the potential is of the form $V(t) = V(\mathbf{r}(t))$. For a uniform translation and a uniform rotation, $\mathbf{r}(t)$ is given by

$$\text{Translation}\quad x(t) = x, \quad y(t) = y, \quad z(t) = z - vt, \tag{1}$$

$$\begin{aligned}\text{Rotation}\quad x(t) &= x\cos(\Omega t) + y\sin(\Omega t),\\ y(t) &= -x\sin(\Omega t) + y\cos(\Omega t),\\ z(t) &= z.\end{aligned} \tag{2}$$

Since the time dependence of the potential is prescribed, our analysis will be applicable only to those cases when the change in the uniform motion of the potential caused by the photon emission (absorption) can be disregarded. This means that we neglect all recoil corrections.

The time dependence of the state vector $\Psi(t)$ is determined by the evolution equation (in atomic units)

$$\begin{aligned} i\partial_t\Psi(t) &= H(t)\Psi(t)\\ &= \left[\frac{1}{2}\mathbf{p}^2 + V(t) + H_F + H_I\right]\Psi(t), \end{aligned} \tag{3}$$

where H_F,

$$\begin{aligned} H_F &= \frac{1}{2}\int d^3x\,(\mathbf{D}^2/\epsilon_0 + \mathbf{B}^2/\mu_0)\\ &= \sum_\lambda \int d^3k\,\omega(\mathbf{k})a_\lambda^\dagger(\mathbf{k})a_\lambda(\mathbf{k})\\ &= \sum_{JM\lambda}\int d\omega\,\omega a_{JM\lambda}^\dagger(\omega)a_{JM\lambda}(\omega), \end{aligned} \tag{4}$$

is the free field Hamiltonian and H_I,

$$H_I = \mathbf{p} \cdot \mathbf{A}(\mathbf{r}) + \frac{1}{2}\mathbf{A}^2(\mathbf{r}), \qquad (5)$$

is the interaction Hamiltonian.

In order to determine the spectrum of emitted photons, we construct the transition amplitude A_{fi},

$$A_{fi} = \langle \Psi_f | U(t, t_0) | \Psi_i \rangle . \qquad (6)$$

The evolution operator $U(t, t_0)$ satisfies the equation

$$i\partial_t U(t, t_0) = H(t) U(t, t_0), \qquad (7)$$

and the initial condition $U(t_0, t_0) = 1$. Next, we introduce a unitary transformation $\exp(iGt)$ to get rid of the time dependence of the Hamiltonian $H(t)$, transforming it back to time $t = 0$,

$$H(0) = e^{iGt} H(t) e^{-iGt}. \qquad (8)$$

The operators G are the generators of uniform translations and uniform rotations,

$$\text{Translation} \quad G = vP_z, \qquad (9)$$

$$\text{Rotation} \quad G = \Omega M_z, \qquad (10)$$

where P_z and M_z denote the operators of the z component of the *total* linear momentum and angular momentum (including the electromagnetic field),

$$P_z = p_z + \sum_\lambda \int d^3k\, k_z a_\lambda^\dagger(\mathbf{k}) a_\lambda(\mathbf{k}), \qquad (11)$$

$$M_z = xp_y - yp_x + \sum_{JM\lambda} \int d\omega\, M a_{JM\lambda}^\dagger(\omega) a_{JM\lambda}(\omega). \qquad (12)$$

Note that the Hamiltonians H_F and H_I are invariant under translations and rotations and, therefore, they are not changed by the unitary transformations (8). One may check that the solution of the evolution equation (7) satisfying the initial condition has the form

$$U(t, t_0) = e^{-iGt} e^{-i(H_0 + H_I)(t - t_0)} e^{iGt_0}, \qquad (13)$$

where

$$H_0 = \frac{1}{2}\mathbf{p}^2 + V(0) + H_F - G. \qquad (14)$$

In the first order of perturbation theory, we obtain

$$U(t, t_0) \approx e^{-iGt} e^{-iH_0(t - t_0)} e^{iGt_0} - ie^{-iGt} \int_{t_0}^{t} d\tau\, e^{-iH_0(t - \tau)} \times \mathbf{p} \cdot \mathbf{A}(\mathbf{r}) e^{iH_0(t_0 - \tau)} e^{iGt_0}. \qquad (15)$$

Upon inserting this expression into the formula (6) for the transition amplitude, we get

$$A_{fi} \approx \langle \Phi_f | e^{-iH_0(t - t_0)} | \Phi_i \rangle - i\langle \Phi_f | \int_{t_0}^{t} d\tau\, e^{-iH_0(t - \tau)} H_I e^{iH_0(t_0 - \tau)} | \Phi_i \rangle, \qquad (16)$$

where

$$|\Phi_i\rangle = e^{iGt_0} |\Psi_i\rangle, \qquad |\Phi_f\rangle = e^{iGt} |\Psi_f\rangle. \qquad (17)$$

The transition amplitude (16) will lead to the Fermi Golden Rule for the transition rate provided the auxiliary vectors $|\Phi_{i,f}\rangle$ are chosen as eigenvectors of H_0. In order to study spontaneous emission, we shall assume that the initial state comprises the excited state of the electron and the vacuum state of the field and the final state comprises the ground state of the electron and a one photon state. The corresponding eigenvectors $|\Phi_{i,f}\rangle$ have a product form and they satisfy the following eigenvalue equations:

$$H_0 |\phi_e(\mathbf{r})\rangle |0\rangle = E_e |\phi_e(\mathbf{r})\rangle |0\rangle, \qquad (18)$$

$$H_0 |\phi_g(\mathbf{r})\rangle |1_{ph}\rangle = (E_g + \omega - \Delta) |\phi_g(\mathbf{r})\rangle |1_{ph}\rangle, \qquad (19)$$

where $\phi_e(\mathbf{r})$ and $\phi_g(\mathbf{r})$ are the eigenfunctions of the electronic part H_E of H_0,

$$\text{Translation} \quad H_E = \frac{1}{2}\mathbf{p}^2 + V(0) - vp_z, \qquad (20)$$

$$\text{Rotation} \quad H_E = \frac{1}{2}\mathbf{p}^2 + V(0) - \Omega(xp_y - yp_x), \qquad (21)$$

corresponding to the eigenvalues E_e and E_g. The shift Δ is an eigenvalue of the photonic part of the generator G [cf. Eqs. (11) and (12)] and is equal to vk_z in the case of translation or to ΩM in the case of rotation. The vectors $|\Phi_{i,f}\rangle$ represent stationary states; the electronic probability densities $|\phi_{e,g}(\mathbf{r})|^2$ are time independent. The original vectors $|\Psi_{i,f}\rangle$, related to the eigenvectors $|\Phi_{i,f}\rangle$ by the transformation $\exp(-iGt)$ represent nonstationary states. Since the transformation $\exp(-iGt)$ acting on the electronic wave function replaces its argument $\mathbf{r}$ by $\mathbf{r}(t)$, the electronic wave functions depend on time not only through a phase factor $\exp(-iEt)$ but also through $\mathbf{r}(t)$. These states may be called quasistationary since the corresponding electronic probability densities $|\psi_{e,g}(\mathbf{r}(t))|^2$ undergo a uniform motion according to the formulas (1) and (2).

The spectrum of the emitted photons can be determined from the standard formula for the decay rate w_{if} applied to the amplitude (16)

$$w_{if} = 2\pi |\langle \Phi_f | H_I | \Phi_i \rangle|^2 \delta(E_e - E_g - \omega + \Delta). \qquad (22)$$

Hence the frequency of the emitted photon is shifted by Δ. For a uniform translation, Δ represents the standard Doppler shift and for a uniform rotation, Δ represents the *rotational frequency shift*,

$$\text{Translation (Doppler)} \quad \omega = E_e - E_g + vk_z, \qquad (23)$$

$$\text{Rotation (RFS)} \quad \omega = E_e - E_g + \Omega M. \qquad (24)$$

In the case of a uniform translation the electronic energy difference $E_e - E_g$ is exactly the same as for a system at rest because the electronic part of the Hamiltonian H_0 is unitarily equivalent (up to a constant) to the corresponding Hamiltonian at rest

$$e^{ivz}\left[\frac{1}{2}\mathbf{p}^2 + V(0) - vp_z\right]e^{-ivz} = \frac{1}{2}\mathbf{p}^2 + V(0) - \frac{1}{2}v^2. \qquad (25)$$

Therefore for a uniform translation the only change in the spectrum of the emitted photons is the Doppler shift. This is simply the result of the Galilean invariance of nonrelativistic quantum mechanics. The situation is quite different for a uniform rotation. The spectrum of the photons emitted by a rotating source is not only shifted by $\Delta = \Omega M$ but it is also, in general, modified due to changes of the energy differences $E_e - E_g$.

This modification is especially simple for rotationally invariant potentials when M_z commutes with the Hamiltonian H_0. In this case the wave functions $\phi_e(\mathbf{r})$ and $\phi_g(\mathbf{r})$ can be chosen as eigenfunctions of the z component of the electronic angular momentum belonging to certain eigenvalues m_e and m_g. Then the change in the energy difference caused by the rotation is equal to $\Omega(m_e - m_g)$. This change cancels *completely* the RFS owing to the conservation of angular momentum and no shift is observed in the laboratory frame. Therefore, the RFS cannot be seen in isolated atoms but it might occur, in principle, in isolated molecules or in atoms placed in an environment that destroys the rotational symmetry.

Even when the potential is not rotationally invariant, the consequences of the RFS are not always significant. In particular, when the rotating body is of macroscopic dimensions the RFS is much smaller, in general, than other effects. For example, in the experiment [1] designed to measure the transverse Doppler shift the emitter and the absorber were mounted on a rotating disk. Even though the angular velocity was large enough to observe the quadratic Doppler shift, the RFS calculated for the conditions of this experiment could not have been seen (it is only about 10^{-4} of the quadratic Doppler shift). The RFS is so small in this case because the ratio of the wavelength to the linear dimensions of the rotating system is exceedingly small. The RFS might be seen for macroscopic rotating systems when the emitted radiation is in the radio-wave range, as is the case for transitions involving nuclear magnetic moments. Taking the frequency of such a transition in the range of 10^8 Hz and assuming an ultrafast centrifuge rotating at 10^3 Hz, one obtains the RFS shift of the order of 10^{-5} of the resonance frequency. Such shifts might be detectable. In order to observe the RFS in the optical domain, the rotational frequency must be much higher and that can be achieved only when the rotating systems are of atomic dimensions.

The simplest model of a rotating atomic system with broken rotational symmetry is an electron bound by a rotating asymmetric harmonic potential in two dimensions.

$$V(t) = \frac{1}{2}[\mu x^2(t) + \nu y^2(t)], \tag{26}$$

where μ and ν are two arbitrary real parameters and the dependence of x and y on time is given by the formulas (2). An especially interesting case is when one of the two oscillators is inverted (one of the parameters μ or ν is negative). Such a model has been recently used to predict the existence and to describe the main features of Trojan wave packets of Rydberg electrons in atoms [2] and in rotating molecules with large electric dipole moments [3]. Since all more elaborate numerical calculations [4–7] fully confirmed the validity of the estimates based on the harmonic approximation, we shall use the model with the potential (26) to study the significance of the RFS in realistic situations.

The Trojan states of Rydberg electrons in hydrogenic atoms are described by nonspreading localized wave packets circling on a large orbit around the nucleus. Their stability is due to an interplay between the centrifugal force and the Coriolis force acting in the rotating frame and the electric field of a circularly polarized electromagnetic wave propagating in the direction perpendicular to the orbit. They were named by us the Trojan states because their mathematical description is similar to the description of the orbits of Trojan asteroids in the Sun-Jupiter system [2]. The Trojan states of electrons offer a perfect example of an atomic system that will exhibit the RFS since in the laboratory frame the interaction with the electric field of the wave is described (in the dipole approximation) by the rotating potential $V(t) = \mathcal{E}[x\cos(\Omega t) + y\sin(\Omega t)]$. The description of the Trojan states in the harmonic approximation [2] leads to the Hamiltonian in which the parameters μ and ν are determined by the frequency of the wave Ω and by the parameter q that measures the ratio between the Coulomb force and the centrifugal force, $\mu = -2q\Omega^2$, $\nu = q\Omega^2$. The frequencies $\omega_\pm$ in this case are

$$\omega_\pm = \Omega\sqrt{(2 - q \pm \sqrt{9q^2 - 8q})/2}. \tag{27}$$

They are real for $8/9 \le q \le 1$.

We shall use now the harmonic oscillator model to exhibit the main features of the RFS for realistic atomic systems. The electronic part H_E of the time independent Hamiltonian for the potential (26) is in this case

$$H_E = \frac{1}{2}\mathbf{p}^2 + \frac{1}{2}(\mu x^2 + \nu y^2) - \Omega(xp_y - yp_x). \tag{28}$$

This Hamiltonian describes two harmonic oscillators with the following characteristic frequencies

$$\omega_\pm = \sqrt{(\mu + \nu + 2\Omega^2 \pm \sqrt{(\mu - \nu)^2 + 8(\mu + \nu)\Omega^2})/2} \tag{29}$$

that determine the spacing between the energy levels. These frequencies are real when the parameters μ, ν, and Ω satisfy the conditions

$$\mu \ge \Omega^2 \quad \text{and} \quad \nu \ge \Omega^2, \tag{30}$$

or

$$-3\Omega^2 \le \mu \le \Omega^2, \qquad -3\Omega^2 \le \nu \le \Omega^2,$$
$$\text{and} \quad (\mu - \nu)^2 + 8(\mu + \nu)\Omega^2 \ge 0. \tag{31}$$

In a degenerate case of an isotropic potential, when $\mu = \nu$, the frequencies $\omega_\pm$ are linear functions of the angular velocity $\omega_\pm = \sqrt{\mu} \pm \Omega$. This is a special example of a rotationally invariant potential, when the RFS is exactly canceled by the dynamical modification of the electronic energy difference. For anisotropic potentials the energy differences are nonlinear functions of Ω and a complete cancellation between the RFS and the dynamical modifications of the energy differences is impossible.

The RFS that occurs in the emission of photons during transitions between the atomic Trojan states has dramatic consequences. The frequencies (24) of the photons emitted during the transitions between the neighboring energy levels of the oscillators, as seen by the observer in the laboratory frame, are

$$\omega = \omega_{\pm} + \Omega M . \qquad (32)$$

The observed photon frequency is drastically shifted by the RFS; the shift *exceeds* the original transition frequency in the rotating frame. This huge effect is due to the fact that the rotation is not just a small modification of the dynamics, but the mere existence of the Trojan states depends on it. Since both frequencies $\omega_{\pm}$ are smaller than Ω, the RFS has a very large effect in this case. The emission of photons with $M = -1$ is forbidden. This means that the photons emitted spontaneously by the Trojan electrons have the same projection of the angular momentum on the z axis as the photons constituting the strong wave that is driving the system. At this point we would like to stress that the RFS and the transfer of the angular momentum to the emitted photon must not in any direct way be attributed to the photons that make the circularly polarized wave. The same frequency shift is obtained for all rotating potentials, regardless of their nature. For example, it would appear for nonspherical rotating nuclei, provided one can treat the rotating nuclear potential as fully prescribed.

In order to illustrate these general results with specific numbers, we shall consider a hydrogen atom driven by the circularly polarized wave with the frequency of 200 GHz and the field amplitude of 1930 V/m, the same as in Ref. [2]. In this case, the separations between the neighboring oscillator levels in the rotating frame are 1.92×10^{11} Hz and 0.66×10^{11} Hz. The frequencies of the circularly polarized radiation observed in the laboratory frame will be shifted by 2×10^{11} Hz and they will be equal to 3.92×10^{11} Hz and 2.66×10^{11} Hz, respectively. Both frequencies are quite different from the frequency of the driving wave.

Our derivation of the RFS depends crucially on the validity of the no-recoil approximation. This assumption is satisfied very well for the Doppler shift but the effect of recoil in the rotational motion is much more significant than in the translational motion. Usually, one may disregard the perturbation of the translational motion by the emission of the photon because the relative change in the velocity is of the order of 10^{-4}. In contrast, the relative change of the angular velocity of a molecule caused by an emission of a photon, as argued below, is several orders of magnitude larger. The principle of energy equipartition gives the relation between the angular velocity and the linear velocity $\Omega = v/R$, where R is the effective radius of the molecule. Therefore the relative changes of the angular velocity and the linear velocity are related by a large factor, the ratio of the wave length to the molecular radius,

$$\delta\Omega/\Omega \approx (1/kR)(\delta v/v), \qquad (33)$$

where k is the photon wave vector. This means that for an isolated molecule the angular recoil should be taken into account. The whole molecule, not just the electron undergoing the radiative transition, must be considered as a radiating system. For the molecule, treated as a closed system, the RFS is unobservable because the total angular momentum is conserved and we end up in the category of rotationally invariant systems. However, for strongly driven atoms or molecules, as in the case of atomic Trojan states, a Mössbauer-like regime is reached since the external field sustains the uniform rotational motion. Under such conditions, the angular recoil can be disregarded since the angular momentum of the emitted photon is provided by the macroscopic field.

The standard explanation of the Doppler shift is based on the kinematics of a special theory of relativity [8,9]. The RFS may also be explained in a similar vein but one must apply general theory of relativity because the rotating coordinate frame is not inertial. One may use in this case the known transformation properties of the Maxwell equations under general coordinate transformations to relate the frequency of the photon emitted in the rotating frame to the frequency observed in the laboratory frame. However, the RFS is always accompanied by the modifications of the emitter dynamics due to rotation and the net frequency shift can be observed *only for quantum mechanical systems* when the electronic Hamiltonian does not commute with the generator of rotations.

This work has been supported by the KBN Grant No. 2P30B01309.

[1] W. Kündig, Phys. Rev. **129**, 2371 (1963).
[2] I. Bialynicki-Birula, M. Kalinski, and J. H. Eberly, Phys. Rev. Lett. **73**, 1777 (1994).
[3] I. Bialynicki-Birula and Z. Bialynicka-Birula, Phys. Rev. Lett. **77**, 4298 (1996).
[4] D. Delande, J. Zakrzewski, and A. Buchleitner, Europhys. Lett. **32**, 107 (1995).
[5] J. Zakrzewski, D. Delande, and A. Buchleitner, Phys. Rev. Lett. **75**, 4015 (1995).
[6] D. Farrelly, E. Lee, and T. Uzer, Phys. Lett. A **204**, 359 (1995).
[7] C. Cerjan, D. Farrelly, and T. Uzer (to be published).
[8] W. Pauli, *Theory of Relativity* (Pergamon, London, 1958).
[9] J. D. Jackson, *Classical Electrodynamics* (Wiley, New York, 1975).

PAPER 6.5

Azimuthal Doppler shift in light beams with orbital angular momentum

L. Allen [a], M. Babiker [a], W.L. Power [b]

[a] *Department of Physics, University of Essex, Colchester CO4 3SQ, UK*

[b] *Optics Section, Blackett Laboratory, Imperial College, London SW7 2BZ, UK*

Received 20 June 1994; revised version received 15 August 1994

Abstract

We show that an atom moving in a light beam with orbital angular momentum experiences an azimuthal shift in the resonant frequency in addition to the usual axial Doppler and recoil shifts. For a Laguerre-Gaussian beam characterised by an orbital angular momentum quantum number l, the shift is lV_ϕ/r where r is the radial atomic position and V_ϕ the azimuthal component of velocity. The predicted shift could play a significant role in interactions between atoms and standing light fields in cooling experiments as well as in ion traps.

There has been a good deal of interest in the mechanical effects that light can exert on material systems. Most attention in the last decade or so [1] has been addressed to the *translational* effects of light.

The best known experiment on the *rotational* aspects of light is that of Beth [2], who measured the torque experienced by a birefringent plate subject to a beam of circularly polarised light. In Beth's experiment the relevant effect is attributable to the spin angular momentum. Recent emphasis on rotational effects has stemmed from work by Allen et al. [3] who suggested that paraxial beams, in particular Laguerre-Gaussian beams, possessing orbital angular momentum should influence the angular motion of a rigid body. They proposed an experiment that exploits the transfer of angular momentum in the process of converting a paraxial beam of angular momentum quantum number l into a Hermite-Gaussian beam which has no orbital angular momentum. A subsequent report by Beijersbergen et al. [4] discussed the actual design, and demonstrated the operation, of a converter which could experience a resultant torque.

There has been speculation recently about the possible rotational effects that atoms may experience when immersed in light beams possessing orbital angular momentum [3,5]. A moving atom interacting with light experiences changes in its gross motion due to an associated radiation force. Such effects have in the past only been considered in the context of plane waves. The way in which the intrinsic orbital angular momentum of the beam manifests itself in the behaviour of the atoms is the main concern of this article. We focus on a motion-induced effect that lends itself to direct experimental study, namely the shift in the atomic resonance. The aim is to quantify the changes that occur in the Doppler effect for atoms interacting with light beams with orbital angular momentum.

The salient features of the theory may be discussed with respect to a two-level atom, of resonant frequency ω_0, interacting with a light mode of frequency ω. In the electric dipole and rotating wave approximations the Hamiltonian may be written [6]

$$H = \hbar\omega_0 \pi^\dagger \pi + P^2/2M + U(\boldsymbol{R}) + \hbar\omega a^\dagger a - i\hbar[\pi^\dagger a f(\boldsymbol{R}) - f^*(\boldsymbol{R}) a^\dagger \pi] \,, \qquad (1)$$

where $\boldsymbol{P}$ and $\boldsymbol{R}$ are the momentum and position vectors of the centre of mass with total mass M, π and $\pi^\dagger$ are ladder operators characterising the internal two-level system, and a and $a^\dagger$ are, respectively, the annihilation and creation operators of the light field. The inclusion of the potential $U(\boldsymbol{R})$ allows the general consideration of the atom, whether free or confined. The operator $f(\boldsymbol{R})$ stems from the electric dipole interaction $-\boldsymbol{d} \cdot \boldsymbol{E}(\boldsymbol{R})$ and we may write

$$f(\boldsymbol{R}) = N\,(\boldsymbol{D}_{12} \cdot \boldsymbol{\mathcal{E}}(\boldsymbol{R}))\,, \qquad (2)$$

where $\boldsymbol{\mathcal{E}}$ is the electric field vector associated with the mode, N is a normalisation factor and $\boldsymbol{D}_{12}$ is the dipole matrix element.

The above Hamiltonian (Eq. (1)) is the basis for a direct quantum mechanical derivation of the Doppler shift [7]. The time evolution of the operator $f(\boldsymbol{R}(t))$ stems from its Heisenberg equation of motion which, together with Eq. (1), gives

$$\frac{df}{dt} = \frac{i}{\hbar}\left[H, f(\boldsymbol{R}(t))\right] = \frac{1}{2M}\left(\boldsymbol{P} \cdot \nabla f(\boldsymbol{R}(t)) + \nabla f(\boldsymbol{R}(t)) \cdot \boldsymbol{P}\right). \qquad (3)$$

As $U(\boldsymbol{R})$ commutes with $f(\boldsymbol{R})$, Eq. (3) holds irrespective of the explicit form of the potential $U(\boldsymbol{R})$.

The solution of Eq. (3) can only be obtained if the time dependence of both $\boldsymbol{R}$ and $\boldsymbol{P}$ have been specified. However we need only evaluate f to leading order in the interaction, which amounts to depending on the initial values of $\boldsymbol{R}$ and $\boldsymbol{P}$ of the atomic operators. We therefore obtain the leading solution of Eq. (3) in the form

$$f(\boldsymbol{R}, t) \approx f_0 \exp\left[\left(\frac{(\boldsymbol{P} \cdot \nabla f + \nabla f \cdot \boldsymbol{P})}{2Mf}\right)_0 t\right] = f_0 \exp(i\delta t)\,, \qquad (4)$$

where the subscript zero denotes operators at the initial time $t = 0$. The change in energy due to the atomic gross motion is identified in terms of the corresponding frequency operator δ appearing in Eq. (4)

$$\delta = -\frac{i}{2M}\left(\frac{\boldsymbol{P} \cdot \nabla f + \nabla f \cdot \boldsymbol{P}}{f}\right)_0 . \qquad (5)$$

In what follows it is convenient to drop the zero subscript from the operators $\boldsymbol{R}$ and $\boldsymbol{P}$, but retain it in f_0.

By restricting attention to the case of a linearly polarised plane wave the frequency shift δ may be readily shown to contain the familiar form of the Doppler shift as well as the recoil shift. For a plane wave of wavevector $\boldsymbol{k}$ and polarisation $\boldsymbol{\epsilon}$ we have $f_0 = N(\boldsymbol{D}_{12} \cdot \boldsymbol{\epsilon}) \exp(-i\boldsymbol{k} \cdot \boldsymbol{R})$ and we then obtain from Eq. (5) the well known result

$$\delta = -\boldsymbol{k} \cdot \boldsymbol{V} + \frac{\hbar k^2}{2M}\,, \qquad (6)$$

where $\boldsymbol{V} = \boldsymbol{P}/M$. The main influence of this effect is tomodify the detuning parameter from $\Delta_0 = \omega - \omega_0$ to Δ, where $\Delta = \Delta_0 + \delta$.

If we assume a large atomic mass M, then we may ignore the recoil shift. The Doppler shift is a maximum for atoms at a given velocity travelling along the light beam direction and zero for those moving transverse to the beam direction. This is the reason for the well known crossed-beams configuration commonly adopted in experiments where an attempt is made to minimise the influence of the Doppler shift in controlling the detuning parameter.

We shall investigate the detailed nature of the shift given by Eq. (5) by considering a beam with azimuthal dependence. Many types of beam exhibiting azimuthal dependence, for example Laguerre-Gaussian laser modes or Bessel beams [8], may be shown to possess orbital angular momentum. We may note, however, that the beams need not be paraxial. Barnett and Allen [9] have shown that full beam solutions of the Maxwell equations which reduce to Laguerre-Gaussian beams in the paraxial limit, also possess intrinsic orbital angular momentum.

We consider the particular case of Laguerre-Gaussian beams which can be readily produced [4]. Such a beam is characterised by two integer indices l and p where l is the orbital angular momentum quantum number. Analytical expressions for the electric field vector of Laguerre-Gaussian modes in the paraxial approximation are well known [10]. The form that will be of concern to us here is essentially equivalent to light linearly polarised along the x-axis with an additional component of field along the direction of propagation. However, circularly polarised light may also be readily represented in the same formalism.

In order to highlight the angular momentum properties of the interaction of atoms with such Laguerre-Gaussian modes, it is sufficient to consider the special case of a dipole orientation along the x-axis such that $\boldsymbol{D}_{12} = D\hat{\boldsymbol{x}}$. More formally, the z-component of the electric field induces a component of the dipole moment in the z-direction; but this is almost everywhere in phase with the x-component and does not affect the result obtained in this special case. Only the x-component $\mathcal{E}_x$ of the electric field vector is involved in this case and we then have in cylindrical coordinates $\boldsymbol{R} = (r, \phi, z)$

$$f_0^l(\boldsymbol{R}) = ND\mathcal{E}_x(\boldsymbol{R}) = -\mathrm{i}\omega ND F_l(r,z)\exp(-\mathrm{i}kz)\exp(-\mathrm{i}l\phi)\,, \tag{7}$$

where $F_l(r, z)$ contains no azimuthal dependence. The explicit form of F_l [11] is

$$F_l(r,z) = (-1)^p \frac{C}{w(z)}\left(\frac{r\sqrt{2}}{w(z)}\right)^l L_p^l(2r^2/w^2) \times \exp[-\mathrm{i}(2p+l+1)\psi(z)] \times \exp[-\mathrm{i}(kr^2/2\bar{z})]\exp[-(r^2/w^2)]\,, \tag{8}$$

where L_p^l are associated Laguerre polynomials and we have set

$$\bar{z} = \frac{z_R^2 + z^2}{z}, \quad \tfrac{1}{2}kw^2(z) = \frac{z_R^2 + z^2}{z_R}, \quad \psi(z) = \tan^{-1}(z/z_R). \tag{9}$$

The quantities $\bar{z}$, $w(z)$ and $\psi(z)$ appearing in Eq. (8) are expressed in Eq. (9) in terms z_R, the Rayleigh range, which is defined by $z_R = \pi w_0^2/\lambda$ where w_0 is the beam waist. Thus f_0^l is, in general, a complex function of $\boldsymbol{R}$ and it is convenient to reformulate it as

$$f_0^l(\boldsymbol{R}) = G_l(\boldsymbol{R})\exp[\mathrm{i}\Theta_l(\boldsymbol{R})]\,, \tag{10}$$

where $G_l(\boldsymbol{R})$ and $\Theta_l(\boldsymbol{R})$ are two real functions, easily deducible from Eqs. (7) and (8).

The corresponding shift arising from Eq. (5), denoted by δ_{LG}, is obtained as

$$\delta_{LG} = -\mathrm{i}\frac{\nabla f_0^l \cdot \boldsymbol{V}}{f_0^l} - \frac{\hbar\nabla^2 f_0^l}{2Mf_0^l}. \tag{11}$$

We shall assume that the atomic mass M is large and concentrate in what follows on the first term, ignoring the recoil term. For convenience we continue to refer to δ_{LG} as the Doppler shift. We obtain, on substitution from Eq. (10),

$$\delta_{LG} = \nabla\Theta_l \cdot \boldsymbol{V} - \mathrm{i}\frac{\nabla G_l \cdot \boldsymbol{V}}{G_l} = \delta_{LG}^{(1)} - \mathrm{i}\delta_{LG}^{(2)}. \tag{12}$$

To leading order, the imaginary part $\delta_{LG}^{(2)}$ can be shown to be due to a change in the electric field amplitude arising from a change in the position of the atom from $\boldsymbol{R}$ to $\boldsymbol{R}+\boldsymbol{V}t$. The real part $\delta_{LG}^{(1)}$ can be expressed in cylindrical coordinates in terms of the components of the atomic velocity vector

$$\delta_{LG}^{(1)} = \nabla\Theta_l \cdot \boldsymbol{V} = \left[-k + \frac{kr^2}{2(z^2+z_R^2)}\left(\frac{2z^2}{(z^2+z_R^2)} - 1\right) - \frac{(2p+l+1)z_R}{(z^2+z_R^2)}\right]V_z - \left(\frac{kr}{\bar{z}}\right)V_r - \left(\frac{l}{r}\right)V_\phi\,, \tag{13}$$

where V_z, V_r and V_ϕ are the axial, radial and azimuthal velocity components of the atom. This is the main result of this article.

The axial Doppler shift is obtained from Eq. (13) as the first term (proportional to V_z). It is easy to see that this consists of the usual leading term $-kV_z$ plus terms accounting for the non-uniformity of the beam. The second contribution to the Doppler shift in Eq. (13) is the radial component and is given by the term proportional to V_r. This is again a property of the non-uniformity of the beam. A Hermite-Gaussian beam would also show such effects [12] and they are not the main concern of this report.

A much more significant effect resides in the final contribution. The corresponding azimuthal Doppler shift, denoted δ_{LG}^ϕ, is given by the V_ϕ term in Eq. (13),

$$\delta_{LG}^\phi = -\frac{V_\phi l}{r} = -l\omega_r\,. \tag{14}$$

The important features of this azimuthal shift are that it is directly proportional to the orbital angular momentum quantum number l of the Laguerre-Gaussian mode and that it is inversely proportional to the radial coordinates of the position of the atom.

A simple physical explanation of the azimuthal and radial Doppler shifts may be given in terms of the

Poynting vector. In a Laguerre-Gaussian mode the Poynting vector at any point is a tangent vector to a helix which winds around the beam axis [3]. At such a point the wave may be regarded as a local plane wave with its wavefronts normal to the Poynting vector. All the axial as well as non-axial components of the shift are directly related to the corresponding components of the local Poynting vector.

The main prediction of this article is that atoms with azimuthal velocity components relative to a Laguerre-Gaussian beam axis experience a frequency shift that is proportional to the orbital angular momentum quantum number l. The cylindrical symmetry of the cross-section of such beams would make them potentially attractive alternatives to the (0,0)-modes usually employed. Laguerre-Gaussian modes may readily be produced with either left- or right-handed orbital angular momentum for beams with l as big as 7 or more. It appears likely, therefore, that the azimuthal shift evaluated here may play an exploitable role in atom-field interactions, especially in experiments on the trapping and cooling of atoms and ions.

The effect may be particularly amenable to experimental measurement for the case of ions in a Penning trap, where, with an appropriate potential $U(\boldsymbol{R})$, the ions can be kept in a circular path of a fixed radius r with a constant velocity V_ϕ and undergo only minor perturbations in orbit due to interaction with the electromagnetic field of the light. In this very special case, the azimuthal shift can be given a transparent, albeit non-rigorous, interpretation. The relevant phase factor of the Laguerre-Gaussian mode $\exp(-il\phi)$ becomes $\exp[-il(\phi + V_\phi t/r)]$ after time t has elapsed, due to a change in the phase angle by an amount equal to the arc length $V_\phi t$ divided by the radius r. This phase factor immediately yields the azimuthal shift when written in the form $\exp(-il\phi - il\omega_r t)$ with $\omega_r = V_\phi/r$. In the case of atoms in a Penning trap the velocity V_ϕ is such that ω_r is typically of the order of a few MHz. The azimuthal shift is then several MHz, which is sufficiently large for experimental detection.

While an azimuthal shift could in principle be determined classically, we have chosen to use a quantum mechanical approach which proves to be very convenient. This has led to a quantum mechanical expression for all components of the shift in the resonant frequency arising from an arbitrary field distribution, including atom recoil. More importantly, perhaps, we have estalished the basis of a theory for evaluating the interaction of the atom with Laguerre-Gaussian light beams where the explicit state of excitation of the system can be included by use of the optical Bloch equations.

The new component in the Doppler shift is just one of the manifestations of the orbital angular momentum effects of non-uniform beams. It is a potentially observable characteristic of the internal motion of the atom arising from its interaction with the light beam. Elsewhere we shall report on the angular momentum and torque due to such beams that are, by contrast, measurable properties of the gross motion of the atom.

We are grateful to Professor Rodney Loudon, FRS and to Dr. Richard Thompson for useful discussions. LA is pleased to acknowledge the award of a Leverhulme Research Fellowship and WLP acknowledges a studentship from the UK Engineering and Physical Science Research Council.

References

[1] For recent reviews, see Fundamental Systems in Quantum Optics, Proc. International School of Physics, Les Houches, session LIII, 1990, eds. J. Dalibard, J.-M. Raimond and J. Zinn-Justin (Elsevier, Amsterdam, 1992).

[2] R.A. Beth, Phys. Rev. 50 (1936) 115.

[3] L. Allen, M.W. Beijersbergen, R.J.C. Spreeuw and J.P. Woerdman, Phys. Rev. A 45 (1992) 8185.

[4] M.W. Beijersbergen, L. Allen, H.E.L.O. van der Veen and J.P. Woerdman, Optics Comm. 96 (1993) 123.

[5] S.J. van Enk and G. Nienhuis, J. Mod. Optics 41 (1994) 963.

[6] V.E. Lembessis, M. Babiker, C. Baxter and R. Loudon, Phys. Rev. A 48 (1993) 1594.

[7] C. Cohen-Tannoudji, B. Diu and F. Laloë, Quantum Mechanics, Vol. II (Wiley, New York, 1977) p. 1366. The authors suggest an equivalent derivation of the plane wave Doppler shift based on the minimal coupling interaction $-e\boldsymbol{A} \cdot \boldsymbol{p}/m$ in the dipole approximation.

[8] J. Durnin, J. Opt. Soc. Am. A 4 (1987) 651.

[9] S.M. Barnett and L. Allen, Optics Comm. 110 (1994) 670.

[10] H.A. Haus, Waves and Fields in Optoelectronics (Prentice-Hall, Englewood Cliffs, NJ, 1984) p. 126.

[11] A.E. Siegman, Lasers (University Science Books, Mill Valley, CA, 1986).

[12] V.G. Minogin and V.S. Letokhov, Laser Light Pressure on Atoms (Gordon and Breach, New York, 1986).

PAPER 6.6

Rotational Frequency Shift of a Light Beam

J. Courtial, D. A. Robertson, K. Dholakia, L. Allen, and M. J. Padgett
School of Physics & Astronomy, University of St. Andrews, St. Andrews, Fife, KY16 9SS, United Kingdom
(Received 28 May 1998)

We explain the rotational frequency shift of a light beam in classical terms and measure it using a mm-wave source. The shift is equal to the total angular momentum per photon multiplied by the angular velocity between the source and observer. This is analogous to the translational Doppler shift, which is equal to the momentum per photon multiplied by the translational velocity. We show that the shifts due to the spin and orbital angular momentum components of the light beam act in an additive way. [S0031-9007(98)07769-2]

PACS numbers: 32.70.Jz, 03.65.Bz, 42.25.Ja, 42.79.Nv

The nonrelativistic, translational Doppler shift is a well-known phenomenon. The relative velocity v between light source and observer leads to a frequency shift $\Delta\omega = vk$, where $\hbar k$ is the linear momentum of each photon. The nonrelativistic rotational frequency shift identified in this paper can be described analogously by $\Delta\omega = \Omega J$, where Ω is the angular velocity between source and observer and $\hbar J$ is the total angular momentum per photon. We show that this result can be understood easily in terms of the rotational symmetry associated with the transverse electric field distribution of circularly polarized Laguerre-Gaussian beams.

It should be emphasized that the rotational frequency shift is completely different from the translational Doppler shift observed for a rotating source that arises from a linear velocity component in the plane of rotation (Fig. 1). The rotational shift is maximal when observed at right angles to the plane of rotation where the translational shift is zero.

It is well known that each photon in a light beam carries a spin angular momentum $\sigma\hbar$, where $\sigma = \pm 1$ corresponds to left- and right-handed circularly polarized light. Less well known is that light beams can also have an orbital angular momentum associated with the phase structure of the beam. In 1992 Allen *et al.* [1] showed for beams with an azimuthal phase term of $\exp(il\phi)$, such as Laguerre-Gaussian laser modes, that the orbital angular momentum is $l\hbar$ per photon. Laguerre-Gaussian modes have helical wave fronts with a phase discontinuity on the axis of the beam and an intensity distribution of one or more concentric rings [2]. These modes form a complete basis set; consequently, any arbitrary light beam can be described by a superposition of Laguerre-Gaussian modes.

For *spin* angular momentum alone, the rotational frequency shift has a number of precedents. In 1981 Garetz [3] attributed phenomena such as rotational Raman spectra and the frequency shift previously observed with a rotating half-wave plate [4] to an "angular Doppler effect." In each case, a circularly polarized light beam rotating with an angular velocity Ω with respect to the observer produces a frequency shift of $\Omega\sigma$. Bialynicki-Birula and Bialynicka-Birula [5] predicted in a quantum mechanical calculation that for an atomic system subject to a rotating potential a polarized transition experiences a frequency shift of ΩJ. We have shown recently [6] that the rotation of beams with an $\exp(il\phi)$ phase term, and so an *orbital* angular momentum of $l\hbar$ per photon, results in a frequency shift of Ωl. This shift can be related to previous predictions for the azimuthal Doppler shift of atomic transition frequencies [7] and to the frequency shift introduced by rotating cylindrical lenses [8]. In this paper we unify previous studies to include both spin and orbital angular momentum simultaneously. We find that the spin and orbital contributions behave in an additive way, such that it is the *total* angular momentum per photon, $\hbar J = \hbar(l + \sigma)$, that gives a rotational frequency shift of $\Delta\omega = \Omega J$.

In our experiment we rotate a circularly polarized beam possessing helical wave fronts and directly measure the corresponding frequency shift. The rotation of a source or detector without the introduction of unwanted off-axis

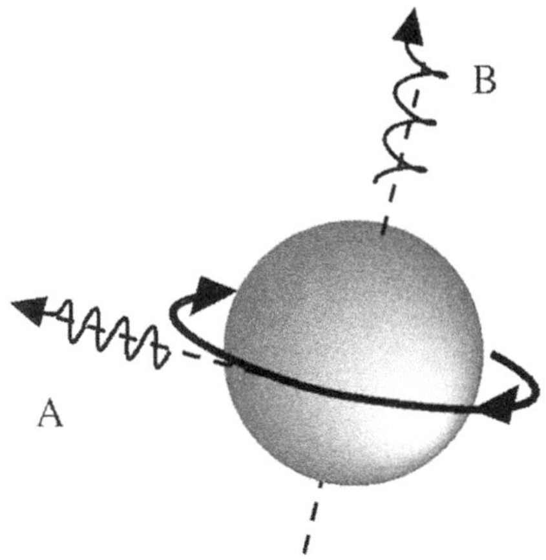

FIG. 1. Both the rotational frequency shift and the translational Doppler shift can be observed in the same rotating source. The frequency of light beam *A* is shifted due to the translational Doppler shift; light beam *B* is frequency shifted due to the rotational frequency shift.

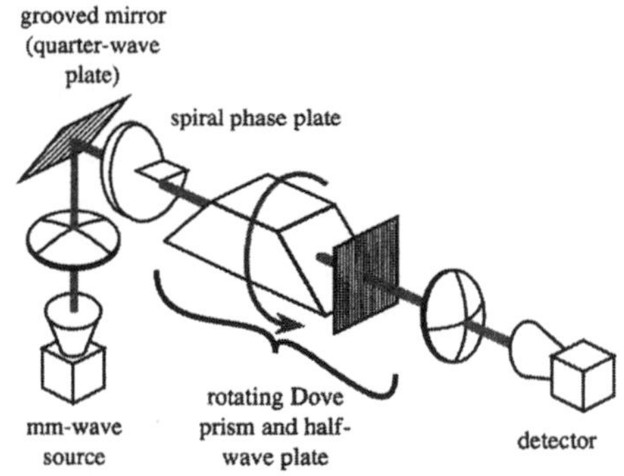

FIG. 2. Schematic diagram of the experimental apparatus for the observation of the rotational frequency shift.

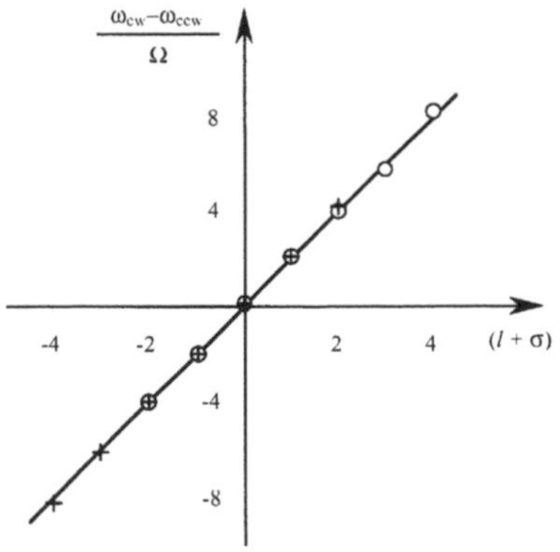

FIG. 3. Observed rotational frequency shift for various values of $(l + \sigma)$. The crosses and circles correspond to σ of -1 and $+1$, respectively. The beam angular velocity Ω was 1.25 Hz. The error associated with each measurement is 0.2 Hz.

movement is difficult to achieve. As an alternative, we use a Dove prism as an image rotator; the transmitted image and associated phase structure rotate at twice the angular velocity of the prism. Similarly, a half-wave plate rotates the electric field vector of a transmitted beam at twice the angular velocity of the wave plate. The combination of a Dove prism and a half-wave plate therefore rotates the whole beam while leaving the source stationary. To minimize difficulties in alignment of the prism with the beam axis, our experiment is performed in the mm-wave regime, using a Gunn diode source oscillating at 94 GHz with a corresponding wavelength of approximately 3 mm. Figure 2 shows the experimental arrangement. The source frequency is phase locked to a high harmonic of a crystal oscillator giving a short-term frequency stability of approximately 1 Hz. A corrugated feed horn couples approximately 10 mW into a linearly polarized fundamental Gaussian beam. This beam is collimated using a polyethylene lens and converted into a beam with helical wave fronts by transmission through a polyethylene phase plate whose thickness increases with azimuthal angle [9]. The beam is circularly polarized by reflection from a surface that has $\lambda/8$ deep grooves aligned at 45° to the incident linear polarization [10]. The frequency of the beam after transmission through the rotating polyethylene Dove prism and half-wave plate [10] is measured directly using a mm-wave frequency counter which is phase referenced to the source. The rotation of the beam introduces a shift in frequency. Reversing either the rotation direction or the sign of J changes the sense of the shift. Consequently, we expect the frequency difference between clockwise and counterclockwise rotation to be given by $\omega_{cw} - \omega_{ccw} = 2\Omega J = 2\Omega(l + \sigma)$. Figure 3 shows the observed frequency difference for various combinations of l and σ. The experimental observations agree with our predictions.

The rotational frequency shift could be interpreted in terms of an energy exchange with the rotating optical element [8,11]. However, it is more illuminating to explain the shift in terms of the field distribution of circularly polarized Laguerre-Gaussian beams. As the electric field evolves in time, its distribution rotates. In each case the field distribution has $(l + \sigma)$-fold rotational symmetry (Fig. 4) and the electric field undergoes a change of $l + \sigma$ cycles for each rotation of the beam.

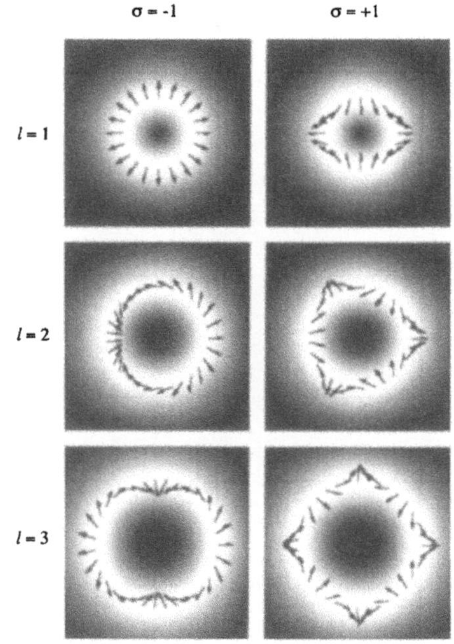

FIG. 4. Vector plots of the transverse electric field for circularly polarized Laguerre-Gaussian beams, showing the $(l + \sigma)$-fold rotational symmetry.

For circularly polarized light the electric field rotates at the optical frequency, and therefore rotation of the beam results in a frequency shift. The direction of this shift depends on whether the sense of the beam rotation adds or subtracts to the rotation of the electric field. This is completely analogous to the translational Doppler shift which also depends upon a change in the number of cycles observed.

The translational Doppler shift is proportional both to the relative velocity and to the unshifted frequency. Although the rotational frequency shift is proportional to the angular velocity, it is independent of the unshifted frequency. This allowed us to perform our experiments in the mm-wave regime without reducing the magnitude of the shift. Unlike the translational Doppler shift, the rotational frequency shift depends on the distribution of the transverse electric field of the beam. Only for a pure circularly polarized Laguerre-Gaussian beam is there a single frequency shift and not a splitting. However, any arbitrary field distribution can be decomposed into a superposition of Laguerre-Gaussian beams; the resulting spectrum will consequently consist of sidebands at shifts proportional to the J value of each component.

We have described classically the rotational frequency shift, analogous to the translational Doppler shift, which arises from an angular velocity Ω between source and observer. For circularly polarized Laguerre-Gaussian beams the total angular momentum per photon is $\hbar J = \hbar(l + \sigma)$. The resulting rotational frequency shift is given by $\Delta\omega = \Omega J$. Any field distribution can be expressed as a superposition of circularly polarized Laguerre-Gaussian beams and, if rotated, results in a frequency spectrum consisting of sidebands about the unshifted frequency. Specific combinations of Laguerre-Gaussian beams allow the generation of particular frequency distributions, which may have applications for ranging, communication systems or optical guiding of atoms.

[1] L. Allen, M. W. Beijersbergen, R. J. C. Spreeuw, and J. P. Woerdman, Phys. Rev. A **45**, 8185 (1992).

[2] A. E. Siegman, *Lasers* (University Science Books, Mill Valley, CA, 1968).

[3] B. A. Garetz, J. Opt. Soc. Am. **71**, 609 (1981).

[4] B. A. Garetz and S. Arnold, Opt. Commun. **31**, 1 (1979).

[5] I. Bialynicki-Birula and Z. Bialynicka-Birula, Phys. Rev. Lett. **78**, 2539 (1997).

[6] J. Courtial *et al.*, Phys. Rev. Lett. **80**, 3217 (1998).

[7] L. Allen, M. Babiker, and W. L. Power, Opt. Commun. **112**, 141 (1994).

[8] G. Nienhuis, Opt. Commun. **132**, 8 (1996).

[9] G. A. Turnbull *et al.*, Opt. Commun. **127**, 183 (1996).

[10] A. H. F. van Vliet and T. de Graauw, Int. J. Infrared Millim. Waves **2**, 465 (1981).

[11] F. Bretenaker and A. Le Floch, Phys. Rev. Lett. **65**, 2316 (1990).

Section 7

ANGULAR MOMENTUM IN NONLINEAR OPTICS

In 1993 Basistiy *et al.* (7.1) examined the interaction of beams containing optical vortices with nonlinear materials and with computer generated holograms. In each case additional vortices, of either the same or opposite handedness, were found to appear in the beam. It was noted that when frequency doubled a beam with a first order screw dislocation, that is $l = 1$, acquired an additional screw dislocation.

The first work in which the role of orbital angular momentum was explicitly considered in a nonlinear process was that of Dholakia *et al.* (7.2). They demonstrated that the frequency doubling of an LG mode gave rise to another LG beam with its azimuthal mode index doubled. They were able to explain this in terms of the phase matching conditions in the crystal which dictated that the Poynting vector should maintain the same spiral trajectory through the crystal. The corresponding doubling of the orbital angular momentum per output photon meant that the total orbital angular momentum was conserved within the light field. The same group generalised their work to include LG modes with different radial as well as different azimuthal mode indices (7.3, **Paper 7.1**). The frequency doubled LG mode with $p = 0$ is another LG mode with index $2l$ and $p = 0$. But, for $p > 0$ the amplitude distribution of the resulting mode, the square of that of the input mode, is a superposition of LG beams of different mode order. Although still conserving orbital angular momentum, the beam has a form that changes upon propagation and which only reproduces the beam waist distribution in the far-field.

Second harmonic generation is a special case of sum frequency mixing in which two input fields of frequencies ω_1 and ω_2 produce a third field $\omega_3 = \omega_1 + \omega_2 = 2\omega_1$. Berzanskis *et al.* (7.4) experimented with sum frequency generation for beams with a variety of l-values between 1 and 3 and interpreted the results in terms of topological charge. Conservation of orbital angular momentum would dictate that $l_3 = l_1 + l_2$.

For high incident intensities, the possibility of additional non-linear processes needs to be considered. For sufficient intensity, both second- and third-order nonlinearities can give rise to self-defocusing and self-focusing leading to fragmentation of the beam. In 1992 Swartzlander and Law demonstrated that, in a self-defocusing medium, an optical vortex could be stable and be used to guide a probe beam (7.5). In 1997, Torner and Petrov (7.6) predicted that in a second-order material an intense input LG_0^1 beam would, because of self-focusing, break up into three. Firth and Skryabin (7.7, **Paper 7.2**) later predicted that subject to a second- or third-order non-linearity, a single annular ring can break up into $2|l|+1$ or $2|l|$ fragments respectively. Because of their soliton properties, these fragments can retain their transverse localisation. The fragments resemble Newtonian particles flying off at a tangent to the initial ring; a vivid demonstration of the conservation of angular momentum. Under different conditions, Soljačić and Segev (7.8, **Paper 7.3**) have predicted that the individual fragments can rotate in a manner akin to a rigid body.

Second harmonic generation and sum frequency mixing are examples of three-wave interactions, where the phase of the two input fields define the phase of the third output beam. Parametric down conversion is also a three-wave interaction and again there is a fixed phase relation between the input pump and output signal and idler fields. However, neither the signal nor idler have an externally set phase and it is their phase sum that is defined by the pump. Classically, the signal and the idler fields are both spatially incoherent (7.9) and neither, therefore, have phasefronts with a well defined azimuthal index. It follows that for classical fields neither signal nor idler beams have a unique orbital angular momentum (7.10). However, at the single-photon level, the phase relationship does imply a fixed relationship between signal and idler beams. Recently, Mair *et al.* (7.11, **Paper 8.1**) have shown that although the orbital angular momentum of the signal and idler photons have a range of values, each photon pair conserves the orbital angular momentum; that is, $l_3 = l_1 + l_2$.

REFERENCES

7.1 IV Basistiy, V YuBazhenov, MS Soskin and MV Vasnetsov, 1993, *Opt. Commun.* **103** 422.

7.2 K Dholakia, NB Simpson, L Allen and MJ Padgett, 1996, *Phys. Rev. A* **54** R3742.

7.3 J Courtial, K Dholakia, L Allen and MJ Padgett, 1997, *Phys. Rev. A* **56** 4193.

7.4 A Berzanskis, A Matijosius, A Piskarskas, V Smilgevicius and A Stabinis, 1997, *Opt. Commun.* **140** 273.

7.5 GA Swartzlander and CT Law, 1992, *Phys. Rev. Lett.* **69** 2503.

7.6 L Torner and DV Petrov, 1997, *J. Opt. Soc. B* **14** 2017.

7.7 WJ Firth and DV Skryabin, 1997, *Phys. Rev. Lett.* **79** 2450.

7.8 M Soljačić and M Segev, 2001, *Phys. Rev. Lett.* **86** 420.

7.9 R Ghosh, CK Hong, ZY Ou and L Mandel, 1986, *Phys. Rev. A* **34** 3962.

7.10 J Arlt, K Dholakia, L Allen and MJ Padgett, 1999, *Phys. Rev. A* **59** 3950.

7.11 A Mair, A Vaziri, G Weihs and A Zeilinger, 2001, *Nature* **412** 313.

PAPER 7.1

Second-harmonic generation and the conservation of orbital angular momentum with high-order Laguerre-Gaussian modes

J. Courtial, K. Dholakia, L. Allen, and M. J. Padgett
School of Physics and Astronomy, University of St. Andrews, Fife KY16 9SS, Scotland
(Received 16 June 1997)

Laguerre-Gaussian modes of various order are frequency doubled. The azimuthal phase structure of the second-harmonic light is measured directly by interfering the beam with its mirror image. We show that the orbital angular momentum per photon is doubled, so conserving the orbital angular momentum in the light beam. The frequency-doubled output beam is shown to have a Gegenbauer-Gaussian amplitude distribution at the beam waist. The beam can be described as a summation of Laguerre-Gaussian modes that interfere so that it changes form with propagation, but the distribution at the beam waist is reproduced in the far field. [S1050-2947(97)03311-8]

PACS number(s): 42.65.Ky, 42.60.Jf

It was predicted in 1992 that monochromatic beams with an azimuthal phase term $e^{il\phi}$, of which Laguerre-Gaussian laser modes are an example, have a well-defined orbital angular momentum of $l\hbar$ per photon [1]. This orbital angular momentum is associated with the azimuthal component of the Poynting vector [2] and is quite distinct from the spin angular momentum associated with circular polarization. The transfer of this orbital angular momentum to a microscopic particle has been demonstrated recently [3]. Microscopic particles held within a Laguerre-Gaussian laser beam have been rotated and the orbital angular momentum quantified by a comparison of this rotation to that induced by the spin angular momentum. The results confirm that the orbital angular momentum is $l\hbar$ per photon [4,5]. Other work has explored theoretically the interaction of these beams with atomic systems [6,7].

Laguerre-Gaussian modes [8] are characterized by two indices l and p, where l is the number of 2π cycles in phase around the circumference and $p+1$ the number of radial nodes. The amplitude u_p^l of such a mode in cylindrical coordinates is given by

$$u_p^l(r,\phi,z)\propto \exp(-ikr^2/2R)\exp(-r^2/w^2)$$
$$\times\exp[-i(2p+l+1)\psi]$$
$$\times\exp(-il\phi)(-1)^p(r\sqrt{2}/w)^l L_p^l(2r^2/w^2), \quad (1)$$

where r is the distance from the beam axis, ϕ the azimuthal angle, z the distance from the beam waist, k the wave number of the light, w the radius for which the Gaussian term falls to $1/e$ of its on-axis value, z_r is the Rayleigh range, $L_p^l(x)$ an associated Laguerre polynomial, and $(2p+l+1)\psi$ is the Gouy phase, where $\psi=\arctan(z/z_r)$.

In 1996 we reported the mode transformation that occurs when a $p=0$ Laguerre-Gaussian mode is frequency doubled [9]. Modes with $p=0$ have a single-annular-ring intensity distribution. They frequency double to give a pure Laguerre-Gaussian mode also with $p=0$, but with an azimuthal index of $2l$, that is, with twice the orbital angular momentum per photon. This mode transformation is readily understood in terms of the spiraling of the Poynting vector, which has the same form for both the fundamental and second-harmonic beams, and was shown to be consistent with conservation of orbital angular momentum within the light beams. In our earlier work the azimuthal phase structure of the frequency-doubled mode was not measured directly. It was deduced by converting the second-harmonic mode into the corresponding Hermite-Gaussian mode by means of a cylindrical lens mode converter [10].

In this paper we confirm our earlier results by the direct measurement of the azimuthal phase structure of the frequency-doubled beams and extend the results to include frequency doubling of the multiringed, $p>0$, Laguerre-Gaussian modes. We show that this results in beams that possess a well-defined orbital angular momentum, but are not simple Laguerre-Gaussian modes.

At the beam waist, $z=0$, the amplitude of a Laguerre-Gaussian mode simplifies to

$$u_p^l(r,\phi,z=0)\propto e^{-r^2/w_0^2}e^{-il\phi}(-1)^p(r\sqrt{2}/w_0)^l L_p^l(2r^2/w_0^2). \quad (2)$$

For $p=0$, the associated Laguerre polynomial is a constant and independent of r. In second-harmonic generation, the amplitude of the frequency-doubled field is proportional to the square of the incident field [11]. It follows that for a $p=0$ mode the second-harmonic beam is also a Laguerre-Gaussian mode that has undergone the following transformations: $k\rightarrow 2k$, frequency doubling; $w_0\rightarrow w_0/\sqrt{2}$, reduction of the beam waist; $p=0\rightarrow p=0$, the amplitude distribution remains single ringed; and $l\rightarrow 2l$, the angular momentum per photon is doubled. When $p>0$ the square of the incident field can no longer be described in terms of a single Laguerre-Gaussian mode. However, the azimuthal phase structure is still of the form $\exp(il\phi)$.

The experimental arrangement for generating the frequency-doubled beams, as well as the subsequent analysis of their intensity and phase structure, is shown in Fig. 1. An intracavity cross wire is used to generate a variety of Hermite-Gaussian modes with indices m and n from a diode-pumped Nd:YAG laser (where YAG denotes yttrium aluminum garnet) operating at 1064 nm producing a linearly polarized output power of ≈ 100 mW. Each mode is converted

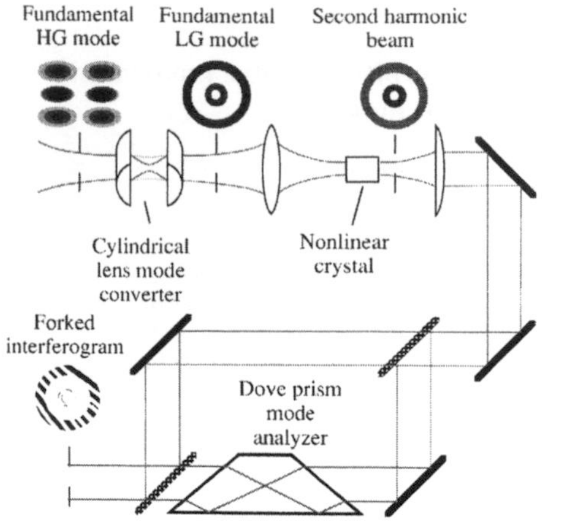

FIG. 1. Experimental apparatus for the generation and analysis of the second-harmonic beams.

into the corresponding Laguerre-Gaussian mode by means of a cylindrical lens mode converter, giving the transformations $l=m-n$ and $p=\min(m,n)$ [10]. The Laguerre-Gaussian modes are then frequency doubled using a 10-mm-long crystal of potassium titanyl phosphate (KTP), angle tuned to give phase matching for the second harmonic at 532 nm. The efficiency of the process is maximized by focusing the incident beam such that its Rayleigh range is comparable to the length of the crystal [12]. Filters placed after the crystal allow either the fundamental or the second-harmonic beam to be selected and imaged onto a charge coupled device array detector. The azimuthal phase structure of the beams can be measured directly using a mode analyzer [13] based on a Dove prism, which allows the interference pattern between the beam and its own mirror image to be obtained. The azimuthal phase component of the beam gives rise to forked interference fringes and the l index of the beam can be inferred directly from the number of fringes on either side of the fork by dividing the number of additional fringes by 2.

Figure 2 shows the forked interferograms obtained for a variety of fundamental Laguerre-Gaussian beams and their second-harmonic counterparts. As in our previous work [9], these results confirm that the azimuthal index l of the beam is doubled in the second-harmonic process. However, here the azimuthal phase has been measured directly and we observe that the doubling of l in the second-harmonic process holds for Laguerre-Gaussian modes of any order of l and p. Again, this is consistent with the conservation of orbital angular momentum within the light beams.

The less than perfect mode converter introduces residual astigmatism into the beam, which manifests itself as a slight ellipticity in the observed images. Corrected radial profiles for the fundamental and second-harmonic beams are obtained by scaling the images to make them symmetrical and averaging the profiles over 80 azimuths. These profiles are then fitted to the predicted field distributions of a Laguerre-Gaussian and its square, respectively, with the amplitude and beam diameter as the fit variables. Figure 3 shows that the corrected profiles are in good agreement with those predicted. The square of a Laguerre polynomial is a Gegenbauer

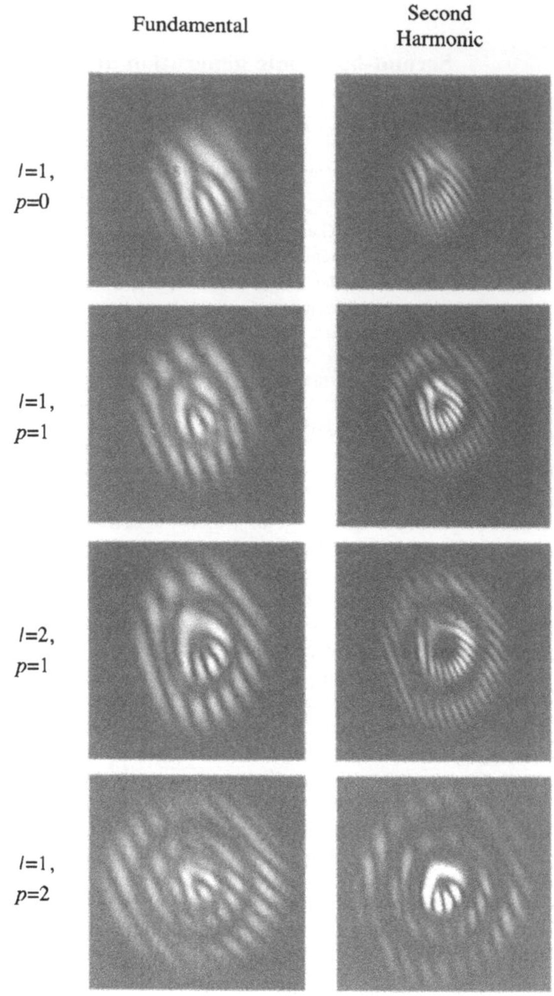

FIG. 2. Forked interferograms derived from a variety of Laguerre-Gaussian beams and their second-harmonic counterparts.

polynomial; consequently, we refer to the amplitude distribution of a frequency-doubled Laguerre-Gaussian mode at the beam waist as a Gegenbauer-Gaussian distribution.

Pure Laguerre-Gaussian modes propagate without changing their form, with a beam divergence dictated by the size of the beam waist and the corresponding Rayleigh range. As discussed, the second-harmonic beam for $p>0$ can no longer be described as a simple Laguerre-Gaussian mode. Insight as to how it can be described can, however, be gained from the experimental evidence: We find that the distribution of intensity in the far field is the same as that in the plane of the nonlinear process at $z=0$. For a monochromatic beam the far-field amplitude distribution is simply the Fourier transform of the distribution at the beam waist [14]. When the Fourier transform of light consisting of the square of Laguerre-Gaussians at z is taken, it is found that the resulting distribution can be described as a superposition of a number

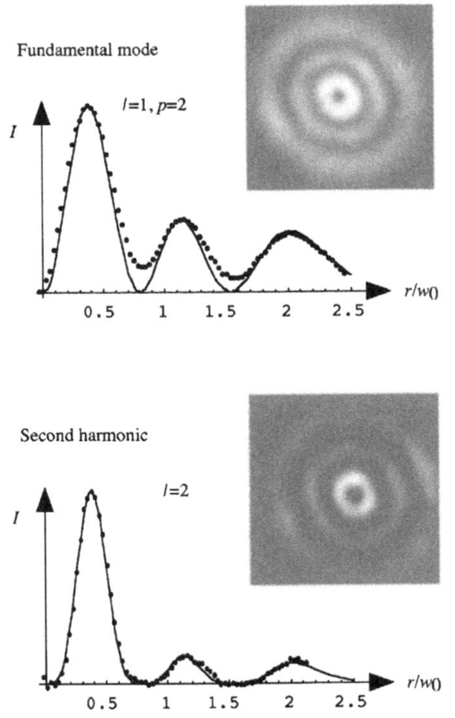

FIG. 3. Corrected radial intensity profiles for the fundamental and second-harmonic beams compared with theory, together with photographs of the observed images recorded in the plane of the frequency-doubling crystal.

of Laguerre-Gaussian modes all with the same index $2l$ but with $p^{(2\omega)}=0,2,\ldots,2p$. For $z=0$, this summation reduces to the square of a Laguerre-Gaussian of the same form as in Eq. (2).

Just as a Laguerre-Gaussian mode is a solution of the paraxial Helmholtz equation, so, too, is any sum of Laguerre-Gaussian modes, and as the second-harmonic beam propagates, interference occurs between the constituent modes. Although the modes all have the same Rayleigh range, the p indices give rise to a differing Gouy phase shift between the modes, which leads to an intensity distribution that changes form with propagation. It is only in the far field, where all the Gouy phase shifts differ by multiples of 2π, that the Gegenbauer-Gaussian distribution is reproduced. We see that the behavior of the $p=0$ modes analyzed in our previous paper [9] is simply a limiting case of the general behavior; when $p=0$ there is only one Laguerre polynomial involved and so only one Laguerre-Gaussian distribution.

Rather than investigate the explicit summation of Laguerre-Gaussian modes to determine the beam distribution as it propagates, we have implemented an algorithm based on the Fourier expansion of the beam in k space into a series of plane waves [15]. The propagation of each plane-wave component to a subsequent plane results in a well-defined change in phase. After propagation a summation of the individual plane-wave components followed by an inverse transform gives the new phase and amplitude distribution of the

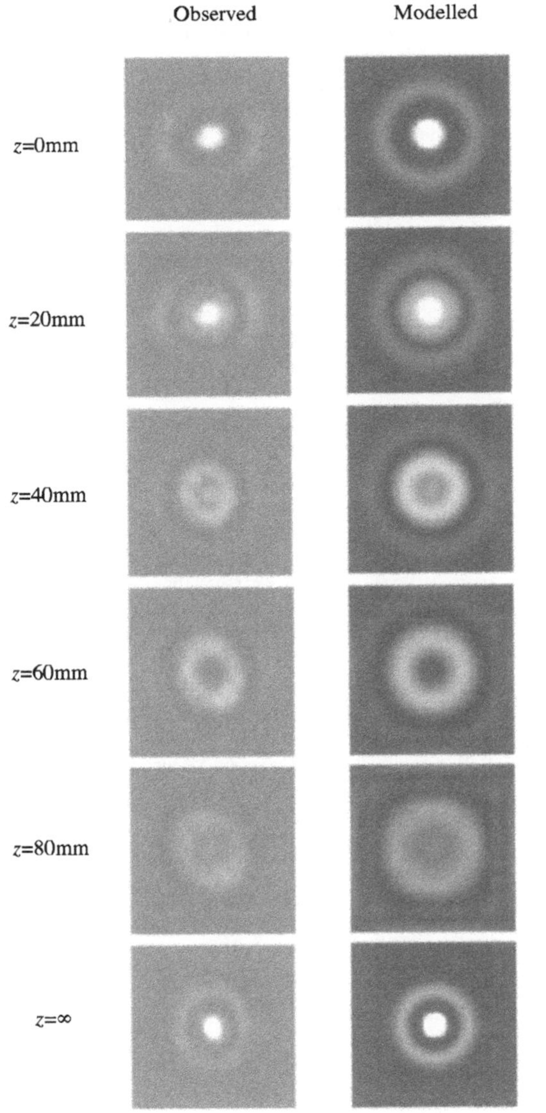

FIG. 4. Observed and modeled intensity distribution of a frequency-doubled Laguerre-Gaussian ($p=1$, $l=0$) mode as it propagates from the beam waist.

beam. The algorithm is limited only by the array size of the Fourier transforms, which restricts the maximum allowed divergence and physical diameter of the beam. Figure 4 shows the observed and predicted intensity distributions for a frequency-doubled Laguerre-Gaussian mode with $p=1$ and $l=0$ as it propagates from the beam waist to the far field. We see that the intensity distribution of the second-harmonic beam reproduces itself in the far field, but varies at all intermediate positions. We observe similar behavior for all frequency-doubled Laguerre-Gaussian modes with $p>0$.

In this paper we extend our earlier work on the frequency doubling of Laguerre-Gaussian modes to $p>0$ and show that the previously investigated case of $p=0$ is a limiting case of the general behavior. The azimuthal phase of the frequency-doubled modes is measured directly and the azimuthal phase index is shown to become doubled for all modes. This corresponds to a doubling of the orbital angular momentum per photon during the second-harmonic generation process. The second-harmonic amplitude distribution may be described as a Gegenbauer-Gaussian beam that, while reproducing its intensity distribution in the far field, varies its distribution at all intermediate positions. Nevertheless, the frequency-doubled beams are still well defined and symmetrical about the beam axis. The nonlinearity of the second-harmonic generation process means that these beams have a radial intensity distribution in which a higher proportion of the energy is closer to the beam axis than in Laguerre-Gaussian modes. Such beams therefore allow the use of lower-aperture optical components in experiments involving the orbital angular momentum of light.

This work was supported by EPSRC. K. D. acknowledges the financial support of the Royal Society of Edinburgh. M. J. P. would like to thank the Royal Society (London) for financial support. It is a pleasure to express our thanks to Stephen M. Barnett for his analytic solution of the relevant Fourier transform.

[1] L. Allen, M. W. Beijersbergen, R. J. C. Spreeuw, and J. P. Woerdman, Phys. Rev. A **45**, 8185 (1992).
[2] M. Padgett and L. Allen, Opt. Commun. **121**, 36 (1995).
[3] H. He, M. E. J. Friese, N. R. Heckenberg, and H. Rubinsztein-Dunlop, Phys. Rev. Lett. **75**, 826 (1995).
[4] M. E. J. Friese, J. Enger, H. Rubinsztein-Dunlop, and N. R. Heckenberg, Phys. Rev. A **54**, 1593 (1996).
[5] N. B. Simpson, K. Dholakia, L. Allen, and M. J. Padgett, Opt. Lett. **22**, 52 (1997).
[6] L. Allen, M. Babiker, W. K. Lai, and V. E. Lembessis, Phys. Rev. A **54**, 4259 (1996).
[7] W. K. Lai, M. Babiker, and L. Allen, Opt. Commun. **133**, 487 (1997).
[8] A. E. Siegman, *Lasers* (University Science Books, Mill Valley, CA, 1986), Sec. 17.5.
[9] K. Dholakia, N. B. Simpson, M. J. Padgett, and L. Allen, Phys. Rev. A **54**, R3742 (1996).
[10] M. W. Beijersbergen, L. Allen, H. E. L. O. van der Veen, and J. P. Woerdman, Opt. Commun. **96**, 123 (1993).
[11] A. Yariv, *Optical Electronics*, 3rd ed. (Holt, Rinehart and Winston, New York, 1985), Chap. 8.
[12] G. D. Boyd and D. A. Kleinmann, J. Appl. Phys. **39**, 3597 (1968).
[13] M. Harris, C. A. Hill, and J. M. Vaughan, Opt. Commun. **106**, 161 (1994).
[14] E. Hecht and A. Zajac, *Optics* (Addison-Wesley, Reading, MA, 1974), Sec. 11.3.3.
[15] E. A. Sziklas and A. E. Siegman, Appl. Opt. **14**, 1874 (1975).

Optical Solitons Carrying Orbital Angular Momentum

W. J. Firth and D. V. Skryabin*
Department of Physics and Applied Physics, John Anderson Building, University of Strathclyde, 107 Rottenrow, Glasgow, G4 0NG, United Kingdom
(Received 10 April 1997; revised manuscript received 1 July 1997)

We predict a new kind of ring-profile solitary wave in nonlinear optical media, with finite orbital angular momentum. During propagation these fragment into fundamental solitons. Like free Newtonian particles, these fly off tangential to the ring, vividly demonstrating conservation of orbital angular momentum in soliton motion. [S0031-9007(97)04148-3]

PACS numbers: 42.65.Tg, 03.40.Kf, 42.65.Ky

Solitons are important in many branches of science [1]. They are nonlinear waves which possess several mechanical and other attributes more commonly associated with particles. This analogy is usually developed in relation to their response to external "forces," i.e., in the context of linear momentum; see, e.g., [2]. Here we describe phenomena which are neatly interpreted as solitons carrying *orbital angular* momentum, arising from motion oblique to the centroid of the system.

These solitons are produced in the fragmentation of a new kind of "doughnut soliton." They have well-defined angular momentum which is transformed into orbital angular momentum of the daughter solitons. Below we demonstrate this phenomenon in two rather different nonlinear optical systems. The close correspondence of the dynamics in the two cases lead us to believe that this scenario should be quite general in solitonic systems with enough dimensions to exhibit angular momentum effects.

Our first model, $\chi^{(3)}$, describes a beam propagating in a saturable self-focusing medium. In the paraxial approximation the evolution in z of the field envelope $E_1(x,y,z)$ obeys the following dimensionless equation [3]

$$i\partial_z E_1 + \frac{1}{2}\vec{\nabla}_\perp^2 E_1 + E_1|E_1|^2/(1+\alpha|E_1|^2) = 0\,. \quad (1)$$

Here $\vec{\nabla}_\perp = \vec{i}\partial_x + \vec{j}\partial_y$. For a pure Kerr medium ($\alpha = 0$) this equation is the well-known nonlinear Schrödinger equation (NLS). With the y-dimension suppressed (1D case) the NLS is integrable, with exact solutions—solitons—of sech profile. In 2D it has solitonlike solutions which are unstable, collapsing to a singularity [3]. Saturation, described by a finite positive α, prevents this collapse [3]. Thus, though it is not integrable, Eq. (1) possesses stable solitary wave solutions, localized in 2D, which we will term "solitons."

Our second model, $\chi^{(2)}$, is physically different, corresponding to the coupled propagation of an optical field and its second harmonic in a quadratically nonlinear medium. Their field envelopes E_1 and E_2 can be described by the following system of rescaled equations [4]:

$$i\partial_z E_1 + \frac{1}{2}\vec{\nabla}_\perp^2 E_1 + E_1^* E_2 = 0\,,$$
$$i\partial_z E_2 + \frac{1}{4}\vec{\nabla}_\perp^2 E_2 + \frac{1}{2}E_1^2 = \beta E_2\,. \quad (2)$$

A recent review [5] provides a good link to experimental parameters, and to complications such as field polarization and walk-off which we neglect here. The parameter β is the phase mismatch. If β is large enough, it can be considered to dominate the derivative terms, then solving for E_2 and substituting into the equation for E_1 gives the NLS. Again Eqs. (2) are not integrable, but stable solitary solutions persist even for small β, far from the NLS limit. We present results for $\beta = 0$ in the following: the phenomena we describe are not very sensitive to β. These solitons are now well known both experimentally and theoretically; see [4–6] and *op cit.*

All quantities in Eqs. (1) and (2) are dimensionless, and these scaled units are used throughout the text and in the figures. Note that both equations have a Galilean invariance, and so the 2D spatial soliton centered on $(x = 0, y = 0)$ generates a family of equivalent solitons which "move" with constant velocity in the (x,y) plane as the beam propagates. It is such moving solitons which are important in the following.

Both models are Hamiltonian, and possess phase, translational, and rotational symmetries. As a consequence, both conserve the energy integral Q, transverse momentum $\vec{P}$, and transverse angular momentum $\vec{L}$, defined as follows: $Q = \int dxdy(|E_1|^2 + 2|E_2|^2)$, $\vec{P} = \int dxdy\,\vec{p} = \int dxdy\frac{i}{2}[E_1(\vec{\nabla}_\perp \cdot E_1^*) + E_2(\vec{\nabla}_\perp \cdot E_2^*) - \text{c.c.}]$, $\vec{L} = \int dxdy\,\vec{r} \times \vec{p}$, where $\vec{p}$ is the transverse momentum density and $\vec{r}$ is the transverse radius vector. (For the $\chi^{(3)}$ model set E_2 to zero.)

The angular momentum carried by light beams has attracted much recent interest. It has been predicted, and proved experimentally, that Laguerre-Gaussian beams with azimuthal model index l carry orbital angular momentum $l\hbar$ per photon [7]. Frequency doubling such a beam has been shown [8] to generate a second harmonic with doubled azimuthal mode index $2l$.

We now show that both our models admit nondiffracting solitary wave solutions with finite angular momentum. Such "doughnut solitons" are ringlike solutions of (1) or (2) with intensity independent of z. Equations (1) and (2) are reduced by the substitution $E_m(z, r, \theta) = A_m(r)\exp[im(\kappa z + l\theta)]$ ($m = 1, 2$) to ordinary differential equations which we solve numerically using finite differences. Here $r = \sqrt{x^2 + y^2}$, θ is the polar angle, κ is a real and free parameter, as is l, which we restrict to integer values to ensure azimuthal periodicity. $A_{1,2}(r)$ are real functions which satisfy zero boundary conditions: $A_{1,2}(0) = A_{1,2}(+\infty) = 0$. For the $\chi^{(3)}$ model, consider $m = 1$ only (here and below).

For each l, $\chi^{(3)}$ doughnut solitons exist for all positive κ (which is required for soliton confinement), while for $\chi^{(2)}$ the existence condition is $\kappa > \max(0, -\beta/2)$. Doughnut solitons obey $\vec{P} = 0$ and $|\vec{L}| = |l|Q$ in both models. Typical spatial profiles are presented in Figs. 1(a) and 1(b).

To investigate doughnut soliton stability under propagation, we initialize (1) or (2) with a doughnut soliton (plus noise) and simulate the subsequent evolution on both Cartesian and polar grids, using split-step algorithms. Both approaches give the same results. As a further check, conservation of the energy, momentum, and angular momentum was monitored during the simulations. We find that these doughnuts usually break up into several solitons, which move off at constant "velocity" (angle to the axis of propagation) along paths tangent to the initial ring. The breakup is due to an azimuthal modulational instability. Before discussing the asymptotic motion of the "daughter" solitons, therefore, we outline our doughnut soliton stability analysis.

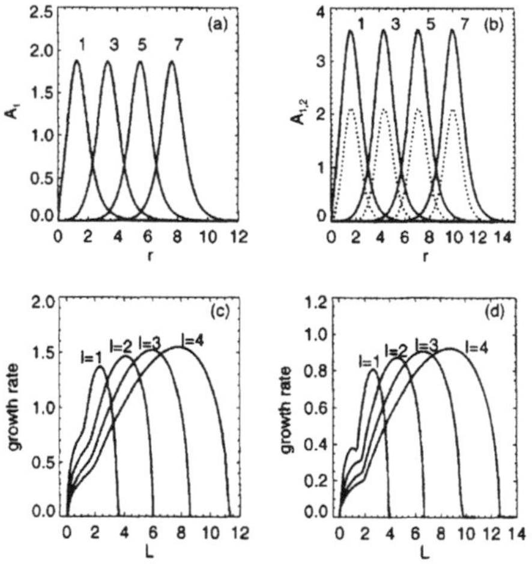

FIG. 1. (a) One-ring solitary wave solutions of Eq. (1): plots of the field amplitude $A_1(r)$ for $\kappa = 1$ and $l = 1, 3, 5, 7$. (b) Corresponding plots of $A_1(r)$ (full) and $A_2(r)$ (dotted) for Eqs. (2). (c) Perturbation growth rate Re$\tilde{\lambda}$ vs L for different values of l, for the $\chi^{(3)}$ case: $\kappa = 1$, $\alpha = 0.1$. (d) Corresponding plots for the $\chi^{(2)}$ case: $\kappa = 1$, $\beta = 0$.

As regards radial perturbations, the criterion $\partial_\kappa Q > 0$ [9] for stability applies to the doughnut solitons, and is satisfied over most of the existence domain. However, $\partial_\kappa Q > 0$ does not imply stability with respect to azimuthal perturbations [10], which break the cylindrical symmetry. We therefore considered azimuthally perturbed doughnut solitons in the general form $E_m = [A_m(r) + \epsilon_m^+(r, z)e^{iL\theta} + \epsilon_m^-(r, z)e^{-iL\theta}]e^{im(\kappa z + l\theta)}$ ($m = 1, 2$). Substitution into (1) or (2) and linearization results in partial differential equations for $\epsilon_{1,2}^{\pm}(r, z)$ which we solve numerically using a Crank-Nicholson scheme. We integrate along z until the perturbation growth rate becomes stationary and then average it over a further propagation distance. This procedure gives us an estimate for the real part of the maximally unstable eigenvalue $\tilde{\lambda}$. Alternatively, setting $\epsilon_{1,2}^{\pm} = [u_{1,2}^{\pm}(r) + iv_{1,2}^{\pm}(r)]e^{\lambda z}$ yields a boundary value problem whose spectrum we find by a finite difference method. The two methods give identical results.

Dependencies of the perturbation growth rate Re$\tilde{\lambda}$ on L for various values of l with other parameters fixed are presented in Figs. 1(c) and 1(d). Physically, L must be an integer to ensure azimuthal periodicity, but it appears in the linearized equations as a real parameter, and extra insight can be gained by studying growth rates for arbitrary positive L. In every case there is a positive growth rate over a range of L values, with a well-defined global maximum for $L = L_{\max}$. We find that azimuthal instability ($L \neq 0$) always dominates radial instability which corresponds to $L = 0$. On propagation, we expect the initially uniform field amplitude around the doughnut to develop N minima and N maxima, where N is the integer closest to $L_{\max}$. As a consequence, the doughnut should break up into N solitons. In the cases illustrated below $N = 2|l|$ for the $\chi^{(3)}$ model and $2|l| + 1$ for $\chi^{(2)}$.

We present results of direct numerical simulation of Eqs. (1) and (2) in Figs. 2 and 3, respectively, for the cases $l = 1, 2, 3$. Figures 2(a)–2(c) and 3(a)–3(c) show the real part of the perturbation eigenmode corresponding to $L = N$ computed from the stability analysis—the real part determines the field amplitude modulation pattern which develops around the initial ring. Figures 2(d)–2(f) and 3(d)–3(f) show the field intensity after propagating far enough for the modulational instability to develop. Each of the N peaks of the eigenmode evolves into an intensity peak (a protosoliton). (For the $\chi^{(2)}$ case the field amplitudes E_1, E_2 are localized in the same region of space due to their nonlinear coupling.)

Figures 2(g)–2(j) and 3(g)–3(j) show the real part of E_1, indicating that adjacent protosolitons are out of phase (this is less obvious in Fig. 3 because N is odd).

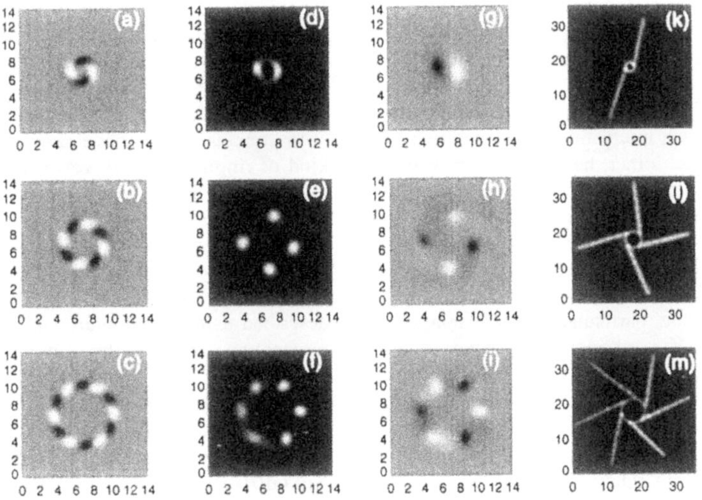

FIG. 2. Azimuthal modulational instability development and soliton trajectories, for $\chi^{(3)}$ case: $\alpha = 0.1$, $\kappa = 1$, with $l = 1, 2, 3$. (a)–(c) Real part of the perturbation field pattern with maximal growth rate. (d)–(f) Numerically computed field intensity $|E_1|^2$ at a point where protosolitons have developed. (g)–(i) Real part of E_1 at the same point, showing relative phases of protosolitons. (k)–(m) Superimposed images of the transverse intensity distribution at different z values, showing soliton trajectories tangential to the initial ring. Note the change of scale.

It has been noted that out-of-phase solitons repel each other, while in-phase attract; see, e.g., Ref. [11]. By symmetry, the resultant force on each protosoliton should thus be *radial* and outward. In Figs. 2(k)–2(m) and 3(k)–3(m) we superimpose a succession of images at different z, to show the daughter soliton trajectories. Far from being radial, these are *tangential* to the initial ring.

Our interpretation is that the intersoliton forces are actually negligible, and that the solitons are behaving like free Newtonian particles, flying off tangential to the doughnut soliton ring, and carrying away its angular momentum *via* the obliquity of their paths. The "mass" of the nth soliton is its energy Q_n, and its angular momentum is $\vec{r}_n \times \vec{p}_n$, where Q_n and $\vec{p}_n$ are now defined *locally* around the nth soliton's position $\vec{r}_n$. Using conservation of energy and angular momentum and the Galilean invariance of Eqs. (1) and (2) we calculate the transverse speed of the solitons to be $|\vec{v}| = |l|/R$, where

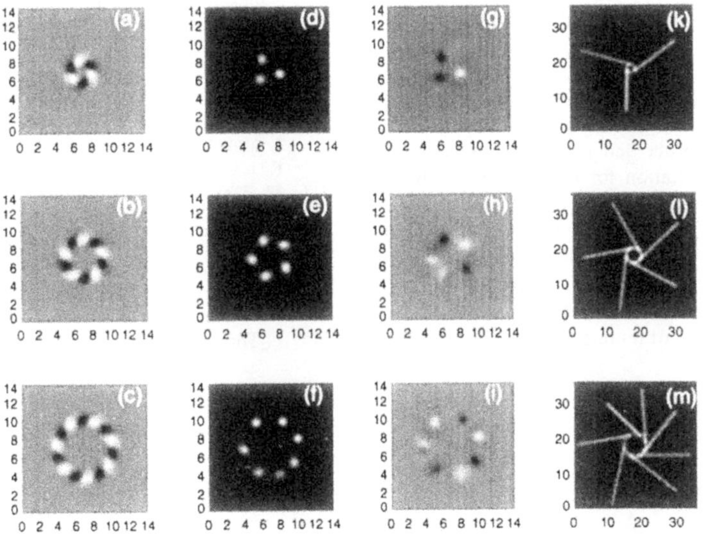

FIG. 3. Same as Fig. 2, but for field E_1 in the $\chi^{(2)}$ model: $\beta = 0$, $\kappa = 2$.

R is the radius of the initial doughnut (the same speed for each daughter soliton, independent of its energy).

This estimate is in good agreement with the numerical results shown in Figs. 2 and 3, with losses (due to nonsoliton "radiation") at most 10%. Thus, in contrast to recent work [11] in which interacting photorefractive solitons spiraled around each other, here we have nearly free solitons, with dynamics dominated by angular momentum conservation. Interactions forces may play a minor role in partitioning the energy among the protosolitons, but the exponential localization of the daughter solitons means they rapidly cease to interact.

Generalizing from the particular cases displayed in Figs. 2 and 3, we find very similar behavior over a wide range of parameter values. For both models the sign of l, which defines the direction of the angular momentum, fixes the orientation of the soliton trajectories. The number of daughter solitons depends strongly on the azimuthal index l, but relatively weakly on all other parameters.

While we believe that our results are conceptually and pedagogically interesting, independent of direct experimental verification, it is naturally of interest to address the question of experimental observation of these or related phenomena. We already mentioned the work of Shih *et al.* [11], and several other recent papers report experiments which relate in some way to our scheme and are thus generally encouraging, though in each case there are important and interesting differences.

Tikhonenko *et al.* report experiments in rubidium vapor (a saturable self-focusing medium) [12,13]. They observed, and confirmed in simulations, fragmentation of beams with a phase dislocation. They did not consider the possibility of doughnut solitons. Their experimental results were quite strongly affected by the lack of azimuthal symmetry of their input beam, especially in the case $l = 2$ [13]. For the case of $l = 1$ their results [12] are quite similar to ours, with two daughter solitons produced. A side view indicates straight-line trajectories with no evidence of interaction forces leading to spiraling.

In the $\chi^{(2)}$ case, Torner and Petrov [14] recently described numerical simulation for the case where the input was a Laguerre-Gaussian for E_1 with $E_2 = 0$. They concentrated on the case $l = 1$. Typically three solitons were output, broadly in accord with our findings, though for quite different initial conditions. Stationary solutions such as doughnut solitons, we would argue, are the natural starting point for studies of more general input conditions. Finally, Fuerst *et al.* recently reported an experiment [6] quite close to our model, except that the input is "unrolled" to form an intense line focus. The observed filamentation into up to six stable solitons is encouraging for similar experiments with finite angular momentum.

In conclusion, we have shown the existence of a new kind of ringlike solitary wave, with finite orbital angular momentum. These are unstable on propagation, breaking into filaments which become solitons, whose number is strongly dependent on the input angular momentum. The solitons fly out tangentially from the initial ring, like free Newtonian particles, and their motion is accurately described using Newtonian conservation laws for energy, momentum, and angular momentum. This scenario is rather robust, with broadly similar phenomena occurring over a wide range of parameters, both in quadratically nonlinear and in self-focusing media.

This work was partially supported by EPSRC Grant No. GR/L 27916. D. V. S. acknowledges financial support from ORS award scheme.

---

*Electronic address: dmitry@phys.strath.ac.uk

[1] A. C. Newell, *Solitons in Mathematics and Physics* (SIAM, Philadelphia, 1985).

[2] A. Aceves *et al.*, J. Opt. Soc. Am. B **7**, 963 (1990); D. E. Edmundson and R. H. Enns, Phys. Rev. A **51**, 2491 (1995).

[3] J. J. Rasmussen and K. Rypdal, Phys. Scr. **33**, 481 (1986).

[4] A. V. Buryak, Y. S. Kivshar, and V. Steblina, Phys. Rev. A **52**, 1670 (1995).

[5] G. I. Stegeman, D. J. Hagan, and L. Torner, Opt. Quantum Electron. **28**, 1691 (1996).

[6] R. A. Fuerst *et al.*, Phys. Rev. Lett. **78**, 2756 (1997).

[7] L. Allen *et al.*, Phys. Rev. A **45**, 8185 (1992); N. R. Heckenberg *et al.*, Opt. Lett. **17**, 221 (1992); N. B. Simpson *et al.*, Opt. Lett. **22**, 52 (1997).

[8] K. Dholakia *et al.*, Phys. Rev. A **54**, R3742 (1996).

[9] M. G. Vakhitov and A. A. Kolokolov, Sov. Radiophys. **16**, 783 (1973); D. E. Pelinovsky, A. V. Buryak, and Y. S. Kivshar, Phys. Rev. Lett. **75**, 591 (1995).

[10] J. M. Soto-Crespo *et al.*, Phys. Rev. A **44**, 636 (1991); J. Atai, Y. Chen, and J. M. Soto-Crespo, Phys. Rev. A **49**, R3170 (1994).

[11] M. Shih, M. Segev, and G. Salamo, Phys. Rev. Lett. **78**, 2551 (1997).

[12] V. Tikhonenko, J. Christou, and B. Luther-Davies, J. Opt. Soc. Am. B **12**, 2046 (1995).

[13] V. Tikhonenko, J. Christou, and B. Luther-Davies, Phys. Rev. Lett. **76**, 2698 (1996).

[14] L. Torner and D. V. Petrov, Electron. Lett. **33**, 608 (1997).

Integer and Fractional Angular Momentum Borne on Self-Trapped Necklace-Ring Beams

Marin Soljačić[1] and Mordechai Segev[2]
[1]*Physics Department, MIT, Cambridge, Massachusetts 02139*
[2]*Physics Department, Technion-Israel Institute of Technology, Haifa 32000, Israel*
(Received 14 March 2000; revised manuscript received 8 September 2000)

We present self-trapped necklace-ring beams that carry and conserve angular momentum. Such beams can have a fractional ratio of angular momentum to energy, and they exhibit a series of phenomena typically associated with rotation of rigid bodies and centrifugal force effects.

DOI: 10.1103/PhysRevLett.86.420 PACS numbers: 42.65.Tg, 05.45.Yv

Many nonlinear-wave systems can be described by the cubic nonlinear Schrödinger equation (NLSE). Solitons in the (1 + 1)D self-focusing version of this equation are stable, displaying interesting physics and applications, yet solitons of (2 + 1)D NLSE are highly unstable [1]. Recent papers [2,3] have proposed self-trapped (2 + 1)D beams that propagate in a stable fashion in a self-focusing Kerr medium: the necklace-ring beams. Necklace beams are shaped like rings whose thickness w is much smaller than their radius R and whose intensity is azimuthally periodically modulated (Fig. 1). Such beams exhibit stable self-trapped propagation for many (>50) physical diffraction lengths [2,3], even though circular (2 + 1)D solitons are inherently unstable in self-focusing Kerr media [1]. In necklace beams, it is the interaction between the spots that stabilizes the structure as a whole. As shown in [3], an isolated individual spot is highly unstable. Furthermore, removing a single spot from the necklace renders the entire necklace unstable. Necklace beams can be thought of as a superposition of two rings carrying equal but opposite topological charge. For such a superposition to exhibit stable propagation, the thickness of the ring must be significantly smaller than its radius, and the thickness must be larger than the azimuthal period. Moreover, the ring has to propagate as one entity, or else the spots walk off each other. In contrast to necklace beams, a single charge-carrying (ring) beam is highly unstable in any self-focusing medium: In a Kerr material it disintegrates, while in saturable nonlinear media it breaks into a number of solitons that can interact with one another [4] or fly off like free particles [5]. This behavior occurs also in quadratic media [6]. Yet necklace beams stay intact and display stable propagation, in Kerr as well as in saturable self-focusing media, if their parameters are chosen properly [2,3]. The stability of self-trapped necklaces (with properly chosen parameters) is unique in soliton science: A superposition of bound solutions (the rings with equal but opposite charge that make up the necklace) is stable, but its individual constituents are *unstable* [7]. Experiments with self-trapped necklaces have already been reported [8]. Here, we present self-trapped necklace beams that carry angular momentum. It is a rare case of self-trapped scalar bright beams that carry angular momentum [9]. In contrast to all known solitons, the angular momentum borne on such necklaces can be a noninteger multiple of the energy. We demonstrate angular momentum and centrifugal force effects.

Solitons that carry angular momentum have been studied in many systems: vortex solitons [10], 3D spiraling of solitons [11], composite solitons [12], quadratic solitons [13,14], and ring solitons in Kerr [15] and cubic-quintic [9] media. Experiments involving transfer of angular momentum carried by light to other forms of angular momentum have been performed [16]. The angular momentum of optical beams is typically associated with azimuthal phase modulation of $\exp(iM\theta)$ [17], which provides angular velocity to every part of the beam with respect to the beam center. The field of solitons is continuous wherever the amplitude is nonzero, that is, everywhere except for the origin. This is because a field discontinuity where the amplitude is nonzero renders the soliton highly unstable, even in a self-defocusing medium. For this reason of stability, for all vortex solitons [having $\exp(iM\theta)$], M is an integer [14]. For the same reason, for all other forms of single solitons carrying angular momentum, Mu is an integer. In the (2 + 1)D cubic self-focusing NLSE,

$$i\frac{\partial\psi}{\partial z}+\frac{1}{2}\left\{\frac{\partial^2\psi}{\partial r^2}+\frac{1}{r}\frac{\partial\psi}{\partial r}+\frac{1}{r^2}\frac{\partial^2\psi}{\partial\theta^2}\right\}+|\psi|^2\psi=0, \quad (1)$$

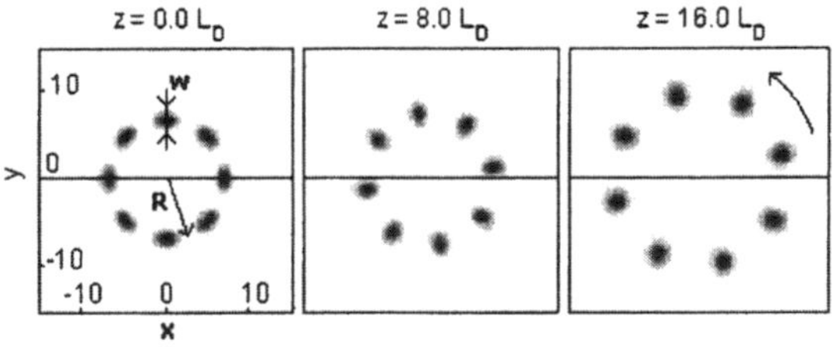

FIG. 1. A rotating necklace with integer L/E. We launch a necklace close to the self-trapped shape. It "breathes" for a short distance, until it reaches the equilibrium shape. This necklace slowly rotates as it propagates. Every necklace slowly expands as it propagates; nevertheless, as seen here, this does not stop the rotation. The input shape is $\sim\psi(r,\theta,z=0)=\mathrm{sech}(r-6.83)\cos(4\theta)\exp(i\theta)$. Dark means high intensity.

the energy is $E = \iint |\psi|^2\,dx\,dy$ and the angular momentum is $L\hat{\mathbf{z}} = \frac{i}{2}\iint \mathbf{r} \times \{\psi\nabla\psi^* - \psi^*\nabla\psi\}\,dx\,dy$. Generically, any beam that can be written as $\psi(r,\theta) = f(r)\exp(iM\theta)$ has $L/E = M$. Since definitions of neither L or E depend on the nonlinearity, this holds for all nonlinearities. Thus, all optical solitons found so far carry an integer L/E. The angular momentum carried by such a beam, when averaged over the number of photons, is exactly $M\hbar$ per photon [17]. If the light is circularly polarized, the total angular momentum is modified by the spin contribution of $\pm\hbar$ per photon.

The fact that L/E is an integer can be intuitively understood by comparison with quantum mechanics (QM). All (paraxial) optical solitons are described by a Schrödinger-type equation. In QM, the solutions of this equation have a quantized angular momentum which is an integer multiple of $\hbar$, and the total probability $\iint |\psi|^2\,dx\,dy$ is normalized to 1. In classical optics $\iint |\psi|^2\,dx\,dy$ is the total power, which is proportional to the average number of photons. The quantization of L in QM resembles the fact that L/E is an integer for all *solitons* found thus far of a *classical* (2 + 1)D normalized NLSE. But L/E is an integer for solitons not due to quantization reasons, but because a noninteger L/E typically leads to a field discontinuity where the intensity is nonzero; such a discontinuity is thought to be unstable in self-focusing/defocusing media. Here we find self-trapped structures carrying noninteger L/E yet stably propagating for many diffraction lengths: necklace-ring quasisolitons that carry noninteger per-photon angular momentum.

Consider a necklace beam whose input shape is approximately $\psi(r,\theta,z=0) = f(r)\cos(\Omega\theta)$ [18], and add to it angular momentum by multiplying it by $\exp(iM\theta)$ (Ω, M integers). As long as M is reasonably smaller than Ω, we find (numerically) that this necklace is stable for more than 50 diffraction lengths L_D [19]. After 50 L_D we reach our computational limits, but it is plausible that the necklaces are stable for much larger distances. Such shapes have $L/E = M$. Since the symmetry between the spots must be preserved in a stable propagation, the angular momentum is manifested by the rotation of the entire necklace as it propagates (Fig. 1). We find numerically that quantization of L/E means that, for a necklace whose parameters are all fixed (except for its M), only certain angular velocities ω are allowed; these ω's are given by $\approx M/R^2$. Two necklaces that differ only in their radii (have the same M) differ in their ω's by a squared ratio of their radii. The fact that the allowed L/E's are quantized shows a connection between solitons and bound states in QM. Both of these systems are described by very similar wave equations and display several similar properties. The fact that some wave quantities that relate solitons and particles are necessarily quantized in optics was not appreciated so far.

Self-trapped necklace beams slowly expand as they propagate [3,4]. The expansion is a consequence of the net radial force exerted on each spot in the necklace. However, even though the necklace slowly expands, the dynamics is very different (and much slower) than diffractive dynamics: It is uniform and it preserves the shape of the necklace. Once angular momentum is added to a necklace beam, it expands faster. The expansion is still highly dominated by the internal dynamics of the necklace. Nevertheless, for two necklaces that differ only in their M's, the one with larger M expands noticeably faster. This implies that what we observe is actually a centrifugal force in a solitonic system. Furthermore, as a necklace beam expands, its L and E are conserved, implying that ω (the angular velocity) cannot be conserved. This is similar to a skater on ice: If she extends her hands while rotating, her ω decreases. We observe this with necklace beams (Fig. 2). Analytically, the angular phase has to be conserved; otherwise, because ω is quantized, the phase would discontinuously jump from, say, $\exp(i\theta)$ to $\exp(i2\theta)$, which is not physical in the continuous evolution describing the necklace propagation. Thus, as the necklace expands, ωR^2 is conserved. Our numerics confirm this prediction. One can develop a moment of inertia formulation for this system. The moment of inertia for necklaces is $I \approx ER^2$. Since $L = I\omega$ and E and L are conserved, ω has to go down with R^2. This is the first prediction of a "skater on ice" effect, which is so obvious in Newtonian mechanics but is unobserved yet in solitonic systems: the slowing down of angular velocity due to conservation of energy and angular momentum.

Given that, in analogy to QM, only integer L/E values are allowed, we recall that there are objects that carry angular momentum (in the form of spin) in multiples of $\hbar/2$ also. However, such spin is an internal degree of freedom and cannot be reproduced as a manifestation of a spatial property of a wave function. Nevertheless, even in QM, the expectation value of angular momentum can be a noninteger multiple of $\hbar$. We build on this idea to construct stable self-trapped beams that carry noninteger L/E. The necklaces described above have $\psi(r,\theta,z=0) = f(r)\{\exp[i(\Omega + M)\theta] + \exp[-i(\Omega - M)\theta]\}/2$. To create a necklace

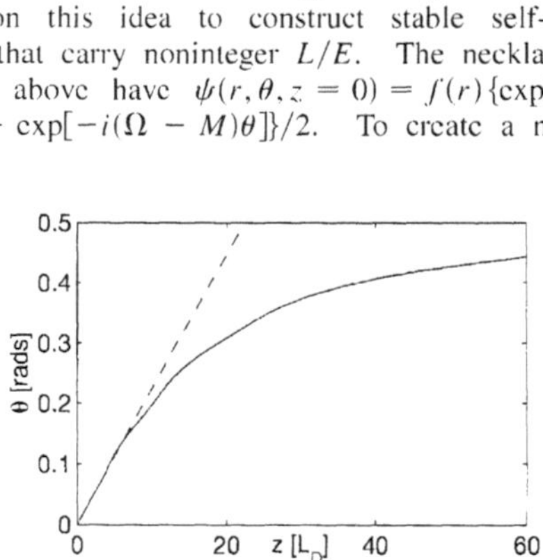

FIG. 2. The rotation angle of the expanding necklace of Fig. 1, as a function of the propagation distance. This particular necklace expands significantly as it propagates. The solid line represents the true instantaneous angle of rotation (measured numerically), whereas the dashed line represents what the instantaneous angle of rotation would have been if the angular velocity were a conserved quantity.

carrying noninteger L/E, we launch $\psi(r,\theta,z=0) = f(r)\{\exp[i(\Omega+M)\theta] + \exp[-i(\Omega)\theta]\}/2$, as in Fig. 3. Such a necklace has $L/E = M/2$. For an odd M, this necklace has a noninteger L/E. Its intensity is given by $f^2(r)\{1+\cos[(M+2\Omega)\theta]\}/2$. In contrast to necklaces that have an even number of spots [3,4], a necklace that carries noninteger L/E has an odd number of spots. Furthermore, in a necklace with an integer (or zero) L/E, adjacent spots are mutually π out of phase (this is why such necklaces expand). This is not the case here since there is an odd number of spots. In order to preserve symmetry between the spots, the angular momentum is manifested in rotation of the necklace, and $\omega = M/(2R^2)$; thus ω is twice slower than for the corresponding necklaces of the previous paragraphs, keeping M and other parameters fixed. Our numerics confirm this prediction, and these necklaces are as stable as the usual necklaces: for many tens of L_D's. We investigate the stability of these necklaces using the same methods as in [2,3].

Another surprising feature is that, although the probability to find photons is not azimuthally symmetric (hence, the nonuniform azimuthal intensity), the local expectation value of L/E is azimuthally symmetric [20]. That is, we calculate analytically the ratio $L(\theta)/E(\theta)$ at $z=0$ and find this ratio to be independent of θ, both in the case of the necklace with an even number of pearls (when it equals M) and in the case of a necklace with an odd number of pearls (when it is $M/2$). Because these necklaces rotate as rigid bodies we expect the $L(\theta)/E(\theta)$ not to change significantly during propagation. Thus, in a necklace with an odd number of spots, each photon contributes exactly $M\hbar/2$ the expectation value of the total angular momentum. One might think that a noninteger per-photon angular momentum is because different regions of the beam have different ratios of $L(\theta)/E(\theta)$, but this is not the case; since the shape of each spot is fixed as the necklace propagates, each part of the beam has the same angular velocity with respect to the center of the necklace.

Next, we construct a necklace carrying an *arbitrary* real per-photon angular momentum. Consider a necklace with $\psi(r,\theta,z=0) = f(r)\{a\exp(iM\theta) + b\exp(iN\theta) + c\exp(-iP\theta) + d\exp(-iQ\theta)\}$; it has $L/E = (a^2M + b^2N - c^2P - d^2Q)/(a^2+b^2+c^2+d^2)$, which can take *any real value*. Not all such necklaces are stable, but one can construct necklaces that are stable for many L_D's. Setting $N = P$, M, and Q to have similar values as N and P, b to be similar to c, and $a, d \ll b, c$, the necklace looks like a "usual" necklace, but with its envelope slightly azimuthally modulated. Small azimuthal perturbations do not destabilize necklaces: we find many such necklaces that are stable for more than $20L_D$, which is plenty for experimental observations. In Fig. 4, we show a necklace that has $d=0$, $N=P=8$, $M=15$, $a=1$, $b=7$, and $c=8$. Therefore, $L/E = -35/38$ for this necklace. As shown in Fig. 5, this necklace indeed has a shape similar to a usual necklace, but with a small azimuthal perturbation. It is interesting to note that in these necklaces the angular momentum is not manifested just in rotation of the necklace, but also in circulation of the modulation of the azimuthal envelope (Fig. 4): Neighboring spots exchange energy and perform a circulation of energy around the necklace, and this is the primary means of transporting the angular momentum upon the propagating beam. The reason for this distinctly different behavior of this necklace from the necklaces with integer or $M/2$ L/E values is symmetry. For a necklace with integer or $M/2$ L/E, the symmetry between the spots is conserved. Thus, if a symmetric necklace is to stay stable, the only way the angular momentum can be manifested is the rotation of the necklace as a whole (Figs. 1 and 3). In contrast, for a necklace described in this paragraph, the symmetry between spots is broken. Thus, spots are allowed to exchange energy and thereby carry angular momentum without a significant rotation of the necklace. Indeed, the frame of the necklace appears stationary (Fig. 4), yet the spots circulate the energy in a preferential direction corresponding to the sign and value of L/E.

The necklace of Figs. 4 and 5 has $L(\theta)/E(\theta)$ which depends on θ. Since in this necklace the spots are not

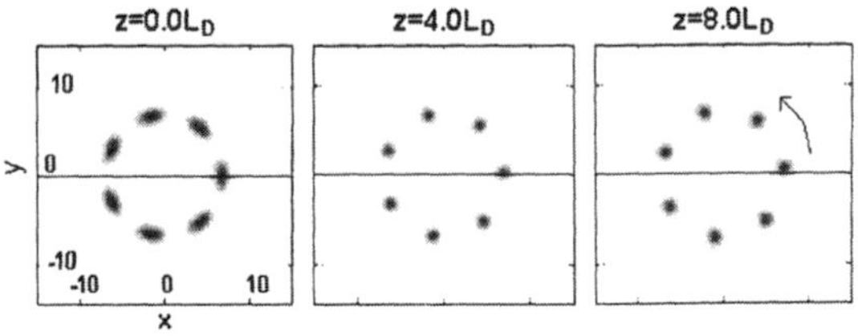

FIG. 3. A necklace with $L/E = 1/2$. This necklace slowly rotates as it propagates. The input shape is approximately $\psi(r,\theta,z=0) = \text{sech}(r-6.83)\{\exp(i4\theta) + \exp(-i3\theta)\}/2$.

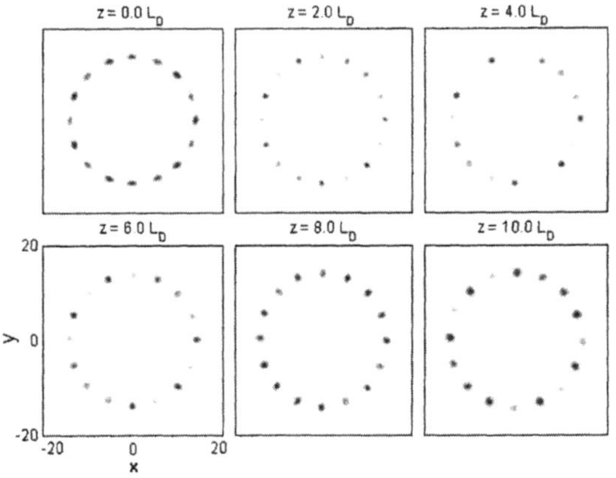

FIG. 4. A necklace with $L/E = -35/38$. This necklace is stable for $8L_D$. Necklaces with better stability are such that the energy exchange between the spots is slow, so it is not visible in a gray-level figure: for example, a necklace with $L/E = 261/1634$ that is stable for at least $50L_D$.

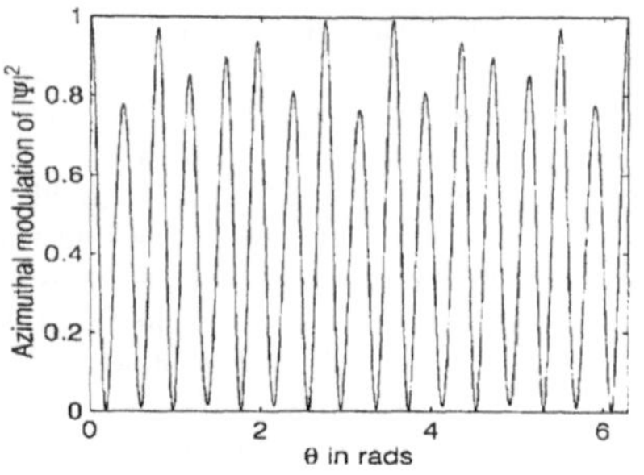

FIG. 5. Azimuthal intensity profile of the necklace of Fig. 4.

"rigid" as with the necklaces from Figs. 1 and 3, different parts of the necklace can have different angular velocities. Thus, the noninteger L/E of these beams does not imply that the angular momentum per photon is a noninteger multiple of $\hbar$. This is in contrast with the $M/2$ case, where the angular momentum per photon is $\hbar M/2$ everywhere.

Note that all necklaces described here are indeed self-trapped. If we start with a shape that is close to the equilibrium necklace shape, as long as this necklace is stable, both L and E are conserved: the L and E carried away by radiation are negligible compared to the initial L and E values. The self-trapped necklace conserves its initial L/E to even much better accuracy, although a tiny fraction of L and E is carried away through radiation.

To the best of our knowledge the necklaces described here are the only self-trapped shapes that have a non-integer per-photon angular momentum (in units of $\hbar$). Necklace beams can be constructed in many nonlinear wave equations. For example, one might think about converting the fractional angular momentum per particle carried by a necklace with $L/E = M/2$ into the angular momentum carried by the spin. This will imply rotation of the polarization state in optics, or spin-orbit interaction in a coherent system, such as a Bose-Einstein condensate. Such a conversion should be even more interesting when the necklaces are made of few photons only [21] (as opposed to a macroscopic number of photons [22]). Another exciting possibility is to investigate atomic necklaces in Bose-Einstein condensates.

We acknowledge enlightening discussions with Professor Meir Orenstein of Technion, Israel. This work was supported by the MURI Project on Optical Spatial Solitons.

---

[1] Circular beams in Kerr media undergo catastrophic collapse above a specific power. See N. N. Akhmediev, Opt. Quantum Electron. **30**, 535 (1998). In reality, when the beam becomes narrow enough, it is no longer represented by a scalar equation. It has been argued that the vectorial nature of the propagation arrests the collapse. Here we discuss only cases for which the physics is contained in the scalar (2 + 1)D cubic self-focusing NLSE.

[2] M. Soljačić, S. Sears, and M. Segev, Phys. Rev. Lett. **81**, 4851 (1998).

[3] M. Soljačić and M. Segev, Phys. Rev. E **62**, 2810 (2000).

[4] V. Tikhonenko, J. Christou, and B. Luther-Davies, Phys. Rev. Lett. **76**, 2698 (1996).

[5] W. J. Firth and D. V. Skryabin, Phys. Rev. Lett. **79**, 2450 (1997). There, the breakup (soliton) products fly off tangentially and do not interact with one another, as opposed to our case, which is in Kerr media where individual solitons are unstable. It is only through the mutual interaction of the spots (which must be present during the propagation) that our necklace is stable [3].

[6] L. Torner and D. V. Petrov, Electron. Lett. **33**, 608 (1997); D. V. Petrov *et al.*, Opt. Lett. **23**, 1444 (1998).

[7] The opposite case, where a superposition of stable solitons is also stable, is a key feature of (1 + 1)D Kerr solitons.

[8] A. Barthelemy, C. Froehly, and M. Shalaby, Proc. SPIE Int. Soc. Opt. Eng. **2041**, 104 (1993).

[9] M. Quiroga-Teixeiro and H. Michinel, J. Opt. Soc. Am. B **14**, 2004 (1997) describe spinning solitons in a qubic-quintic nonlinearity. Recently, D. Mihalache *et al.* have shown [Phys. Rev. E **61**, 7142 (2000)] that these solitons are stable only for a finite distance. Apart from the necklaces, the cubic-quintic solitons are the only stable scalar self-trapped bright beam carrying angular momentum.

[10] G. A. Swartzlander and C. T. Law, Phys. Rev. Lett. **69**, 2503 (1992); B. Luther-Davies *et al.*, Opt. Lett. **19**, 1816 (1994); Z. Chen *et al.*, Phys. Rev. Lett. **78**, 2948 (1997).

[11] M. Shih, M. Segev, and G. Salamo, Phys. Rev. Lett. **78**, 2551 (1997).

[12] Z. Musslimani *et al.*, Phys. Rev. Lett. **84**, 1164 (2000).

[13] J. P. Torres *et al.*, Opt. Commun. **149**, 77 (1998); G. Molina-Terriza *et al.*, Opt. Commun. **158**, 170 (1998).

[14] In self-defocusing Kerr media the only stable vortex solitons are with $M = \pm 1$. Multiply charged vortex solitons are unstable, although the instability is suppressed by saturation [A. Dreischuh *et al.*, Phys. Rev. E **60**, 6111 (1999)].

[15] V. V. Afanasjev, Phys. Rev. E **52**, 3153 (1995).

[16] R. A. Beth, Phys. Rev. **50**, 115 (1936); T. W. Hansch and A. L. Schawlow, Opt. Commun. **13**, 68 (1975); A. Ashkin, Science **210**, 1081 (1980); H. He *et al.*, Phys. Rev. Lett. **75**, 826 (1995).

[17] L. Allen *et al.*, Phys. Rev. A **45**, 8185 (1992) showed the angular momentum per photon in a Laguerre-Gauss beam is an integer multiple of $\hbar$. If one takes $u(r,\theta) = f(r)\exp(iM\theta)$ with an arbitrary $f(r)$ and use Eq. (6) in Allen's paper, it follows that the ratio of angular to linear momentum is $M\lambda/2\pi$. Since the linear momentum per photon is h/λ, then the angular momentum per photon is $M\hbar$. This breaks down if paraxiality is not valid.

[18] $f(r)$ must go at least as fast as r^2 close to the origin, so no singularities appear when this ψ is substituted into the NLSE; this issue is explained in [3,4].

[19] The diffraction length L_D equals 1 normalized unit. If the nonlinearity is turned "off" in Eq. (1), then a beam of $\psi(x, z = 0) = \exp(-x^2/2)$ expands by $\sqrt{2}$ within $1L_D$.

[20] The expectation value of the angular momentum of each photon is the same, anywhere on the necklace.

[21] S. E. Harris and L. V. Hau, Phys. Rev. Lett. **82**, 4611 (1999).

[22] J. E. Avron *et al.*, Eur. J. Phys. **20**, 153 (1999).

Section 8

ENTANGLEMENT OF ANGULAR MOMENTUM

Entanglement in quantum theory is a consequence of the superposition principle for probability amplitudes. The term relates to a pair of quantum systems whose state cannot be expressed as the product of the single state of each system. The simplest example is the singlet, or zero angular momentum, state of a pair of spin-half particles. This has the form $(1/\sqrt{2})(|+\rangle_1|-\rangle_2 - |-\rangle_1|+\rangle_2)$ where $|+\rangle_i$ is the spin up, and $|-\rangle_i$ the spin down, state for particle i.

Entangled states have remarkable properties. A measurement on one of the entangled particles appears to modify instantaneously the state of its partner, an effect which occurs irrespective of the distance between the particles. The apparent conflict between this result and the predictions of relativity is perhaps the most startling feature of modern quantum theory (8.1).

The bizarre consequences of entanglement were exposed to experimental test through the theoretical contributions of Bell (8.2). He showed that the combination of locality and reality, that is the absence of superluminal influences and the fact that properties exist even if we choose not to measure them, lead to an experimentally testable inequality. Quantum mechanics predicts that this inequality can be violated and so there is a conflict between quantum theory and "local realism". Aspect and co-workers demonstrated a violation of Bell's inequality in experiments on the pairs of photons emitted in a $J = 0$ to $J = 1$ to $J = 0$ radiative cascade in calcium (8.3). The polarisations of the two emitted photons are entangled, with a state in the form $(1/\sqrt{2})(|R\rangle_1|R\rangle_2 + |L\rangle_1|L\rangle_2)$. This is a state of entangled spin angular momentum. Because circular polarisation is assigned relative to the direction of propagation, this state of two counter-propagating photons includes two left-handed photons and two right-handed and is thus a state of zero total angular momentum. The evidence of these and subsequent experiments is strongly supportive of quantum mechanics and provides convincing evidence of the existence of entangled states. "Non-local" phenomena associated with entangled polarisation states are an active field of study and play an important role in the emerging field of quantum information (8.4).

Entanglement of orbital angular momentum has recently been demonstrated in spontaneous parametric down-conversion (8.5, **Paper 8.1**). This experiment demonstrated both the conservation of orbital angular momentum in individual down-conversion events at the single-photon level and that the signal and idler photon pair are formed in an entangled superposition of orbital angular momentum states. The form of this entangled state can be derived from the phase matching conditions, or from linear momentum conservation, for the signal and idler photons (8.6, **Paper 8.2**). A simplified version of this derivation will serve to illustrate the idea. Consider a plane wave pump field propagating in the z-direction. Momentum conservation in the transverse (x–y) plane means that the dependence of the wavefunction for the signal (s) and idler (i) photons must be of the form $\delta(k_{s.x} + k_{i.x})\delta(k_{s.y} + k_{i.y})$. A Fourier transform allows this to be written in the position representation as:

$$\delta(x_s - x_i)\delta(y_s - y_i) = \frac{1}{r_s}\delta(r_s - r_i)\delta(\phi_s - \phi_i) = \frac{1}{2\pi r_s}\delta(r_s - r_i)\sum_{l=-\infty}^{\infty}\exp(il\phi_s)\exp(-il\phi_i),$$

where r and ϕ are the radial and azimuthal coordinates of the position of emission of the photons. The exponential terms are eigenstates of orbital angular momentum and the state, derived on the basis of phase matching, is an entangled state comprising a superposition of states in which the signal and idler orbital angular momenta sum to zero. Naturally, the experimental system is more complicated than this, but the principle that orbital angular momentum entanglement arises from phase matching remains valid. A more complicated picture may occur when extending this idea to more complex crystal geometries (8.7, 8.8, 8.9) beyond the paraxial approximation.

Entanglement is a precious resource for quantum information applications. For this reason it is desirable to have an efficient method for measuring entangled states. A multi-channel analyser capable of determining the orbital angular momentum, l, for a single photon has been built and is described in (8.10, **Paper 8.3**).

While spin angular momentum and polarised light can be characterised by two orthogonal states, orbital angular momentum is higher dimensional (8.11, **Paper 2.4**); multi-dimensional entanglement of orbital angular momentum states might well find interesting applications in quantum information and cryptography.

REFERENCES

8.1 A Einstein, B Podolsky and N. Rosen, 1935, *Phys. Rev.* **47** 777.

8.2 Bell's papers are collected and reprinted in JS Bell, 1987, *Speakable and unspeakable in quantum mechanics* (Cambridge: Cambridge University Press).

8.3 A Aspect, P Grangier and G Roger, 1981, *Phys. Rev. Lett.* **47** 460; 1982 *ibid.* **49** 91.

8.4 MA Nielsen and IL Chuang, 1999, *Quantum Computation and Quantum Information* (Cambridge: Cambridge University Press).

8.5 A Mair, A Vaziri, G Weihs and A Zeilinger, 2001, *Nature* **412** 313.

8.6 S Franke-Arnold, SM Barnett, MJ Padgett and L Allen, 2002, *Phys. Rev. A* **65** 033823.

8.7 HH Arnaut and GA Barbosa, 2000, *Phys. Rev. Lett.* **85** 286.

8.8 ER Eliel, SM Dutra, G Nienhuis and JP Woerdman, 2001, *Phys. Rev. Lett.* **86** 5208.

8.9 HH Arnaut and GA Barbosa, 2001, *Phys. Rev. Lett.* **86** 5209.

8.10 J Leach, MJ Padgett, SM Barnett, S Franke-Arnold and J Courtial, 2002, *Phys. Rev. Lett.* **88** 257901.

8.11 L Allen, J Courtial and MJ Padgett, 1999, *Phys. Rev. E* **60** 7497.

Entanglement of the orbital angular momentum states of photons

Alois Mair*, Alipasha Vaziri, Gregor Weihs & Anton Zeilinger

Institut für Experimentalphysik, Universität Wien, Boltzmanngasse 5, 1090 Wien, Austria

Entangled quantum states are not separable, regardless of the spatial separation of their components. This is a manifestation of an aspect of quantum mechanics known as quantum non-locality[1,2]. An important consequence of this is that the measurement of the state of one particle in a two-particle entangled state defines the state of the second particle instantaneously, whereas neither particle possesses its own well-defined state before the measurement. Experimental realizations of entanglement have hitherto been restricted to two-state quantum systems[3–6], involving, for example, the two orthogonal polarization states of photons. Here we demonstrate entanglement involving the spatial modes of the electromagnetic field carrying orbital angular momentum. As these modes can be used to define an infinitely dimensional discrete Hilbert space, this approach provides a practical route to entanglement that involves many orthogonal quantum states, rather than just two Multi-dimensional entangled states could be of considerable importance in the field of quantum information[7,8], enabling, for example, more efficient use of communication channels in quantum cryptography[9–11].

Multi-dimensional entanglement is a way—in addition to multi-particle entanglement—to extend the usual two-dimensional two-particle state. There have been suggestions[12,13] (and only a proof-of-principle experiment[14]) as to how to realize higher-order entanglement via multiport beam splitters. Here we present an experiment in which we used the spatial modes of the electromagnetic field carrying orbital angular momentum to create multi-dimensional entanglement. The advantage of using these modes to create entanglement is that they can be used to define an infinitely dimensional discrete (because of the quantization of angular momentum) Hilbert space.

The experimental realization proceeded in the following two steps. First, we confirmed that spontaneous parametric down-conversion conserves the orbital angular momentum of photons. This was done for pump beams carrying orbital angular momenta of $-\hbar$, 0 and $+\hbar$ per photon, respectively. Second, we showed that the state of the down-converted photons can not be explained by assuming classical correlation—in the sense that the photon pairs produced are just a mixture of the combinations allowed by conservation of angular momentum. We proved that, in contrast, they are a coherent superposition of these combinations, and hence they have to be considered as entangled in their orbital angular momentum. After completion of the experimental work presented here, related theoretical work was brought to our attention[15,16].

We will now discuss in order the two steps mentioned above. For paraxial light beams, Laguerre–gaussian (LG) modes (Fig. 1) define a possible set of basis vectors. As predicted[17] and observed[18], LG modes carry an orbital angular momentum for linearly polarized light that is distinct from the intrinsic angular momentum of photons associated with their polarizations. This external angular momentum of the photon states is the reason why they have been suggested for gearing micromachines, and it has been shown that they can be used as optical tweezers[19–21].

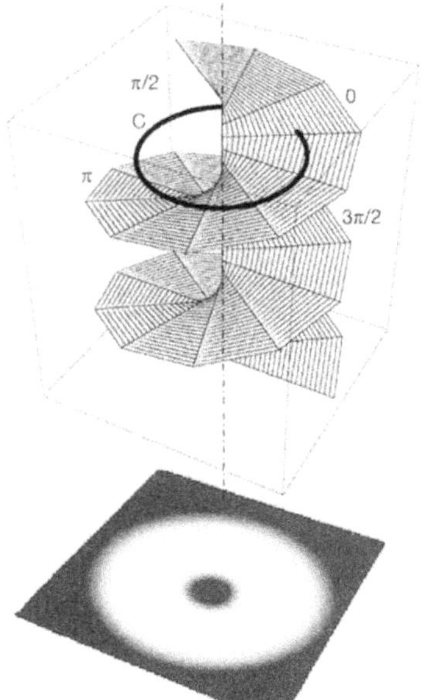

Figure 1 The wave front (top) and the intensity pattern (bottom) of the simplest Laguerre–gaussian (LG_p^l) or 'doughnut' mode. The index l is referred to as the winding number, and $(p+1)$ is the number of radial nodes. Here we only consider cases of $p = 0$. The customary gaussian mode can be viewed as an LG mode with $l = 0$. The handedness of the helical wave fronts of the LG modes is linked to the sign of the index l, and can be chosen by convention. The azimuthal phase term $\exp il\phi$ of the LG modes results in helical wave fronts. The phase variation along a closed path C around the beam centre is $2\pi l$. Therefore, in order to fulfil the wave equation, the intensity has to vanish in the centre of the beam.

* Present address: Harvard-Smithsonian Center for Astrophysics, 60 Garden Street, Cambridge, Massachusetts 02138, USA.

To demonstrate the conservation of the orbital angular momentum carried by the LG modes in spontaneous parametric down-conversion, we investigated three different cases—for pump photons possessing orbital angular momenta of $-\hbar$, 0 and $+\hbar$ per photon, respectively. As a pump beam, we used an argon-ion laser (wavelength 351 nm) which we could operate either with a simple gaussian mode profile ($l = 0$) or in the first-order LG modes ($l = \pm 1$) after astigmatic mode conversion (for a description of this technique, see ref. 22). Spontaneous parametric down-conversion was done in a 1.5-mm-thick BBO (β-barium borate) crystal cut for type-I phase matching (that is, both photons carry the same linear polarization). The crystal cut was chosen so as to produce down-converted photons at a wavelength of 702 nm at an angle of 4° off the pump direction.

The mode detection of the down-converted photons was performed for gaussian and LG modes. The gaussian mode ($l = 0$) was identified using mono-mode fibres (Fig. 2) in connection with avalanche detectors. All other modes have a larger spatial extension, and therefore cannot be coupled into the mono-mode fibre. The LG modes ($l \neq 0$) were identified using mode detectors consisting of computer-generated holograms and mono-mode optical fibres (Fig. 2).

Computer-generated holograms have often been exploited for creating LG modes of various orders[23]. Our holograms were phase gratings 5 mm × 5 mm in size, with 20 lines mm^{-1}, which we first recorded on holographic films and bleached afterwards to increase the transmission efficiency (Fig. 2). We made holograms that had one or two dislocations in the centre, and designed them to have their maximum intensity in the first diffraction order. This enabled us to distinguish between LG modes $l = -2, -1, 0, 1, 2$ using all holograms in the first diffraction order—for which they have been blazed. For analysing an LG mode with a negative index, the holograms were rotated by 180° around the axis perpendicular to the grating lines. The total transmission efficiency of all our holograms was about 80%, and they diffracted 18% of the incoming beam into the desired first order. These characteristics were measured at 632 nm wavelength because a laser source at 702 nm was not available.

The diffraction efficiency is not the only loss that occurs. We also have to account for Fresnel losses at all optical surfaces (95% transmission), imperfect coupling into the optical fibres (70% for a gaussian beam), non-ideal interference filters (75% centre transmission), and the efficiency of the detectors (30%). A conservative estimate of all the losses yields an overall collection efficiency of 2–3%. Comparing the unnormalized ($l_{pump} = l_1 = l_2 = 0$) coincidence rates of about 2,000 s^{-1} to the single count rates of about 100,000 s^{-1} we deduce an efficiency of 2%, in agreement with the above estimation.

The mode analysis was performed in coincidence for all cases where mode filter 1 was prepared for analysing LG modes $l_1 = 0, 1, 2$ and mode filter 2 for those with $l_2 = -2, -1, 0, 1, 2$. For analysing an LG mode with mode index $l = 0$—that is, a gaussian mode—the dislocation of the hologram was shifted out of the beam path. Thus the beam was sent through the border of the hologram where it acts as a conventional grating without changing the photons' angular momentum. The results are shown in Fig. 3 for different values of orbital angular momenta of the pump beam. Within experimental accuracy, coincidences were only observed in those cases where the sum of the orbital angular momenta of the down-converted photons was equal to the pump beam's orbital angular momentum. However, the absolute count rates of these cases are not equal. This is probably due to unequal emission probabilities of the photons into the different modes in the down-conversion process.

These results confirm conservation of the orbital angular momentum in parametric down-conversion. The signal-to-noise ratios achieved were as high as $V = 0.976 \pm 0.038$ and $V = 0.916 \pm 0.009$ for pump beams with and without orbital angular momentum, respectively. V is defined as $(I_{out} - I_{in})/(I_{out} + I_{in})$, where I_{in} and I_{out} denote the maximum and the minimum of the coincidences with the dislocation of the hologram respectively in and out of the beam.

It is only by using a coincidence measurement that we could show that the conservation of the orbital angular momentum holds for each single photon pair. In contrast, cumulative detection methods using many photons result in an incoherent pattern[24], because

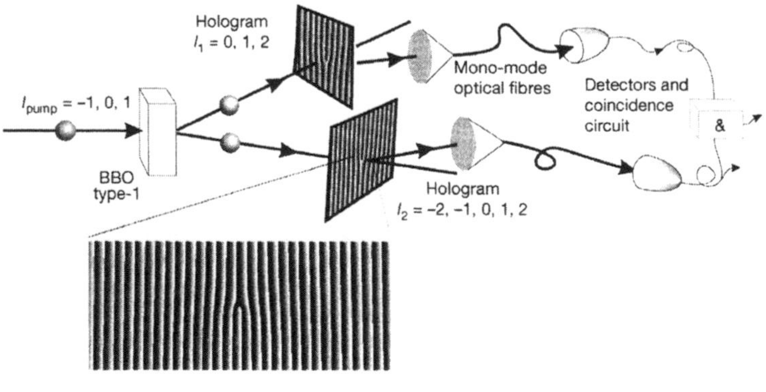

Figure 2 Experimental set-up for single-photon mode detection. After parametric down-conversion, each of the photons enters a mode detector consisting of a computer-generated hologram and a mono-mode optical fibre. By diffraction at the hologram, the incoming mode undergoes a mode transformation in such a way that an LG mode can be transformed into a gaussian mode. As it has a smaller spatial extension than all LG modes, only the gaussian mode can be coupled into the mono-mode fibre. Thus observation of a click projects the mode incident on the fibre coupler into the gaussian mode. The hologram is a phase grating with Δm dislocations in the centre blazed for first-order diffraction. An incoming gaussian laser beam passing through the dislocation of the hologram is diffracted by the grating, and the nth diffraction order becomes an LG mode with an index $l = n\Delta m$ and vice versa. Intuitively speaking, the phase dislocation exerts a 'torque' onto the diffracted beam because of the difference of the local grating vectors in the upper and lower parts of the grating. This 'torque' depends on the diffraction order n and on Δm. Consequently the right and left diffraction orders gain different handedness. Reversing this process, a photon with angular momentum $\Delta m\hbar$ before the grating can be detected by the mono-mode fibre detector placed in the first diffraction order. A photon with zero angular momentum (gaussian mode) is detected by diffracting the beam at the border of the hologram far away from the dislocation. All our measurements were performed in coincidence detection between the two down-converted photons.

each beam from parametric down-conversion by itself is an incoherent mixture. Therefore some previous workers[24] using these classical detection methods—which are in principle unsuitable at the single photon level—were led to believe that the orbital angular momentum is not conserved in spontaneous parametric down-conversion.

Given this experimental verification of the conservation of orbital angular momentum, entanglement between the two photons produced in the conversion process might be expected. But to explain the conservation of the orbital angular momentum, the photons do not necessarily have to be entangled: it would be sufficient to assume classical correlation. But further experimental results (see below) showed that the two-photon state goes beyond classical correlation, and indeed, we were able to prove the entanglement for photon states with phase singularities.

To confirm entanglement, we have to demonstrate that the two-photon state is not just a mixture but a coherent superposition of product states of the various gaussian and LG modes which obey angular momentum conservation. For simplicity, we restricted ourselves to superpositions of two basis states only. An important distinction between coherent superposition and incoherent mixture of gaussian and LG modes is that the latter possess no phase singularity. This is because adding the spatial intensity distributions of these two modes will yield a finite intensity everywhere in the resulting pattern. In contrast, in a coherent superposition the amplitudes are added, and therefore the phase singularity must remain and is displaced to an eccentric location (Fig. 4). It will appear at that location where the amplitudes of the two modes are equal, with opposite phase. Therefore the radial distance of the singularity from the beam centre is a measure of the amplitude ratio of the gaussian to the LG components, whereas the angular position of the singularity is determined by their relative phase. Intuitively speaking, the position of the dislocation with respect to the beam is equivalent to the orientation of a polarizer.

Superpositions of LG and gaussian modes can be realized experimentally by shifting the dislocation of the hologram out of the centre of the beam by a certain (small) amount. Hence in order to detect a photon having an orbital angular momentum that is a superposition of the gaussian and the LG mode, the hologram was placed in a position such that the dislocation was slightly displaced from the beam centre. In the intensity pattern these modes possess an eccentric singularity (Fig. 4). To demonstrate the entanglement, we therefore shifted one of the holograms and scanned the gaussian mode filter on the other side while recording the coincidences.

The results shown in Fig. 4 verify the correlation in superposition bases of the LG ($l = \pm 2$) and gaussian ($l = 0$) modes. A closer analysis shows that there are two conditions necessary to obtain the measured curves. First, the shifted hologram has to work as described above, and second, the source must emit an angular-momentum-entangled state. Assume that the source only emits classically correlated but not entangled singularities. Then on the side with the shifted hologram, the various terms of the classical mixture would be projected onto a state with displaced singularity leaving the total state again in a mixture. Respecting the conservation of angular momentum we would then have to sum the probabilities of the various components on the other side, resulting

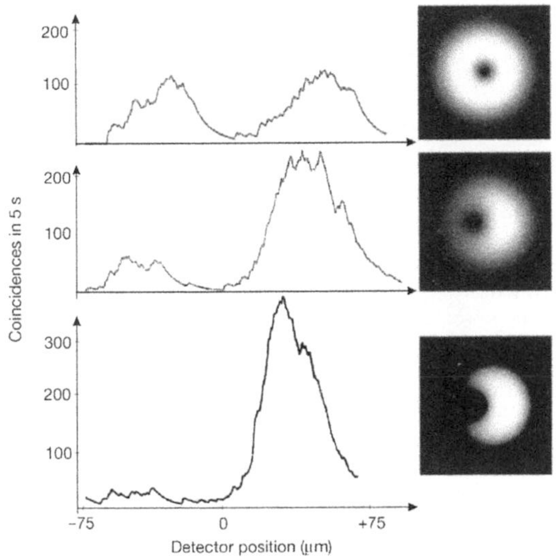

Figure 4 Experimental evidence (left; right, simulation) of entanglement of photon states with phase singularities. The dislocation of the hologram in the beam of photon 1 is shifted out of the beam centre step by step (top, middle, bottom). In these positions, this hologram—together with the mono-mode fibre detector—projects the state of photon 1 into a coherent superposition of LG and gaussian modes. The mode filter for photon 2 with the hologram taken out makes a scan of the second photon's intensity distribution (detector position) in order to identify the location of its singularity with respect to the beam centre. The coincidences show that the second photon is also detected in a superposition of LG and gaussian modes. Classical correlation would yield a coincidence picture which is just a mixture of gaussian and LG modes. In that case, the intensity minimum would remain in the beam centre but would become washed out. In the experiment a hologram with two dislocations in the first diffraction order was used.

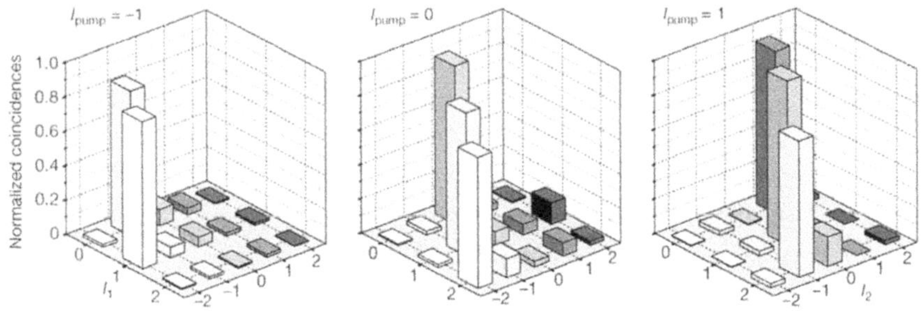

Figure 3 Conservation of orbital angular momentum. Coincidence mode detections for photon 1 and photon 2 in 15 possible combinations of orthogonal states were performed. This was done for a pump beam having an orbital angular momentum of $-\hbar$, 0 and $+\hbar$ per photon, respectively. Coincidences were observed in all cases where the sum of the orbital angular momenta of the down-converted photons was equal to the pump beam's orbital angular momentum. The coincidence counts for each fixed value of the orbital angular momentum of photon 1 were normalized by the total number of coincidences varying the orbital angular momentum of photon 2.

in a coincidence pattern not containing any intensity zeroes. Such a coincidence pattern would also be observed if a shifted hologram together with a mono-mode detector were not able to analyse for superposition states.

An entangled state represents correctly both the correlation of the eigenmodes and the correlations of their superpositions. Having experimentally confirmed the quantum superposition for $l = 0$ and $l = \pm 2$, it is reasonable to expect the quantum superposition will also occur for the other states. Nevertheless, ultimate confirmation of entanglement will be a Bell inequality experiment generalized to more states[25]. Such an experiment will be a major experimental challenge, and we are preparing to perform it.

For a pump beam with zero angular momentum, the emitted state must then be represented by

$$\psi = C_{0,0}|0\rangle|0\rangle + C_{1,-1}|1\rangle|-1\rangle + C_{-1,1}|-1\rangle|1\rangle + C_{2,-2}|2\rangle|-2\rangle + C_{-2,2}|-2\rangle|2\rangle + \ldots\ldots \tag{1}$$

as the LG modes form an infinite dimensional basis. Here the numbers in the brackets represent the indices l of the LG modes, and the $C_{i,j}$ denote the corresponding probability amplitude for measuring $|i\rangle|j\rangle$. The state (1) is a multi-dimensional entangled state for two photons, which in general will also contain terms with radial mode index $p \neq 0$. It means neither photon in state (1) possesses a well-defined orbital angular momentum after parametric down-conversion. The measurement of one photon defines its orbital angular momentum state, and projects the second one into the corresponding orbital angular momentum state.

It is conceivable that these states could in the future be extended to multi-dimensional multi-particle entanglement. A growing body of theoretical work calls for entanglement of quantum systems of higher dimensions[7,8]. These states have applications in quantum cryptography with higher alphabets and in quantum teleportation. As such states increase the flux of information, it is conceivable that they could be important for many other applications in quantum communication and in quantum information. The possibility of using these photon states to drive micromachines, and the application of these states as optical tweezers, make them versatile and potentially suitable for future technologies[19–21]. □

Received 12 March; accepted 5 June 2001.

1. Schrödinger, E. Die gegenwärtige Situation in der Quantenmechanik. *Naturwissenschaften* **23,** 807–812; 823–828; 844–849 (1935).
2. Schrödinger, E. Discussion of probability relations between separated systems. *Proc. Camb. Phil. Soc.* **31,** 555–563 (1935).
3. Bouwmeester, D., Pan, J.-W., Daniell, M., Weinfurter, H. & Zeilinger, A. Observation of a three-photon Greenberger-Horne-Zeilinger state. *Phys. Rev. Lett.* **82,** 1345–1349 (1999).
4. Pan, J.-W., Bouwmeester, D., Daniell, M., Weinfurter, H. & Zeilinger, A. Experimental test of quantum nonlocality in three-photon Greenberger-Horne-Zeilinger entanglement. *Nature* **403,** 515–519 (2000).
5. Sackett, C. A. *et al.* Experimental entanglement of four particles. *Nature* **404,** 256–259 (2000).
6. Pan, J.-W., Daniell, M., Gasparoni, S., Weihs, G. & Zeilinger, A. Experimental demonstration of four-photon entanglement and high-fidelity teleportation. *Phys. Rev. Lett.* **86,** 4435–4438 (2001).
7. DiVincenzo, D. P., More, T., Shor, P. W., Smolin, J. A. & Terhal, B. M. Unextendible product bases, uncompletable product bases and bound entanglement. Preprint quant-ph/9908070 at ⟨http://xxx.lanl.gov⟩ (1999).
8. Bartlett, S. D., de Guise, H. & Sanders, B. C. Quantum computation with harmonic oscillators. Preprint quant-ph/0011080 at ⟨http://xxx.lanl.gov⟩ (2000).
9. Bechmann-Pasquinucci, H. & Peres, A. Quantum cryptography with 3-state systems. *Phys. Rev. Lett.* **85,** 3313–3316 (2000).
10. Bechmann-Pasquinucci, H. & Tittel, W. Quantum cryptography using larger alphabets. *Phys. Rev. A* **61,** 62308–62313 (2000).
11. Bourennane, M., Karlsson, A. & Björk, G. Quantum key distribution using multilevel encoding. *Phys. Rev. A* (in the press).
12. Reck, M., Zeilinger, A., Bernstein, H. J. & Bertani, P. Experimental realization of any discrete unitary operator. *Phys. Rev. Lett.* **73,** 58–61 (1994).
13. Zukowski, M., Zeilinger, A. & Horne, M. Realizable higher-dimensional two-particle entanglements via multiport beam splitters. *Phys. Rev. A* **55,** 2564–2579 (1997).
14. Reck, M. *Quantum Interferometry with Multiports: Entangled Photons in Optical Fibers.* Thesis, Univ. Innsbruck (1996).
15. Arnaut, H. H. & Barbosa, G. A. Orbital and angular momentum of single photons and entangled pairs of photons generated by parametric down-conversion. *Phys. Rev. Lett.* **85,** 286–289 (2000).
16. Franke-Arnold, S., Barnett, S. M., Padgett, M. J. & Allen, L. Two-photon entanglement of orbital angular momentum states. *Phys. Rev. A* (in the press).
17. Allen, L., Beijersbergen, M. W., Spreeuw, R. J. C. & Woerdman, J. P. Orbital angular momentum of light and the transformation of laguerre-gaussian laser modes. *Phys. Rev. A* **45,** 8185–8189 (1992).
18. He, H., Fries, M., Heckenberg, N. & Rubinsztein-Dunlop, H. Direct observation of transfer of angular momentum to absorbtive particles from a laser beam with a phase singularity. *Phys. Rev. Lett.* **75,** 826–829 (1995).
19. Simpson, N. B., Dholakia, K., Allen, L. & Padgett, M. J. Mechanical equivalence of spin and orbital angular momentum of light: An optical spanner. *Opt. Lett.* **22,** 52–54 (1997).
20. Galaida, P. & Ormos, P. Complex micromachines produced and driven by light. *Appl. Phys. Lett.* **78,** 249–251 (2001).
21. Friese, M. E. J., Enger, J., Rubinsztein-Dunlop, H. & Heckenberg, N. Optical angular-momentum transfer to trapped absorbing particles. *Phys. Rev. A* **54,** 1593–1596 (1996).
22. Bejersbergen, M. W., Allen, L., van der Veen, H. E. L. O. & Woerdman, J. P. Astigmatic laser mode converters and transfer of orbital angular momentum. *Opt. Commun.* **96,** 123–132 (1993).
23. Arlt, J., Dholakia, K., Allen, L. & Padgett, M. J. The production of multiringed laguerre-gaussian modes by computer-generated holograms. *J. Mod. Opt.* **45,** 1231–1237 (1998).
24. Arlt, J., Dholakia, K., Allen, L. & Padgett, M. Parametric down-conversion for light beams possessing orbital angular momentum. *Phys. Rev. A* **59,** 3950–3952 (1999).
25. Kaszlikowski, D., Gnacinski, P., Zukowski, M., Miklaszewski, W. & Zeilinger, A. Violation of local realism by two entangled n-dimensional systems are stronger than for two qubits. *Phys. Rev. Lett.* **85,** 4418–4421 (2000).

Acknowledgements

This work was supported by the Austrian Fonds zur Förderung der wissenschaftlichen Forschung (FWF).

Correspondence and requests for materials should be addressed to A.Z. (e-mail: anton.zeilinger@univie.ac.at).

Two-photon entanglement of orbital angular momentum states

Sonja Franke-Arnold and Stephen M. Barnett
Department of Physics and Applied Physics, University of Strathclyde, Glasgow G4 0NG, Scotland

Miles J. Padgett and L. Allen
Department of Physics and Astronomy, University of Glasgow, Glasgow G12 8QQ, Scotland
(Received 12 January 2001; published 26 February 2002)

We investigate the orbital angular momentum correlation of a photon pair created in a spontaneous parametric down-conversion process. We show how the conservation of the orbital angular momentum in this process results from phase matching in the nonlinear crystal.

DOI: 10.1103/PhysRevA.65.033823 PACS number(s): 42.65.Ky, 42.50.Dv, 03.65.Ta

Entanglement is one of the most puzzling and powerful properties of quantum theory and has received undiminished attention since the earliest days of quantum mechanics. Entangled states play a crucial role in the investigation of the EPR paradox [1] and in the evaluation of Bell's inequality [2], which distinguishes between local and nonlocal formulations of quantum mechanics. However, two-photon entanglement not only poses fundamental questions to our understanding of quantum theory, but it also plays an important role in applications of quantum mechanics including quantum cryptography, teleportation [3,4], and quantum images [5–7].

Parametric down-conversion has proved a reliable tool for the generation of pairs of entangled photons [8,9]. The resulting spatially separated photons, usually named the "signal" and "idler," are entangled in their arrival times at the respective detectors [10] and in their transverse positions [6]. They can also be entangled in their polarization states [9,11]. All of these effects have been studied experimentally as well as theoretically. Here we will concentrate on a further entangled property, the orbital angular momentum of the two generated photons [13,14]. The time entanglement arises from the energy conservation in the down-conversion process, which is expressed by the frequency matching condition. Similarly, the entanglement of the transverse position in the far field arises from the momentum conservation expressed by the phase-matching condition.

In this paper, we will show that the orbital angular momentum entanglement, a transverse property of the beam, follows also directly from the phase-matching condition and is related to the conservation of orbital angular momentum. Our approach differs from a recent theoretical paper [12] that dealt with the possible conservation of the combined spin and orbital angular momentum in relation to the susceptibility of the down-conversion crystal. The nonlinear electric susceptibility of the crystal determines the polarization properties of the down-converted photons. In the case of type I down-conversion, the polarization of the signal and the idler photons will be identical, its direction being determined by the polarization of the pump. In the case of type II down-conversion, the two down-converted photons have orthogonal polarizations. In specific setups, this can result in polarization entanglement in each pair of signal and idler photons. For each photon pair, be it generated in type I or type II parametric down-conversion, phase-matching is fulfilled. Although the nature of the nonlinear susceptibility constrains the polarization of the three interacting waves, it in no way determines the orbital angular momentum of any of the beams. The orbital angular momentum is determined solely by the phase structure of each beam. Our analysis provides the theoretical background to a recent experiment, in which the conservation of the orbital angular momentum in parametric down-conversion has been observed for the first time [13,14].

While the spin angular momentum describes the intrinsic photon spin and corresponds to the optical polarization of light, the orbital angular momentum is associated with the transverse phase front of a light beam. Light with an azimuthal phase dependence $\exp(il\varphi)$ carries a well-defined orbital angular momentum of $l\hbar$ per photon [15]. The associated phase discontinuity produces an intensity null on the beam axis. Such light beams are conveniently described in terms of Laguerre-Gaussian modes, characterized by the mode indices l and p, where $p+1$ gives the number of radial nodes, and $2p+l$ the mode order N.

Laguerre-Gaussian beams can be generated by using holograms which have the form of distorted diffraction gratings with an l-pronged fork dislocation on the beam axis. The first-order diffracted beam then has l intertwined helical wavefronts. Alternatively, such holograms can be used in reverse to detect modes of a particular angular momentum number. In this configuration, if the number of dislocations in the hologram matches the angular momentum number of the incident beam, then the first-order diffracted beam has an on-axis intensity which can be detected. If, however, the number of forks does not match the number of dislocations, then no on-axis intensity results.

The correlation between Laguerre-Gaussian modes was studied in a recent down-conversion experiment [13,14]. For a pump beam with $l_{\text{pump}}=0$, nonvanishing coincidence count rates were measured for the detection of $l_{\text{signal}}=2$ and $l_{\text{idler}}=-2$, whereas no significant coincidence rates were found, for example, for the detection of $l_{\text{signal}}=\pm 2$ and $l_{\text{idler}}=0$. This suggests that the angular momentum is conserved in the down-conversion process, so that $l_{\text{signal}}+l_{\text{idler}}=l_{\text{pump}}=0$.

In the following, we derive the correlation between general transverse modes of the signal and idler photon. We then quantify these correlations for the special case of Laguerre-

Gaussian modes and show how these are related to the conservation of the orbital angular momentum.

Here we are concerned with the transverse-mode correlations between the signal and idler beams. Hence we describe the modes of the pump, signal, and idler by the normalized transverse mode functions $\Phi_{0,1,2}(\mathbf{x})$, where $\mathbf{x}$ is a two-dimensional vector in the plane perpendicular to the propagation direction of the light; we work in the paraxial limit. Quantities and operators concerning the pump, signal, and idler are indicated by the subscripts 0, 1, and 2, respectively. For simplicity, we suppress the explicit beam propagation and assume frequency matching to be fulfilled.

We first determine the state of the light generated by a parametric down-conversion process [16,17]. In the Schrödinger picture, the two photons in the signal and the idler mode are created by applying the operator $\hat{a}^\dagger(\mathbf{k}_1)\hat{a}^\dagger(\mathbf{k}_2)$ to the initial vacuum state $|0\rangle$, where $\mathbf{k}_{1,2}$ is the transverse component of the signal and idler wave vector, respectively. Similarly, $\mathbf{k}_0$ denotes the transverse component of the pump beam. We assume polarization states that are in agreement with the condition imposed by the nonlinear susceptibility of the crystal. For photons in such polarization states, phase-matching between the down-converted photons and the pump is satisfied. It can be described by a sinc function $\Pi_{j=x,y}\mathrm{sinc}[(\mathbf{k}_0-\mathbf{k}_1-\mathbf{k}_2)_j L_j/2]$, where L_j is the length of the crystal in the directions transverse to the beam propagation. For simplicity, we assume that the transverse dimension of the crystal is sufficiently large so that the phase-matching condition can described by a two-dimensional delta function $\delta^{(2)}(\mathbf{k}_0-\mathbf{k}_1-\mathbf{k}_2)$ and that all other modes are damped out [16,17]. Moreover, the fact that $\mathbf{k}_0-\mathbf{k}_1-\mathbf{k}_2=0$ is a good approximation for fields carrying orbital angular momentum has already been experimentally verified in the frequency up-conversion of Laguerre-Gaussian modes [18]. While the δ function constrains the sum of the transverse signal and idler wave vectors, their absolute difference $|\mathbf{k}_1-\mathbf{k}_2|$ cannot be arbitrarily large either. This is immediately apparent within the paraxial limit, the regime in which most down-conversion experiments are operating. More generally one can argue that large values of $\mathbf{k}_1-\mathbf{k}_2$ require large values of $\mathbf{k}_1$ and/or $\mathbf{k}_2$ and hence of the signal and idler frequencies ω_1 and ω_2. Increasing $|\mathbf{k}_1-\mathbf{k}_2|$ to an arbitrarily large value, therefore, would lead to a violation of the frequency-matching condition $\omega_0=\omega_1+\omega_2$ associated with energy conservation. Clearly, energy conservation requires $|\mathbf{k}_1-\mathbf{k}_2|\leq 2\pi/\lambda$, where λ is the pump wavelength. We include this constraint in our analysis by means of the purely geometrical function $\Delta(\mathbf{k}_1-\mathbf{k}_2)$, normalized such that $\int d\mathbf{k}|\Delta(\mathbf{k})|^2=1$. This function will be zero for large values of its argument so as to satisfy the requirement of energy conservation. In practice, other constraints including the sizes of apertures used and crystal geometry will prescribe the precise form of $\Delta(\mathbf{k}_1-\mathbf{k}_2)$. We note that this function will result only in a scaling factor for the single and coincidence count rates. It cancels in the normalized count rates and plays no part in the derivation of orbital angular momentum conservation.

The two-photon wave function of the signal and idler then takes the form [19]

$$|\Psi\rangle=\int d\mathbf{k}_0\int d\mathbf{k}_1\int d\mathbf{k}_2\Phi_0(\mathbf{k}_0)\hat{a}_2^\dagger(\mathbf{k}_2)\hat{a}_1^\dagger(\mathbf{k}_1)\times\Delta(\mathbf{k}_1-\mathbf{k}_2)\delta^{(2)}(\mathbf{k}_0-\mathbf{k}_1-\mathbf{k}_2)|0\rangle. \quad (1)$$

Here

$$\Phi(\mathbf{k}_{0,1,2})=\frac{1}{2\pi}\int d\mathbf{x}\Phi_{0,1,2}(\mathbf{x}_{0,1,2})\exp(i\mathbf{k}_{0,1,2}\cdot\mathbf{x}_{0,1,2}) \quad (2)$$

denote the normalized mode functions in Fourier space. Similarly, the Fourier transformation of the creation operators is given by

$$\hat{a}_{1,2}^\dagger(\mathbf{k}_{1,2})=\frac{1}{2\pi}\int d\mathbf{x}\hat{a}_{1,2}^\dagger(\mathbf{x}_{1,2})\exp(-i\mathbf{k}_{1,2}\cdot\mathbf{x}_{1,2}). \quad (3)$$

Using these equations, we can write the two-photon wave function (1) in the position representation:

$$|\Psi\rangle=\int d\mathbf{x}_1\int d\mathbf{x}_2\Phi_0\left(\frac{\mathbf{x}_1+\mathbf{x}_2}{2}\right)\times\Delta(\mathbf{x}_1-\mathbf{x}_2)\hat{a}_1^\dagger(\mathbf{x}_1)\hat{a}_2^\dagger(\mathbf{x}_2)|0\rangle. \quad (4)$$

This is the normalized wave function of the combined system of the signal and idler generated by parametric down-conversion. It contains all necessary information about the outcome of single or coincidence measurements. The count rates of such measurements are proportional to the probabilities of detecting a photon of the signal or idler in the desired mode. In order to calculate these probabilities, we need to find the overlap of the two-photon state with the normalized one-photon state of the signal or idler,

$$|\Psi_{1,2}\rangle=\int d\mathbf{x}_{1,2}\Phi_{1,2}(\mathbf{x}_{1,2})\hat{a}_{1,2}^\dagger(\mathbf{x}_{1,2})|0\rangle, \quad (5)$$

associated with detecting a photon in the mode $\Phi_{1,2}$.

The coincidence probability for finding one photon in the signal mode Φ_1 and one photon in the idler mode Φ_2 is then given by

$$P(\Phi_1,\Phi_2)=|\langle\Psi_2,\Psi_1|\Psi\rangle|^2=\left|\int d\mathbf{x}_1\int d\mathbf{x}_2\Phi_1^*(\mathbf{x}_1)\Phi_2^*(\mathbf{x}_2)\Phi_0\left(\frac{\mathbf{x}_1+\mathbf{x}_2}{2}\right)\Delta(\mathbf{x}_1-\mathbf{x}_2)\right|^2. \quad (6)$$

If we average this expression over all possible signal modes, then we obtain the probability for finding a photon in the idler mode,

$$P(\Phi_2)=|\langle\Psi_2|\Psi\rangle|^2$$

$$=\int d\mathbf{x}_1\int d\mathbf{x}_2\int d\mathbf{x}_2'\Phi_2(\mathbf{x}_2')\Phi_2^*(\mathbf{x}_2)\Phi_0^*\left(\frac{\mathbf{x}_1+\mathbf{x}_2'}{2}\right)$$

$$\times\Phi_0\left(\frac{\mathbf{x}_1+\mathbf{x}_2}{2}\right)\Delta^*(\mathbf{x}_1-\mathbf{x}_2')\Delta(\mathbf{x}_1-\mathbf{x}_2), \quad (7)$$

and a similar expression describes the probability of finding a single photon in the signal mode.

We recall that $\Delta(\mathbf{k})$ is a very broad function that cuts out modes with large transverse wave vectors. The mode functions will therefore vary very little over the region where $\Delta(\mathbf{x}_1-\mathbf{x}_2)$, the Fourier transform of $\Delta(\mathbf{k})$, does not vanish. Under this assumption, Eq. (6) and (7) can be written as

$$P(\Phi_1,\Phi_2)=\left|\int d\mathbf{y}\Delta(\mathbf{y})\right|^2\left|\int d\mathbf{x}\Phi_1^*(\mathbf{x})\Phi_2^*(\mathbf{x})\Phi_0(\mathbf{x})\right|^2, \quad (8)$$

$$P(\Phi_2)=\left|\int d\mathbf{y}\Delta(\mathbf{y})\right|^2\int d\mathbf{x}|\Phi_2^*(\mathbf{x})\Phi_0(\mathbf{x})|^2. \quad (9)$$

The factor $|\int d\mathbf{y}\Delta(\mathbf{y})|^2$ will in general be very small and limit the count rates. This is because the down-converted photons are emitted into a wide spatial range.

The normalized coincidence probability, however, is independent of $\Delta(\mathbf{y})$:

$$P(\Phi_1,\Phi_2)^N$$

$$=\frac{P(\Phi_1,\Phi_2)}{\sqrt{P(\Phi_1)P(\Phi_2)}}$$

$$=\frac{\left|\int d\mathbf{x}\Phi_1^*(\mathbf{x})\Phi_2^*(\mathbf{x})\Phi_0(\mathbf{x})\right|^2}{\sqrt{\int d\mathbf{x}|\Phi_2^*(\mathbf{x})\Phi_0(\mathbf{x})|^2}\sqrt{\int d\mathbf{x}|\Phi_1^*(\mathbf{x})\Phi_0(\mathbf{x})|^2}}. \quad (10)$$

This probability can take values between 0 and 1. It becomes 1 if signal and idler are perfectly correlated so that the detection of the signal in mode Φ_1 implies that the idler is in mode Φ_2, and it vanishes if signal and idler are anticorrelated.

We now need to connect the idler mode function Φ_0 with the mode functions of the down-converted signal and idler $\Phi_{1,2}$. More precisely, we want to determine the idler mode for a given pump mode and detected signal mode. The state of the idler photon collapses into $|\Psi_2\rangle=\langle\Psi_1|\Psi\rangle$ if the signal photon was found to be in state $|\Psi_1\rangle$. We can calculate this state from Eqs. (4) and (5),

$$|\Psi_2\rangle=\int d\mathbf{y}\Delta(\mathbf{y})\int d\mathbf{x}_2\Phi_1^*(\mathbf{x}_2)\Phi_0(\mathbf{x}_2)\hat{a}^\dagger(\mathbf{x}_2)|0\rangle, \quad (11)$$

where we have evaluated the function $\Delta(\mathbf{y})$ in the same way as previously. In comparison with Eq. (5), we find that the mode function of the idler can be expressed as the product of the mode functions of the signal and the pump,

$$\Phi_2=\left(\int d\mathbf{y}\Delta(\mathbf{y})\right)\Phi_0\Phi_1^*, \quad (12)$$

where the integral enforces the normalization of the idler mode. This may be interpreted "backwards" or retrodictively [20]: The measured mode function Φ_2 is "reflected" at the $\chi^{(2)}$ crystal, where it interacts with the pump mode Φ_0 and is modified into the idler mode function.

So far our considerations have been valid for any transverse mode function, including Hermite or Laguerre-Gaussian modes. In the following, we will concentrate on Laguerre-Gaussian modes and specifically investigate the correlation of the orbital angular momentum between the signal and the idler. The normalized Laguerre-Gaussian modes in polar coordinates are given by

$$\Phi_{p,l}(r,\varphi)=\sqrt{\frac{2p!}{\pi(|l|+p)!}}\sqrt{\frac{1}{w}}\left(\frac{r\sqrt{2}}{w}\right)^{|l|}$$

$$\times L_p^{|l|}\left(\frac{2r^2}{w^2}\right)e^{-r^2/w^2}e^{-il\varphi}, \quad (13)$$

where the z-dependent phase was omitted and $L_p^{|l|}$ denotes the associated Laguerre polynominal,

$$L_p^l=\sum_{m=0}^{p}(-1)^m C_{p-m}^{p+l}r^m/m!. \quad (14)$$

Laguerre-Gaussian modes with an integer orbital angular momentum of $\hbar l$ per photon are orthonormal solutions of the paraxial wave equation. We note that these modes can be defined in the same way for fractional values of l. These fractional modes are still normalized, but form an overcomplete set. They can be written as sums of stable Laguerre-Gaussian modes with different integer l's. As each of these modes has a different Gouy phase, the resulting beams with fractional l's are unstable and do not maintain their amplitude distribution upon propagation. It has been suggested, however, that such beams can be prepared or detected by use of holograms or phaseplates [21] or by using nonlinear optical devices, such as the optical parametric oscillator [22].

By inserting the spatial mode functions $\Phi_{p_{0,1,2},l_{0,1,2}}$ in Eq. (12) and comparing the phase factors, we find that the orbital angular momentum of the idler is given by $l_2=l_0-l_1$. This signifies that the orbital angular momentum must be conserved,

$$l_0=l_1+l_2. \quad (15)$$

We want to stress that we have derived this equation solely from the phase-matching condition for Laguerre-Gaussian modes or other modes with the same azimuthal phase dependence. Our derivation is independent of the nature of the nonlinear susceptibility of the down-conversion crystal.

One example of the conservation of the orbital angular momentum can be seen in Fig. 1, where the mode profile of the idler mode function is shown as the product of the complex pump mode and the signal mode according to Eq. (12). Our derivation of Eq. (15) is valid for all modes that can be

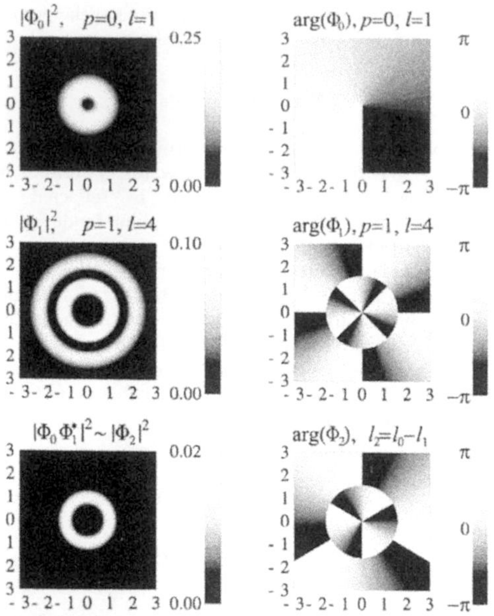

FIG. 1. Intensity $|\Phi|^2$ and phase arg(Φ) of the pump, signal, and idler mode. If the pump and signal beams are prepared/measured in the indicated L-G modes, the idler collapses into a mode with an amplitude proportional to $\Phi_0\Phi_1^*$. The phase structure of the idler corresponds to $l_2=l_0-l_1$ and an orbital angular momentum of $3\hbar$.

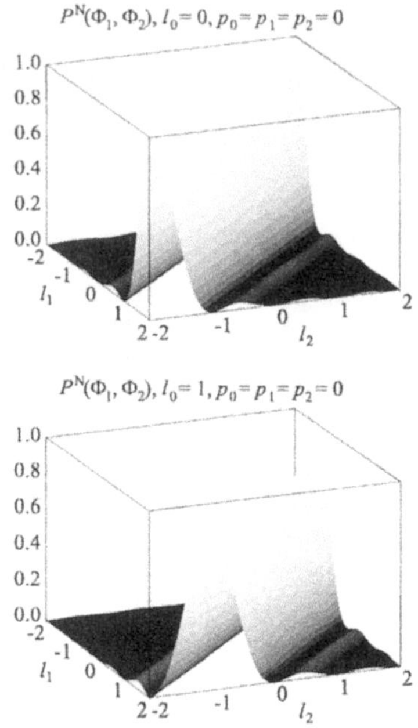

FIG. 2. The normalized coincidence probability $P^N(\Phi_1,\Phi_2)$ for the orbital angular momenta of the signal and idler photons for $l_0=0$ and 1. The ratio of idler and signal waists to that of the pump is $W=0.4$.

expressed in the form (13). We note that the conservation of orbital angular momentum can be fulfilled not only for integer quantities $l_{0,1,2}$ but also for continuous values of the orbital angular momentum numbers.

Equation (15) is in agreement with the experimental observation of orbital angular momentum conservation reported in [13,14]. To our knowledge, this is the only relevant experiment at the single-photon level. Classical signal and idler beams consist of many photon pairs, for each of which the conservation $l_0=l_1+l_2$ is satisfied. Both signal and idler beams will then exhibit modes with mixtures of various l's rather than pure Laguerre-Gaussian modes [23]. These mixed signal and idler modes will still be correlated. However, unless we can observe each pair of photons, their correlation will be lost as there is no coherence between photons generated from different pump photons. This makes it very difficult to demonstrate the validity of Eq. (15) in the classical regime. In contrast to this, the orbital angular momentum of each photon created by a frequency-doubling process is uniquely determined as $l_{2\omega}=2l_\omega$, where l_ω denotes the orbital angular momentum of the incoming photons. Consequently, the conservation of orbital angular momentum can be observed in classical up-conversion experiments [18].

The l conservation entangles the angular momentum modes of the signal and idler for a given pump mode. We set the waist of the idler equal to the waist of the signal, $w_1=w_2=w$, and denote the fraction of the signal and idler waist to the pump waist by $W=w/w_0$. By inserting the form of Laguerre-Gaussian modes (13) into the normalized coincidence probability (10), we find

$$P^N(\Phi_1,\Phi_2)=\mathrm{sinc}^2[(l_1+l_2-l_0)\pi]\frac{|R_{12}|^2}{\sqrt{R_2R_1}}. \tag{16}$$

Here we have denoted

$$R_{12}=\int_0^\infty dr\, r^{(|l_1|+|l_2|+|l_0|)/2}e^{-r(1+W^2/2)} \times \mathrm{L}_{p_1}^{|l_1|}(r)\mathrm{L}_{p_1}^{|l_2|}(r)\mathrm{L}_{p_0}^{|l_0|}(rW^2), \tag{17}$$

$$R_1=\frac{(p_1+|l_1|)!}{p_1!} \times \int_0^\infty dr\, r^{|l_2|+|l_0|}e^{-r(1+W^2)}[\mathrm{L}_{p_1}^{|l_1|}(r)]^2[\mathrm{L}_{p_0}^{|l_0|}(rW^2)]^2, \tag{18}$$

and similarly for R_2. Figure 2 shows the normalized coincidence probability as a function of the continuous orbital angular momenta of the signal and idler photon.

The sinc function in Eq. (15) becomes maximal if the orbital angular momentum is conserved and vanishes if $(l_0-l_1-l_2)$ is a nonzero integer. It is therefore most likely to detect a pair of signal and idler modes that conserve the pump angular momentum and it is impossible to find combinations of signal and idler modes that violate angular momentum conservation by an integer. Secondary maxima of the sinc function near half-integer values of $(l_0-l_1-l_2)$ seem to imply that there is a violation of the conservation of orbital angular momentum. This is not the case, as the beams of fractional l may be expressed in terms of modes with integer l, some of which satisfy the conservation requirement. This is due to the fact that the fractional l modes are overcomplete. While this structure is determined by the phase dependence of the L-G modes, we find additional dips along the diagonals of maximum coincidence at $l_1+l_2=l_0$. These are related to the radial mode profile in Eq. (16) and again arise from the finite overlap of the fractional modes (12). There is no equivalent to this in polarization correlations. However, it should be stressed that this additional structure arises from the radial structure of the mode. If this radial structure is ignored, as in a detection process that measures only l, then this detailed structure disappears.

Compared with the spin angular momentum, orbital angular momentum offers a far richer structure. The spin angular momentum can only take values between -1 and 1 for right and left circularly polarized light, respectively. The orbital angular momentum is in principle unlimited, and there exists an infinite number of Laguerre-Gaussian modes. While the spin angular momentum modes are defined in a two-dimensional Hilbert space, the orbital angular momentum modes occupy different Hilbert spaces depending on the mode order $N=l+2p$ with dimensionality $N+1$ [24].

Orbital angular momentum states with mode order $N=1$ are exactly analogous to spin angular momentum states. In each case they can be depicted on the Poincaré sphere [25]. Any two diametrically opposed points on the Poincaré sphere correspond to orthogonal states. For the polarization states these are left and right circular polarization, and horizontal and vertical linear polarization. For the orbital angular momentum states these are Laguerre-Gaussian modes with positive and negative phase, vertical and horizontal Hermite-Gaussian modes, or superpositions of such modes. It is well known [26] that maximum violation of Bell's inequality occurs between any two measurements made for polarization states separated by 45° on the Poincaré sphere. For orbital angular momentum, states, we would therefore anticipate a maximum violation of Bell's inequality to be obtained for measurements of superpositions of Laguerre-Gaussian and Hermite-Gaussian modes of mode order 1 separated by 45° degree on the Poincaré sphere. For larger values of orbital angular momentum, this analogy breaks down due to the higher number of possible states. The quantum correlations, however, persist and offer the prospect of novel demonstrations and applications of entanglement.

Another approach has been proposed by Mair *et al.* [14], whereby a hologram displaced from its on-axis position no longer produces or detects a pure Laguerre-Gaussian mode but a complicated superposition of modes with differing indices. This situation can be modeled by a numerical evaluation of Eq. (10), which we will report elsewhere [27]. In general, the coincidence rate is maximized for hologram offsets, which result in mode combinations with high contributions by mode pairs which satisfy the conservation of orbital angular momentum (15), namely $l_0=l_1+l_2$.

In this paper, we have considered the correlation between Laguerre-Gaussian modes of the signal and idler with varying amounts of orbital angular momentum. The finite mode overlap between modes differing by a fractional number of l is to some extent reminiscent of the finite overlap between nonorthogonal polarization states, as they are used for testing Bell's inequality. This suggests the possibility of observing a violation of Bell's inequality for measurements of angular momentum states differing by fractional values of l. Such measurements may be possible with the use of holograms that impose a fractional change of the orbital angular momentum by means of a fractional change of the phase step.

We are grateful to Alois Mair for useful discussions and for providing us with invaluable information about his experiment on orbital angular momentum conservation. We would like to acknowledge the enthusiastic encouragement of the late Alan J. Duncan. This work was supported by the TMR program of the Commission of the European Union through the Quantum Structures Network and by the Leverhulme Trust.

[1] A. Einstein, B. Podolsky, and N. Rosen, Phys. Rev. **47**, 777 (1935).

[2] J.S. Bell, Physics (Long Island City, N.Y.) **1**, 195 (1964); *Speakable and Unspeakable in Quantum Mechanics* (Cambridge University Press, Cambridge, England, 1987).

[3] A. Ekert, Phys. Rev. Lett. **67**, 661 (1991).

[4] D. Bouwmeester, J.V. Pan, K. Mattle, M. Eible, H. Weinfurter, and A. Zeilinger, Nature (London) **390**, 575 (1997).

[5] L.A. Lugiato and A. Gatti, Phys. Rev. Lett. **70**, 3868 (1993); A. Gatti and L.A. Lugiato, Phys. Rev. A **52**, 1675 (1995); A. Gatti, H. Wiedermann, L.A. Lugiato, I. Marzoli, G.-L. Oppo, and S.M. Barnett, *ibid.* **56**, 877 (1997).

[6] T.B. Pittman, Y.H. Shih, D.B. Strekalov, and A.V. Sergienko, Phys. Rev. A **52**, R3429 (1995).

[7] C.H. Monken, P.H. Souto Ribeiro, and S. Pádua, Phys. Rev. A **57**, 3123 (1998).

[8] D.N. Klyshko and D.P. Krindach, Pis'ma Zh. Éksp. Teor. Fiz. **54**, 697 (1968) [JETP Lett. **17**, 371 (1968)].

[9] P.G. Kwiat, K. Mattle, H. Weinfurter, A. Zeilinger, A.V. Sergienko, and Y. Shih, Phys. Rev. Lett. **75**, 4337 (1995).

[10] J.G. Rarity, P.R. Tapster, E. Jakeman, T. Larchuk, R.A. Campos, M.C. Teich, and B.E.A. Saleh, Phys. Rev. Lett. **65**, 1348 (1990); Z.Y. Ou, X.Y. Zou, L.J. Wang, and L. Mandel, Phys. Rev. A **42**, 2957 (1990); Y.H. Shih and A.V. Sergienko, Phys. Lett. A **186**, 29 (1994).

[11] Y.H. Shih, A.V. Sergienko, M.H. Rubin, T.E. Kiess, and C.O. Alley, Phys. Rev. A **50**, 23 (1994).

[12] H.H. Arnaut and G.A. Barbosa, Phys. Rev. Lett. **85**, 286 (2000); E.R. Eliel *et al.*, *ibid.* **86**, 5208 (2001).

[13] A. Mair and A. Zeilinger, *Vienna Circle Institute Yearbook 7/1999*, edited by A. Zeilinger *et al.* (Kluwer, Dordrecht, 1999).

[14] A. E. Mair, Ph.D. thesis, Leopold Franzens Universität Innsbruck (2000) (in German); A. E. Mair, A. Vaziri, G. Weihs, and Anton Zeilinger, Nature (London) **412**, 313 (2001).

[15] L. Allen, M.W. Beijersbergen, R.J.C. Spreeuw, and J.P. Woerdman, Phys. Rev. A **45**, 8185 (1992); M.W. Beijersbergen, L. Allen, H.E.L.O. van der Veen, and J. Woerdman, Opt. Commun. **96**, 123 (1992); L. Allen, M.J. Padgett, and M. Babiker, Prog. Opt. **39**, 291 (1999).

[16] C.K. Hong and L. Mandel, Phys. Rev. A **31**, 2409 (1985).

[17] See e.g., L. Mandel, and E. Wolf, *Optical Coherence and Quantum Optics* (Cambridge University Press, New York, 1995), p.1069f.

[18] K. Dholakia, N.B. Simpson, M.J. Padgett, and L. Allen, Phys. Rev. A **54**, R3742 (1996); J. Courtial, K. Dholakia, L. Allen, and M.J. Padgett, *ibid.* **56**, 4193 (1997).

[19] We note that, if we wanted to include polarization properties in our calculation, we would need to consider the nonlinear susceptibility tensor in Eq. (1).

[20] D.T. Pegg and S.M. Barnett, J. Opt. B: Quantum Semi-Classical Opt. **1**, 442 (1999), and references therin.

[21] M.W. Beijersbergen, R.P.C. Coerwinkel, M. Kristensen, and J.P. Woerdman, Opt. Commun. **112**, 321 (1994).

[22] G.-L. Oppo, A.J. Scroggie, and W.J. Firth, Phys. Rev. E **63**, 066209 (2001).

[23] J. Arlt, K. Dholakia, L. Allen, and M.J. Padgett, Phys. Rev. A **59**, 3950 (1999).

[24] L. Allen, J. Courtial, and M.J. Padgett, Phys. Rev. E **60**, 7497 (1999).

[25] M.J. Padgett and J. Courtial, Opt. Lett. **24**, 430 (1999).

[26] See, e.g., A.J. Duncan, *Progress in Atomic Spectroscopy* (Plenum Press, New York, 1987), pp. 477–505.

[27] M.J. Padgett, J. Courtial, L. Allen, S. Franke-Arnold, and S.M. Barnett, J. Mod. Opt. (to be published).

Measuring the Orbital Angular Momentum of a Single Photon

Jonathan Leach,[1] Miles J. Padgett,[1] Stephen M. Barnett,[2] Sonja Franke-Arnold,[2] and Johannes Courtial[1,*]
[1]*Department of Physics and Astronomy, University of Glasgow, Glasgow, Scotland*
[2]*Department of Physics and Applied Physics, University of Strathclyde, Glasgow, Scotland*
(Received 21 January 2002; published 5 June 2002)

We propose an interferometric method for measuring the orbital angular momentum of single photons. We demonstrate its viability by sorting four different orbital angular momentum states, and are thus able to encode two bits of information on a single photon. This new approach has implications for entanglement experiments, quantum cryptography and high density information transfer.

DOI: 10.1103/PhysRevLett.88.257901 PACS numbers: 03.67.–a, 42.50.Ct

It is well known that photons can carry both spin and orbital angular momentum (OAM) [1,2]. The spin is associated with polarization and the OAM with the azimuthal phase of the complex electric field. Each photon of a beam with an azimuthal phase dependence of the form $\exp(il\phi)$, for example, carries an OAM of $l\hbar$. The polarization of a single photon is described by a state in a two-dimensional space. For this reason, photon polarization provides a useful physical realization of a single qubit and has been widely employed in demonstrations of quantum key distribution [3,4]. As recently pointed out [5], the infinite number of orthogonal states of OAM places no limit on the number of bits that can be carried by a single photon. Moreover, the ability to create states with different OAM and superpositions of these allows the realization of quNits, that is quantum states in an N-dimensional space, with single photons. The problem addressed in this Letter is the realization of a multichannel device for determining the OAM of a single photon. Such a device will allow us to take advantage of the increase in information capacity associated with orbital rather than spin angular momentum.

Previous work has shown that the OAM of a laser beam containing many photons in the same mode can be measured. For example, interfering an $\exp(il\phi)$ beam with its mirror image produces an interferogram with $2l$ radial spokes [6,7] [Fig. 1(a)]. Although this technique can discriminate between an arbitrarily large number of states, the state of one single photon cannot be measured as many photons are required to form the full interference pattern.

For individual photons, computer-generated holograms [8,9], when used in combination with a pinhole in a setup similar to the one shown in Fig. 1(b), can determine the particular OAM state [10]. Holograms are often used in the generation of $\exp(il\phi)$ beams, where the fork dislocation in the hologram introduces helical phase fronts in the diffracted beam. When operated in reverse, the hologram "flattens" the helical phase fronts. The beam, now with planar phase fronts, can be focused through a pinhole and detected. However, this technique allows photons to be tested only for one particular state. A more complex computer-generated hologram was demonstrated that can detect several different l states but with an efficiency that cannot exceed the reciprocal of the number of different l values [11]. This low efficiency means that this method is not likely to be useful for quantum information applications. In a different approach, the difference in Gouy phase was used within an interferometer to distinguish between modes of two different orders (the mode order is dependent on l) [12]. Although working at the single-photon level such a scheme sorts between only two states.

In this Letter we describe an interferometric technique that can distinguish individual photons in arbitrarily many OAM states with a theoretical efficiency of 100%. We demonstrate this principle with a device that simultaneously sorts four different OAM states, corresponding to

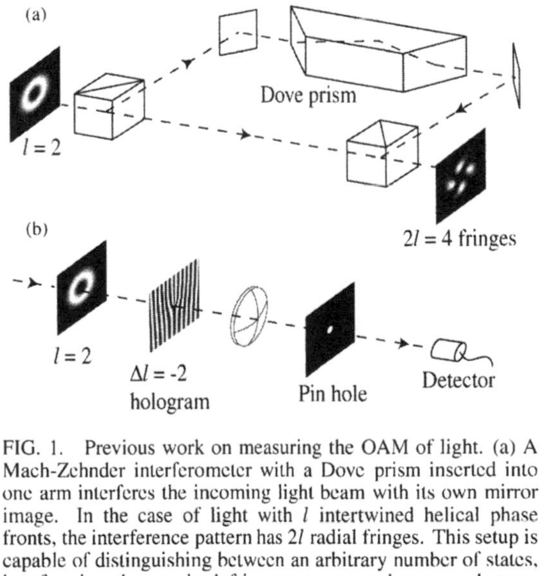

FIG. 1. Previous work on measuring the OAM of light. (a) A Mach-Zehnder interferometer with a Dove prism inserted into one arm interferes the incoming light beam with its own mirror image. In the case of light with l intertwined helical phase fronts, the interference pattern has $2l$ radial fringes. This setup is capable of distinguishing between an arbitrary number of states, but forming the required fringe pattern needs many photons. (b) A hologram can be used to "flatten" the phase fronts of light modes with specific values of l, which makes it possible to focus these modes (but no others) through a pinhole, behind which they can be detected. While this latter setup works with individual photons, it can only test for one particular OAM state.

two bits of information that can be transferred with a single photon.

Our device relies on the $\exp(il\phi)$ form of the transverse modes. On rotation of the beam through an angle α, this phase dependence becomes $\exp[il(\phi + \alpha)]$. This corresponds to a phase shift of $\Delta\psi = l\alpha$ [13], which is a manifestation of a geometrical phase [14]. For particular combinations of l and α, the rotated beam may be either in or out of phase with respect to the original. For example, when $\alpha = \pi$, a beam with even l is in phase with the original but a beam with odd l rotated by the same angle is out of phase by π (Fig. 2). If such a rotation is incorporated into the arms of a two-beam interferometer, then the phase shift between the two arms becomes l dependent. It follows that for different angles of rotation, constructive and destructive interference occurs for different values of l.

This concept can be realized in the form of a Mach-Zehnder interferometer with a Dove prism inserted into each arm (Fig. 3). A Dove prism flips the transverse cross section of any transmitted beam [15]. Two Dove prisms, rotated with respect to each other through an angle $\alpha/2$, rotate a passing beam through an angle α. In the example shown in Fig. 3, $\alpha/2 = \pi/2$ and hence the relative phase difference between the two arms of the interferometer is $\Delta\psi = l\pi$. By correctly adjusting the path length of the interferometer we can ensure that photons with even l appear in port A1 and photons with odd l appear in port B1. If the input state is a mixture of even and odd l components, then these components are "sorted" into an even channel A1 and an odd channel B1.

Our principle can be extended further to enable us to test for an arbitrarily large number of OAM states. This is achieved by cascading additional Mach-Zehnder interferometers with different rotation angles (Fig. 4). (Note that the scheme outlined in Ref. [12] could be extended in an analogous fashion.) The first interferometer, stage 1, sorts photons with even and odd values of l into ports A1 and B1, respectively. Photons with even l are then passed into the second stage where they are sorted further. The angle between the Dove prisms of the second stage is $\alpha/2 = \pi/4$ corresponding to $\Delta\psi = l\pi/2$. Therefore, modes with $l = 4n$, where n is an integer, go into port A2 and beams with a phase term of $l = 4n + 2$ go into port B2. Unfortunately, there is no rotation angle that allows us to unambiguously sort odd-l photons in the same way. We solve this problem by placing a hologram in front of one interferometer of the second stage so that we can increase the azimuthal phase of the odd-l photons by 1, thereby making their l values even. An additional interferometer with $\alpha = \pi/2$ will now separate the original odd-l photons in the same way as the even-l photons were sorted. Figure 4 outlines the first three sorting stages, which allow discrimination between eight different values of l. By adding further stages, this procedure can be extended to allow an arbitrarily large number of OAM states to be distinguished. It should be noted that, in the absence of holograms, a scheme similar to that illustrated in Fig. 4 can be constructed to sort beams where l takes on the values of 0 or 2^n, where n is an integer.

To demonstrate the viability of our proposed mechanism, three Mach-Zehnder interferometers were built to

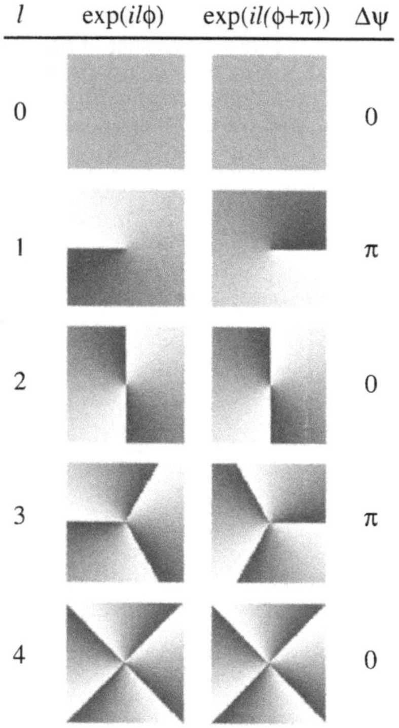

FIG. 2. Gray-scale representations of phase profiles of nonrotated and rotated beams with an $\exp(il\phi)$ phase structure. After a rotation through π, a beam with even l is unchanged, while one with odd l is out of phase by π with the nonrotated beam. Interfering an l beam with a rotated copy of itself therefore results in constructive interference for even l and destructive interference for odd l.

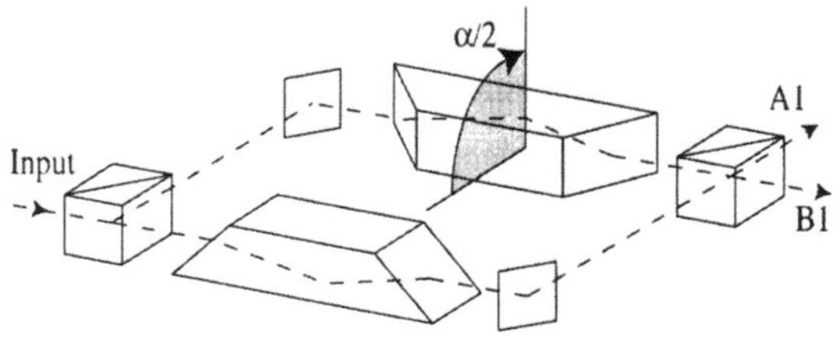

FIG. 3. First stage of our OAM sorter. A Mach-Zehnder interferometer with a Dove prism placed in each arm. The beams in the two arms are rotated with respect to each other through an angle α, where $\alpha/2$ is the relative angle between the dove prisms. In the example shown, $\alpha/2 = \pi/2$, this device sorts photons with even values of l into Port A1 and those with odd values of l into Port B1.

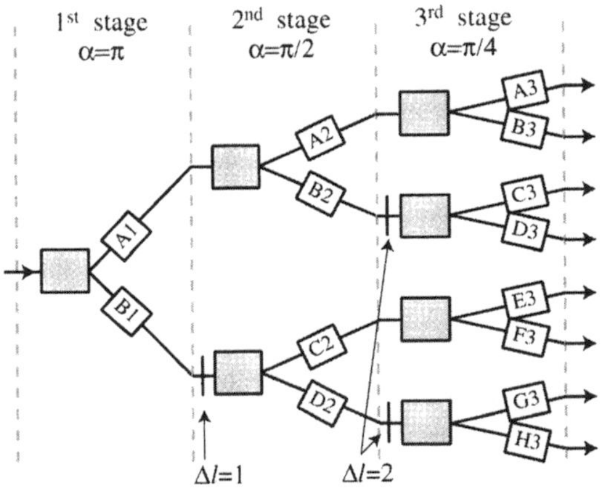

FIG. 4. First three stages of a general sorting scheme. The gray boxes each represent an interferometer of the form shown in Fig. 3 with different angles between the Dove prisms. The first stage introduces a phase shift of $\alpha = \pi$ and so sorts multiples of 2: even ls into Port A1 and odd ls into Port B1. The odd-l photons then pass though an $\Delta l = 1$ hologram so that they become even-l photons. The second stage introduces a phase shift of $\alpha = \pi/2$, so it sorts even-l photons into even and odd multiples of 2. The $\Delta l = 2$ hologram is required before the photons are sorted further in the third stage.

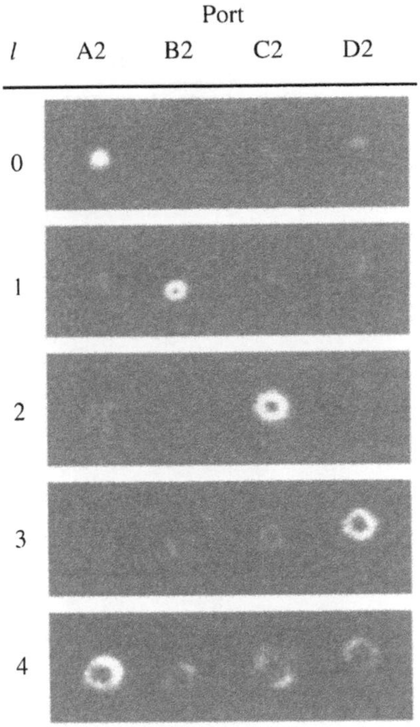

FIG. 5. Experimental results of a 2-stage sorting scheme. The four different output ports correspond to $\exp(il\phi)$ modes with the values $l(\mathrm{mod}4) = 0, 1, 2, 3$, respectively.

form the first two stages of the general OAM sorter outlined in Fig. 4. The light source used in this experiment was a helium-neon laser with a power of <1 mW. An intracavity cross wire introduced rectangular symmetry to the laser cavity and forced the laser to oscillate in high-order Hermite-Gaussian ($HG_{m,n}$) modes. Such modes are characterized by the indices m and n which correspond to zeros of intensity in the electric field in the x and y directions, respectively. The Hermite-Gaussian modes were then converted to Laguerre-Gaussian modes by passing them through a $\pi/2$ mode converter based on cylindrical lenses [16]. The resulting Laguerre-Gaussian modes have an $\exp(il\phi)$ phase structure and corresponding OAM of $l\hbar$ per photon. This conversion of Hermite-Gaussian ($HG_{m,n}$) beams gives Laguerre-Gaussian (LG_p^l) beams characterized by $l = |m - n|$ and $p = \min(m, n)$. Adjustments to the intracavity cross wire allowed us to generate $HG_{m,0}$ modes with $m = 0, 1, 2, \ldots$, which in turn gave rise to LG_0^l beams with $l = 0, 1, 2, \ldots$. The interferometers had an arm length of approximately 30 cm and were built from standard optical components. The $\Delta l = 1$ hologram was manufactured using standard photographic techniques [17]. Note that such a hologram increases the l value of any $\exp(il\phi)$ mode by 1 [18]. The four ports were directed onto a screen so a camera could take an image of the output.

Figure 5 shows the output from the two-stage sorting process. As can be seen, we succeeded in sorting modes from $l = 0$ to $l = 4$ into different ports. The $l = 4$ mode appears in the same port as the $l = 0$ beam, as one would expect. In this first experiment, the overall efficiency of the OAM sorter was limited by the poor optical efficiency of the particular hologram used to approximately 10%.

To demonstrate that our device works at the single-photon level, a further experiment was carried out at intensities so low that on average less than one photon was present in each interferometer at any one time. This was achieved by inserting neutral-density filters to attenuate the power of the laser beam to <0.3 nW. This experiment used a 1-stage interferometer. The output ports of the interferometer were directed into a camera that averaged over a number of frames. As anticipated, this interferometer still sorts between odd and even ls with an efficiency limited only by the quality of the optical components (Fig. 6). Although the device proposed in this Letter can sort individual photons according to their OAM, we did not detect photons individually. We plan to do this in the future with the use of a single photon source and single photon detectors.

Our OAM sorter is the analog of the polarizing beam splitter in that it selects the optical path on the basis of OAM, one path for each of the distinguishable states. In this way, our sorter can be used to generate entanglement between the optical path and OAM in the same way that a polarizing beam splitter can create entanglement between the optical path and polarization [19]. This will make it

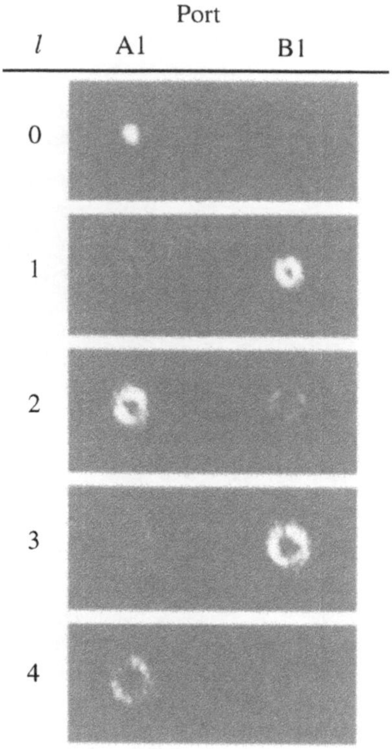

FIG. 6. Experimental results of a 1-stage (even-odd) sorting scheme at the single-photon level. Neutral density filters were used to reduce the number of photons so that the intensity corresponded to one photon in the interferometer at one time.

useful in generating highly entangled states and extending the optical realization of quantum logic elements to OAM quNits [20].

We have demonstrated experimentally that a single photon in an OAM eigenstate can be measured in any one of a number of different orthogonal states corresponding to different values of l (the OAM in units of $\hbar$). Our approach is in principle 100% efficient, limited only by the efficiency of the components.

The ability to measure a single photon to be in any one of an arbitrarily large number of orthogonal states has a number of potential implications for quantum information processing. The efficient measurement of the OAM of a single photon allows us access to a larger state space than that associated with optical polarization. This provides the possibility of a greater density of information transfer along with the generation and analysis of entanglement involving large numbers of states [10]. The implications of this work for entanglement based applications such as superdense coding [21], teleportation [22], and quantum computation [23] remain to be explored.

This work was supported by the Glasgow-Strathclyde University Synergy fund, the Royal Society, the Leverhulme Trust, the Royal Society of Edinburgh, and the Scottish Executive Education and Lifelong Learning Department.

*Electronic address: j.courtial@physics.gla.ac.uk

[1] L. Allen, M. W. Beijersbergen, R. J. C. Spreeuw, and J. P. Woerdman, Phys. Rev. A **45**, 8185 (1992).
[2] L. Allen, M. J. Padgett, and M. Babiker, in *Progress in Optics XXXIX*, edited by E. Wolf (Elsevier Science B. V., New York, 1999), pp. 291–372.
[3] D. Bouwmeester, A. Ekert, and A. Zeilinger, *The Physics of Quantum Information* (Springer, Berlin, Germany, 2000).
[4] S. Pheonix and P. Townsend, Contemp. Phys. **36**, 165 (1995).
[5] G. Molina-Terriza, J. P. Torres, and L. Torner, Phys. Rev. Lett. **88**, 013601 (2002).
[6] M. Harris, C. A. Hill, P. R. Tapster, and J. M. Vaughan, Phys. Rev. A **49**, 3119 (1994).
[7] M. J. Padgett, J. Arlt, N. B. Simpson, and L. Allen, Am. J. Phys. **64**, 77 (1996).
[8] V. Y. Bazhenov, M. V. Vasnetsov, and M. S. Soskin, JETP Lett. **52**, 429 (1990).
[9] N. R. Heckenberg, R. McDuff, C. P. Smith, and A. G. White, Opt. Lett. **17**, 221 (1992).
[10] G. Weihs and A. Zeilinger, Nature (London) **412**, 313 (2001).
[11] V. V. Kotlyar, V. A. Soifer, and S. N. Khonina, J. Mod. Opt. **44**, 1409 (1997).
[12] M. V. Vasnetsov, V. V. Slyusar, and M. S. Soskin, Quantum Electron. **31**, 464 (2001).
[13] J. Courtial, D. A. Robertson, K. Dholakia, L. Allen, and M. J. Padgett, Phys. Rev. Lett. **81**, 4828 (1998).
[14] M. J. Padgett and J. Courtial, Opt. Lett. **24**, 430 (1999).
[15] M. Born and E. Wolf, *Principles of Optics* (Pergamon Press, Oxford, 1980), 6th ed.
[16] M. W. Beijersbergen, L. Allen, H. E. L. O. van der Veen, and J. P. Woerdman, Opt. Commun. **96**, 123 (1993).
[17] J. Arlt, K. Dholakia, L. Allen, and M. J. Padgett, J. Mod. Opt. **45**, 1231 (1998).
[18] If the incident mode is an LG mode, which is of the form $\exp(il\phi)$, the transmitted beam is no longer a pure LG mode, but is still of the form $\exp(il'\phi)$ (with $l' = l + 1$). There is no fundamental reason limiting the holographic conversion of $\exp(il\phi)$ modes, but in practice they currently convert at best about 85% of the incident power.
[19] T. J. Herzog, P. G. Kwiat, H. Weinfurter, and A. Zeilinger, Phys. Rev. Lett. **75**, 3034 (1995).
[20] N. J. Cerf, C. Adami, and P. G. Kwiat, Phys. Rev. A **57**, R1477 (1998).
[21] C. H. Bennett and S. J. Wiesner, Phys. Rev. Lett. **69**, 2881 (1992).
[22] C. H. Bennett, G. Brassard, R. Jozsa, A. Peres, and W. K. Wootters, Phys. Rev. Lett. **70**, 1895 (1993).
[23] M. A. Nielsen and I. L. Chuang, *Quantum Computation and Quantum Information* (Cambridge University Press, Cambridge, United Kingdom, 2000).

Milton Keynes UK
Ingram Content Group UK Ltd.
UKHW051948071024
449327UK00026B/2216

9 780367 578534